HORMONES AND SIGNALING

VOLUME I

HORMONES AND SIGNALING

ASSOCIATE EDITORS

Lutz Birnbaumer
Los Angeles, California

James Darnell
New York, New York

Ronald Evans
La Jolla, California

Tony Hunter
La Jolla, California

Anthony Means
Durham, North Carolina

Wylie Vale
La Jolla, California

HORMONES AND SIGNALING

VOLUME I

Editor-in-Chief

Bert W. O'Malley

ACADEMIC PRESS
San Diego   London   Boston   New York   Sydney   Tokyo   Toronto

QP
517
.C45
H675

ADI-9960

This book is printed on acid-free paper. ∞

Copyright © 1998 by ACADEMIC PRESS

All Rights Reserved.
No part of this publication may be reproduced or transmitted in any form or by any means, electronic or mechanical, including photocopy, recording, or any information storage and retrieval system, without permission in writing from the Publisher.
The appearance of the code at the bottom of the first page of a chapter in this book indicates the Publisher's consent that copies of the chapter may be made for personal or internal use of specific clients. This consent is given on the condition, however, that the copier pay the stated per copy fee through the Copyright Clearance Center, Inc. (222 Rosewood Drive, Danvers, Massachusetts 01923), for copying beyond that permitted by Sections 107 or 108 of the U.S. Copyright Law. This consent does not extend to other kinds of copying, such as copying for general distribution, for advertising or promotional purposes, for creating new collective works, or for resale. Copy fees for pre-1998 chapters are as shown on the title pages. If no fee code appears on the title page, the copy fee is the same as for current chapters.
1094-0103/98 $25.00

Academic Press
a division of Harcourt Brace & Company
525 B Street, Suite 1900, San Diego, California 92101-4495, USA
http://www.apnet.com

Academic Press Limited
24-28 Oval Road, London NW1 7DX, UK
http://www.hbuk.co.uk/ap/

International Standard Book Number: 0-12-312411-5

PRINTED IN THE UNITED STATES OF AMERICA
97 98 99 00 01 02 EB 9 8 7 6 5 4 3 2 1

Contents

Contributors xi
Preface xiii

Glucocorticoids and Oxysterols in Lymphoid Apoptosis

E. Brad Thompson

    I. Introduction 2
   II. Mechanisms of Action of the Glucocorticoid Receptor 3
       A. Control of Transcription 3
       B. Posttranscriptional Control 5
       C. Regulation of Growth Factors 5
  III. Actions of Oxysterols on Lymphoid Cells 6
  IV. Apoptosis 7
       A. DNA Lysis 8
       B. Calcium Flux 8
       C. Proto-oncogenes 8
       D. Cell Shrinkage 10
       E. Induced Lethality Genes 10
       F. Fas and the Cysteine Proteases 11
       G. Summary 12
   V. Glucocorticoids and Oxysterols Cause Apoptosis of CEM Cells 12
       A. Derivation of CEM Cell Clones 12
       B. Both Glucocorticoids and Oxysterols Cause CEM Cell Death by Apoptosis 14

C. The Glucocorticoid Pathway to Apoptosis 16
   D. The Role of the GR in Apoptosis: Mapping the GR for Cell
      Death Domains 20
   E. The Oxysterol Path to Apoptosis in CEM Cells 27
VI. Conclusions 31
   References 32

RAGE: A Receptor with a Taste for Multiple Ligands and Varied Pathophysiologic States

Ann Marie Schmidt, Jean-Luc Wautier, David Stern, and Shi Du Yan

   I. Introduction 42
  II. Identification and Characterization of RAGE 42
 III. Expression and Functions of RAGE: Endothelial Cells and
      Mononuclear Phagocytes 47
      A. Endothelial Cells 47
      B. Mononuclear Phagocytes 49
  IV. RAGE and Diabetes 51
   V. RAGE and Amphoterin 55
  VI. RAGE and Alzheimer's Disease 56
 VII. Hypothesis 60
      References 61

The Function and Regulation of the G-Protein-Coupled Receptor Kinases

Alexander D. Macrae and Robert J. Lefkowitz

   I. Introduction 66
  II. The Regulation of Cellular Signaling 66
 III. The G-Protein-Coupled Receptor Kinases 69
      A. Cloning and Structure of the GRKs 69
      B. Expression of the GRKs 70
  IV. The Regulation of the G-Protein-Coupled Receptor Kinases 71
      A. Regulation by Membrane Association 71
      B. Regulation by Membrane Lipids 73
      C. Microsomal Membrane Association 75
      D. Regulation by PKC 75
   V. Kinase Enzymology 76
  VI. Receptor–Kinase Specificity 76
 VII. Animal Models 79
      A. Animal Studies by Transgenic Gene Expression 79
      B. Animal Studies by Gene Deletion 80

VIII. Physiological Significance   81
IX. Conclusions   83
   References   83

Characteristics and Function of the Novel Estrogen Receptor β

George G. J. M. Kuiper, Stefan Nilsson, and Jan-Åke Gustafsson

I. Introduction   90
II. Cloning of ERβ cDNA   91
III. Ligand-Binding Characteristics of ERβ Protein   95
IV. Transactivation Function of ERβ Protein   99
V. Expression of ERβ mRNA   101
VI. Physiological Role of ERβ Protein   103
   A. ERβ in the ERα Knock-Out Mouse   103
   B. ERβ and the Prostate   105
   C. ERβ and Environmental Endocrine Disruptors   106
VII. Concluding Remarks   106
   References   107

EGF Family Receptors and Their Ligands in Human Cancer

Careen K. Tang and Marc E. Lippman

I. Introduction   113
II. EGF Family Receptors   114
   A. Overview   114
   B. Structure, Expression, and Transforming Potential of EGF Family Receptors   116
III. EGF-like Growth Factors   121
   A. Overview   121
   B. Common Structure of EGF like Growth Factor Family Members   121
   C. Function of EGF-like Growth Factors   122
   D. Expression and Biological Role of EGF-like Growth Factors   123
IV. Activation of EGF Family Receptors   130
   A. Interaction between the EGF Family Receptors and Their Ligands   130
   B. The Mechanisms of Receptor Activation   130
   C. Heterodimerization, Transphosphorylation   134

V. Clinical Significance of EGF Family Receptors   136
   A. EGFR   136
   B. ErbB-2   137
   C. ErbB-3   138
   D. ErbB-4   139
VI. Potential Clinical Application by Targeting of EGF Family Receptor Members and Ligands   139
   A. Immuno (Antibody) Therapy   140
   B. Coupling of Receptor Antibodies or Ligands to Toxin Molecules   141
   C. Antisense Strategies   143
   D. Receptor Tyrosine Kinase Inhibitors (EGFR PTK Inhibitors)   144
VII. Conclusion   145
   References   146

Fertilization: Common Molecular Signaling Pathways across the Species

Timothy A. Quill and David L. Garbers

I. Introduction   167
II. Oviductal Transport   168
III. Motility Modulation   170
   A. Motility Stimulation (Chemokinesis)   170
   B. Egg Peptide Receptors   175
   C. Chemotaxis   180
IV. Gamete Adhesion   183
V. Acrosome Reaction   185
VI. Egg Plasma Membrane Interactions and Egg Activation   192
   References   195

The JNK Family of MAP Kinases: Regulation and Function

Audrey Minden and Michael Karin

I. Introduction   210
II. Regulation of c-Jun Expression and Activity   212
III. Phosphorylation of the c-Jun Amino-Terminal Sites by JNK   213
IV. Signal Transduction Pathways Leading to Activation of JNK and Other MAP Kinases   216
V. Other Substrates for the JNKs   222

VI. Biological Functions of JNK and Other MAP Kinase
       Pathways   224
VII. Conclusions and Future Considerations   227
       References   228

PPARα: Tempting Fate with Fat
Pallavi R. Devchand and Walter Wahli

  I. PPARα as a Ligand-Activated Transcription Factor   236
 II. Lipid Homeostasis: How Complex Can It Be?   237
      A. Cell and Lipid Type   238
      B. Peroxisomes Are Multifunctional Organelles   238
      C. Cross-Talk at the Transcription Factor Level   241
III. PPARα Expression during Development and Adulthood   242
IV. PPARα and Its Activators: Open Relationships?   243
      A. Assays for PPAR Activators   243
      B. Fatty Acids   245
      C. Channeling Arachidonic Acid   245
      D. Peroxisome Proliferators   249
      E. Thiazolidinediones   249
      F. Species Differences   250
      G. Summary of PPARα Activation Profile   250
 V. PPARα Ligands   251
VI. PPARα Functions   251
      A. Evaluation of PPARα Function *in Vitro*   251
      B. PPARα Knock-Out Mice   253
VII. Conclusion   255
      References   255

Molecular Mechanisms of Neuronal Survival and Apoptosis
Sandeep Robert Datta and Michael E. Greenberg

   I. The Neurotrophic Theory   259
  II. Apoptosis—A Means of Neuronal Death   261
 III. Activity, Intracellular Calcium, and Survival   262
 IV. The Calcium Set-Point Hypothesis   263
  V. Mechanisms of Survival: Neurotrophin Receptors and
        Second Messengers   264
 VI. The Ras–MAPK Pathway   266
VII. The Phosphatidylinositide-3′-OH Kinase/Akt Pathway   270
VIII. Protein Kinase C   273

IX. Neurotrophin-Dependent Pathway Cross-Talk   274
X. Calcium-Dependent Kinase Pathways   274
XI. Interaction of Calcium and Neurotrophin Signaling in Survival   276
XII. Mechanisms of Death   278
XIII. Upstream Events in Neuronal Death and p75   278
XIV. Death and Second Messengers: Ceramide and JNK/p38   281
XV. Signal Integration: The Balance between Life and Death   285
XVI. Caspases and the Bcl-2 Family   288
XVII. Cell-Extrinsic Signaling Meets Cell-Intrinsic Survival and Death Mechanisms   291
XVIII. Frontiers in Neuronal Survival and Death   292
References   292

Nuclear Orphan Receptors: The Search for Novel Ligands and Signaling Pathways

Patricia J. Willy and David J. Mangelsdorf

I. Introduction   308
   A. Background   308
   B. Definition of an Orphan Receptor   311
II. Discovery of Orphan Receptor Ligands   315
   A. The Hunt for Ligands   315
   B. The Retinoid X Receptors   316
III. Orphan Receptors That Function as RXR Heterodimers   319
   A. RXR Heterodimer Partners with Ligands and/or Activators   319
   B. RXR Heterodimer Partners with No Known Ligands or Activators   322
   C. Receptors That Mediate Retinoid Signaling through Heterodimerization with RXR   323
IV. Orphan Receptors That Function Independently of Dimerization with RXR   327
   A. Receptors That Generally Function as Transcriptional Activators   327
   B. Receptors That Generally Function as Transcriptional Repressors   334
   C. Receptors with Unknown Activation Functions   339
V. Summary and Perspective   340
References   342

Index   359

Contributors

Numbers in parentheses indicate the pages on which the authors' contributions begin.

Sandeep Robert Datta (257) Greenberg Lab, Division of Neuroscience, Children's Hospital, Harvard Medical School, Boston, Massachusetts 02115

Pallavi Devchand (235) Institut de Biologie Animale, Bâtiment de Biologie, Université de Lausanne, CH-1015 Lausanne, Switzerland

David L. Garbers (167) Department of Pharmacology and Howard Hughes Medical Institute, University of Texas Southwestern Medical Center, Dallas, Texas 75235

Michael E. Greenberg (257) Division of Neuroscience, Children's Hospital, Harvard Medical School, Boston, Massachusetts 02115

Jan-Åke Gustafsson (89) Center for Biotechnology, Karolinska Institute, NOVUM, S-14186 Huddinge, Sweden

Michael Karin (209) Department of Pharmacology, Program in Biomedical Sciences, UCSD School of Medicine, La Jolla, California 92093

George G. J. M. Kuiper (89) Center for Biotechnology, Karolinska Institute, NOVUM, S-14186 Huddinge, Sweden

Robert J. Lefkowitz (65) Duke University Medical Center, Howard Hughes Medical Institute, Durham, North Carolina 27710

Marc E. Lippman (113) Lombardi Cancer Center, Georgetown University Medical School, Washington, DC 20007

Alexander D. Macrae (65) Duke University Medical Center, Howard Hughes Medical Institute, Durham, North Carolina 27710

David J. Mangelsdorf (307) Howard Hughes Medical Institute and Department of Pharmacology, University of Texas Southwestern Medical Center, Dallas, Texas 75235

Audrey Minden (209) Department of Biological Sciences, Columbia University, New York, New York 10027

Stefan Nilsson (89) Center for Biotechnology, Karolinska Institute, NOVUM, S-14186 Huddinge, Sweden

Timothy A. Quill (167) Department of Pharmacology, University of Texas Southwestern Medical Center, Dallas, Texas 75235

Ann Marie Schmidt (41) Department of Surgery and Medicine, College of Physicians and Surgeons, Columbia University, New York, New York 10032

David M. Stern (41) Department of Surgery and Medicine, College of Physicians and Surgeons, Columbia University, New York, New York 10032

Careen K. Tang (113) Lombardi Cancer Center, Georgetown University Medical School, Washington, DC 20007

E. Brad Thompson (1) Department of Human Biological Chemistry and Genetics, University of Texas Medical Branch, Galveston, Texas 77555

Walter Wahli (235) Institut de Biologie Animale, Bâtiment de Biologie, Université De Lausanne, CH-1015 Lausanne, Switzerland

Jean-Luc Wautier (41) Laboratory of Vascular Disease Research, University-7 of Paris, Paris, France

Patricia J. Willy (307) Howard Hughes Medical Institute and Department of Pharmacology, University of Texas Southwestern Medical Center, Dallas, Texas 75235

Shi Du Yan (41) Department of Surgery and Medicine, College of Physicians and Surgeons, Columbia University, New York, New York 10032

Preface

The fields of molecular endocrinology and signal transduction have undergone explosive expansion over the past decade. The cloning of new receptors and intermediates in the respective pathways of action has brought intense excitement to the quest to understand cellular regulation of gene expression. Pathways for steroid and peptide hormone action, intracellular phosphorylation cascades for biological response, cell-cycle intermediates and regulators, generic transcription factors and their attendant coactivators and corepressors, cell–cell communication mechanisms, differentiation determinants and signals for senescence, oncogenes, and tumor repressors have all been elucidated substantially within the approximate past decade. The biology of these gene products and pathways has been advanced by the combined technologies of recombinant DNA methodology, mutational analyses and cell transfection, yeast-based mutagenesis and two-hybrid analyses, structural biology and crystallization, molecular endocrinology technologies, transgenic and targeted gene interruption methodologies, and cell-free interaction methods. So what has all of this advanced technological development yielded? The answer is: a great deal.

Recent efforts to unravel the mechanisms of initiation of transcription have revealed amazing complexity, but at least the molecular players are rapidly becoming characterized. At present, well over 50 polypeptides participate in the GTF–polymerase preinitiation complex at the promoter (TATA box). Combinatorial assembly of these myriad proteins, stabilized via multiple weak protein–protein interactions, leads to transcription that can be specific for promoter context. Regulation occurs primarily by regulatory influences generated at upstream enhancers, which in the case of the endocrine system are the hormone response elements where receptors and other transcription factors are driven by the chemical signals.

We now have a much better appreciation of the complexities and a rudimentary understanding of the signaling mechanisms of early development. We understand that embryo development depends in large part on gradients of growth factors/hormones that utilize many of the same signaling pathways of the adult state to effect axis polarity and cellular differentiation. These same pathways play a continuing role throughout early postnatal development and adult life. Activation begins at two major sites: the cellular membrane receptors and the intracellular/nuclear receptors. Hormones acting at the membrane include the growth factors, peptide hormones, catecholamines, prostaglandins, and neurotransmittors, among others. Representative receptors from each of the major pathways have now been cloned and sequenced. Combinations of receptors are employed to effect enzyme activation on the inner membrane, leading to the generation of small second messengers, which induce a myriad of kinases and/or phosphatases to effect intracellular cascades. The downstream phosphorylation targets activated by phosphorylation/dephosphorylation are allosterically activated or multimerize for enzymatic function or relocate to another cellular compartment such as the nucleus, where they interact with collaborator proteins at the DNA level to effect gene regulation.

In contrast, the intracellular receptors are a large superfamily of structurally related transcription factors, now over 50 in number and which include the steroid receptors, thyroid hormone receptors, and receptors for certain vitaminoids such as vitamin D_3 and retinoids. The "orphan" receptors, termed as such because their ligands are currently unknown, as are many of their target genes, constitute the largest subgroup within the family and now number more than 40. Already they have been shown to regulate many of our more important developmental pathways, and some of their putative ligands have been uncovered recently. Ligands bind to their cognate receptors within the cell, usually in the nucleus, and allosterically activate these transcription factors so that they multimerize, bind to DNA, recruit a complex set of coactivators/corepressors, destabilize and remodel the local nucleosome structure at the target gene, and seduce and stabilize general transcription factors at the promoter via protein–protein interactions. Both the membrane receptor and the nuclear receptor pathways usually end their response at the level of DNA, where life in general begins and ends.

Molecular endocrinology is vibrant at this time, with new receptors and new hormones continuing to be published every few months. Pathways are becoming known in much greater detail, and new hormonal target genes are published with regularity. Mice resulting from targeted gene deletions are providing important new information on mammalian development that holds great promise for future studies of the pathobiology of disease. It is clear that the horizon holds a significant payback for the money spent by NIH and other organizations on basic research. Many new drugs are entering the pipeline at pharmaceutical and biotechnology companies based on the explosion of information just in the past decade. The field typified by *Hormones and Signaling* will make a substantial impact on the needs of society, with special contributions to nutrition, mammalian development, aging, endocrinology, oncology, reproductive biology, genetics, and neurobiology, among others.

The intention of the *Hormones and Signaling* series is to critically evaluate the field of regulatory biology and select the very best of the ongoing work for presentation. We have assembled a broad and representative group of leaders in this field as editors, whose job is to select scientists to contribute chapters on topics that both summarize their own cutting edge research and provide a perspective of the state of the art in their areas of expertise. Readers should find the reviews topical, critical, and informative. The volumes in this series will be best used as comprehensive and insightful summaries of topics for graduate students, postdoctoral fellows, active investigators, and teachers. They will be good reference sources for scientists and should tempt individuals to collect every volume in the series.

Bert W. O'Malley

E. Brad Thompson

Department of Human Biological Chemistry and Genetics
The University of Texas Medical Branch at Galveston
Galveston, Texas 77555

Glucocorticoids and Oxysterols in Lymphoid Apoptosis

Glucocorticoid-induced death of lymphoid cells is a classic model for apoptosis. Oxysterols, a different class of steroids, also produce lymphoid apoptosis. Conflicting proposals have been advanced for a general mechanism of apoptosis, and many initiating events may eventually activate a "final common pathway." This hypothesis remains open for testing in specific systems worked out in detail, to allow meaningful comparisons and contrasts. After briefly reviewing mechanisms of glucocorticoid and oxysterol actions, and some major themes in the control of apoptosis, this chapter summarizes our work comparing the effects of glucocorticoids and oxysterols on CEM cells, a line of human leukemic T cells. When continuously present, each class of steroids kills CEM cells after a delay of about 24 hr. Both cause a rapid, deep reduction in cMyc, which may be critical for subsequent apoptosis. Both steroid classes cause DNA nicking, then overt fragmentation into distinctive large and small size classes. Nicking appears to begin a few hours before cell death; fragmentation is approximately concomitant with cell death. The glucocorticoid receptor (GR) is required for the glucocorticoid-induced pathway, and transient transfections of GR mutants suggest that when the ligand binding domain (LBD) is intact, the DNA binding domain, but not the amino-terminal region, is required to initiate apoptosis. When the LBD is removed, the residual GR fragment constitutively causes cell death, possibly by a novel mechanism involving protein:protein interactions, *e.g.,* with cJun. Oxysterols do not utilize the GR; their receptor remains unknown. These two independent steroid classes may regulate events which converge to produce similar morphologic and biochemical end points.

I. Introduction

Glucocorticoids exert profound effects on lymphoid tissue. Thymic involution following *in vivo* administration of glucocorticoids to young rodents was an early observation (1), and later, *in vitro* administration of glucocorticoids to freshly dispersed thymocytes showed the effect to occur directly on the thymic cells, with suppression of macromolecular synthesis and overt cell death (2). In the early 1970s, binding assays in thymocytes were used to demonstrate the existence of glucocorticoid receptors (GRs), and the lethal potency of many glucocorticoids corresponded to their affinity for GRs (3). It was also observed that glucocorticoid treatment acutely reduced thymocyte glucose uptake and cellular ATP levels (3). Demonstration that mouse lymphoma cell lines in tissue culture also were killed by glucocorticoids, with correlated GR binding (4), allowed application of somatic cell genetic techniques. Selection for survivors from steroid treatment showed that nearly all appeared to be GR mutants (5,6). We demonstrated the presence of GRs in a leukemic lymphoid cell line sensitive to glucocorticoids (7) and showed that selection of these human cells for resistance also produced mostly GR mutants (8–10). By transfecting clonal resistant cells that expressed only a mutant GR with a plasmid expressing normal GR, we demonstrated that the GR is both necessary and sufficient for glucocorticoid-induced lymphoid cell death (11). Along with many other studies, such work provided powerful arguments for a central role of the lymphoid cell GR in the cell death evoked by glucocorticoids (12–14). The further question is, by what processes does ligand-activated GR cause its classic apoptosis? Decoding the mechanism of glucocorticoid-evoked lymphoid cell death should help clarify general apoptotic mechanisms. Also to be explained by any general model for glucocorticoid actions on lymphoid cells are: the differential sensitivity of various subclasses of lymphoid cells despite equivalent contents of GR, the disparity in time course between cell death in thymocytes (hours) and transformed lymphoid cells (days), and the antagonism between glucocorticoidal and negative selection-induced death, when both independent apoptotic stimuli are concurrent (15).

Though immature T cells are often discussed as if they were the only lymphoid targets of glucocorticoid-evoked cell death, certain immature B cells, leukemias, and lymphomas also are sensitive (16–19). More knowledge of the make-up of thymic cell subgroups has revealed that several classes of developing lymphoid cells are killed by glucocorticoids, especially CD4$^+$, CD8$^+$ thymocytes, prior to further differentiation. Mature, differentiated peripheral lymphoid cells are much less sensitive, unless first provoked to cell division (20). In general, myeloid cells and their oncogenically transformed counterparts are relatively resistant to death caused by glucocorticoids, though the steroids certainly can regulate genes in myeloid cells, and eosinophils at least are killed by glucocorticoids (21). In all cases, a primary interaction between the steroid and the GR is accepted as an essential component in the entrainment of cell death. Therefore, it is necessary to appreciate the various actions and interactions of the GR.

II. Mechanisms of Action of the Glucocorticoid Receptor

A. Control of Transcription

A simple initial model for the molecular mechanism of action of the ligand-activated GR followed the important discovery (22–24) of the palindromic Glucocorticoid Response Element (GRE) in the DNA sequence of the long terminal repeat (LTR) that forms the transcriptional regulatory region of the genome-integrated mouse mammary tumor virus (MMTV). In this model, a homodimer of the GR binds the GRE, with the transactivation domain of the GR directly contacting the general transcription factor complex to stimulate transcription. Consistent with this model, *in vitro* transcription experiments have shown that the activated GR enhances the stable preinitiation transcription complex in GRE-driven constructs (25). However, this clearly cannot be the full explanation of all glucocorticoid/GR transcriptional effects, for instance the control of genes lacking palindromic GREs or the many cases of negative gene regulation. Chromatin structure may be involved; *e.g.*, as the GR interacts with a key GRE in the MMTV LTR, the structure of a phased nucleosome is altered, altering the access of other transcription factors (26–28). *In vitro* transcription studies also showed influence of other factors and their sites, *e.g.*, NF1 (29). In fact, activated GR bound to the GRE that lies poised on the critical phased nucleosome in the MMTV LTR appears to make available an otherwise masked NF1 site. Thus, even at "simple" GRE sites, the GR seems to interact with or influence other proteins to alter transcription. The term glucocorticoid response unit (GRU) has been suggested to describe complexes of interactive regulatory molecules (30) on DNA. How the GR interacts with other factors, on chromatin, to regulate transcription remains an active area of interest (27,28,31).

Glucocorticoids down-regulate certain genes, and in some genes negative or nGREs have been proposed as the basis for such action (32), but no universal nGRE has been found. Negative (as well as positive) regulation in other cases appears to be carried out by interactions between the GR and other regulatory proteins (27,31,33, and references therein, 34). In fact, the importance of the ability of the GR to bind other proteins is only now emerging. Without ligand present, cytoplasmic GR appears to be complexed with several heat shock proteins, some of which at least may contribute to the high-affinity ligand binding site (35). When ligand binds, the GR sheds these proteins, is transferred to the nucleus, and interacts with other proteins to cause positive or negative alterations in transcription of specific genes. For example, the GR can interact with cJun or cFos (which together form the AP-1 transcription factor). The balance between the concentrations of cJun, GR, and cFos (and perhaps other members of the AP-1 family) can alter the level of transcription of genes under AP-1 control (31,33). Another important node of interaction appears to be NFκB, an important regulator of lymphoid cell function. The GR can bind NFκB and also can increase the expression of IκB, the cytoplasmic molecule that sequesters NFκB from its nuclear site of action. By two mechanisms, therefore, active GR can influence the function of this factor (36). As is AP-1, NFκB is involved in the regulation of many cytokines. Many examples involving other important transcription factors have been found. The full extent and physiologic importance of these GR interactions with heterologous transcription factors are not known, but it seems likely that they are a common mechanism in gene control.

Other protein interactions may play a role in GR actions. A transcription transactivation domain in estrogen was found toward the extreme carboxy-terminal end of the molecule (37). Homologous sequences exist in other members of this receptor family, including the GR. Recently, receptor class-specific proteins have been discovered or proposed that bind to this region (38,39). In test systems, these proteins act as coactivators and corepressors of the transcription regulated by their appropriate receptors.

Other signal transduction pathways impinge on that of the corticosteroid/GR system. For example, a long-standing observation in endocrinology is that glucocorticoids are "permissive" for activities controlled by cyclic AMP (40). The interactions between the cAMP and GR pathways have long been of interest, and it has been shown that cyclic nucleotides as well as glucocorticoids can cause cytolysis of lymphoid cells (41–43). Recent work has shown that ligands and drugs known to stimulate the cAMP pathway can, in some cells and tissues, activate certain steroid hormone receptors (44). In lymphoid cells, there is synergistic interaction between the cAMP path and the GR to evoke cell kill. We recently found a startling instance of such synergy: activation of protein kinase A by forskolin can restore glucocorticoid-induced cell death in a line of GR+ but otherwise glucocorticoid-resistant cells (43, and see below). The detailed mechanisms for these interactions are as yet unknown. Many possibilities present themselves, including altered phosphorylation of the GR (though the data so far lend little support to this), interactions with pathway intermediates, and interactions with the ultimate transcription factors controlled by the various pathways, with coactivators and corepressors, resulting in combined mutual control over critical genes.

B. Posttranscriptional Control

While transcription control by the activated GR is certainly a major mechanism of action, it should not be forgotten that posttranscriptional regulation by steroid hormones, including glucocorticoids, has been recorded or implied in a number of systems (45–50). Several of these are specifically relevant here, because they involve reductions in cytokines important to maintenance of the health and viability of hematopoietic cells. The quantitative contribution of this level of regulation *in vivo* to the physiological or pharmacological actions of glucocorticoids needs to be established. In experimental systems posttranscriptional regulation can be a major factor in, or even the dominant level of, control. Often the mechanism of such regulation appears to involve controlling the stability of mRNAs, but other posttranscriptional steps sometimes have been implicated. For a valuable general review of posttranscriptional mRNA regulation, including discussion of some steroid-regulated systems, see Ross (51).

C. Regulation of Growth Factors

Not all glucocorticoidal regulation of the growth, life, differentiation, and death of lymphoid cells is exerted by direct effects of the steroids on the affected cells. Paracrine effects brought about by steroids are important in their ultimate control of a given cell type. In the case of glucocorticoids and lymphoid cells, especially normal lymphoid cells, the glucocorticoidal regulation of lymphokines seems to be of great importance. Glucocorticoids exert negative regulatory effects on the production of many lymphokines, upon which lymphoid and other hematopoietic cell growth and viability depend (18,21,52–56). In primary cultures of glucocorticoid-sensitive hematopoietic cell types, varying the level of essential lymphokines results in varying sensitivity to glucocorticoid-evoked cell death: more lymphokine, less glucocorticoid sensitivity. Omission of all appropriate lymphokines from primary cultures of hematopoietic cells often leads to spontaneous cell death, which is hastened by addition of glucocorticoids (21,57,58). The exact molecular interplay of transduction pathways due to the effects of these two classes of control molecules—steroids and lymphokines—in regulating the viability of lymphoid cells is still unknown. In sum, indirect regulation of the viability of lymphoid and several other classes of hematopoietic cells by glucocorticoids is carried out through the steroids' control of production of essential lymphokines. This control may be both transcriptional and posttranscriptional.

Many oncogenically transformed lymphoid cells have reduced or no dependence on exogenous lymphokines for growth and survival. Yet malignant cells may be quite susceptible to the apoptotic effects of glucocorticoids, presumably due to their direct lethal effect. Fundamental aspects of the direct apoptotic effects of glucocorticoids on lymphoid cells may therefore be uncovered by studies of transformed cell lines. In addition, the widespread use of glucocorticoids in the therapy of leukemias and lymphomas makes understanding the mechanisms by which glucocorticoids kill such cells of crucial importance. Such understanding not only provides hope for unravelling basic mechanisms of cell death, but also gives promise for the development of novel new therapies.

III. Actions of Oxysterols on Lymphoid Cells

Oxysterols were discovered by Kandutsch to be extremely potent negative regulators of cholesterol synthesis (59). Acting at both transcriptional and posttranscriptional levels, they strongly reduce 3-hydroxy-3-methylglutaryl coenzyme A reductase (HMGCAR), the rate-limiting enzyme in the cholesterol synthetic pathway (60). They also regulate transcription of several other important genes involved in cholesterol homeostasis, including the LDL receptor (61,62). Though they can be found in lipoprotein particles, oxysterols do not require the LDL/LDL receptor system to carry out these regulatory effects. Their precise modes of action in cholesterol homeostasis are unknown, but they may control the processing or function of site-specific transcription factors (SREBPs) that regulate genes of the system (63). A general overview of oxysterol action is available (64).

In the 1970s, it was noted that adding oxysterols to fibroblastoid or lymphoid cells in culture stopped their growth. When oxysterols are added to their culture medium, growing lymphoid cells are blocked in the G0/G1 phase of the cell cycle, and longer incubation with the sterols leads to cell death with the characteristics of apoptosis (65–67). Comparison of normal and malignant lymphoid cells has suggested that the latter are more sensitive, raising the possibility that oxysterols might have therapeutic value (68).

The detailed mechanisms of action of oxysterols have not been worked out as well as those of glucocorticoids. Oxysterols are found in the circulation, largely bound to albumin but also, when esterified, in LDL particles (69,70). Oxysterols can be generated by auto-oxidation, and are intracellular products of the synthesis and metabolism of cholesterol. From their extracellular location, oxysterols freed from albumin probably enter cells directly, forming the major extracellular source. The esterified oxysterols in LDL particles can enter cells through the LDL receptor mechanism as well. Experiments that add oxysterols to cells incubated in serum-free medium containing only a little carrier albumin show clearly that free oxysterols rapidly enter cells and carry out regulatory functions. The added oxysterols localize mostly in cell membranes, with a small fraction found bound to an oxysterol-specific cytoplasmic Oxysterol Binding Protein (OBP) (71,72). The affinities of a large number of oxysterols for OBP correlate well with their potency in regulating HMGCAR and other genes in the cholesterol homeostatic system (60,61). For this reason, OBP has been proposed as a receptor for oxysterols (73). Occupancy of OBP by 25-hydroxycholstrol also correlates with the concentrations of the sterol that evoke cell death (74). However, OBP is not primarily a nuclear protein and does not seem to be a transcription factor (75), so its function has not been defined. Quite recently, a specific subset of oxysterols have been shown to bind to LXRα, a member of the steroid/thyroid/retnoid receptor family (76). This binding pattern and subsequent transactivation potency seem unique, and will have to be reconciled with other structure:potency patterns shown by oxysterols.

IV. Apoptosis

"Apoptosis" was coined to designate a form of cell death observed during microscopic examination of many tissues, and its classic description (77) provides definitive morphologic characteristics. This intriguing designation for the distinctive appearance of certain dead and dying cells has led to the hypothesis of regulated responses deliberately designed to cause relatively unobtrusive cell loss. Certainly this occurs during ontogeny, and certainly something with closely similar appearance can be triggered by many exogenous ligands and treatments. Many efforts have been made to uncover the biochemical basis of these characteristic morphologic changes. While these efforts have uncovered many important correlates, many apoptosis-causing ligands, and many gene products that regulate the apoptotic pathway, as yet no universal step-by-step biochemical pathway of apoptosis has been defined. There is disagreement as basic as whether induction of specific "death genes" ever occurs, or whether mere activation of constitutive proteins is sufficient to cause apoptosis. A popular current hypothesis is that many pathways initiate apoptosis, all eventually activating a universal common and final set of events. Some current major areas under study are the roles of DNA lysis, Ca^{2+} and other ion fluxes, ICE-like proteases, Myc, Bcl2, other protooncogenes, Fas/Apo1 and Fas ligand, steroids and other hormones or hormone-like growth factors, ceramide and sphingomyelin break-down, cell shrinkage, reactive oxygen, and the tumor suppressors p53 and Rb.

Each of these topics has been widely studied, discussed, and reviewed, but a brief mention of certain issues is relevant here. The reader is referred to the many excellent reviews (see below) on these topics for more details and an introduction into the extensive primary literature.

A. DNA Lysis

Lysis of DNA is a very frequent accompaniment of apoptosis, and while it may not always be a necessary step, its close correlation with apoptosis, particularly after glucocorticoids (78), has made it the basis for many assays used to detect or follow the process (79). During apoptosis at least two distinctive types of DNA fragments appear: large pieces of about 50 and 300 kb, and small pieces that are multiples of about 180 bp (80). It has been noted that the large pieces correspond to the sizes of DNA associated with the loops and rosettes found in functioning chromatin, though it has not been proven conclusively that these are in fact the source of the DNA fragments seen. The 180-bp multiples suggest cuts occurring in internucleosomal DNA. The exact endonucleases involved are not defined, although considerable work has produced some candidates (81–84). With respect to T-lymphocyte apoptosis, it has been suggested that the large fragments are precursors for the small pieces, and that dexamethasone-induced apoptosis can occur even without the 180-bp fragments developing, if the latter process is blocked with Zn^{2+} (85).

B. Calcium Flux

Calcium ion influx and/or shifts in intracellular Ca^{2+} pools have been noted in many cases of apoptosis. This fact, taken together with the knowledge that several endonucleases and proteases are Ca^{2+}/Mg^{2+}-activated, and that addition of calcium ionophores to cells can produce apoptosis, has led to the hypothesis that alterations in Ca^{2+} concentrations are a central step in the apoptotic program. This hypothesis has yet to explain the cases in which Ca^{2+} fluxes do not seem to be involved, nor has it led to the discovery of *specific* paths to apoptosis that are Ca^{2+}-activated. Work remains active in this area, and tests of the hypothesis which are more rigorous than heretofore can be expected. For a recent review that reevaluates the Ca^{++} hypothesis, see McConkey and Orrenius (86).

C. Proto-oncogenes

Altered expression of several proto-oncogenes has been found to be tied closely to apoptosis in various systems. Space here permits mention of only a few that are particularly relevant to CEM cells.

1. cMyc

One of the proto-oncogenes is *c-myc*, whose product, the transcription factor cMyc, is required for cell cycle progression.

a. Overexpression of Myc Overexpression of Myc is seen frequently in malignancies, where it takes part in the oncogenic process. But in cultured fibroblasts, epithelial, and myeloid cells that have been prevented from growing by being deprived of some essential growth factor, deliberate overexpression of transfected c-*myc* causes apoptosis. This paradox led to the hypothesis that Myc signals for both growth and death, and that the death signal must be blocked by specific intracellular countersignals to avert apoptosis (87). As yet, these hypothetical specific countersignals have not been identified. An interesting target of Myc regulation was noted recently, however, when it was learned that Cdc25A, a cell cycle phosphatase, is induced by cMyc (88). Overexpression of Cdc25A in serum-starved cells also produced apoptosis. An alternative interpretation and hypothesis from the c-*myc* overexpression work holds that Myc overexpression in a cell blocked from cycling causes a nonspecific imbalance in the signals to grow and stop growing, and that the cell somehow senses this imbalance (13,89).

b. Reduction in c-myc expression This reduction correlates with several cases of lymphoid apoptosis. In continuous cultures of transformed lymphoid cells treated with glucocorticoids, a contrasting role of c-*myc* has been seen. There, rather than overexpression, profound reduction in c-*myc* expression is associated with apoptosis (90–95). Forced continued expression of c-*myc* alone, or together with cyclin D3, protects against the otherwise lethal steroids (93,96), and continued expression of c-*myc* protects the β-lymphoma lines WEHI-2131 and CH31B from death due to treatment with anti-μ (97).

Understanding these contrasting effects of Myc will require further research. It has been suggested that the balance between Myc and several other related proteins, such as Max, may account for cellular life or death. While the overexpression of these proteins from transfected genes may well result in cell death or escape from cell death, whether the regulation of their relative amounts naturally present in a given cell can alone explain cell death or viability remains an open question. Over/underexpression of other genes that are involved in cell cycling, including cJun, cFos, and several cytokines, also has been associated with cell death (87,98,99).

2. p53

The tumor suppressor protein p53 sometimes can act as an apoptotic signal (100) and at other times to simply halt the cell cycle. Apparently cell growth is stopped, at least in part, by p53 activating transcription of the gene for a 21-kDa protein inhibitor of cyclin/cdks (101). This connects the action of p53 with that of the tumor suppressor Rb, since p21 prevents the cyclin/cdk complex from hyperphosphorylating Rb, resulting in continued sequestration of translation factor E2F with Rb. These findings do not, however, completely explain the role of either p53 or Rb in apoptosis, since an increase in p53 levels sometimes leads to cell death and sometimes simply to cell cycle arrest. Other regulatory functions involving p53 and the Rb protein may be relevant (102,103). In addition, p53 may control other genes, including those of the Bcl2 family, and p53 seems to be required for the apoptosis caused by cMyc/Cdc25 overexpression (104). The orderly, regulated expression of cyclins, cdks, and their inhibitors is essential for proper progression through the cell growth cycle, and many apoptotic agents lead to the misregulation of the genes of this system. Hence, additional cell cycle regulatory proteins may be involved in control of cell death (96,105–108).

3. Bcl2

Bcl2, a protein found associated with several intracellular membranes, notably those of mitochondria, the inner surface of the plasma membrane, and the outer nuclear membrane, can attenuate or even block completely the apoptotic effects initiated by various signals (109,110). Among these are damage correlated with oxidative challenge, altered cellular Ca^{2+} pools, and expression of certain components of the cyclin system. In some cases of apoptosis, however, Bcl2 has only temporary or no protective effect. Bcl2 is a member of a family of related proteins of unknown primary biochemical action which appear to be involved in the regulation of cell death pathways. Bax, a smaller homolog of Bcl2, seems to have apoptosis-promoting activity. The balance between intracellular concentrations of Bcl2 and Bax, and analogous interactions between various other members of the family, have been suggested as controlling elements in the hypothetical "final common pathway." Instances in which Bcl2 fails to protect against apoptosis raise the possibility that there is more than one such pathway. Proponents of the universality of Bcl2 involvement argue that other family members' expression levels may overcome this apparent difficulty. We await better understanding of the biochemical actions of Bcl2, which should help in deducing its anti-apoptotic activity.

D. Cell Shrinkage

While this remains a cardinal feature of apoptosis, the mechanisms behind this effect have not been defined. Whether loss of cell volume is merely another correlate of the apoptotic process or is one of the forces driving it remains unknown (111). The effects of steroidal apoptotic agents on membrane ion pumps are actively being investigated (112).

E. Induced Lethality Genes

One mechanism of apoptosis originally proposed was that specific genes, *e.g.*, special endonucleases, were turned on to drive the process (113). Consistent with a need for gene expression—at least for many cases of apoptosis—

new macromolecular synthesis often appears necessary, since inhibitors of protein and/or RNA synthesis can often block or delay the appearance of cell death, or at least of some easily measured biochemical correlate. This is specifically so for glucocorticoid-evoked lymphoid apoptosis. Efforts have been made to clone the putative induced lethality genes (114,115).

F. Fas and the Cysteine Proteases

Recently, important apoptotic regulatory systems have been uncovered that operate by activation of preexisting molecules. One of these is the Apo1/Fas system (116). Apo1 or Fas, related to the TNF receptor group, is the receptor for the Fas ligand, a 40,000-kDa membrane peptide. Ligand binding, or interaction of Fas with anti-receptor antibodies, causes rapid apoptosis of lymphoid and other cells, without a requirement for further macromolecular synthesis. Part of the cytoplasmic portion of the transmembrane Fas protein contains a "death" sequence, to which other molecules bind. Activation of Fas by antibody or ligand causes these molecules to transmit a lethal signal to the cell, presumably through activation of certain proteases and/or kinases. The PITSLRE protein kinases have been postulated to play such a role (117). The cell death brought about by activation of Fas is noteworthy for its rapidity, taking place in a few hours, rather than the many hours or even days required for certain other systems.

Specific cysteine proteases have been shown to participate in the cell death brought about by some activators of apoptosis. The first such protease found was termed ICE (Interleukin Converting Enzyme) due to its action in converting pro-IL-1β into the active, smaller IL-1β (118,119). Several ICE-like proteases have since been uncovered (120,121). Activation of these enzymes leads to cytosolic and nuclear changes typical of cell death, including DNA lysis. Their activation in cytoplasts (enucleated cells) has resulted in changes which mimic the cytoplasmic aspects of apoptosis seen in whole cells (122), and addition of cytoplasm containing activated proteases to isolated nuclei has resulted in endonucleolysis (123). The ICE-like proteases cleave a variety of cytoplasmic and nuclear proteins, but no unique substrates specifically responsible for the apoptotic process have been identified as yet. Introducing several varieties of proteases into cells can evoke typical apoptotic changes, opening to question the exclusivity of the ICE-like cysteine proteases (124).

It has been proposed that the Fas and/or ICE-like protease systems explain apoptosis fully, doing away with the need to consider a requirement for new macromolecular synthesis, gene regulation, or any primary involvement of nuclear events. This interpretation has the problem of reconciling the large number of experiments which show that for many systems, blocking macromolecular synthesis at either the transcriptional or translational level blocks apoptosis. It also encounters the problem of explaining the fact that not all apoptosis occurs quickly. One possible way to reconcile all the data is to hypothesize that whether or not blocking macromolecular synthesis prevents apoptosis depends on the half-life of the preformed, critical proteolytic enzymes in that particular cell system, and/or whether they must be induced. Biochemical evidence concerning this point is scarce. Whether or not protease activation precipitates the final fulminant apoptotic catastrophe, there remains the problem of defining the prior ligand- and cell-specific events that result in apoptosis when a considerable time elapses between stimulus and the final cell death outcome. To utilize cell-specific apoptotic pathways—in tumor therapy, for example—will require such definition.

G. Summary

Scores of events and ligands that initiate apoptosis have been identified. Several molecules stand out as being frequently involved in the apoptotic processes. These include the Bcl2 and Myc families of proteins, the Fas/TNF receptors, certain protein kinases, ICE-like proteases, p53, Rb, cJun, and others not dealt with here. Two hypotheses explaining the mechanism have emerged. One holds that regulation of gene expression is an important, even an essential, step in apoptosis. The other states that no macromolecular synthesis is required, only activation of preexisting receptors and/or proteases. Whether future research will show that in fact both these interpretations of the data are compatible in a single theory of apoptosis, or will prove that there is more than one way to evoke similar cellular morphology in death remains to be seen. It may be that the mechanism used will depend on the stimulus applied.

In such circumstances, it is most important for specific apoptotic systems to be studied in detail, so that each complete apoptotic pathway can be understood and compared. We are attempting to do so in the CEM cell culture system.

V. Glucocorticoids and Oxysterols Cause Apoptosis of CEM Cells

A. Derivation of CEM Cell Clones

CEM cells are T-lymphoid cells grown from the blood of a female child with late-stage acute lymphoblastic leukemia. To provide greater uniformity of results and allow genetic analysis, we have worked with clones of CEM cells. *One clone, CEM-C7, has provided the basis for most of our work.* CEM-C7 cells are killed outright by concentrations of glucocorticoids that are sufficient to occupy their intracellular GRs. CEM-C7 cells express p53, Bcl2, and Bax constitutively. A resistant clone, CEM-C1, contained functional GR (125) and therefore seemed blocked in apoptosis downstream from the GR's actions. Other resistant clones were obtained from the steroid-sensitive CEM-C7 clone by selection in high concentrations of dexamethasone (dex). Nearly all of these appeared to be GR mutants, generally with one of two phenotypes: little or no specific, high-affinity GR binding of steroid, or reduced binding and impaired GR function (126–128). The rate of occurrence of spontaneous, GR-based resistance was consistent with a haploid genomic state for the receptor. This was confirmed after the cloning of the GR, when it was proved that CEM-C7 cells possessed one normal and one mutant GR allele. Loss of the normal allele due to spontaneous or induced mutation accounted for the loss of sensitivity to glucocorticoids. The endogenous mutant allele was a point mutant leu753phe in the LBD, and the receptor coded by this mutant gene was shown to lack proper function (127–129). The GR-positive, lysis-resistant clone CEM-C1 proved to have the same GR alleles as its sensitive CEM-C7 sister clone, and it appears that over time, C1 cells also can lose or reduce expression of their normal allele (130,131).

CEM cells proved susceptible to lysis by oxysterols, and from CEM-C7 we also have isolated subclones of cells resistant to high levels of 25-hydroxycholesterol (74,132). This cell system thus has provided a set of closely related clones in which somatic cell genetics and biochemistry can be combined to study the mechanisms by which steroids of two distinct types bring about cell death. Table I provides an outline of the derivation and properties of several useful clones.

TABLE I Properties of CEM Clones Mentioned Herein

	GR		c-Myc[a]			Apoptosis		
Clone	Genotype	Phenotype	Basal	+Dex	+OHC[b]	Basal	+Dex	+OHC
C7	+/753[c]	Active	+++	±	+	0[d]	+++	+++
H10[e]	+/753	Active	+++	±	ND[f]	0	+++	ND
C1	+/753	Active	+++	+++	ND	0	0	+++
ICR27	Δ/753	Inactive	+｜｜	｜++	ND	0	0	+++
4R4	−/753	Activation labile	｜｜｜	+++	ND	0	0	+++
M1OR5	+/753	Active	+++	±	+++	0	+++	0

[a] c-Myc refers to mRNA and/or protein; +++, strong positive signal; ±, low or not detected.
[b] Dex, dexamethasone, OHC, 25-hydroxycholesterol.
[c] + Refers to wild-type; 753 refers to the point mutation leu753phe; Δ means allele deleted; − means lack of expression allele, precise mutation not yet defined.
[d] 0, ≤5% nonviable cells; +++, death of ≥99% of cells in freshly subcloned culture (7–10,65,66,74,92–94,125–129,132,135,138,144,150).
[e] H10 is a somatic cell hybrid between ICR27 and C1 (92 and ref. therein).
[f] ND, not determined.

B. Both Glucocorticoids and Oxysterols Cause CEM Cell Death by Apoptosis

Simple observation of cultures of CEM-C7 cells treated with either agonist glucocorticoids or oxysterols provides convincing evidence that both types of steroid kill the cells. The central question remains, what processes cause that cell death? Several classic tests show that both glucocorticoids and oxysterols cause apoptosis (7,65,66,74,133,134). Cell shrinkage is seen within hours after either glucocorticoid or oxysterol treatment of sensitive, but not resistant, clones. As discussed above, oxysterols readily enter and remain in cell membrane lipid layers and also bind a cytoplasmic protein with high affinity, leading to the hypothesis that it is a receptor. Recently, they also have been shown to bind to the $LxR\alpha$ receptor. They quickly reduce cholesterol synthesis in many cells, including CEMs, and also have been shown to regulate expression of genes not involved in cholesterol synthesis. Since cell shrinkage is one of the classic identifying features of apoptosis, it will be important to determine its mechanism. *In sum, both glucocorticoids and oxysterols promptly cause cell shrinkage of CEM cells, one of the classic morphologic hallmarks of apoptosis. The mechanism remains unknown, but for the glucocorticoids, the classic intracellular GR seems involved, since cells lacking functional GR do not shrink in the presence of dex.*

Other morphologic hallmarks of apoptosis are seen after addition of either glucocorticoids or oxysterols to the sensitive CEM-C7 cells, including loss of microvilli, pericentric nuclear accumulation of condensed chromatin, appearance of cytoplasmic vacuoles, and, eventually, development of typical, small, dense cells or cell fragments, containing nuclear compartments completely filled with heterochromatin. Some of these changes are depicted in Fig. 1. After either class of steroid is added to logarithmic-phase cultures containing $\geq 95\%$ viable cells, significant numbers of cells begin to manifest such changes following a delay of about 24 hr, and before actual cell death can be detected. Thereafter, with the steroid present, increasing numbers of cells develop increasingly severe morphologic alterations during the subsequent 2–3 days. This subsequent progression roughly coincides with loss of viable cells from the culture as measured by counts of trypan-blue-excluding cells or by clonogenicity assays. One can conclude that morphologic changes characteristic of apoptosis are brought about in CEM-C7 cells by both oxysterols and glucocorticoids. Cell shrinkage occurs well before the actual death of the cells, while other changes begin shortly before overt cell death.

A frequent correlate of apoptosis is endonucleolysis of DNA within cells that are still recognizable as membrane-enclosed entities. This event can be identified and quantified in various ways, *e.g.*, by FACS analysis, electrophoresis of DNA, Elisa assay, or end labeling with terminal deoxynucleotide

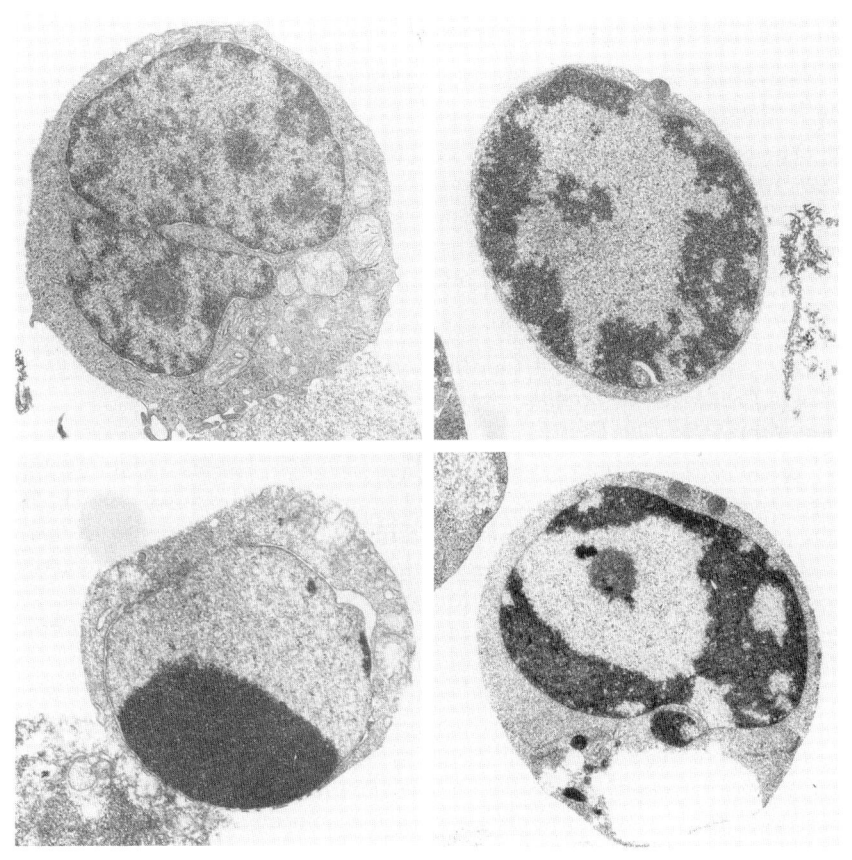

FIGURE 1 Both glucocorticoids and oxysterols evoke typical apoptotic morphology in CEM-C7 human lymphoid leukemia cells. Transmission electron micrographs of CEM-C7 cells in various stages of apoptosis after addition of 1 μM dex or 300 nM 25-hydroxycholesterol. Upper left, control cell from a log-phase culture, ×15,000. Upper right and lower right, 24 and 48 hr 1 μM dex, ×15,000. Lower left, 48 hr 1 μM 25-hydroxycholesterol, ×10,400. The steroid-treated cells show varying degrees of shrinkage, loss of microvilli, centrifugal heterochromatization, and vacuole formation.

transferase (TUNEL assay). Exposure of CEM-C7 cells to oxysterol or to dex caused endonucleolysis, and produced large DNA fragments of ~50 and 300 kbp as well as smaller pieces that were multiples of 180 bp (65,66). The time course and pattern of fragmentation differed slightly between the steroids. For example, the TUNEL assay, which measures nicked ends of DNA, was more markedly positive after oxysterol treatment than after dex, and oxysterol-treated cells showed evidence of DNA breakage occurring several hours before overt cell death demonstrated by dye exclusion assays. In general, however, both classes of steroids cause DNA lysis characteristic of apoptosis.

C. The Glucocorticoid Pathway to Apoptosis

1. The Glucocorticoid Receptor Is a Required Part of the Pathway

The data from CEM as well as other lymphoid cells and cell lines strongly support the conclusion that glucocorticoids must interact with their intracellular receptor to initiate apoptosis (4–11,95,126–131,135–137). To understand how the activated GR transduces the steroids' signal to cause apoptosis, one must define the stepwise events after addition of glucocorticoid to CEM-C7 cells.

2. The Kinetics of the CEM Cell Response to Glucocorticoids Show a Prolonged Reversible Phase, Followed by an Irreversible Commitment to Cell Death

When cells in midlog growth are exposed to concentrations of dex that fully occupy GR, there follows a period of at least 24 hr during which the cells remain fully viable, while events critical for eventual apoptotic death occur. Expression of several genes changes, the cells shrink in volume, and evidence of DNA nicking can be found. Nonetheless, during this time the cells continue unchecked in logarithmic growth, and removal of steroid (138) or addition of a glucocorticoid antagonist (135) completely prevents cell death, permitting the cells to continue growing as well as untreated control cultures. Beyond 24 hr in the presence of enough agonist glucocorticoid to occupy the GR, the cells begin to collect in the G0/G1 phase of the cell cycle and, roughly in parallel, to lose viability (138). During this period, increasing DNA lysis into specific-sized pieces occurs (65,66), thymidine kinase activity falls, polysomes disaggregate (7), and the morphologic manifestations of apoptosis become apparent in increasing numbers of cells. Over the succeeding few days, apoptosis spreads to virtually all the cells, sparing only those few resistant mutants that have accumulated in the culture. Figure 2 diagrams this time course.

3. DNA Lysis Is Prominent during Glucocorticoid-Evoked Apoptosis

One of the most common correlates of apoptosis is DNA lysis. It has been proposed that this is a leading, even causative, event in the eventual cell death; considerable effort has been made to isolate one or more specific endonucleases hypothesized to be induced by apoptotic agents (78–84), although no unique endonucleases have as yet been demonstrated compellingly. However, instances of apoptosis have been found in which endonucleolysis appears to have little or no role (139). Hence, the *universality* of endonucleolysis as the primary driving force for apoptosis is in doubt. Whether or not it is a primary cause of death, DNA lysis into specific sizes is a prominent part of glucocorticoid action on lymphoid cells. Pulse-field

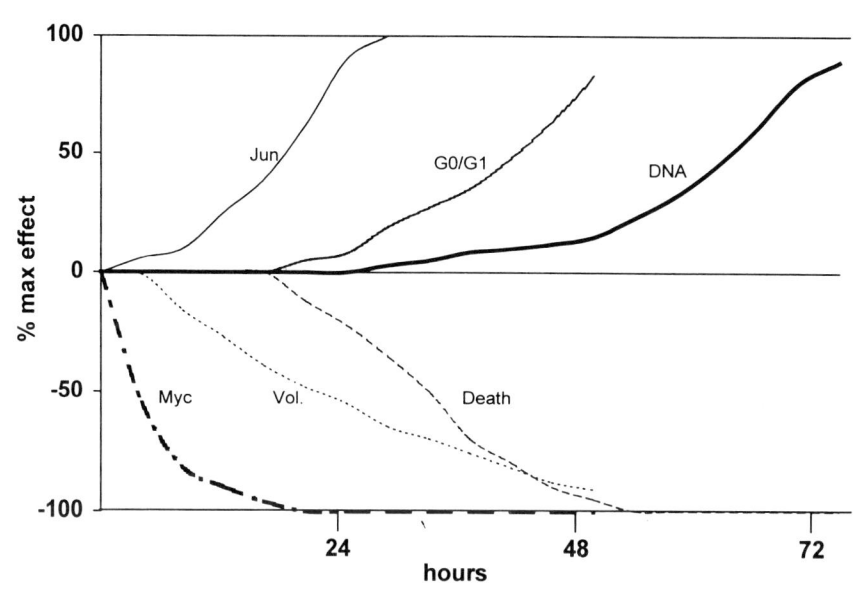

FIGURE 2 Sequence of dex effects on CEM-C7 cells. Diagrammatic representation of some events in CEM-C7 cells following addition of glucocorticoid (dexamethasone). Each parameter is plotted as a percentage of its own maximum (minimum) effect. For example, in absolute terms, Myc levels fall to about 10% of basal. From data in Thompson et al. (7,66,92,93,99,138).

gel electrophoresis of DNA from dex-treated CEM-C7 cells shows very large (~50 kb) DNA fragments and a continuum of medium-sized fragments with a maximum at about 15 kb. This "smear" of DNA is evident on the gels about 24 hr after the addition of dex, with the 50-kb band appearing hours later (66). The traditional "ladder" of DNA fragments, representing multiples of ~180 bp—presumably due to cuts occurring between nucleosomes—is not a prominent feature of apoptosis in CEM-C7 cells, but can be demonstrated, after ≥36 hr in dex. It has been suggested that the large (50-kb and greater) fragments are precursors to the DNA ladders (85). Our data do not rule out a precursor–product relationship between the big and small DNA fragments, though at the intervals we have studied, we see no precedence in time for the larger fragments. The appearance of typical, extensive, apoptotic DNA lysis into both large and small fragments occurs in CEM-C7 cells near to or concomitantly with final collapse and death.

4. Essential Events Occur during the Reversible Phase of Glucocorticoid-Evoked Apoptosis

We hypothesize that during the reversible period that precedes cell death a sequence of reversible but essential switches are thrown, eventually placing the cell in a state from which it cannot recover, manifested by fulminant apoptosis. Testing the hypothesis requires identification of the critical, early, essential but reversible changes. From screens of regulatory molecules of potential importance, we have begun to see the outline of such a sequence. One of its later stages is the DNA lysis just discussed. The first early regulatory molecule found to be important in the pathway was cMyc.

a. Repression of the Growth-Regulatory Transcription Factor cMyc Is Important for Apoptosis of CEM Cells Lymphoid cells' activation depends on increased expression of a number of genes which control and determine progression through the cell cycle. Among these is the proto-oncogene c-*myc*. Overexpression of c-*myc*, usually without structural mutation, is associated with oncogenic transformation of lymphoid and many other cell types (140). This transformation usually occurs as c-*myc* acts in conjunction with other oncogenes, notably c-*ras*. In 1986 it was observed in the mouse lymphoma cell line S-49 that incubation with dex caused rapid down-regulation of c-*myc* mRNA (among other genes), prior to cell death (90). When we examined the effect of dex on several proto-oncogenes in clones of mutant CEM cells, we found exact correlation only between c-*myc* down-regulation and cell death (92). That is, c-*myc* was greatly reduced by the steroid in sensitive clones CEM-C7 and H10 (a dex-sensitive somatic cell hybrid between CEM-C1 and ICR-27), but was not affected in resistant clones ICR-27 or CEM-C1. While it expresses some mutant GR protein, ICR-27 shows little or no GR binding by whole cell assay and is completely resistant to dex; the CEM-C1 cells tested contained functional GR (able to induce glutamine synthetase) but were resistant to the apoptotic effects of glucocorticoids (125). Thus, among 20 or so genes tested, c-*myc* was the first gene we identified that proved to be unaffected in dex-resistant CEM cells whether GR^- or GR^+. Negative regulation of c-*myc* by dex in CEM-C7 cells has been confirmed independently (94). We also showed that the loss of c-*myc* mRNA following addition of dex was matched by loss of cMyc protein. Expression of the Myc binding partner Max (often proposed as an apoptotic factor) did not vary from its basal level in any clone of dex-treated CEM cells tested (EBT, unpublished results). We hypothesized that down-regulation of c-*myc* is an essential and early effect for CEM cell apoptosis and have carried out several tests of the predictions imposed by this hypothesis (93,135).

The down-regulation of cMyc is rapid and can be seen to begin by an hour after adding the steroid. Levels continue to fall until reaching a minimum of about 10% of controls after 12–18 hr, well before any loss of cell viability. The hypothesis predicts that preventing the fall in cMyc will block dex-evoked apoptosis. We tested this prediction by constructing plasmids expressing c-*myc* under the control of three different promoters not downregulated by dex. When any of the three was transiently transfected into CEM-C7 cells, Myc levels remained significantly higher in the face of dex treatment, and there was a significant reduction in dex-evoked apoptosis.

The hypothesis further predicts that reducing myc sufficiently by an agent other than dex will lead to CEM cell death. As predicted, when CEM-C7 cell Myc levels were reduced by use of antisense c-*myc* RNA, the result was cell death. If the effects of dex are removed during the reversible window preceding overt apoptosis, c-*myc* levels should rebound quickly, before the culture starts to resume growth. We reversed dex action after various times by adding the GR-occupying glucocorticoid antagonist RU38486 and found that the prediction was confirmed. Very shortly after adding RU38486, c-*myc* recovered to control levels and cell growth was restored (135). Finally, we discovered that restoration of dex sensitivity to the GR$^+$ but dex-resistant clone CEM-C1 by simultaneous treatment with forskolin and dex (see below) resulted in a decrease in c-*myc* mRNA to an extent similar to that seen in CEM-C7 cells treated with dex alone (43; R. Medh, F. Saeed, and E. B. Thompson, unpublished results).

Negative regulation of c-*myc* during glucocorticoid-induced cell death has been observed in other lymphoid cell systems. The mouse lymphoid line P1798 is growth-arrested and eventually killed by glucocorticoids, and c-*myc* reduction is an early event in this process. The cells can be rescued by coexpression of c-*myc* and cyclin D$_3$, another gene negatively regulated by steroids in those cells (96). Cells of the Jurkat human lymphoid line lack functional GR, but when transfected with a vector expressing GR they become sensitive to dex-evoked cell death. In these circumstances, c-*myc* is rapidly down-regulated after addition of the steroid (95). Although it is not a direct glucocorticoid effect, further evidence supporting the importance of c-*myc* for lymphoid cell viability comes from the WEHI cell system. Treatment of WEHI-231 and CH31 cells with anti-μ causes a brief, abrupt increase in c-*myc* expression, followed by a profound and sustained decrease. Apoptosis ensues. Addition of antisense c-*myc* in this system prevents basal c-*myc* levels from changing significantly, and apoptosis is prevented (97). In sum, down-regulation of c-*myc* by glucocorticoids appears to be a critical early step in the glucocorticoidal apoptotic pathway in cultures of transformed lymphoid cells.

b. cJun Induction Is Important for Glucocorticoid-Evoked CEM-C7 Cell Apoptosis A few hours following the addition of dexamethasone to CEM-C7 cells, well after the decrease in Myc is underway, induction of c-*jun* occurs. Induction of both c-*jun* mRNA and Jun protein can be seen as early as 6–12 hr after adding dexamethasone, and by 24 hr Jun levels reach a plateau six- to eightfold over basal levels, where they remain until apoptosis overwhelms the culture (99). Thus sustained c-*jun* induction, like c-*myc* reduction, precedes glucocorticoid-induced apoptosis of CEM-C7 cells. Using methods analogous to those employed in our Myc studies, we have tested the hypothesis that this induction of cJun also is a requisite step. First, from the lack of such a response in the GR-deficient, resistant clone ICR-

27, we concluded that the induction required functioning GR. In the GR$^+$ resistant clone CEM-C1, we found that there was induction of c-*jun*, but from a basal level of expression it was so much lower than that in clone C7 that after full induction, cJun levels in CEM-C1 cells only approached the basal levels found in CEM-C7 cells. We reasoned that sustaining cJun at a level well above this might be important for apoptosis to ensue. Second, we expressed antisense c-*jun* in CEM-C7 cells, blocking the induction of cJun without greatly affecting its basal levels. In such transfected cells, apoptosis was blocked (99). Thus, it seems that sustained expression of cJun above a certain threshold is part of the apoptotic pathway evoked by glucocorticoids in CEM cells. It may play a role in the apoptotic process in more than this one type of cell, since in other cell systems overexpression of c-*jun* can lead to or is associated with apoptosis and inhibition of its elevation blocks apoptosis [see references in Zhou and Thompson (99)]. The mechanism of cJun induction is unclear. Its delay suggests that it is not a simple, direct induction by activated GR; rather, some intervening step is needed. Preliminary data suggest that both transcriptional and posttranscriptional regulation are involved (E. B. Thompson and F. Zhou, preliminary results). Early work on c-*myc* suggested that an AP-1 site in its regulatory region exerted a negative effect on its expression (141). If this is so, the increase in cJun might contribute to the sustained repression of c-*myc*.

These experiments in CEM cells provide the beginnings of an outline of the events that define glucocorticoid-induced apoptosis in this lymphoid cell line. Figure 2 diagrams that sequence. The apoptotic response of CEM cells can be thought of as a continuum that results from two functionally different periods: an early sequence of events which are critical for eventual apoptosis but which are fully reversible if agonist steroid is removed in time, and a later set of changes which either closely precede or are part and parcel of the apoptotic collapse of the cell. The earliest event observed so far is the onset of repression of c-*myc*, followed shortly by cell shrinkage and induction of c-*jun*. (In the same interval the GR itself and glutamine synthetase are induced; however, neither of these seems to be critical for the apoptotic process.) The goal of future work is to identify other critical steps following the changes in c-*myc* and c-*jun* and antedating the ultimate apoptotic catastrophe. But G0/G1 arrest, gross DNA lysis into easily seen fragments of several sizes, and evidence of failure of major macromolecular synthetic systems all occur later, as the full morphologic evidence of overt apoptosis becomes manifest.

D. The Role of the GR in Apoptosis: Mapping the GR for Cell Death Domains

I. Regulation When the GR LBD Is Intact

In the early models of glucocorticoid-evoked apoptosis, it was presumed that the steroid-activated GR caused the induction of some lethal gene or genes. The discovery that reduced c-*myc* expression is important opened the question of which GR functions are required for apoptosis, because some GR domains required for gene induction are not needed for gene repression. Nevertheless, since a glucocorticoid agonist must be present for many hours in order for CEM cell apoptosis to occur, the activated, ligand-bearing GR must be required continually. Mutational analysis has identified a number of functional domains within this molecule, including ligand- and DNA-binding domains, nuclear translocation signals, and regions important for activation of genes. Also identified to some extent are the binding sites for certain heat shock proteins and other factors. A recent review of GR structure and mutations gives a useful analysis of the domain concept and its limitations (142). GR domains essential for activation of genes need not play a role in gene repression, which often seems to be the result of interactions between the GR and other transcription factors. Since it was not obvious that direct gene induction by the GR was responsible for the death of CEM cells, we employed transfections of the dex-resistant clone ICR-27 that contains only a GR lack of function mutant to map the regions of the human GR for delivery of a ligand-dependent apoptotic signal (11). First, electroporation conditions were found which resulted in transfection of a reasonable fraction (~40%) of the cells without any loss of cell viability (143). Then it was shown that restoration of the holoGR to these cells resulted in restoration of the apoptotic response to dex with its usual time course. About 25–30% of the cells were killed when glucocorticoid was supplied following transfection of the wild-type holoreceptor. Standard assays for specific competitive binding sites on the receptor showed that the transfected culture contained an average 3000–4000 sites per cell, with affinity typical of the natural GR. The parental CEM-C7 clone contains about 10,000 sites per cell; considering that some 40% of the ICR-27 cells were transfected, it seems likely that there was not gross overexpression of the transfected GR gene. Our observation that ~25% of the transfected cells were killed following addition of dex is consistent with this view and with the interpretation that a reasonable fraction of the transfected cells expressed and retained sufficient GR to transmit the lethal response. It is well documented that increasing the intensity of electroporation conditions results in higher transfection efficiency, but also in the death of many cells. When we intensified the electroporation conditions while transfecting the holoGR, we found we could raise the proportion of surviving cells which could be killed subsequently by dex to virtually 100% (M.El-Nagy and E. B. Thompson, unpublished results).

Using the milder electroporation conditions to avoid any confounding lethal effect of electroporation itself, we mapped the GR for regions essential to deliver dex-evoked cell death (11,144). Two domains proved necessary— the DNA Binding Domain (DBD) and the Ligand Binding Domain (LBD). Mutations that deleted either of the zinc fingers of the DBD or amino acid substitutions in critical sites within the amino-terminal zinc finger all completely blocked the lethal response. There seemed to be a specific requirement for the GR DBD, since a receptor containing a mutated DBD, altered to bind solely to an Estrogen Response Element (ERE), resulted in loss of lethal activity, whereas a similar DBD that recognized a GRE as well was active in causing death.

The LBD mutations showed an absolute requirement for ligand binding, but the ligand did not have to be a glucocorticoid, since a chimeric molecule in which the DBD of the GR replaced that of the estrogen receptor, leaving intact the estrogen-specific LBD, conveyed estrogen-specific binding and lethality to the ICR-27 cells. This experiment provided the first data suggesting that the amino-terminal portion of the human GR, with its potent transactivation domain, was not at all necessary. Other transfections, using GR mutants lacking the GR amino-terminal transactivation domain, showed that in this system, there is no absolute requirement for the transfected GR to have this, or indeed most of its amino-terminal region, in order to convey a steroid-activated lethality signal. This contrasts with results from the S49 mouse lymphoma cell system, where one of the classic dex-resistant mutants proved to contain an amino-terminal truncation of the GR that destroyed the transactivation domain. Recent experiments using stable transfectants of S49 cells have confirmed that result and given evidence that in the S49-based system, an intact amino-terminal transactivation domain is essential for evoked cell death (136). Why the differing results in the CEM and S49 cells? There are numerous differences between the systems: transient or stable transfections, human or mouse cells and GR, lymphoid cells in somewhat differing stages of development, GR mutations at nonequivalent loci, and dissimilar basal mutant GRs expressed in the two cell lines, to name a few. Understanding the molecular basis for the difference in the basic result, however, will be important in unravelling the general mechanisms of glucocorticoid-evoked lymphoid cell apoptosis.

In the two human cell systems tested so far, however, the data suggest that transactivation by the GR is not important for provoking apoptosis (95,143,144). The view that gene repression, not gene induction, is the critical apoptotic function of the GR—in human lymphoid cells, at least—is consistent with our results showing that down-regulation of c-*myc* expression is an essential and very early effect of glucocorticoids in CEM cells. Studies on the GR-deficient human Jurkat cell line showed that transfection of a GR mutant incapable of gene transactivation nevertheless caused cell death (95). Like ICR-27 cells, Jurkat cells contain some residual mutant GR, so the existence of some obscure cooperative complementation of function between the endogenous mutant GRs and the transfected GRs has not been ruled out. As mentioned above, c-*myc* reduction following addition of glucocorticoids, and well before the onset of overt apoptosis, has been observed in both mouse and human cell lines. As to mechanism, in mouse

P1798 cells this negative regulation is a direct transcriptional effect (145). Although one report on the subject found the level of control in CEM cells to be posttranscriptional (146), our results on CEM-C7 cells indicate strong inhibition of c-*myc* transcription following dex (F. Zhou and E. B.Thompson, unpublished results). In sum, the weight of evidence in the human systems at present strongly favors the conclusion that the transcription transrepressive, not the transactive, function of the GR is responsible for evoking apoptosis. This is not to say that no gene induction is required. As we have pointed out (93), c-*myc* is a transcription factor with the capacity to both induce and repress genes. Reduction of cMyc could therefore relieve the repression of other genes, some of which are perhaps essential for apoptosis. The results of Wood *et al.* (94) suggest that this prediction may be true, since they show inhibition of dex-evoked apoptosis of CEM cells by the protein synthesis inhibitor cycloheximide. This interpretation would also fit with our observation that cJun induction, which occurs after cMyc repression, is essential for CEM cell apoptosis.

The effects of mutations in the human GR on its ability to cause apoptosis in the human lymphoid cell lines CEM and Jurkat are shown in Fig. 3. The conclusions from these results are: (1) that when the LBD of the GR is intact, the GRE-specific DBD but not the GR amino-terminal portion—which includes the major transactivation domain—is essential for cell death, and (2) that ligand-dependent gene repression is a primary action of the human GR in human leukemic lymphoid cell apoptosis. As a consequence, genes originally repressed through factors down-regulated by the ligand–GR complex may be induced.

2. Constitutively Lethal Fragments of the GR: A Ligand-Independent Pathway to Cell Death

Among the GR mutants tested for domains essential for cell death were several altogether lacking the LBD. When cells of the ICR-27 clone were transfected with these, a remarkable result was observed—a reduction in cells equivalent to that occurring after transfection of holoreceptor and administration of steroid ligand. Double mutants of the GR, with deletions of both the amino-terminal transactivation domain and the SBD, were nearly equally effective (143,144). The reduction in viable cells by these LBD-mutant forms of the GR followed strikingly different kinetics than those seen after holoGR and added steroid. Without the LBD, the transfected GR gene fragments acted within the first 24 hr, following which no further effect was seen. This differed from the holoGR with steroid, which killed cells only following a 24-hr delay. Figure 4 summarizes these results.

One singular mutant GR lacking a LBD was caused by a frame-shift beginning at amino acid 465 and termed "465*" (147). The frame shift predicts a novel 21-amino-acid sequence followed by a termination codon beginning just after the *ileu* following the 2nd invariant *cys* of the second

No loss of function

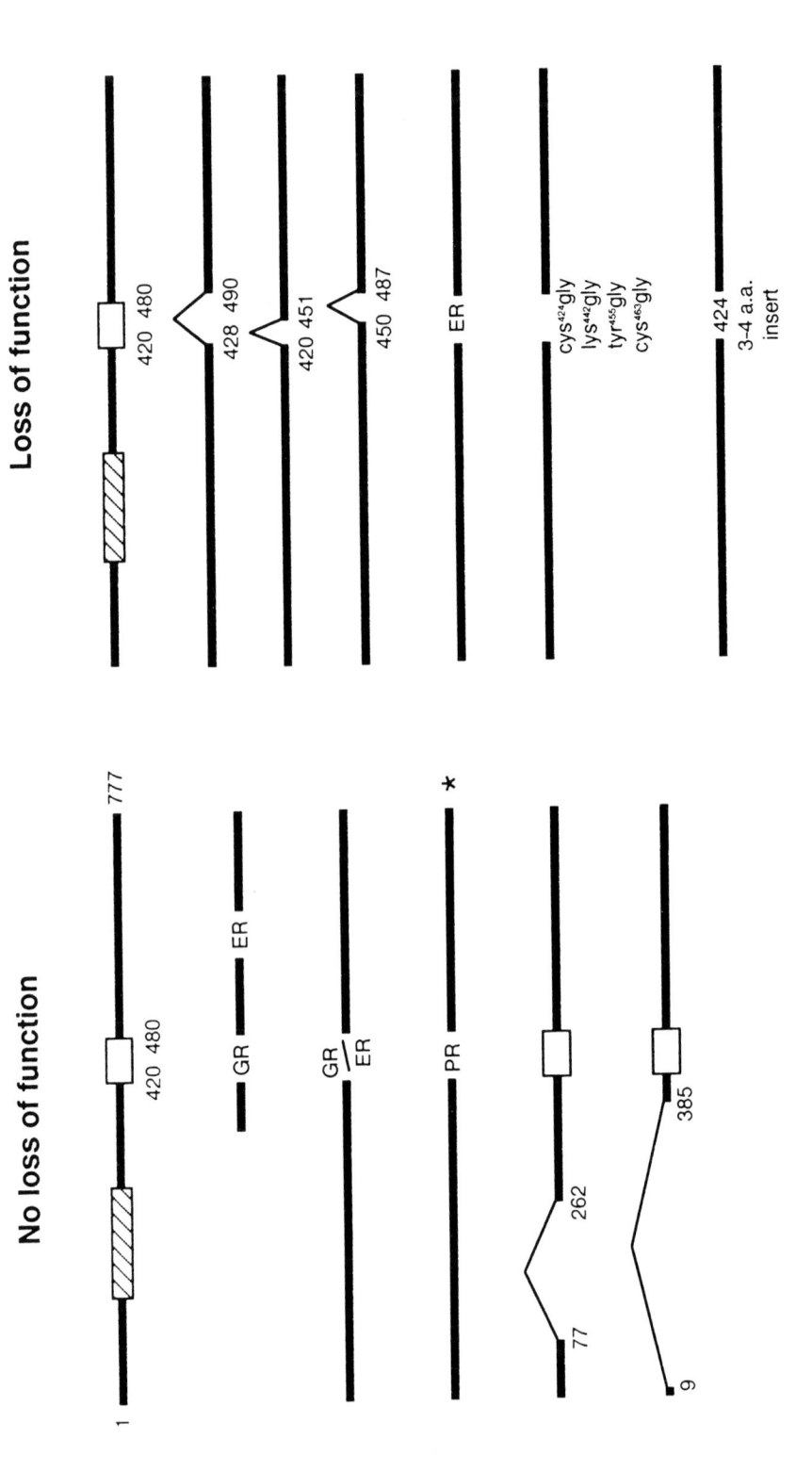

Loss of function

zinc finger. Expression of this construct in a baculovirus system has produced a protein of the expected size (G. Srinivasan and E. B. Thompson, unpublished results). Transactivation assays had shown 465* to have virtually no ability to provoke GRE-driven transcription (147), but the expression plasmid bearing GR mutant 465* was as effective in cell kill as several others that lacked the SBD but retained the complete, normal DBD of the GR (144). As with hologR, when stronger electroporation conditions were used, these mutants lacking the SBD killed increasing proportions of the surviving cells, and conditions were found at which most of the cells were killed, compared to controls. We have subsequently removed most of the amino-terminal amino acids as well from 465* with only a little loss of activity. The resultant gene (398–465*), coding for only amino acids 398–465 plus the predicted 21-amino-acid frame-shifted extension, causes loss of transfected cells (143). Removal of the c-terminal 21 amino acids causes loss of activity (E. B.Thompson, unpublished results).

To see whether these mutant fragments of the GR were universally toxic to hematopoietic cells, we compared the effects of transfected expression vectors carrying mutants 465* and 398–465* on several lines of lymphoid T- and B-derived as well as myeloid-derived cells (143). The cell lines tested contained varying levels of GR, but all were resistant to glucocorticoid-induced apoptosis. Transfection efficiency proved to be about equal in the CEM clones tested and in the myeloid cells. All the lymphoid cell lines were found to be susceptible to the 465* mutants, but both myeloid lines tested were resistant. This could be due to differential expression or stability of p465*, or to a basic difference in the nature of myeloid and lymphoid cells. The data to date suggest that for whatever reason, these GR mutants are not universal cell toxins but display some cell selectivity. It is therefore possible that such GR fragments could be useful in the therapy of certain hematopoietic malignancies, since in principle they provide a way to kill leukemic cells without the systemic effects of glucocorticoids. Determining

---

FIGURE 3 Mapping the GR for ligand-dependent apoptosis-evoking activity. Transient transfections of receptor constructs into the dex-resistant CEM-C7 subclone ICR-27 with or without added dex as ligand and following the cultures for cell death led to the conclusions represented here. The normal GR is represented at the top of each column by a bar indicating the usual 1–777 amino acids, with the approximate position of the *tau* 1 major transcription domain indicated by a hatched box and the DBD, by an open box between aa 420 and 480. GR or a box in that position indicates a GRE-specific DBD. GR/ER indicates a DBD capable of binding either a GRE or an ERE; ER means an ERE-specific DBD. PR indicates the progesterone receptor. *The progesterone receptor is not identical in length to the GR. Deletions, single aa changes, and insert mutations are shown in standard fashion. In the "No loss of function" column, all mutants shown exhibited essentially the same degree of ligand-dependent lethal activity. In the "Loss of function" column, all mutants shown failed to cause any cell death, with or without ligand. From data in Thompson *et al.* (11,43,143,144).

Normal GR

```
1 ▬▬▬▬▬▬▬▬▭▬▬777
         420 480
```

Mutants   **Cell Loss Without Ligand**

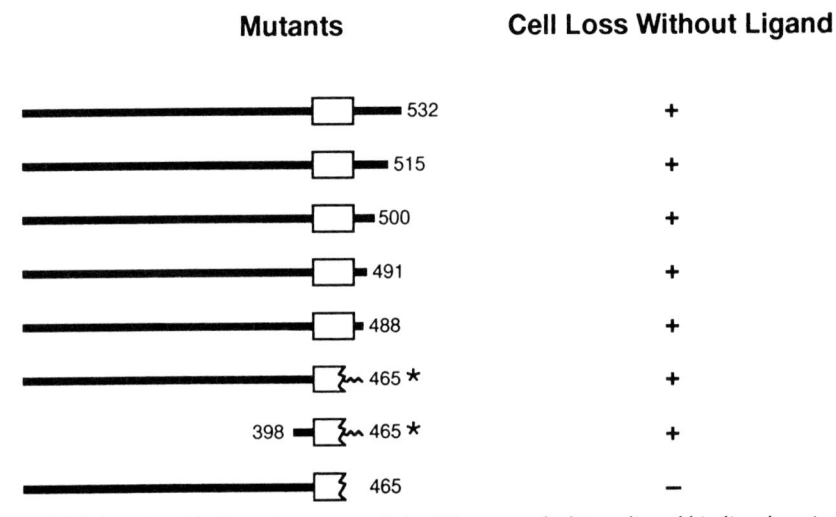

─────────────▭─ 532   +

─────────────▭─ 515   +

─────────────▭─ 500   +

─────────────▭─ 491   +

─────────────▭─ 488   +

─────────────▭∿ 465 *   +

398 ▬▬▭∿ 465 *   +

─────────────▭ 465   −

FIGURE 4 Ligand-independent apoptosis by GR mutants lacking a ligand binding domain. Transient transfections of GR constructs into various cell lines, including ICR-27, led to the conclusions represented here. Diagram of the normal GR at top, as in Fig. 3. Mutants lacking the LBD are shown, aligned by the N-terminal end of their DBD (open box). The approximate position of the *tau* 1 transactivation domain is indicated in the normal GR and is present along with the entire N-terminal sequence in all mutants except 398–465* (11,43,143,144, and E. B. Thompson, unpublished results).

their mechanism of action and their structural requirements for activity therefore will be important.

We have focused on the 465* mutation since its carboxy-terminal sequence is so radically different from that of the normal GR DBD and hinge region. The 465* mutant protein, expressed in the baculovirus system, has been utilized in gel-shift experiments and compared with the recombinant GR peptide 1–500, which codes for the entire normal DBD. (Both genes have equivalent constitutive killing activity in transfection experiments.) DNA binding experiments with radiolabeled consensus GRE oligonucleotide showed peptide 465* to bind far less strongly than peptide 1–500. In cell-free transcription experiments, using a GRE-driven G-free gene, a system previously shown to be highly sensitive to site-specific stimulation by the holoGR (25), we found that 465* had only minimal stimulatory activity.

It did not inhibit transcriptional regulation by the holoGR. It seems improbable that at the levels likely to be expressed in transfected cells, 465* interferes with cell viability by regulating GRE-driven genes or by heterodimerizing with the GR. It also does not seem to interfere with basal transcription machinery (H. Chen, G. Srinivasan, and E. B. Thompson, unpublished results).

When the naturally fluorescent Green Fluorescent Protein (GFP) was attached to the amino terminus of 465*, and the chimeric protein was expressed in transient transfections of CEM clones, fluorescence was seen throughout the cells, but more intensely in the cytoplasmic compartment (H. Chen, M. El-Nagy, and E. B. Thompson, unpublished results).

Collectively, these results suggest that 465* may have its predominant action through heterodimerizations with other proteins. A chimeric protein consisting of glutathione S-Transferase (GST) attached to the amino terminal end of 465* has provided a means of testing this notion. GST:465* was attached to Sepharose beads coated with glutathione. The glutathione:GST link allowed the mutant 465* peptide to remain free to interact with other proteins. When CEM cell extracts were exposed to this GST:465* "trap" a number of proteins were seen to bind preferentially. Several of these will bear investigation, but one already identified is cJun.

When nuclear extracts of CEM cells were exposed to the trap, cJun was among the proteins that adhered. Recombinant cJun also bound to the GST:465* and not to control beads. The ability of peptide 1–465* to interact functionally with cJun was tested in transfection assays. CV-1 cells were cotransfected with plasmids expressing 1–465*, cJun, and constructs containing reporter genes under the control of AP-1 sites (the cJun DNA binding site). As expected, cJun caused induction of the reporter gene. Coexpression of 465* significantly reduced the Jun-driven induction. Addition of excess Jun overcame the 465* effect. Tests of GR peptide 1–500 showed it to be even more strongly inhibitory of cJun-driven transcription (H. Chen and E. B. Thompson, unpublished results). While it seems unlikely that this particular interaction fully explains the inhibitory effects of the mutant peptides on cell growth and viability, the demonstration that these truncated GR forms can interact physically and functionally with important regulatory molecules –for example, with one of importance in the glucocorticoid-induced apoptotic pathway—serves as an important paradigm for their mechanism of action.

E. The Oxysterol Path to Apoptosis in CEM Cells

As described above, oxysterols cause clear-cut apoptosis of CEM cells. Relative to the glucocorticoid path, much less is known about the biochemical steps of the oxysterol apoptotic path, but information is accumulating. After exposure to oxysterols, the cells shrink, show other morphologic changes typical of apoptosis, and undergo marked intracellular lysis of their DNA in a time course resembling that seen following glucocorticoids. Though very similar, the cellular catastrophe brought about by the two different types of steroids is not identical in detail, and certainly differs in its initial steroid:cell interactions. Glucocorticoids require their intracellular receptor to cause apoptosis, but GR mutants resistant to glucocorticoids are fully sensitive to oxysterols (74). Clones selected for resistance to oxysterols are sensitive to glucocorticoids also (66). The molecular basis for resistance in these clones has not yet been determined.

1. Oxysterols May Evoke Apoptosis by Mechanisms Other Than Inhibition of Cholesterol Synthesis and Uptake

Certainly, cells must have cholesterol to grow. One view of how oxysterols might cause apoptosis is that by severely reducing its synthesis, they cause cells to "run out of cholesterol" as the demand for membrane production in growing cells outstrips the supply. This hypothesis fails to explain several observations. First, oxysterols can cause the death of primary cultures of mouse thymocytes and activated cytotoxic lymphocytes (67,148). Second, oxysterols can cause cell death of growing normal or malignant lymphoid cell cultures even in the presence of whole serum, which provides abundant cholesterol in LDL (67,74,148). Under serum-free conditions, the CEM cell death caused by low concentrations of 25-hydroxycholesterol can be reversed by a large excess of cholesterol, but in the presence of this excess cholesterol, raising the oxysterol concentration moderately results in cholesterol-resistant cell death (74). Further, the morphologic changes of apoptosis and the extensive DNA lysis seen in oxysterol-treated cells suggest that processes other than simple rupture/leakage of membranes due to reduced cholesterol content are involved.

2. Oxysterols Can Regulate Genes outside the Cholesterol Homeostasis Pathway

Recently it has been discovered that oxysterols regulate genes not involved in the cholesterol homeostatic pathways (148). One report indicated that 7,25-dihydroxycholesterol reduced production of the cytokine IL-2 (149). We have shown that 25-hydroxycholesterol reduces Cellular Nucleic acid Binding Protein (CNBP) mRNA in CEM-C7 cells (150), and not in cells selected for resistance to oxysterol-induced apoptosis. CNBP was originally suggested as a direct regulator of sterol synthetic genes; however, CNBP-L, the form originally identified, did not seem to control such genes. On the other hand, CNBP has been shown to bind to the CT element of the c-*myc* gene and to induce a CT-driven reporter gene in transfection assays (151). While some isoform of CNPB may still be found to be involved in regulation of cholesterol synthesis, the protein may have a more general regulatory role.

Very recently, we have found oxysterol regulation of several genes important for cell cycling. Among these is c-*myc*. The negative regulation of CNBP by oxysterols may therefore contribute to the reduction in c-*myc* by the sterols. These discoveries reinforce the possibility that regulatory events other than blocking cholesterol synthesis are involved in oxysterol-induced apoptosis.

3. Possible Mechanisms for the Transduction of Oxysterols' Apoptotic Signal

The transduction mechanism for oxysterols' apoptotic action is unknown. One candidate for an intermediate in the system is Oxysterol Binding Protein (OBP), a ubiquitous cytosolic protein with high affinity and specificity for oxysterols. Oxysterols in the blood appear to be bound mostly to albumin, with some found also in LDL particles (69,70). Free oxysterol (and possibly that in LDL particles taken up by LDL receptors) appears to be responsible for the regulatory actions of added, exogenous oxysterols. When exposed to oxysterols in serum-free medium containing a low concentration of carrier albumin, cells rapidly take up the sterols, which distribute into membrane compartments, except for a small proportion that binds to OBP (71). We have shown a correlation between OBP binding and evocation of apoptosis by 25-hydroxycholesterol in CEM cells (74). The function of OBP remains unknown, though its "knockout," *i.e.*, genomic deletion, resulted in extremely early lethality in mouse embryos (M. Brown and J. Goldstein, personal communication), suggesting a vital role in development.

Another explanation of oxysterols' mechanism of action is that they alter the activity of proteolytic enzymes controlling the release of SREBPs. If they do, the mechanism is unknown—maybe by intercalation into the ER, where the concentration of sterols is low and slight changes might make a significant difference, or maybe by interaction with a receptor (OBP, RxRα, or an undiscovered molecule). The possibility that oxysterols may regulate the action of proteolytic enzymes involved in cholesterol homeostasis raises the possibility that oxysterols also regulate proteases involved in apoptosis. Recently it was reported that at least some oxysterols can bind and activate transcription factor LxRα, a member of the steroid/thyroid/retinoid receptor family of proteins (76, 76A). This new finding raises exciting possibilities for understanding oxysterol actions and interactions.

4. The Overall Kinetic Pattern of Oxysterol-Evoked Apoptosis Resembles That Following Glucocorticoids

The mechanistic pathway for oxysterol-evoked apoptosis thus is still sketchy. Figure 5 diagrams some of the known events that follow addition of oxysterols to logarithmically growing CEM cells. Exogenous oxysterol, with or without LDL present, rapidly enters the cell, dissolving in membranes and binding OBP. Within a few hours, HMGCAR activity has fallen to

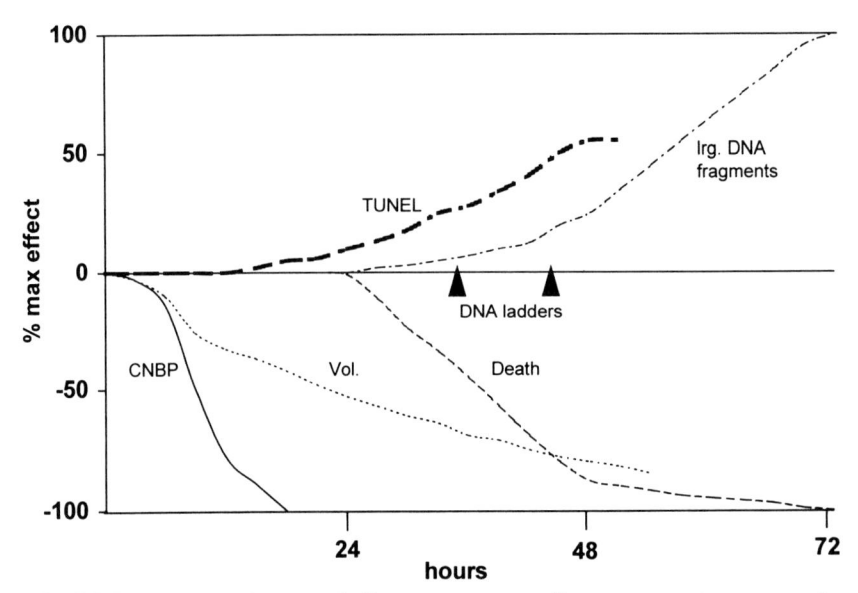

FIGURE 5 Sequence of oxysterol effects on CEM-C7 cells. Diagrammatic representation of events in CEM-C7 cells following addition of lethal concentrations of oxysterol (25-hydroxycholesterol). Each parameter is plotted as a percentage of its own maximum (or minimum) effect. The closed triangles indicate the approximate times at which 180-nt multiple DNA fragment "ladders" were detected. Morphologic characteristics of apoptosis became apparent at ~24 hr.

about 20% of control levels. Cell volume decrease can be documented by 12 hr after addition of oxysterol and progresses continuously thereafter until cell death ensues. DNA nicking can be demonstrated by TUNEL assay as early as 24 hr after adding oxysterol, and specific large-sized DNA fragments show up by 36 hr. From about 48 hr onward, classic DNA "ladders," multiples of ~180 bp, can be shown by gel electrophoresis. Increasing numbers of cells collect in G0/G1 after about 24 hr, at which time CNBP mRNA levels are diminished (150). Morphologic changes of apoptosis at the subcellular level visualized by electron microscopy are rare before 18–24 hr but become increasingly frequent and severe from 24 hr onward. The general picture resembles that seen when glucocorticoids are used as the apoptotic agents: early changes in gene expression and the beginning of cell shrinkage during a "quiet" interval lasting 18–24 hr, then a dramatic set of clearly apoptotic morphologic changes accompanied by severe DNA lysis and general breakdown of cellular synthetic machinery. During the quiet period, and going into the end-stage period, several cell cycle regulatory gene products are increased or diminished, and DNA lysis may begin. The similarity of their final cellular morphology and DNA breakdown suggest that in their concluding phases the oxysterol and glucocorticoid apoptotic pathways ultimately evoke a universal, atavistic set of events.

VI. Conclusions

Remembering that the original definition of apoptosis was histopathological, the number and variety of agents that cause this form of cell death make it obvious that many transduction pathways can produce death with the same appearance. Though morphology is no guarantee of identical biochemical causation, it seems likely that the many starting pathways are converging on a limited number of biochemical steps which produce the final distinctive appearance. The extremes of initiating processes can be appreciated by comparing the Apo1/Fas path, which requires no new macromolecular synthesis with the glucocorticoid-evoked path, wherein changes in expression of certain genes seem essential. Glucocorticoid initiation of thymic cell death is one of the classic models for apoptosis. Determining the exact steps involved in this steroid-induced apoptosis will provide a basis for specific comparisons with other pathways. Recently, oxysterols also have been found to cause apoptosis, particularly of lymphoid cells. Comparison of oxysterol- and glucocorticoid-induced apoptosis should help clarify unique and common points in the apoptotic pathway.

We have employed a well-defined tissue culture system of human lymphoid leukemia T cells to study the apoptosis caused by glucocorticoids and oxysterols. Virtually all the cells used are clonal and all but one are subclones of a single glucocorticoid-sensitive clone, making for greater baseline biochemical similarity when comparisons are made. In the wild-type, glucocorticoid-sensitive mother clone CEM-C7, exposure to glucocorticoids causes rapid loss of cMyc, followed shortly by induction of GR, glutamine synthetase, and cJun, as well as the loss of cell volume. From a variety of experiments, we conclude that the down-regulation of c-*myc* and the induction of c-*jun* are important for the glucocorticoidal apoptosis pathway in these cells. For at least a day after adding glucocorticoids to the cells, the effects on these genes and indeed the entire apoptotic sequence are reversible upon removal of glucocorticoid or its replacement by a glucocorticoid antagonist. During this time of reversible events, the presence of functional, agonist-occupied GR is essential. Determining the further steps in the reversible period is the goal of future research.

Transfection of plasmids expressing various mutants of the human GR into a CEM clone lacking functional GR has allowed determination of GR domains critical for transmitting the signal for cell death. The ligand- and DNA-binding regions seem essential, but the amino-terminal part of the GR, with its potent transactivation domain, is not required (although it may enhance activity somewhat). When the ligand-binding domain of the GR is removed completely, the residual peptide becomes constitutively lethal, with much more rapid kinetics than the holoreceptor plus steroid. Thus, such truncated peptides may be causing cell death by another pathway. Such an effect is of interest for its potential in therapy of leukemias. Questions of optimization of potency and targeted delivery of this GR fragment remain to be addressed.

In cells with functional GR the standard glucocorticoid-evoked death pathway causes cells to become irreversibly trapped in G0/G1, and DNA lysis into specific large and small forms occurs. Eventually polysomes are disrupted, thymidine kinase activity falls, and the overt morphology of apoptosis is seen. We conclude that this represents an orchestrated, final phase of the pathway, set off by the events begun earlier, during the reversible phase. We hypothesize that the earlier events, though reversible, are integral parts of the path.

The oxysterol pathway culminates in very similar events, but is initiated by unknown steps, possibly involving the OBP. Again, one sees a preliminary, reversible, steroid-dependent period during which specific gene expression is altered, followed by overt apoptosis. Early DNA lysis during the reversible phase is more prominent during oxysterol than glucocorticoid exposure, but later DNA fragmentation accompanying apoptosis is similar. Some genes important for apoptosis and controlled by glucocorticoids also are regulated by oxysterols, suggesting some functional overlap in the two initiatory paths.

References

1. Dougherty, T. F., and White, A. (1945). Functional alterations in lymphoid tissue induced by adrenal corticol secretion. *Am. J. Anat.* **77**, 81–116.
2. Makman, M. H., Nakagawa, S., and White, A. (1967). Studies of the mode of action of adrenal steroids on lymphocytes. *Recent Prog. Horm. Res.* **23**, 195–227.
3. Munck, A., and Wira, C. (1971). Glucocorticoid receptors in rat thymus cells. In "Advances in the Biosciences" (G. Raspé, ed.), pp. 301–330. Pergamon, Berlin.
4. Baxter, J. D., Harris, A. W., Tomkins, G. M., and Cohn, M. (1971). Glucocorticoid receptors in lymphoma cells in culture: Relationship to glucocorticoid killing activity. *Science* **171**, 189–191.
5. Sibley, C. H., and Tomkins, G. M. (1974). Isolation of lymphoma cell variants resistant to killing by glucocorticoids. *Cell (Cambridge, Mass.)* **2**, 213–220.
6. Sibley, C. H., and Tomkins, G. M. (1974). Mechanisms of steroid resistance. *Cell (Cambridge, Mass.)* **2**, 221–227.
7. Norman, M. R., and Thompson, E. B. (1977). Characterization of a glucocorticoid-sensitive human lymphoid cell line. *Cancer Res.* **37**, 3785–3791.
8. Schmidt, T. J., Harmon, J. M., and Thompson, E. B. (1980). Activation-labile glucocorticoid-receptor complexes of a steroid-resistant variant on CEM-C7 human lymphoid cells. *Nature (London)* **286**, 507–510.
9. Harmon, J. M., and Thompson, E. B. (1981) Isolation and characterization of dexamethasone-resistant mutants from human lymphoid cell line CEM-C7. *Mol. Cell. Biol.* **1**, 512–521.
10. Thompson, E. B., Zawydiwski, R., Brower, S. T., Eisen, H. J., Simons, S. S., Jr., Schmidt, T. J., Schlechte, J. A., and Moore, D. E. (1983). Properties and function of human glucocorticoid receptors in steroid-sensitive and resistant leukemic cells. In "Steroid Hormone Receptors: Structure and Function" (H. Eriksson and J. A. Gustafsson, eds.), pp. 171-194. Elsevier, Amsterdam.

11. Harbour, D. V., Chambon, P., and Thompson, E. B. (1990). Steroid mediated lysis of lymphoblasts requires the DNA binding region of the steroid hormone receptor. *J. Steroid Biochem.* **35**, 1–9.
12. Cohen, J. J. (1992). Glucocorticoid-induced apoptosis in the thymus. *Immunology* **4**, 363–369.
13. Thompson, E. B. (1994). Apoptosis and steroid hormones. *Mol. Endocrinol.* **8**, 665–673.
14. Schwartzman, R. A., and Cidlowski, J. A. (1994). Glucocorticoid-induced apoptosis of lymphoid cells. *Int. Arch. Allergy Immunol.* **105**, 347–354.
15. King, L. B., Vacchio, M. S., and Ashwell, J. D. (1994). To be or not to be: Mutually antagonistic death signals regulate thymocyte apoptosis. *Int. Arch. Allergy Immunol.* **105**, 355–358.
16. Gomi, M., Moriwaki, K., Katagiri, S., Kurata, Y., and Thompson, E. B. (1990). Glucocorticoid effects on myeloma cells in culture: Correlation of growth inhibition with induction of glucocorticoid receptor mRNA. *Cancer Res.* **50**, 1873–1878.
17. Johnson, B. H., Gomi, M., Jakowlew, S. B., Moriwaki, K., and Thompson, E. B. (1993). Actions and interactions of glucocorticoids and transforming growth factor β on two related human myeloma cell lines. *Cell Growth Differ.* **4**, 25–30.
18. Bateman, A., Singh, A., Kral, T., and Solomon, S. (1989). The immune-hypothalamic-pituitary-adrenal axis. *Endocr. Rev.* **10**(1), 92–112.
19. Galili, U. (1983). Glucocorticoid induced cytolysis of human normal and malignant lymphocytes. *J. Steroid Biochem.* **19**(1B), 483–490.
20. Galili, N., Galili, U., Klein, E., Rosenthal, L., and Nordenskjold, B. (1980). Human T lymphocytes become glucocorticoid-sensitive upon immune activation. *Cell. Immunol.* **50**(2), 440–444.
21. Thompson, E. B. (1997). Glucocorticoids and apoptosis of hemopoietic cells: Transcriptional and post-transcriptional control. In "Inhaled Glucocorticoids in Asthma Mechanisms and Clinical Actions" (R. P. Schleimer, W. W. Busse, and P. M. O'Byrne, eds.), pp. 81–106. Dekker, New York.
22. Lee, F., Mulligan, R., Berg, P., and Ringold, G. (1981). Glucocorticoids regulate expression of dihydrofolate reductase cDNA in mouse mammary tumor virus chimaeric plasmids. *Nature (London)* **294**, 228–232.
23. Ringold, G. M. (1985). Steroid hormone regulation of gene expression. *Annu. Rev. Pharmacol. Toxicol.* **25**, 529–566.
24. Beato, M. (1989). Gene regulation by steroid hormones. *Cell (Cambridge, Mass.)* **56**, 335–344.
25. Tsai, S. Y., Srinivasan, G., Allan, G. F., Thompson, E. B., O'Malley, B. W., and Tsai, M. J. (1990). Recombinant human glucocorticoid receptor induces transcription of hormone response genes *in vitro*. *J. Biol. Chem.* **265**, 17055–17061.
26. Truss, M., Bartsch, J., Schelbert, A., Hache, R. J., and Beato, M. (1995). Hormone induces binding of receptors and transcription factors to a rearranged nucleosome on the MMTV promoter *in vivo*. *EMBO J.* **14**(8), 1737–1751.
27. Beato, M., Gandau, R., Chavez, S., Mows, C., and Truss, M. (1996). Interaction of steroid hormone receptors with transcription factors involves chromatin remodelling. *J. Steroid Biochem. Mol. Biol.* **56**, 47–59.
28. Hager, G. L., Archer, T. K., Fragoso, G., Bresnick, E. H., Tsukagoshi, Y., John, S., and Smith, C. L. (1993). Influence of chromatin structure on the binding of transcription factors to DNA. *Cold Spring Harbor Symp. Quant. Biol.* **58**, 63–71.
29. Allan, G. F., Ing, N. H., Tsai, S. Y., Srinivasan, G., Weigel, N. L., Thompson, E. B., Tsai, M. J., and O'Malley, B. W. (1991). Synergism between steroid response and promoter elements during cell-free transcription. *J. Biol. Chem.* **266**, 5905–5910.
30. Grange, T., Roux, J., Rigaud, G., and Pictet, R. (1989). Two remote glucocorticoid responsive units interact cooperatively to promote glucocorticoid induction of rat tyrosine aminotransferase gene expression. *Nucleic Acids Res.* **17**(21), 8695–8790.

31. Beato, M., Truss, M. and Chavez, S. (1996). Control of transcription by steroid hormones. *Ann. N.Y. Acad. Sci.* **784**, 93–123.
32. Drouin, J., Sun, Y. L., Chamberland, M., Gauthier, Y., De Léan, A., Nemer, M., and Schmidt, T. J. (1993). Novel glucocorticoid receptor complex with DNA element of the hormone-repressed POMC gene. *EMBO J.* **12**(1), 145–156.
33. Liu, W., Hillmann, A. G., and Harmon, J. M. (1995). Hormone-independent repression of AP-1-inducible collagenase promoter activity by glucocorticoid receptors. *Mol. Cell. Biol.* **15**(2), 1005–1013.
34. Alam, T., An, M. R., Mifflin, R. C., Hsieh, C.-C., Ge, X., and Papaconstantinou, J. (1993). *trans*-activation of the α_1-acid glycoprotein gene acute phase responsive element by multiple isoforms of C/EBP and glucocorticoid receptor. *J. Biol. Chem.* **268**(21), 15681–15688.
35. Hutchison, K. A., Scherrer, L. C., Czar, M. J., Stancato, L. F., Chow, Y. H., Jove, R., and Pratt, W. B. (1993). Regulation of glucocorticoid receptor function through assembly of a receptor-heat shock protein complex. *Ann. N.Y. Acad. Sci.* **684**, 35–48.
36. Cato, A. C. B., and Wade, E. (1996). Molecular mechanisms of anti-inflammatory action of glucocorticoids. *BioEssays* **18**(5), 371–378.
37. Parker, M. G. (1995). Structure and function of estrogen receptors. *Vitam. Horm.* **51**, 262–287.
38. Hong, H., Kohli, K., Trivedi, A., Johnson, D. L., and Stallcup, M. R. (1996). GRIP1, a novel mouse protein that serves as a transcriptional coactivator in yeast for the hormone binding domains of steroid receptors. *Proc. Natl. Acad. Sci. U.S.A.* **93**(10), 4948–4952.
39. Nordeen, S. K., Bona, B. J., Beck, C. A., Edwards, D. P., Borror, K. C., and DeFranco, D. B. (1995). The two faces of a steroid antagonist: When an antagonist isn't. *Steroids* **60**(1), 97–104.
40. Thompson, E. B., and Lippman, M. E. (1974). Mechanism of action of glucocorticoids. *Metab. Clin. Exp.* **23**, 159–202.
41. Gruol, D. J., Rajah, F. M., and Bourgeois, S. (1989). Cyclic AMP-dependent protein kinase modulation of the glucocorticoid-induced cytolytic response in murine T-lymphoma cells. *Mol. Endocrinol.* **3**, 2119–2127.
42. Jondal, M., Xue, Y., McConkey, D. J., and Okret, S. (1995). Thymocyte apoptosis by glucocorticoids and cAMP. *Curr. Top. Microbiol. Immunol.* **200**, 67–79.
43. Saeed, M. F. (1995). Synergistic interaction between the glucocorticoid and the cAMP-dependent protein kinase pathways in human leukemic cells. Master of Science Thesis, University of Texas Medical Branch at Galveston.
44. Power, R. F., Mani, S. K., Codina, J., Conneely, O. M., and O'Malley, B. W. (1991). Dopaminergic and ligand-independent activation of steroid hormone receptors. *Science* **254**, 1636–1639.
45. Thompson, E. B., Gadson, P., Wasner, G., and Simons, S. S., Jr. (1988). Differential regulation of tyrosine amino-transferase by glucocorticoids: Transcriptional and post-transcriptional control. *In* "Gene Regulation by Steroid Hormones IV" (A. K. Roy, and J. H. Clark, eds.), pp. 63–77. Springer-Verlag, Rochester, NY.
46. Hurme, M., Siljander, P., and Anttila, H. (1991). Regulation of interleukin-1β production by glucocorticoids in human monocytes: The mechanism of action depends on the activation signal. *Biochem. Biophys. Res. Commun.* **180**, 1383–1389.
47. Amano, Y., Lee, S. W., and Allison, A. C. (1993). Inhibition by glucocorticoids of the formation of interleukin-1α, interleukin-1β, and interleukin-6: Mediation by decreased mRNA stability. *Mol. Pharmcol.* **43**, 176–182.
48. Peppel, K., Vinci, J. M., and Baglioni, C. (1991). The AU-rich sequences in the 3′ untranslated region mediate the increased turnover of interferon mRNA induced by glucocorticoids. *J. Exp. Med.* **173**, 349–355.
49. Dodson, R. E., and Shapiro, D. J. (1994). An estrogen-inducible protein binds specifically to a sequence in the 3′ untranslated region of estrogen-stabilized vitellogenin mRNA. *Mol. Cell. Biol.* **14**, 3130–3138.

50. Pastori, R. L., and Schoenberg, D. R. (1993). The nuclease that selectively degrades albumin mRNA *in vitro* associates with xenopus liver polysomes through the 80S ribosome complex. *Arch. Biochem. Biophys.* **305**, 313–319.
51. Ross, J. (1995). mRNA stability in mammalian cells. *Microbiol. Rev.* **59**(3), 423–450.
52. Paliogianni, F., Raptis, A., Ahuja, S. S., Najjar, S. M., and Boumpas, D. T. (1993). Negative transcriptional regulation of human interleukin 2 (IL-2) gene by glucocorticoids through interference with nuclear transcription factors AP-1 and NF-AT. *J. Clin. Invest.* **91**(4), 1481–1489.
53. Chikanza, L. C., and Panayi, G. S. (1993). The effects of hydrocortisone on *in vitro* lymphocyte proliferation and interleukin-2 and -4 production in corticosteroid sensitive and resistant subjects. *Eur. J. Clin. Invest.* **23**(12), 845–850.
54. Snijdewint, F. G., Kapsenberg, M. L., Wauben-Penris, P. J., and Bos, J. D. (1995). Corticosteroids class-dependently inhibit *in vitro* TH1- and TH2-type cytokine production. *Immunopharmacology* **29**(2), 93–101.
55. Mori, A., Suko, M., Nishizaki, Y., Kaminuma, O., Kobayashi, S., Matsuzaki, G., Yamamoto, K., Ito, K., Tsuruoka, N., and Okudaira, H. (1995). IL-5 production by CD4+ T cells of asthmatic patients is suppressed by glucocorticoids and the immunosuppressants FK506 and cyclosporin A. *Int. Immunol.* **7**(3), 449–457.
56. Kunicka, J. E., Talle, M. A., Denhardt, G. H., Brown, M., Prince, L. A. and Goldstein, G. (1993). Immunosuppression by glucocorticoids: Inhibition of production of multiple lymphokines by *in vivo* administration of dexamethasone. *Cell. Immunol.* **149**, 39–49.
57. Nieto, M. A., and Lopez-Rivas, A. (1992). Glucocorticoids activate a suicide program in mature T lymphocytes: Protective action of interleukin-2. *Ann. N.Y. Acad. Sci.* **650**, 115–120.
58. Lanza, L., Scudeletti, M., Puppo, F., Bosco, O., Peirano, L., Filaci, G., Fecarotta, E., Uidali, G., and Indiveri, F. (1996). Prednisone increases apoptosis in *in vitro* activated human peripheral blood T lymphocytes. *Clin. Exp. Immunol.* **103**(3), 482–490.
59. Kandutsch, A. A., and Chen, H. W. (1973). Inhibition of sterol synthesis in cultured mouse cells by 7α-hydroxycholesterol, 7β-hydroxycholesterol, and 7-ketocholesterol. *J. Biol. Chem.* **248**, 8408–8417.
60. Taylor, F. R., Saucier, S. E., Shown, E. P., Parish, E. J., and Kandutsch, A. A. (1984). Correlation between oxysterol binding to a cytosolic binding protein and potency in the repression of hydroxymethylglutaryl coenzyme A reductase. *J. Biol. Chem.* **259**(20), 12382–12387.
61. Taylor, F. R. (1992). Correlation among oxysterol potencies in the regulation degradation of 3-hydroxy-3-methylglutaryl CoA reductase repression of 3-hydroxy-3-methlglutaryl CoA synthase affinities for the oxysterol receptor. *Biochem. Biophys. Res. Commun.* **186**, 182–189.
62. Takagi, K., Alverez, J. G., Favata, M. F., Trzaskos, J. M., and Strauss, J. F., III. (1989). Control of low density lipoprotein receptor gene promoter activity. Ketoconazole inhibits serum lipoprotein but not oxysterol suppression of gene transcription. *J. Biol. Chem.* **264**, 12352–12357.
63. Wang, X., Sato, R., Brown, M. S., Hua, X., and Goldstein, J. L. (1994). SREBP-1, a membrane-bound transcription factor released by sterol-regulated proteolysis. *Cell* (*Cambridge, Mass.*) **77**, 53–62.
64. Smith, L. L., and Johnson, B. H. (1989). Biological activities of oxysterols. *Free Radical Biol. Med.* **7**, 285–332.
65. Ayala-Torres, S., Moller, P. C., Johnson, B. H., and Thompson, E. B. (1997). Characteristics of 25-hydroxycholesterol-induced apoptosis in the human leukemic cell line CEM. *Exp. Cell Res.* (in press).
66. Johnson, B. H., Ayala-Torres, S., Chan, L.-N. L., El-Naghy, M., and Thompson, E. B. (1997). Glucocorticoid/oxysterol-induced DNA lysis in human leukemia cells. *J. Steroid Biochem. Mol. Biol.* (in press).

67. Christ, M., Luu, B., Mejia, J. E., Moosbrugger, I., and Bischoff, P. (1993). Apoptosis induced by oxysterols in murine lymphoma cells and in normal thymocyts. *Immunology* **78**, 455–460.
68. Hietter, H., Bischoff, P., Beck, J. P., Ourisson, G., and Luu, B. (1986). Comparative effects of 7 β-hydroxycholesterol towards murine lymphomas, lymphoblasts and lymphocytes: Selective cytotoxicity and blastogenesis inhibition. *Cancer Biochem. Biophys.* **9**(1), 75–83.
69. Lin, C. Y., and Morel, D. W. (1995). Distribution of oxysterols in human serum: Characterization of 25-hydroxycholesterol association with serum albumin. *J. Nutr. Biochem.* **6**, 618–625.
70. Lin, C. Y., and Morel, D. W. (1996). Esterification of oxysterols in human serum: Effects on distribution and cellular uptake. *J. Lipid Res.* **37**, 168–178.
71. Kandutsch, A. A., and Thompson, E. B. (1980). Cytosolic proteins that bind oxygenated sterols: Cellular distribution, specificity, and some properties. *J. Biol. Chem.* **255**, 10813–10821.
72. Kandutsch, A. A., Chen, H. W., and Shown, E. P. (1977). Binding of 25-hydroxycholesterol and cholesterol to different cytoplasmic proteins. *Proc. Natl. Acad. Sci. U.S.A.* **74**, 2500–2503.
73. Taylor, F. R., Shown, E. P., Thompson, E. B., and Kandutsch, A. A. (1989). Purification, subunit structure, and DNA binding properties of the mouse oxysterol receptor. *J. Biol. Chem.* **264**, 18433–18439.
74. Bakos, J. T., Johnson, B. H., and Thompson, E. B. (1993). Oxysterol-induced cell death in human leukemic T-cells correlates with oxysterol binding protein occupancy and is independent of glucocorticoid-induced apoptosis. *J. Steroid Biochem. Mol. Biol.* **46**, 415–426.
75. Dawson, P. A., Ridgway, N. D., Slaughter, C. A., Brown, M. S. and Goldstein, J. L. (1989). cDNA cloning and expression of oxysterol-binding protein, an oligomer with a potential leucine zipper. *J. Biol. Chem.* **264**, 16798–16803.
76. Janowski, B. A., Willy, P. J., Devi, T. R., Falck, J. R., and Mangelsdorf, D. J. (1996). An oxysterol signalling pathway mediated by the nuclear receptor LXRα. *Nature (London)* **383**, 728–731.
76a. Lehmann, J. M., Kliewer, S., Moore, L. B., Smith-Oliver, T. A., Oliver, B. B., Su, J-L., Sundseth, S. S., Winegar, D. A., Blanchard, D. E., and Spencer, T. A. (1997). Activation of the nuclear receptor LXR by oxysterols defines a new hormone response pathway. *J. Biol. Chem.* **272**(6), 3137–3140.
77. Kerr, J. F. R., Wyllie, A. H., and Currie, A. R. (1972). Apoptosis: A basic biological phenomenon with wide-ranging implications in tissue kinetics. *Br. J. Cancer* **26**, 239–257.
78. Wyllie, A. H. (1980). Glucocorticoid-induced thymocyte apoptosis is associated with endogenous endonuclease activation. *Nature (London)* **284**, 555–556.
79. Wyllie, A. H., Morris, R. G., Smith, A. L., and Dunlop, D. (1984). Chromatin cleavage in apoptosis: Association with condensed chromatin morphology and dependence on macromolecular synthesis. *J. Pathol.* **142**, 67–77.
80. Cohen, G. M., Sun, S. M., Fearnhead, H., MacFarlane, M., Brown, D. G., Snowden, R. T., and Dinsdale, D. (1994). Formation of large molecular weight fragments of DNA is a key committed step of apoptosis in thymocytes. *J. Immunol.* **153**(2), 507–516.
81. Gaido, M. L., and Cidlowski, J. A. (1991). Identification, purification, and characterization of a calcium-dependent endonuclease (NUC18) from apoptotic rat thymocytes. *J. Biol. Chem.* **266**, 18580–18585.
82. Wyllie, A. H., Arends, M. J., Morris, R. G., Walker, S. W., and Evan, G. (1992). The apoptosis endonuclease and its regulation. *Semin. Immunol.* **4**, 389–397.
83. Khodarev, N. N., and Ashwell, J. D. (1996). An inducible lymphocyte nuclear Ca2+/Mg(2+)-dependent endonuclease associated with apoptosis. *J. Immunol.* **156**(3), 922–931.

84. Shiokawa, D., Ohyama, H., Yamada, T., Takahashi, K., and Tanuma, S. (1994). Identification of an endonuclease responsible for apoptosis in rat thymocytes. *Eur. J. Biochem.* **226**(1), 23–30.
85. Brown, D. G., Sun, X.-M., and Cohen, G. M. (1993). Dexamethasone-induced apoptosis involves cleavage of DNA to large fragments prior to internucleosomal fragmentation. *J. Biol. Chem.* **268**, 3037–3039.
86. McConkey, D. J., and Orrenius, S. (1996). The role of calcium in the regulation of apoptosis. *J. Leukocyte Biol.* **59**, 775–782.
87. Harrington, E. A., Fanidi, A., and Evan, G. I. (1994). Oncogenes and cell death. *Curr. Opin. Genet. Dev.* **4**(1), 120–129.
88. Galaktionov, K., Chen, X., and Beach, D. (1996). Cdc25 cell-cycle phosphatase as a target of c-*myc*. *Nature (London)* **382**, 511–517.
89. Thulasi, R., and Thompson, E. B. (1993). The role of glucocorticoids in cell growth, differentiation and programmed cell death. *In* Protooncogenes and Growth Factors in Steroid Hormone Induced Growth and Differentiation" (S. Khan and G. M. Stancel, eds.), pp. 221–240. CRC Press, Boca Raton, FL.
90. Eastman-Reks, S. B. and Vedeckis, W. V. (1986). Glucocorticoid inhibition of c-*myc*, c-*myb*, c-*rasKi* expression in a mouse lymphoma cell line. *Cancer Res.* **46**, 2457.
91. Forsthoefel, A. M., and Thompson, E. A., Jr. (1987). Glucocorticoid regulation of transcription of the c-*myc* cellular protooncogene in P1798 cells. *Mol. Endocrinol.* **1**, 899–907.
92. Yuh, Y.-S., and Thompson, E. B. (1989). Glucocorticoid effect on oncogene/growth gene expression in human T lymphoblastic leukemic cell line CCRF-CEM: Specific c-*myc* mRNA suppression by dexamethasone. *J. Biol. Chem.* **264**, 10904–10910.
93. Thulasi, R., Harbour, D. V., and Thompson, E. B. (1993). Suppression of c-*myc* is a critical step in glucocorticoid-induced human leukemic cell lysis. *J. Biol. Chem.* **268**, 18306–18312.
94. Wood, A. C., Waters, C. M., Garner, A., and Hickman, J. A. (1994). Changes in c-*myc* expression and the kinetics of dexamethasone-induced programmed cell death (apoptosis) in human lymphoid leukaemia cells. *Br. J. Cancer* **69**, 663–669.
95. Helmberg, A., Auphan, N., Caelles, C., and Karin, M. (1995). Glucocorticoid-induced apoptosis of human leukemic cells is caused by the repressive function of the glucocorticoid receptor. *EMBO J.* **14**, 452–460.
96. Rhee, K., Bresnahan, W., Hirai, A., Hirai, M., and Thompson, E. A. (1995). c-Myc and cyclin D3 (CcnD3) genes are independent targets for glucocorticoid inhibition of lymphoid cell proliferation. *Cancer Res.* **55**(18), 4188–4195.
97. Fischer, G., Kent, S. C., Joseph, L., Green, D. R., and Scott, D. W. (1994). Lymphoma models for B cell activation and tolerance. X. Anti-μ-mediated growth arrest and apoptosis of murine B cell lymphomas is prevented by the stabilization of myc. *J. Exp. Med.* **179**, 221–228.
98. Gruber, J., and Greil, R. (1994). Apoptosis and therapy of malignant diseases of the hematopoietic system. *Int. Arch. Allergy Immunol.* **105**, 368–373.
99. Zhou, F., and Thompson, E. B. (1996). Role pf c-*jun* induction in the glucocorticoid-evoked apoptotic pathway in human leukemic lymphoblasts. *Mol. Endocrinol.* **10**, 306–316.
100. Kastan, M. B. (1993). p53: Evolutionally conserved and constantly evolving. *J. NIH Res.* **5**, 53–57.
101. Marx, J. (1993). How p53 suppresses cell growth. *Science* **262**, 1644–1645.
102. Picksley, S. M., and Lane, D. P. (1994). p53 and Rb: Their cellular roles. *Curr. Opin. Cell Biol.* **6**, 853–858.
103. Wu, X., and Levine, A. J. (1994). p53 and E2F-1 cooperate to mediate apoptosis. *Proc. Natl. Acad. Sci. U.S.A.* **91**, 3602–3606.

104. Hermeking, H., and Eick, D. (1994). Mediation of c-Myc-induced apoptosis by p53. *Science* **265**, 2091–2093.
105. Wang, J., and Walsh, K. (1996). Resistance to apoptosis conferred by Cdk inhibitors during myocyte differentiation. *Science* **273**, 359–361.
106. Meikrantz, W., and Schlegel, R. (1996). Suppression of apoptosis by dominant negative mutants of cyclin-dependent protein kinases. *J. Biol. Chem.* **271**(17), 10205–10209.
107. Kranenburg, O., van der Eb, A. J., and Zantema, A. (1996). Cyclin D1 is an essential mediator of apoptotic neuronal cell death. *EMBO J.* **15**(1), 46–54.
108. Chiarugi, V., Magnelli, L., Cinelli, M., and Basi, G. (1994). Apoptosis and the cell cycle. *Cell. Mol. Biol. Res.* **40**, 603–612.
109. Reed, J. C. (1995). Regulation of apoptosis by bcl-2 family proteins and its role in cancer and chemoresistance. *Curr. Opin. Oncol.* **7**, 541–546.
110. Korsmeyer, S. J. (1995). Regulators of cell death. *Trends Genet.* **11**(3), 101–105.
111. Bortner, C. D., and Cidlowski, J. A. (1997). Cell volume regulation and the movement of ions during apoptosis. *In* "Program Cell Death" (D. D. Scott, Y. B. Shi, Y. Shi, and X. Yu, eds.). Plenum, New York (in press).
112. Benson, R. S. P., Heer, S., Dive, C., and Watson, A. J. M. (1996). Characterization of cell volume loss in Cem-C7A cells during dexamethasone-induced apoptosis. *Am. J. Physiol.* **39**(4), C1190–C1203.
113. Arends, M. J., Morris, R. G., and Wyllie, A. H. (1990). Apoptosis: The role of the endonuclease. *Am. J. Pathol.* **136**, 593–608.
114. Owens, G. P., Hahn, W. E., and Cohen, J. J. (1991). Identification of mRNAs associated with programmed cell death in immature thymocytes. *Mol. Cell. Biol.* **11**(8), 4177–4188.
115. Chapman, M. S., Qu, N., Pascoe, S., Chen, W. X., Apostol, C., Gordon, D., and Miesfeld, R. L. (1995). Isolation of differentially expressed sequence tags from steroid-responsive cells using mRNA differential display. *Mol. Cell. Endocrinol.* **108**(1-2), R1–R7.
116. Nagata, S., and Golstein, P. (1995). The Fas death factor. *Science* **267**, 1449–1455.
117. Lahti, J. M., Xiang, J., and Kidd, V. J. (1995). Cell cycle-related protein kinases and T cell death. *In* "Glycoimmunology" (A. Alavi and J. S. Axford, eds.), pp. 247–258. Plenum, New York.
118. Thornberry, N. A., Bull, H. G., Calaycay, J. R., Chapman, K. T., Howard, A. D., Kostura, M. J., Miller, D. K., Molineaux, S. M., Weidner, J. R., and Aunins, J. (1992). A novel heterodimeric cysteine protease is required for interleukin-1 beta processing in monocytes. *Nature (London)* **356**, 768–774.
119. Miura, M., Zhu, H., Rotello, R., Hartwieg, E. A., and Yuan, J. (1993). Induction of apoptosis in fibroblasts by IL-1β-converting enzyme, a mammalian homolog of the C. elegans cell death gene ced-3. *Cell (Cambridge, Mass.)* **75**, 653–660.
120. Patel, T., Gores, G. J., and Kaufmann, S. H. (1996). The role of proteases during apoptosis. *FASEB J.* **10**(5), 587–597.
121. Kumar, S. (1995). ICE-like proteases in apoptosis. *Trends Biochem. Sci.* **20**(5), 198–202.
122. Nakajima, H., Golstein, P., and Henkart, P. A. (1995). The target cell nucleus is not required for cell-mediated granzyme- or Fas-based cytotoxicity. *J. Exp. Med.* **181**(5), 1905–1909.
123. Chow, S. C., Weis, M., Kass, G. E., Holmstrom, T. H., Eriksson, J. E., and Orrenius, S. (1995). Involvement of multiple proteases during Fas-mediated apoptosis in T lymphocytes. *FEBS Lett.* **364**(2), 134–138.
124. Williams, M. S., and Henkart, P. A. (1994). Apoptotic cell death induced by intracellular proteolysis. *J. Immunol.* **153**, 4247–4255.
125. Zawydiwski, R., Harmon, J. M., and Thompson, E. B. (1983). Glucocorticoid-resistant human acute lymphoblastic leukemic cell line with functional receptor. *Cancer Res.* **43**, 3865–3873.
126. Harmon, J. M., Schmidt, T. J., and Thompson, E. B. (1984). Molybdate-sensitive and molybdate-resistant activation-labile glucocorticoid-receptor mutants of the human lymphoid cell line CEM-C7. *J. Steroid Biochem.* **21**, 227–236.

127. Ashraf, J., and Thompson, E. B. (1993). Identification of the activation-labile gene: A single point mutation in the human glucocorticoid receptor presents as two distinct receptor phenotypes. *Mol. Endocrinol.* **7**, 631–642.
128. Palmer, L. A., and Harmon, J. M. (1991). Biochemical evidence that glucocorticoid-sensitive cell lines derived from the human leukemic cell line CCRF-CEM express a normal and a mutant glucocorticoid receptor gene. *Cancer Res.* **51**, 5224–5231.
129. Powers, J. H., Hillmann, A. G., Tang, D. C., and Harmon, J. M. (1993). Cloning and expression of mutant glucocorticoid receptors from glucocorticoid-sensitive and -resistant human leukemic cells. *Cancer Res.* **53**, 4059–4065.
130. Strasser-Wozak, E. M. C., Hattmannstorfer, R., Hála, M., Hartmann, B. L., Fiegl, M., Geley, S., and Kofler, R. (1995). Splice site mutation in the glucocorticoid receptor gene causes resistance to glucocorticoid-induced apoptosis in a human acute leukemic cell line. *Cancer Res.* **55**, 348–353.
131. Geley, S., Hartmann, B. L., Hala, M., Strasser-Wozak, E. M. C., Kapelari, K., and Kofler, R. (1996). Resistance to glucocorticoid-induced apoptosis in human T-cell acute lymphoblastic leukemia CEM-C1 cells is due to insufficient glucocorticoid receptor expression. *Cancer Res.* **56**, 5033–5038.
132. Ayala-Torres, S., Johnson, B. H., and Thompson, E. B. (1994). Oxysterol sensitive and resistant lymphoid cells: Correlation with regulation of cellular nucleic acid binding protein mRNA. *J. Steroid Biochem. Molec. Biol.* **48**, 307–315.
133. Adebodun, F., and Post, J. F. M. (1993). 19F NMR studies of changes in membrane potential and intracellular volume during dexamethasone-induced apoptosis in human leukemic cell lines. *J. Cell. Physiol.* **154**, 199–206.
134. Abebodun, F., and Post, J. F. (1994). 31P NMR characterization of cellular metabolism during dexamethasone induced apoptosis in human leukemic cell lines. *J. Cell. Physiol.* **158**, 180–186.
135. Thompson, E. B., Thulasi, R., Saeed, M. F., and Johnson, B. H. (1995). Glucocorticoid antagonist RU486 reverses agonist-induced apoptosis and c-*myc* repression in human leukemic CEM-C7 cells. *Ann. N.Y. Acad. Sci.* **761**, 261–275.
136. Dieken, E. S., and Miesfeld, R. L. (1992). Transcriptional transactivation functions localized to the glucocorticoid receptor N terminus are necessary for steroid induction of lymphocyte apoptosis. *Mol. Cell. Biol.* **12**, 589–597.
137. Bourgeois, S., and Newby, R. F. (1977). Diploid and haploid states of the glucocorticoid receptor gene of mouse lymphoid cell lines. *Cell (Cambridge, Mass.)* **11**, 423–430.
138. Harmon, J. M., Norman, M. R., Fowlkes, B. J., and Thompson, E. B. (1979). Dexamethasone induces irreversible G1 arrest and death of a human lymphoid cell line. *J. Cell. Physiol.* **98**, 267–278.
139. Schulze-Osthoff, K., Walczak, H., Droge, W., and Krammer, P. H. (1994). Cell nucleus and DNA fragmentation are not required for apoptosis. *J. Cell Biol.* **127**(1), 15–20.
140. Marcu, K. B., Bossone, S. A., and Patel, A. J. (1992). Myc function and regulation. *Annu. Rev. Biochem.* **61**, 809–860.
141. Hay, N., Takimoto, M., and Bishop, J. M. (1989). A Fos protein is present in a complex that binds a negative regulator of myc. *Genes Dev.* **3**, 293–303.
142. Simons, S. S., Jr. (1994). Function/activity of specific amino acids in glucocorticoid receptors. *Vitam. Horm. (N.Y.)* **49**, 49–130.
143. Nazareth, L. V., Johnson, B. H., Chen, H., and Thompson, E. B. (1996). Transfected glucocorticoid receptor and certain GR fragments evoke cell death in malignant lymphoid, not myeloid cell lines. *Leukaemia* **10**, 1789–1795.
144. Nazareth, L. V., Harbour, D. V., and Thompson, E. B. (1991). Mapping the human glucocorticoid receptor for leukemic cell death. *J. Biol. Chem.* **266**, 12976–12980.
145. Forsthoefel, A. M., and Thompson, E. A., Jr. (1987). Glucocorticoid regulation of transcription of the c-*myc* cellular protooncogene in P1798 cells. *Mol. Endocrinol.* **1**, 899–907.

146. Maroder, M., Martinotti, S., Vacca, A., Screpanti, I., Petrangeli, E., Frati, L., and Gulino, A. (1990). Post-transcriptional control of c-*myc* proto-oncogene expression by glucocorticoid hormones in human T lymphoblastic leukemic cells. *Nucleic Acids Res.* **18**(5), 1153–1157.
147. Giguere, V., Hollenberg, S. M., Rosenfeld, M. G., and Evans, R. M. (1986). Functional domains of the human glucocorticoid receptor. *Cell (Cambridge, Mass.)* **46**, 645–652.
148. Spangrude, G. J., Sherris, D., and Daynes, R. A. (1982). Inhibitory effects of various oxygenated sterols on the differentiation and function of tumor-specific cytotoxic T lymphocytes. *Transplantation* **33**(5), 482–491.
149. Moog, C., Luu, B., Beck, J. P., Italiano, L., and Bischoff, P. (1988). Studies on the immunosuppressive properties of 7,25-hydroxycholesterol. I. Reduction of interleukin productive by treated lymphocytes. *Int. J. Immunopharmacol.* **10**(5), 511–518.
150. Ayala-Torres, S., Johnson, B. H., and Thompson, E. B. (1994). Oxysterol sensitive and resistant lymphoid cells: Correlation with regulation of cellular nucleic acid binding protein mRNA. *J. Steroid Biochem. Mol. Biol.* **48**, 307–315.
151. Michelotti, E. F., Tomonaga, T., Krutzsch, H., and Levens, D. (1995). Cellular nucleic acid binding protein regulates the CT element of the human c-*myc* protooncogene. *J. Biol. Chem.* **270**(16), 9494–9499.

Ann Marie Schmidt
Jean-Luc Wautier*
David Stern
Shi Du Yan

Departments of Medicine, Physiology & Cellular Biophysics, Surgery, and Pathology
College of Physicians and Surgeons of Columbia University
New York, New York 10032
*Laboratory of Vascular Disease Research
University-7 of Paris
Paris, France

RAGE: A Receptor with a Taste for Multiple Ligands and Varied Pathophysiologic States

The classical concept of one receptor with specificity and high affinity for only one ligand has evolved considerably. Furthermore, there are apparently accidental but, nonetheless, pathophysiologically relevant ligands, such as Intercellular Adhesion Molecule-1, which interacts with rhinoviruses to facilitate their entry into cells. RAGE, a member of the immunoglobulin of cell surface molecules, shares such properties. RAGE interacts with different ligands, with varied implications for cellular functions, depending on the physiologic or pathophysiologic setting. For example, during normal development, RAGE interacts with amphoterin, a molecule which promotes neurite outgrowth. In pathophysiologic states such as diabetes or amyloidosis observed in the setting of renal dialysis, RAGE binds nonenzymatically glycated adducts of macromolecules termed Advanced Glycation Endproducts, or AGEs, resulting in perturbation of multiple cellular properties. Alzheimer's disease represents a situation in which RAGE expression increases dramatically, and amyloid-beta peptide, thought to be critical to the pathogenesis of neurodegeneration, is another ligand for RAGE. The diverse circumstances in which these varied ligands interact with RAGE are the subject of intense investigation in order to understand the distinct mechanisms that regulate the temporal and spatial expression of this receptor.

Introduction

The classical notion of one receptor with specificity and high affinity for only one ligand has evolved considerably. Integrins display specific interactions with multiple ligands that bear target arginine–glycine–aspartic acid–serine-containing sequences. In addition, Mac-1, a counterligand for the integrin CD11a/CD18b, binds fibrogens and coagulation Factor X. Furthermore, there are apparently accidental, but nonetheless pathophysiologically relevant, processes such as chemokine receptor recognition of malarial parasites and intercellular adhesion molecule-1 interaction with rhinoviruses, that facilitate their entry into cells. The immunoglobulin superfamily molecule Receptor for Advanced Glycation Endproducts (RAGE) is such a receptor. As outlined in this chapter, RAGE interacts with different ligands, with varied consequences for cellular functions, depending on the physiologic or pathophysiologic setting. During normal development RAGE interacts with amphoterin, a molecule which promotes neurite outgrowth. In diabetes, RAGE binds nonenzymatically glycated adducts of macromolecules termed Advanced Glycation Endproducts (AGEs), resulting in perturbation of multiple cellular properties. Alzheimer's disease represents a situation in which RAGE expression increases dramatically, and amyloid-β peptide, thought to be critical to the pathogenesis of neurodegeneration, is another ligand which interacts with RAGE. In this chapter the biology of RAGE and its ligands recognized to date will be reviewed, based on the hypothesis that expression of RAGE responds to environmental cues and that the nature of the ligands determines effector mechanisms triggered following engagement of the receptor.

Identification and Characterization of RAGE

As the cells which form the luminal vascular surface, endothelia are subject to and immersed in the vascular microenvironment. The importance of endothelial cell regulation of homeostasis is evident from the important functions it serves in the quiescent state, including: maintenance of blood fluidity; formation of a dynamic barrier excluding bulk passage of macromolecules, cells, and fluid, and which selectively transports nutrients and mediators to the tissues; regulation of leukocyte trafficking; and control of vessel tone. An especially prevalent disorder in which each of these vascular mechanisms becomes dysfunctional is diabetes. Although initially manifested as a disorder of glucose utilization, over time, in diabetes the excessive levels of glucose modify macromolecules to form nonenzymatically glycated adducts. The first products of nonenzymatic glycation are the reversible Schiff bases and Amadori products. An example of the latter is hemoglobin A_{1c}, which is used for longer-term monitoring of blood glucose control. Following further molecular rearrangements, the irreversible AGEs form (1). AGEs are heterogeneous, many having in common characteristic fluorescence, a propensity to form cross-links, generation of reactive oxygen intermediates (ROIs), and interaction with cellular receptors (1–7). Two AGEs found in human tissues are carboxymethyllysine (8), the most abundant AGE characterized thus far and pentosidine (9), both of which are under intensive study to determine their role in the accumulation of biologic properties of AGEs present *in vivo*.

The interaction of AGEs with cellular elements modulates function in a manner potentially underlying organ dysfunction accompanying diabetes. This is especially evident in the endothelium. AGEs increase the permeability of endothelial monolayers, enhance expression of Vascular Cell Adhesion Molecule-1 (VCAM-1) and the procoagulant cofactor tissue factor, and have the capacity to quench nitric oxide via their generation of reactive oxygen intermediates (10–12). Furthermore, AGE-mediated perturbation of these cellular properties coincides with concentrations of nonenzymatically glycated adducts which resulted in occupancy of cell surface binding sites. The relevance of these cellular interaction sites for AGEs is emphasized by the distribution of AGEs in tissues, which demonstrates an intimate association with cellular elements. For example, a sector of kidney from a patient with diabetic glomerulosclerosis displays intense staining with affinity-purified anti-AGE IgG (13; probably principally displaying carboxymethyllysine adducts) in vascular walls, in Bowman's capsule, and in the expanded extracellular matrix (Fig. 1; left panel). These considerations stimulated our laboratory to identify cellular interaction sites for AGEs, in order to gain insight into the molecular mechanisms at work when AGEs engaged cellular targets.

At the outset, we anticipated that the cell binding site for AGEs would be a member of the scavenger receptor family of trimeric polypeptides known to interact with acetylated and oxidized low-density lipoproteins via their collagenous domains. This seemed logical, as AGE adducts form as the result of protein or lipid modification/damage, often as a result of oxidative mechanisms. Critical to our isolation procedure was the development of a competitive binding assay to detect AGE binding activity following solubilization of cell membranes. The binding of radioiodinated AGE albumin (^{125}I-AGE albumin) to cultured endothelial cells was dose-dependent and saturable, displaying $K_d \approx 50$ nM (14; Fig. 2). Binding was prevented by pretreating cells with trypsin or exposing cultures to octyl-β-glucoside (Fig. 2, inset). Cell extracts prepared in the presence of octyl-β-glucoside were diluted in high pH buffer and then adsorbed to microtiter wells. Radioligand binding assays with ^{125}I-AGE albumin and cell extract adsorbed to the plastic wells demonstrated specific binding. The binding isotherms obtained using extracts of endothelial proteins immobilized on plastic wells were similar to those observed with intact cells, with respect to both the dose-dependence

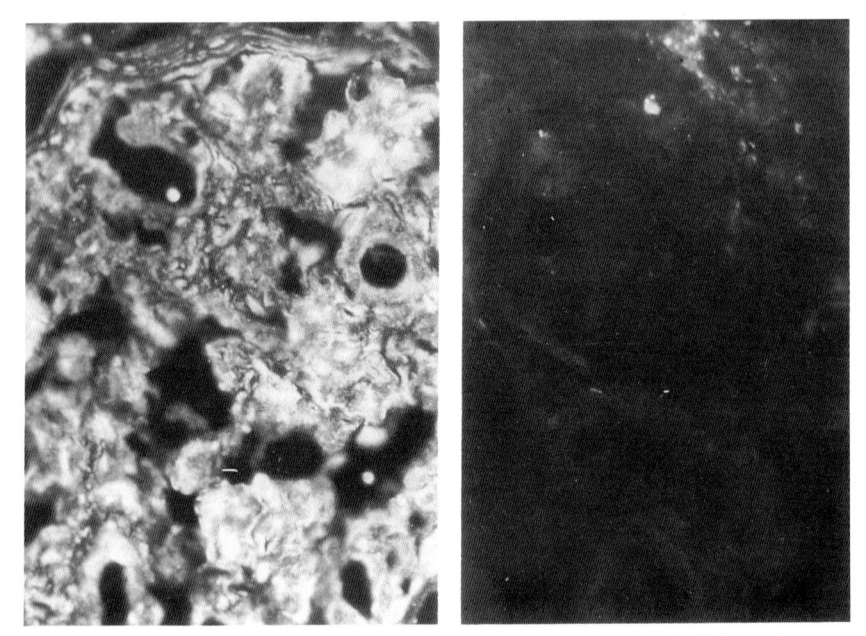

FIGURE 1 Immunostaining of kidney tissue from a 20-year-old man with diabetic glomerulosclerotic disease stained with affinity-purified anti-AGE IgG (left) or nonimmune control IgG (right).

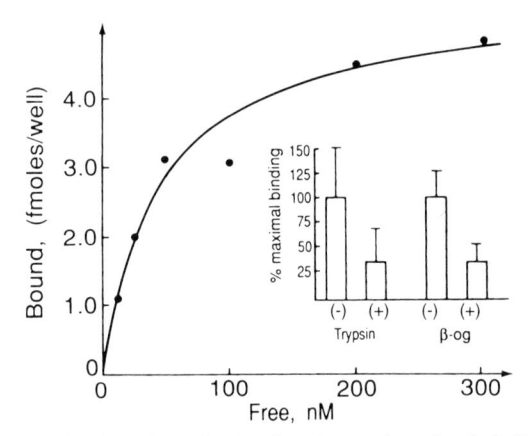

FIGURE 2 Binding of radioiodinated AGE albumin to cultured endothelial cells. Confluent bovine aortic endothelial cell monolayers were incubated with the indicated concentrations of ^{125}I-AGE albumin alone or in the presence of a 20-fold molar excess of unlabeled AGE albumin. Parameters of binding fit to a one-site model were $K_d = 43 \pm 8$ nM and capacity = 5.4 ± 0.3 fmol/well. The inset shows the effect of pretreatment of the monolayers with trypsin or detergent on the subsequent binding of radioiodinated AGE albumin.

of the binding of ^{125}I-AGE albumin and the ability of trypsin to abrogate specific binding. These experiments were next extended to acetone extract of lung; octyl-β-glucoside-solubilized proteins also displayed similar binding of ^{125}I-AGE albumin in the microtiter binding assay.

Based on these data, our purification procedure for isolation of endothelial cell binding proteins for AGEs utilized lung as the starting material. Lung has an especially rich vasculature and, as indicated above, the binding activity for ^{125}I-AGE albumin in lung extract was similar to that observed on intact cultured endothelial cells. Purification steps included chromatography on hydroxylapatite and S Sepharose. Binding activity of fractions was screened using the ^{125}I-AGE albumin binding assay and lung proteins immobilized on microtiter wells. Three polypeptides were purified using this procedures: a novel $\approx$35-kDa polypeptide, lactoferrin, and a high-mobility group I nonhistone chromosome DNA binding protein (14). The $\approx$35-kDa polypeptide was selected for intensive study because of its expression on a range of cell surfaces, and the possibility that it might orchestrate the interaction of AGEs with cellular elements. At the protein level, the $\approx$35-kDa polypeptide was composed of a single chain (on both reduced and nonreduced SDS–PAGE); it was sensitive to degradation by trypsin and bound AGEs with $K_d \approx 50$ nM. Application of the $\approx$35-kDa polypeptide to a column with immobilized AGE albumin resulted in its adsorption; elution was accomplished in the presence of high salt.

Molecular cloning was undertaken to better understand the nature of the $\approx$35-kDa AGE binding protein (15). Analysis of the full-length cDNA showed that the $\approx$35-kDa form was a degradation product of an $\approx$55-kDa polypeptide. Sequencing of tryptic peptides derived from cleavage of the $\approx$35 and $\approx$55-kDa polypeptides has shown that the $\approx$35-kDa form results from proteolytic cleavage of the $\approx$55-kDa molecule releasing the carboxy-terminal portion of the molecule. Computer analysis of the deduced amino acid sequence of the $\approx$55-kDa polypeptide demonstrated a putative extracellular domain composed of three immunoglobulin-like regions: one "V"-type followed by two "C"-type regions (Fig. 3). The latter portion of the

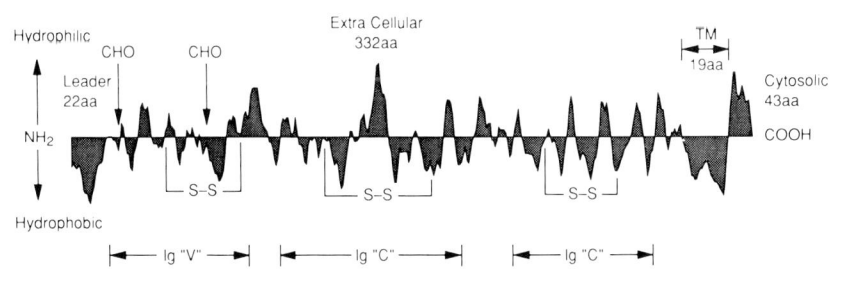

FIGURE 3 Hydrophilicity plot of bovine RAGE was generated from the Hopp and Woods program of intelligenetics.

molecule had cysteines placed at sites which would serve to stabilize immunoglobulin domains, and most closely resembled Neural Cell Adhesion Molecule-1 (NCAM-1) and MUC18. Following the extracellular domain was a single transmembrane spanning domain and a short, highly charged cytosolic tail of 43 amino acids. The cytosolic tail most closely resembled that found in the B cell activation marker CD20. There were no obvious phosphorylation sites or other motifs in the cytosolic tail, suggesting that its capacity to activate signal transduction mechanisms would probably be the result of interactions with cytosolic proteins. Thus, the ≈55-kDa polypeptide appeared to have properties of an integral membrane protein receptor and was termed RAGE. That RAGE was indeed a bona fide cell surface binding site for AGEs was shown by transient transfection of 293 cells with the RAGE cDNA. Cell surface expression of RAGE was observed, and coincided with binding of ^{125}I-AGE albumin ($K_d \approx 100$ nM; Fig. 4); the latter was blocked by monospecific polyclonal antibody to RAGE. Studies on cultured endothelial cells also showed that RAGE was the principal site mediating binding of ^{125}I-AGE albumin to the cell surface.

To gain insights into the biology of RAGE, tissue surveys were performed using monospecific antibody to the receptor in order to evaluate its localization in normal and pathologic samples (16). Quiescent vasculature showed staining for RAGE in endothelial cells and smooth muscle cells. However, RAGE expression was variable and often weak in normal vessels compared with sites of inflammation or other types of pathology (see below), where levels of RAGE were clearly increased (17). The most unexpected finding

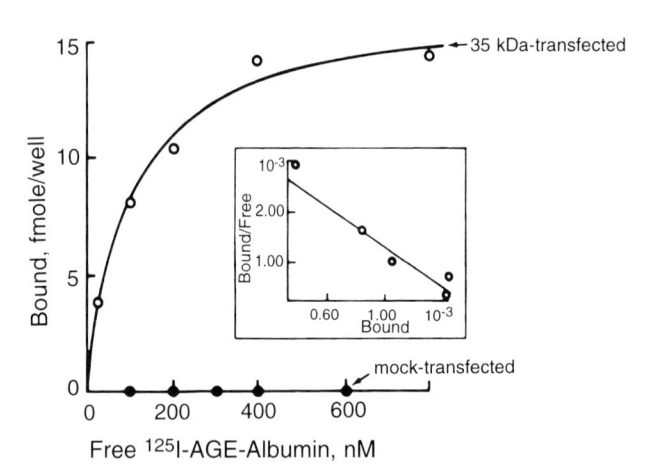

FIGURE 4 Binding of ^{125}I-AGE albumin to transfected 293 cells. A radioligand binding assay was performed on 293 cells transfected with the cDNA for RAGE (open circles) or mock-transfected 293 controls (closed circles) by adding the indicated concentrations of ^{125}I-AGE albumin alone or in the presence of unlabeled AGE albumin (20-fold molar excess). Parameters of binding are $K_d = 100 \pm 20$ nM and capacity = 17 ± 1 fmol/well.

in the tissue survey was the presence of RAGE in neurons. RAGE was present in occasional cortical neurons in normal brain, and spinal motor neurons stained intensely for RAGE. Peripheral nerves also demonstrated expression of RAGE which increased in diabetic neuropathy. Multiple controls were performed to be certain that anti-RAGE IgG was not cross-reacting with NCAM, but the staining appeared to reflect RAGE expression. These data provided our first suggestion that RAGE was a neuronal cell surface protein, leading us to speculate that it might have a role in the physiology of the central nervous system.

Expression and Functions of RAGE: Endothelial Cells and Mononuclear Phagocytes

Endothelial Cells

RAGE is expressed on the endothelial cell surface where it functions as a receptor for AGEs. This was first demonstrated by studies using AGE albumin prepared *in vitro*, and these findings have been extended to AGEs formed *in vivo*. A particularly relevant example of pathophysiologically relevant AGEs is the modification of structures on the surface of diabetic red cells by nonenzymatic glycation (18). In view of the long transit time of red cells in the circulation and their exposure to hyperglycemia, it is not surprising that glycation of red cell components, in addition to hemoglobin, would occur. The significance of cell surface AGEs on diabetic red cells is their potential to support adherence to the endothelium. An earlier study had shown increased adhesivity of diabetic red cells to endothelium, leading us to determine whether this was due to AGE–RAGE interaction (19). The enhanced binding of diabetic red cells to cultured human umbilical vein endothelial cells was blocked by preincubation of diabetic red cells with affinity-purified anti-AGE IgG or by preincubation of endothelium with anti-RAGE IgG (19). Furthermore, addition of a truncated form of the receptor, the extracellular domain (≈ 35 kDa) of RAGE, which is termed soluble RAGE (sRAGE), blocked the diabetic red cell–endothelial interaction. sRAGE functions as a decoy (20), preventing the binding of AGEs to cell surface RAGE, as depicted in Fig. 5. Our *in vitro* observations led us to predict that diabetic red cells infused into normal animals might display a shortened survival due to enhanced binding to the vessel wall. For these studies, rats were rendered diabetic with streptozotocin, and the red cells were infused into syngeneic animals (11). Survival of diabetic erythrocytes was shortened, compared with normal erythrocytes, and this was reversed, in part, by administration of anti-RAGE IgG. These data supported the likelihood that diabetic red cells do interact with the vessel wall *in vivo*, mediated by AGE on the diabetic red cell binding to endothelial RAGE,

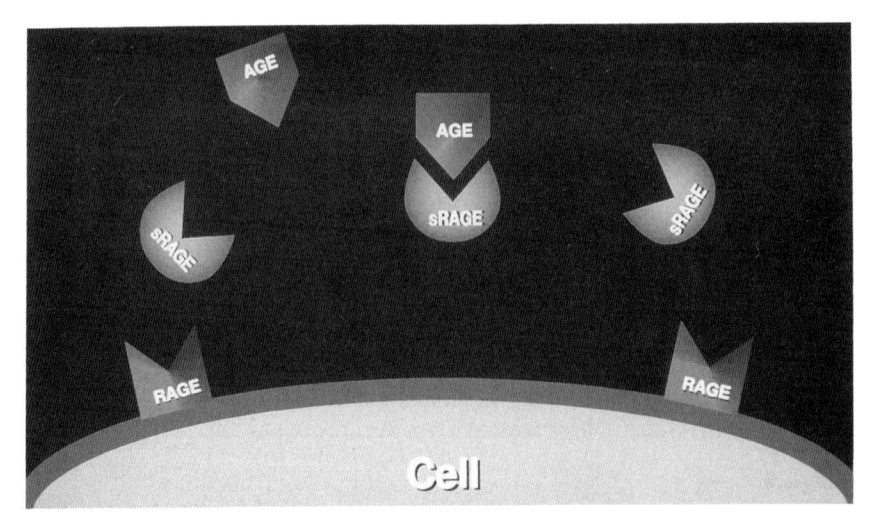

FIGURE 5 sRAGE functions as a decoy interfering with the binding of AGEs to cellular RAGE.

and suggested the importance of characterizing functional consequences of this interaction.

Increased vascular permeability is one of the hallmarks of diabetic vasculopathy. Such vascular dysfunction is most likely multifactorial, reflecting hemodynamic and vessel wall etiologies in patients. Endothelia as the cells which form the blood–tissue interface, have a critical role in barrier function of the vasculature. Incubation of diabetic red cells (harvested from patients with diabetes) with endothelium increased diffusional transit of solutes across the cell monolayer (Fig. 6). This was prevented by anti-RAGE IgG or by sRAGE, indicating that diabetic red cell interaction with endothelial RAGE was critical to the observed perturbation of endothelial function. To determine if diabetic red cells could modulate vascular barrier function *in vivo*, red cells harvested from diabetic rats were infused into normal rats and the changes in vascular permeability were measured using the tissue blood isotope ratio (an index of leakage of ^{125}I-albumin corrected for pooling of ^{51}Cr-labeled red cells in the tissue). Diabetic red cells increased vascular leakage in many organs, compared with normal red cells which did not; this was completely blocked by pretreating the animals with anti-RAGE IgG. The critical test of our hypothesis concerning the ability of AGE–RAGE interaction to mediate vascular dysfunction resulting in hyperpermeability was to perform studies using diabetic rats. Increased vascular permeability was observed in diabetic rats, especially in intestine, skin, and kidney (though in other organs, as well, but to a lesser extent) (Fig. 7). Administration of sRAGE reversed the hyperpermeability completely in intestine and skin, and by about 90% in kidney (Fig. 7). These data support the hypothesis that

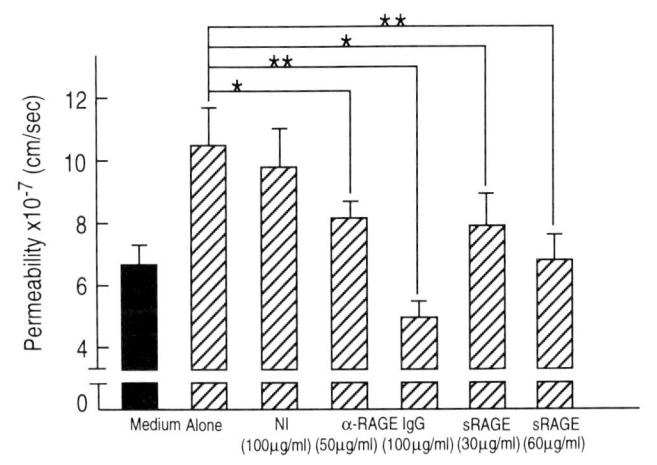

FIGURE 6 Effect of diabetic red blood cells on the barrier function of cultured endothelial cell monolayers: Effect of RAGE blockade. Postconfluent bovine aortic ECs were incubated with medium alone, diabetic (▨) or normal (■) RBC as indicated and permeability to ^{125}I-albumin was determined. sRAGE and anti-RAGE IgG reduced the permeability significantly ($**p < 0.01$ and $*p < 0.05$).

AGE–RAGE interaction is relevant to the pathophysiology of diabetic vasculopathy.

Mononuclear Phagocytes

As professional scavenger cells, mononuclear phagocytes would be expected to have a major role in cellular interactions of AGEs. Monocytes express RAGE, but they also display other receptors reported to bind AGEs,

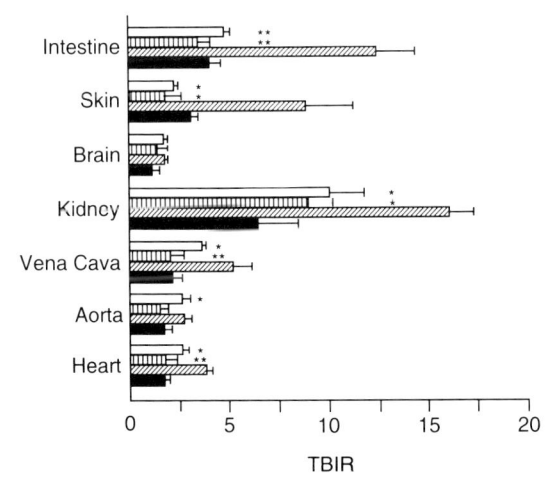

FIGURE 7 Infusion of sRAGE reduces vascular permeability in diabetic rats. Diabetic (▨) and normal (■) rats were infused with two different doses of sRAGE (□, 2.25; ▥, 5.15 mg/kg) and tissue–blood isotope ratio (TBIR) was determined. $*p < 0.05$, $**p < 0.01$.

such as the macrophage scavenger receptor. Thus, it was important to dissect the contribution of AGE–RAGE interaction to AGE-mediated perturbation of monocyte functions. RAGE is expressed by monocytes and mediates the binding of ^{125}I-AGE albumin and an *in vivo*-derived ligand, ^{125}I-AGE-β_2 microglobulin, to the cell surface (13,21). The critical aspect of AGE–RAGE interaction relates to its potential role in monocyte activation. Soluble AGEs, such as AGE albumin, induce directional migration of monocytes down a concentration gradient (13). In contrast, no induction of the migration of polymorphonuclear leukocytes was observed (these cells do not appear to express significant levels of RAGE). However, when monocytes reach a site of immobilized AGEs, as in tissues of a diabetic animal, their migration is halted. Experiments addressing this issue involved adsorbing AGE albumin to the upper surface of chemotaxis chamber membranes, and then adding monocytes to the upper chamber. Cell migration was initiated by adding the chemotactic peptide formyl-methionyl-leucinyl-phenylalanine (fMLP) to the lower chamber. Wells in which the upper surface of the membrane was coated with AGE albumin did not allow monocytes to cross the membranes and emerge below; the monocytes were trapped on the upper surface of the filter. This haptotactic response of monocytes to immobilized AGE albumin was blocked by anti-RAGE F(ab')$_2$ or by sRAGE, indicating that RAGE engagement of the AGE albumin on the membrane was responsible for suppressing monocyte migration. Consistent with these data, monocytes plated on an AGE-albumin-coated surface showed diminished migration using the phagokinetic track assay. In this assay, cells migrate on a defined matrix which has been overlayed with colloidal gold particles; the locomoting cells push the gold particles out of the way, making tracks that are dark, as visualized by dark-field microscopy. Note that the phagokinetic tracks are long and extensive on the native albumin substrate (Fig. 8, left), whereas they are short and small on the AGE albumin substrate (Fig. 8, right). Under normal conditions, monocytes are continuously circulating through the tissues. When they encounter a deposit of immobilized AGEs, their locomotion is arrested. Furthermore, contact of AGE-modified matrix results in cell activation with expression of cytokines and growth factors.

These experiments with monocytes and AGE albumin prepared *in vitro* led us to perform studies with a pathophysiologically relevant AGE-modified protein. Certain patients with renal failure develop an arthropathy termed Dialysis-Related Amyloidosis (DRA). The major component of the amyloid in this disorder is an AGE-modified, acidic form of β_2-microglobulin. The latter accumulates in the joint space and mediates attraction of inflammatory cells underlying the erosive arthropathy that eventuates in destruction of the joint. These observations led us to study the interaction of AGE-β_2-microglobulin, prepared from patient samples, with monocytes. Binding of ^{125}I-AGE-β_2-microglobulin to monocytes was due to interaction with RAGE, as shown by the dose-dependent inhibitory effect of sRAGE or anti-RAGE

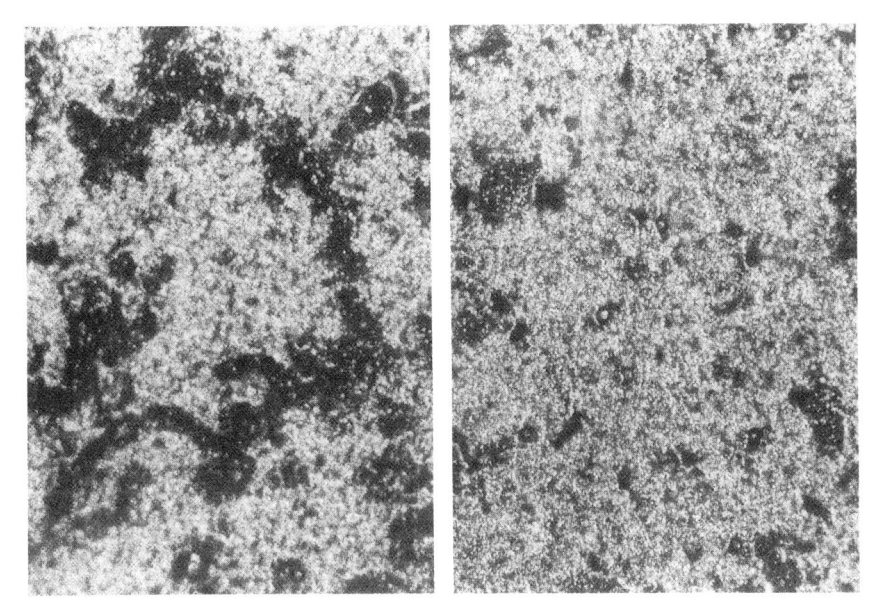

FIGURE 8 Phagokinetic track assay. Mononuclear phagocyte migration was studied on matrices composed of either native albumin (left) or AGE albumin (right) which had been coated with colloidal gold particles.

IgG (21; Fig. 9). Monocytes exposed to AGE-β_2-microglobulin became activated, as illustrated by the increased expression of tumor necrosis factor α. This was also suppressed by blockade of RAGE. These data indicated that althouth there may be several monocyte cell surface structures capable of binding AGEs, RAGE is critical in mediating AGE-induced cell activation. For this reason, we propose that AGE–RAGE interaction is a potentially critical lock-and-key type mechanism, especially in diabetes, for explaining a range of AGE-mediated disturbances in cellular functions (Fig. 10).

RAGE and Diabetes

Hyperglycemia is an obvious situation in which the formation of AGEs is favored. Accumulation of AGEs, as well as increased expression of RAGE, occurs in diabetic vasculature compared with age-matched controls (22; Fig. 11). These data indicate that ligand and receptor are juxtaposed in the diabetic milieu. One mechanism hypothesized to underly diabetic complications is the induction of cellular oxidant stress. Based on the close association of oxidant stress with disturbance of endothelial cell barrier function, we tested whether AGE–RAGE interaction would induce an oxidant stress phenotype. Although AGEs by themselves generate oxygen free radicals, in view of the antioxidant milieu present *in vivo* we reasoned that AGE–RAGE

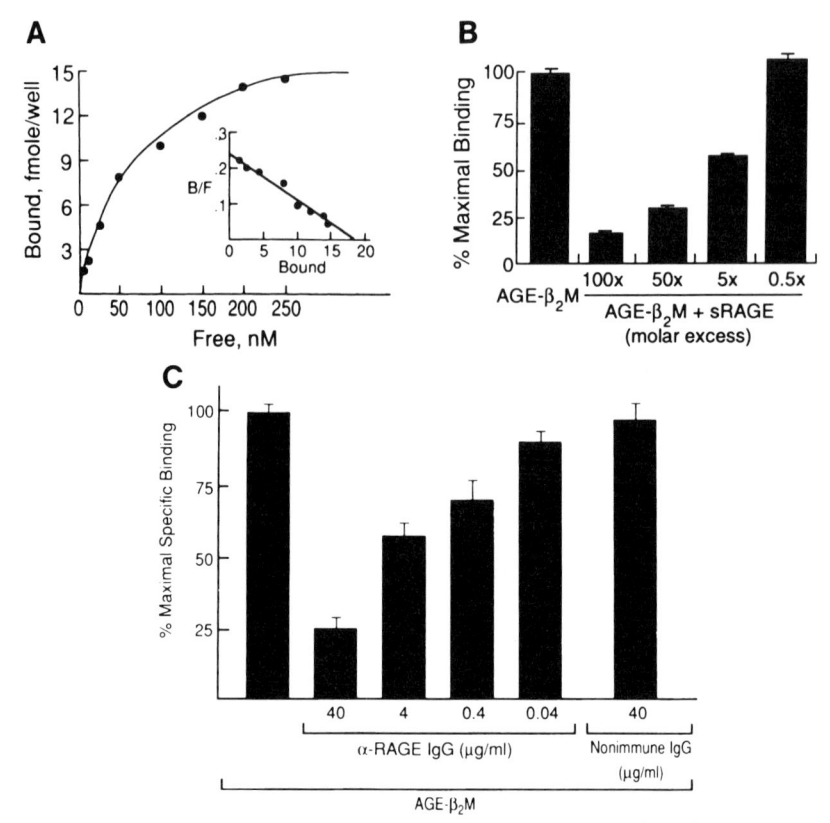

FIGURE 9 Binding of AGE-β_2-microglobulin to monocytes. Mononuclear phagocytes were exposed to ^{125}I-AGE-β2-microglobulin alone or in the presence of excess unlabeled material. Parameters of binding were $K_d = 81.6 \pm 9.92$ nM and capacity = 19.15 ± 0.886 fmol/well. (B and C) The effects of excess sRAGE and anti-RAGE IgG are shown, respectively.

interaction would facilitate the possible induction of an oxidant stress response, as it would bring AGEs in proximity to the cell surface. Infusion of AGE albumin into mice resulted in increased generation of thiobarbituric acid-reactive substances, induction of heme oxygenase mRNA/antigen, and activation of the transcription factor NF-κB (23). These findings, along with their suppression by antioxidants, probucol or N-acetylcysteine, indicated that AGE-mediated oxidant stress was likely to be the critical event underlying their expression. The central role of RAGE in AGE-mediated induction of oxidant stress was supported by studies indicating that anti-RAGE IgG had an inhibitory effect. Further studies have shown that beyond tethering AGEs to the cell surface, AGE ligation of RAGE activates NADPH oxidase in certain cell types (such as endothelium) (24), thereby potentiating the oxidant stress response.

Induction of oxidant stress as a consequence of AGE–RAGE interaction could be an important factor in vascular dysfunction in diabetes. Inactivation

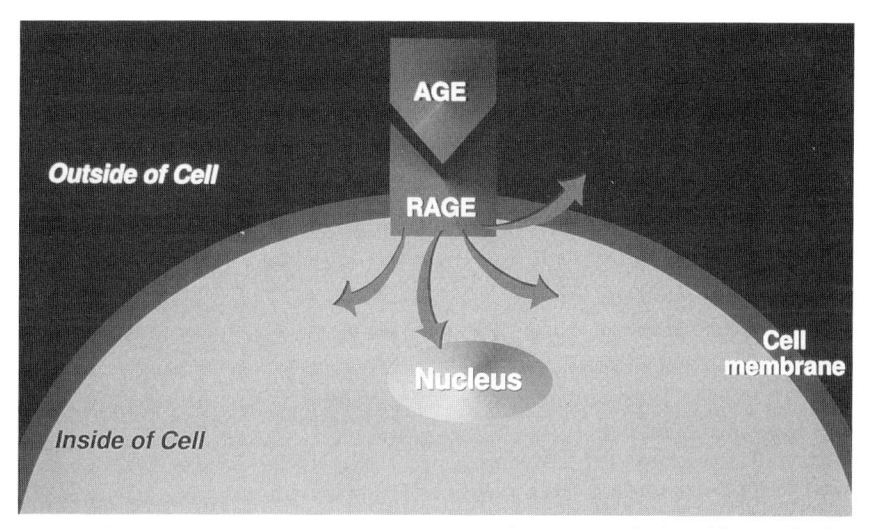

FIGURE 10 AGE–RAGE interaction is a potentially important lock-and-key mechanism, especially in diabetes, for explaining a range of AGE-mediated disturbances in cellular function.

of endothelial-derived nitric oxide is a direct mechanism by which AGE-generated oxygen free radicals might alter regulation of vasomotor tone. AGE–RAGE-mediated oxidant stress could have far-ranging effects on cell function. First, AGE-induced vascular hyperpermeability was found to be due, at least in part, to an oxidant stress mechanism, as it was ameliorated by antioxidants (11). Second, AGE-induced activation of endothelial NF-κB resulted in increased expression of VCAM-1 (12). This cell adhesion molecule, a member of the immunoglobulin superfamily whose expression is closely linked to experimental atherosclerosis in animal models (25,26), was expressed in endothelium exposed to AGEs. The pathway for AGE induction included binding to endothelial RAGE, induction of oxidant stress with activation of NF-κB, increased rate of VCAM-1 transcription, and increased translation and cell surface expression of VCAM-1. Once on the cell surface, VCAM-1 supported binding of Molt-4 cells, indicating that it was a functionally competent adhesion molecule. Infusion of AGE albumin into mice causes vascular expression of VCAM-1, consistent with the *in vivo* relevance of these findings.

These data suggested that monitoring vascular VCAM-1 expression might provide an index of perturbation of the vessel wall in diabetes. As VCAM-1 is also present in the plasma in a soluble form (sVCAM-1), we considered whether its release, as following AGE–RAGE-induction of VCAM-1, might reflect ongoing vascular oxidant stress. Experiments in cell culture showed that, in parallel with AGE induction of cell surface VCAM-1, the soluble form was released into culture supernatants. Although initially we considered that sVCAM-1 might be part of a

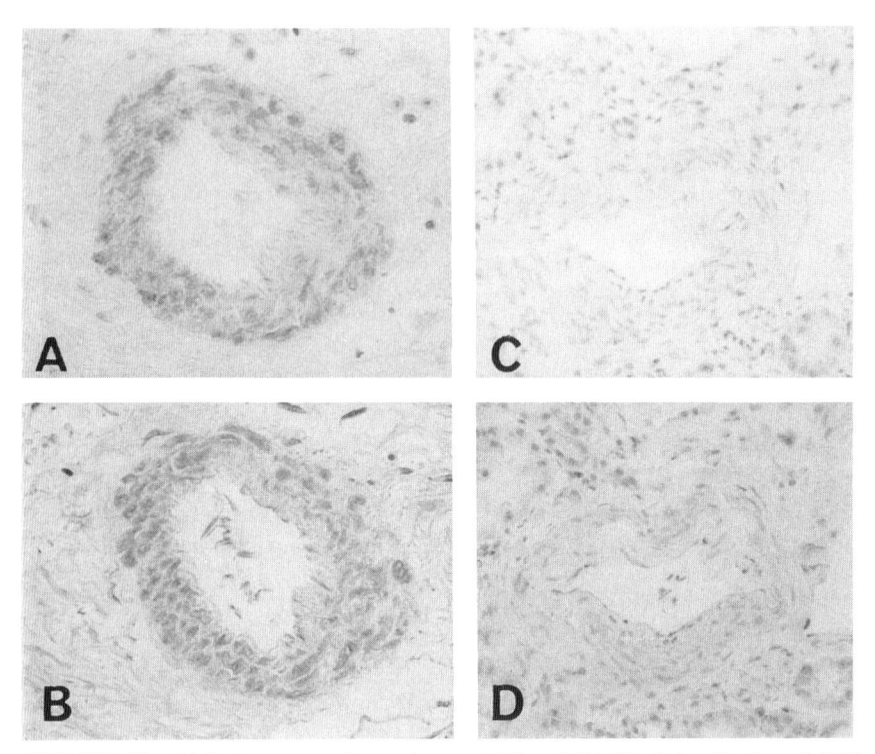

FIGURE 11 Diabetic vs normal vasculature: AGE and RAGE. Colocalization of AGE antigen (A) and RAGE epitopes (C) in adjacent sections of diabetic vasculature (A,C) versus their absence in age-matched control vasculature (B and D, respectively, for AGE and RAGE). Affinity-purified anti-AGE IgG and anti-RAGE IgG were used to visualize the respective antigens.

protective mechanism, preventing ligation of VCAM-1 by cell surface VLA-4, sVCAM-1 at levels present in culture supernatants or *in vivo* (see below) was not an antagonist of VLA-4–VCAM-1 interaction. Because of the association of oxidant stress with vascular dysfunction, especially in diabetes, we sought a patient group at high risk for the development of vascular complications early in their clinical course. Patients with diabetes and microalbuminuria compose such a group, as microalbuminuria has been identified as an independent risk factor for cardiovascular disease in types I and II diabetes. Patients with diabetes and microalbuminuria had higher sVCAM-1 levels (757.7 ± 85 ng/ml) compared with those with diabetes without microalbuminuria (505 ± 55 ng/ml; $p < 0.05$) (27). These findings indicate the importance of determining whether AGE–RAGE-mediated oxidant stress underlies expression of sVCAM-1 in the plasma of patients with diabetes.

RAGE and Amphoterin

The identification of RAGE as a member of the immunoglobulin superfamily of cell surface molecules suggested that this receptor was likely to have functions quite distinct from those of a scavenger receptor or a binding site exclusively for AGEs. We wondered if RAGE might also be involved in cell–cell or cell–matrix interactions, or if it might possibly function as a cytokine or growth factor receptor. Recent data demonstrating that RAGE is on chromosome 6 in the Major Histocompatibility Complex, where genes for cytokines and molecules involved in the immune response are found, was consistent with the possibility that RAGE might be involved in host response mechanisms (28). These data led us to identify other ligands of RAGE present in tissues. Our approach was to employ a column with immobilized sRAGE as the critical affinity purification step for putative RAGE ligands. Using bovine lung as the starting material, two polypeptides adhered to the RAGE column with highest affinity; these were termed p23 and p12 based on their mobility on SDS–PAGE (i.e., ≈23 and ≈12 kDa). First, we have characterized the ≈23-kDa polypeptide in detail (29). Based on N-terminal and internal protein sequence analysis, this molecule was identical to amphoterin. Amphoterin is an amphoteric polypeptide, possessing a basic, lysine-rich N-terminus and an acidic, glutamic acid-rich C-terminus, and is highly expressed in the developing nervous system. *In vitro*, rat embryonic neurons cultured on amphoterin-coated substrates demonstrate outgrowth of neurites, as previously described (30,31). Blockade of amphoterin has been shown to suppress such neurite outgrowth. Thus, amphoterin has been suggested to contribute to neuronal development via regulation of cell–matrix interaction.

Our studies have demonstrated that amphoterin is a ligand for RAGE. Amphoterin binding to RAGE is not mediated by AGEs, but by determinants in the protein backbone. However, AGEs and amphoterin compete for binding to RAGE, suggesting that they recognize identical or overlapping sites on the receptor. Radioligand studies showed dose-dependent binding of ^{125}I-amphoterin to purified RAGE which fit to a one-site model demonstrating K_d of ≈6 nM (this is higher affinity than that of RAGE for AGEs, which is closer to ≈50 nM). Studies on neonatal rat cortical neurons also demonstrated binding of ^{125}I-amphoterin to RAGE. Half-maximal occupancy of cell surface binding sites occurred at ≈9 nM and was blocked by either anti-RAGE IgG or excess sRAGE. Consistent with a role for RAGE in mediating neuronal binding to amphoterin was the observed inhibition of neurite outgrowth by sRAGE or anti-RAGE IgG when neurons were plated on amphoterin-coated substrates (Fig. 12).

Although it is too soon to speculate about the significance of RAGE in neuronal development, our studies in developing rat brain have shown high levels of RAGE in neurons of cerebral cortex. Furthermore, sites of RAGE

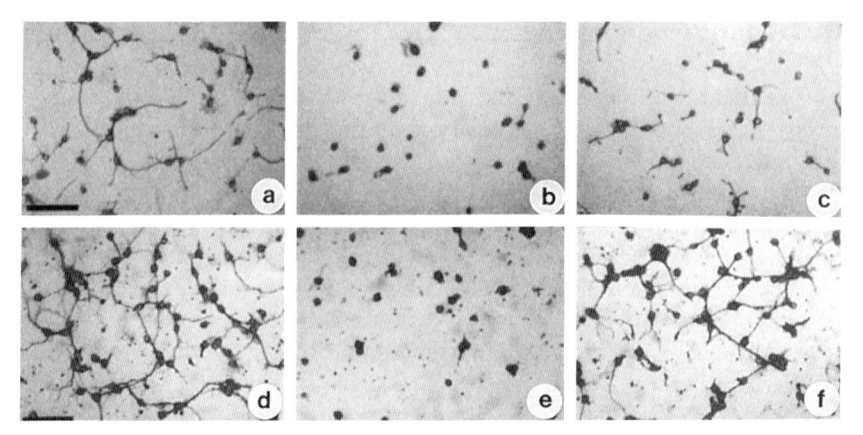

FIGURE 12 Neurite outgrowth assays and the effect of RAGE blockade. Wells were coated with amphoterin for 18 hr. Cortical neuronal cells from E17 rat embryos were fixed with paraformaldehyde and NP-40 and stained with anti-tubulin antibody. Effect of sRAGE. Amphoterin-coated wells and neuronal cells were pretreated with no additive (a), 50 μg/ml sRAGE (b), or 5 μg/ml sRAGE (c). Effect of anti-RAGE F(ab')₂. Amphoterin-coated wells and neuronal cells were pretreated with nonimmune F(ab')₂ (d), 40 μg/ml anti-RAGE F(ab')₂ (e), or 4 μg/ml anti-RAGE F(ab')₂ (f).

expression are proximal to areas where amphoterin is present, suggesting that RAGE–amphoterin interaction might indeed occur. Additional studies are underway to determine if RAGE engagement of amphoterin and/or other ligands provides a pathway which is important in neuronal development.

RAGE and Alzheimer's Disease

The discovery that RAGE is a receptor for amyloid-β peptide (Aβ) was fortuitous. Our initial investigations derived from experiments directed at identifying AGEs in Alzheimer's disease. AGE formation is facilitated by hyperglycemia, but delayed turnover of proteins, especially if they are rich in lysine, also promotes nonenzymatic glycation even with euglycemia. Our first studies focused on the intracellular microtubule-associated protein τ (a protein rich in lysine), as it accumulates in an aggregated form within affected neurons in Alzheimer's disease (32). Paired helical filament τ, the form of τ which largely composes neurofibrillary tangles (one of the hallmarks of Alzheimer's-type pathology), was found to be present in an AGE-modified form. The pathophysiologic significance of these AGEs was suggested by generation of oxygen free radicals by Alzheimer's-derived τ, suppressed in part by anti-AGE IgG. Liposome-mediated introduction of AGE-τ into neuroblastoma cells induced an oxidant stress phenotype with increased expression of heme oxygenase type I and activation of NF-κB (both of which were blocked by antioxidants) (33). These findings suggested the relevance of AGEs as a possible progression factor in Alzheimer's disease and led us to study the extracellular deposits of Aβ. Evidence of AGE formation was difficult to detect, but further studies have suggested that a limited amount of AGEs might be associated with neuritic plaques. To determine the effect of glycation on Aβ, the peptide was incubated with high concentrations of glucose. These studies resulted in rapid identification of an experimental problem: precipitation of AGE-Aβ preventing further experiments. However, we found that nonglycated control Aβ (1–40 or 1–42) was very effective in the induction of cellular oxidant stress, even in the presence of high concentrations of serum. The latter finding led us to speculate that a cell surface receptor might be involved in tethering Aβ to the cell surface, thereby placing it in a privileged environment with respect to inducing oxidant stress in target cells. Our efforts to purify a cell surface Aβ binding protein rapidly led to isolation of ≈35 and ≈55-kDa bands which proved to be identical to RAGE (34). That RAGE was interacting with Aβ itself, and not possible AGEs, was confirmed by two lines of evidence: synthetic peptides (Aβ 1–40 and 1–42) bound to RAGE, and anti-AGE IgG had no effect on binding of synthetic peptides or Aβ purified from brains of patients with Alzheimer's disease. These studies opened a new chapter in the biology of RAGE, in view of its possible role in the neuronal damage associated with Alzheimer's disease.

The binding of binding of Aβ to RAGE was evaluated in parallel studies either by radioiodinating Aβ and immobilizing RAGE on plastic microtiter wells or by labeling RAGE and adsorbing the Aβ to the wells. Because of the propensity of Aβ to aggregate (thus producing an inhomogeneous tracer) and questions concerning possible modification of its biologic activity after radiolabeling, it was necessary to employ both approaches. Furthermore, experiments in which Aβ was adsorbed to the substrate allowed us to use the relatively insoluble form of the peptide, Aβ 1–42, as well as Aβ purified for Alzheimer brain. When ^{125}I-Aβ (synthetic 1–40, freshly prepared) was incubated with RAGE immobilized on the microtiter wells, dose-dependent specific binding was observed which varied with the concentration of RAGE adsorbed to the wells, and, at one RAGE concentration, was proportional to the amount of ^{125}I-Aβ added. The binding of ^{125}I-Aβ fit best to a one-site model, analyzed using nonlinear least squares analysis and the method of Klotz and Hunston (34; K_d ≈50 nM). Binding of Aβ 1–40 was competed by excess unlabeled peptide, either freshly prepared or incubated for 3 days at 37°C to allow aggregates to form. Although Aβ 1–20 was not a competitor of ^{125}I-Aβ binding to RAGE, Aβ 25–35, neither scrambled Aβ 25–35 nor unrelated peptides, such as Arg–Gly–Asp–Ser, were competitors. Binding of ^{125}I-Aβ to RAGE was also blocked by anti-RAGE IgG, but not by nonimmune IgG, and by addition of increasing amounts of sRAGE, but not by a similar soluble form of VCAM-1. These data are consistent with the specificity of Aβ–RAGE interaction, and indicated the importance of assessing whether this would occur on cell surfaces. One important issue concerning the nature of Aβ which binds to RAGE is that our experiments do not address whether the receptor-bound form of Aβ is a monomer, small oligomer, extensive aggregate, or fibril. (Although preparations of unlabeled Aβ which were either predominately monomer or aggregate both competed with ^{125}I-Aβ for binding to RAGE, it is impossible to rule out that this is due to small amounts of oligomers, fibrils or monomers present in these preparations.) As Aβ-induced cell toxicity has been linked to fibril formation (35), this is an important issue for future studies to dissect. Similar binding results were obtained when experiments were performed with ^{125}I-sRAGE and Aβ (1–42,1–40, or that derived from Alzheimer's brain). In fact, replicates in the binding assay were tighter and nonspecific binding was lower in experiments in which sRAGE was the tracer, possibly because it was a more homogeneous tracer than Aβ.

RAGE-transfected COS cells acquired the ability to bind Aβ (Fig. 13). Binding was specific for Aβ–RAGE interaction, since antibodies to RAGE and sRAGE were competitors. In fact, the results of binding studies with multiple competitors, including Aβ-derived peptides, were very similar to those observed in the complementary binding assay using ^{125}I-Aβ and RAGE immobilized on microtiter wells. To extrapolate these results to cells more relevant to the pathophysiology of Alzheimer's disease, experiments were performed on endothelial cells, neuronal-like cells, and microglia. Binding of ^{125}I-Aβ to endothelial cells appeared to be mediated exclusively by RAGE: anti-RAGE IgG and sRAGE completely blocked specific binding. On PC12 cells the situation was somewhat different. Although sRAGE completely blocked the interaction between Aβ and the cell surface, inhibition by anti-

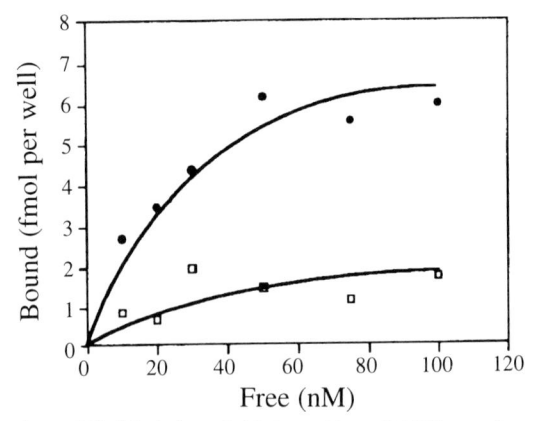

FIGURE 13 Binding of ^{125}I-labeled amyloid-β peptide to RAGE-transfected cos-1 cells (●) vs mock-transfected cos-1 cells (□). $K_d = 25 \pm 6.2$ nM for RAGE-transfected cos-1 cells; mock-transfected cos-1 cells bound ^{125}I-labeled amyloid-β peptide only minimally.

RAGE IgG was 60–80%. These data suggested that in addition to RAGE, another binding site was involved in tethering Aβ to PC12 cells, although once the amyloidogenic peptide was bound to sRAGE, Aβ did not interact with either of these cell surface sites. On microglia or a transformed murine microglial cell line (BV-2 cells), the pattern of binding inhibition was similar to that observed on neurons; complete inhibition by sRAGE and incomplete inhibition by anti-RAGE IgG. As microglia express the macrophage scavenger receptor, which is reported to be a binding site for Aβ, it is not surprising that our data support the presence of another binding site for Aβ on these cells.

Although there appeared to be more than one binding site for AGEs on microglia and neurons, the important issue was to determine the contribution of RAGE to Aβ-mediated perturbation of cellular properties. Addition of Aβ to PC12 cells or cortical neurons resulted in oxidant stress, as reflected by generation of thiobarbituric acid-reactive substances and activation of NF-κB; both of these events were blocked completely by anti-RAGE IgG and sRAGE. In addition, blockade of RAGE was effective in preventing Aβ-induced suppression of PC12 reduction of MTT [3-(4,5-dimethylthiazol-2-yl)-2,5 diphenyl tetrazolium bromide] and Aβ-mediated perturbation of PC12 appearance. Incubation of PC12 cells with Aβ caused retraction of cell processes and rounding up of the cell bodies; this was prevented by blockade of RAGE with either sRAGE or anti-RAGE IgG (34). On microglia, blockade of RAGE prevented both chemotaxis to soluble Aβ, the haptotactic response to immobilized Aβ, and induction of tumor necrosis factor α due to Aβ. These data indicated that although Aβ appears to interact with several sites on neurons and microglia, RAGE has an important role in Aβ-induced changes in cellular function. We hypothesize that Aβ, because of its heterogeneity (due to aggregation), is a multivalent ligand interacting with multiple RAGE molecules, as well as other cell binding sites. It is possible that through the interaction with several sites, the amyloidogenic peptide is tethered to the cell surface with increased affinity.

Further support for the involvement of RAGE in Alzheimer's disease was derived from the study of patient-derived brain tissue. Levels of RAGE antigen were increased by ELISA, and immunohistochemistry demonstrated increased expression of RAGE in neurons close to deposits of Aβ as well as the vasculature, compared with lesser expression in age-matched controls. These were the same neurons which demonstrated evidence of oxidant stress, based on expression of heme oxygenase type I and nuclear localization of the p50 component of the NF-κB family. Our recent studies demonstrating that NF-κB elements in the RAGE promoter regulate expression of the receptor suggest that oxidant stress might also control levels of RAGE. This could result in a potential positive feedback loop whereby Aβ–RAGE interaction induces NF-κB activation which drives RAGE transcription, thereby further enhancing Aβ–RAGE interaction.

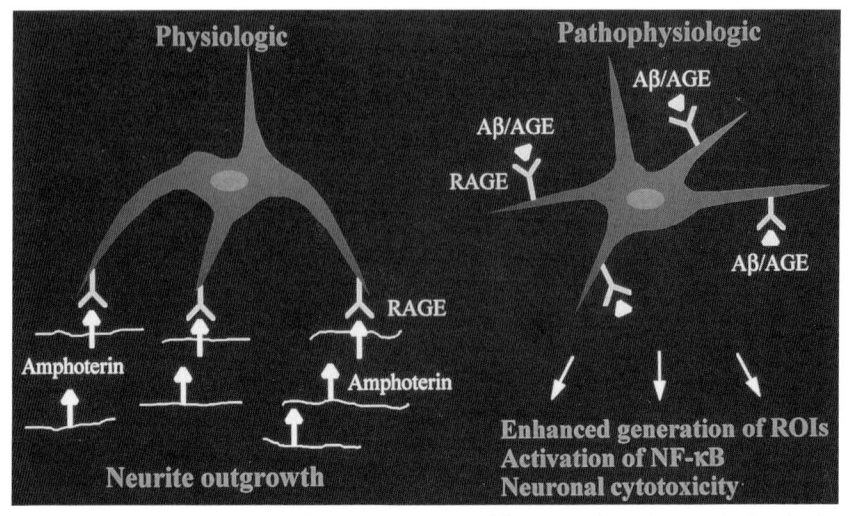

FIGURE 14 Diverse effects of RAGE in neuronal homeostasis and perturbation by its varied ligands. Schematic diagram of how RAGE may contribute to pathophysiologic and physiologic situations in neurons.

Hypothesis

Taken together, these data present a picture of RAGE as a cellular receptor for multiple ligands. How can we assemble these observations into a paradigm for integrating the varied effects of RAGE on cellular functions? Figure 14 depicts our concept of how RAGE may contribute to pathophysiologic and physiologic situations in neurons. During the embryonic period, RAGE is expressed at high levels, especially in neurons, where it contributes to neurite outgrowth and neuronal development via ligands such as amphoterin. Another potentially physiologic ligand for RAGE is low levels of $A\beta$ produced throughout life. In adulthood, levels of RAGE seem to be quite low in vasculature, neurons, and other tissues. The role of RAGE in such homeostatic conditions, if any, is not yet clear. However, when a pathophysiologic process intervenes, such as diabetes or Alzheimer's disease, expression of RAGE is enhanced. Under these conditions, when RAGE engages $A\beta$ or AGEs, the result is cellular stress, and, in certain cases such as neurons, cytotoxicity. These predictions form a framework for our future studies in genetically manipulated mice to rigorously test the role of RAGE in development, homeostasis, and pathophysiologic situations.

Acknowledgments

This work was supported by grants from the USPHS (AG00602, AG00690, HL56881), the New York affiliate of the American Heart Association, American Diabetes Association, American Health Assistance Foundation, and Surgical Research Foundation.

References

1. Ruderman, N., Williamson, J., and Brownlee, M. (1992). Glucose and diabetic vascular disease. *FASEB J.* **6,** 2905–2914.
2. Baynes, J. (1991). Role of oxidative stress in development of complications in diabetes. *Diabetes* **40,** 405–412.
3. Sell, D., and Monnier, V. (1989). Structure elucidation of a senescence cross-link from human extracellular matrix; implication of pentoses in the aging process. *J. Biol. Chem.* **264,** 21597–21602.
4. Brownlee, M., Cerami, A., and Vlassara, H. (1988). Advanced glycosylation end products in tissue and the biochemical basis of diabetic complications. *N. Engl. J. Med.* **318,** 1315–1320.
5. Bucala, R., Makita, Z., Koschinsky, T., Cerami, A., and Vlassara, H. (1993). Lipid advanced glycosylation: Pathway for lipid oxidation in vivo. *Proc. Natl. Acad. Sci. U.S.A.* **90,** 6434–6438.
6. Hicks, M., Delbridge, L., Yue, D., and Reeve, R. (1988). Catalysis of lipid peroxidation by glucose and glycosylated proteins. *Biochem. Biophys. Res. Commun.* **151,** 649–655.
7. Hunt, J., Smith, C., and Wolff, S. (1990). Autooxidative glycosylation and possible involvement of peroxides and free radicals in LDL modification by glucose. *Diabetes* **30,** 1420–1424.
8. Reddy, S., Bichler, J., Wells-Knecht, K. J., Thorpe, S. R., and Baynes, J. W. (1995). Nε-(Carboxymethyl)lysine is a dominant Advanced Glycation End Product (AGE) antigen in tissue proteins. *Biochemistry* **34,** 10872–10878.
9. Miyata, T., Taneda, S., Kawai, R., Ueda, Y., Horiuchi, S., Hara, M., Maeda, K., and Monnier, V. M. (1996). Identification of pentosidine as a native structure for advanced glycation end products in β_2-microglobulin-containing amyloid fibrils in patients with dialysis-related amyloidosis. *Proc. Natl. Acad. Sci. U.S.A.* **93,** 2353–2358.
10. Esposito, C., Gerlach, H., Brett, J., Stern, D., and Vlassara, H. (1989). Endothelial receptor-mediated binding of glucose-modified albumin is associated with increased monolayer permeability and modulation of cell surface coagulant properties. *J. Exp. Med.* **170,** 1387–1407.
11. Wautier, J.-L., Zoukourian, C., Chappey, O., Wautier, M. P., Guillausseau, P. J., Cao, R., Hori, O., Stern, D., and Schmidt, A. M. (1996). Receptor-mediated endothelial cell dysfunction in diabetic vasculopathy: Soluble receptor for advanced glycation endproducts blocks hyperpermeability. *J. Clin. Invest.* **97,** 238–243.
12. Schmidt, A.-M., Hori, O., Chen, J., Crandall, J., Zhang, J., Cao, R., Brett, J., and Stern, D. (1995). Advanced glycation endproducts interacting with their endothelial receptor induce expression of vascular cell adhesion molecule-1 (VCAM-1): A potential mechanism for the accelerated vasculopathy of diabetes. *J. Clin. Invest.* **96,** 1395–1403.
13. Schmidt, A. M., Yan, S. D., Brett, J., Mora, R., Nowygrod, R., and Stern, D. (1993). Regulation of mononuclear phagocyte migration by cell surface binding proteins for advanced glycosylation endproducts. *J. Clin. Invest.* **92,** 2155–2168.
14. Schmidt, A.-M., Vianna, M., Gerlach, M., Brett, J., Ryan, J., Kao, J., Esposito, C., Hegarty, H., Hurley, W., Clauss, M., Wang, F., Pan, Y.-C., Tsang, T. C., and Stern, D. (1992). Isolation and characterization of binding proteins for advanced glycosylation end products from lung tissue which are present on the endothelial cell surface. *J. Biol. Chem.* **267,** 14987–14997.
15. Neeper, M., Schmidt, A.-M., Brett, J., Yan, S.-D., Wang, F., Pan, Y.-C., Elliston, K., Stern, D., and Shaw, A. (1992). Cloning and expression of RAGE: A cell surface receptor for advanced glycosylation end products of proteins. *J. Biol. Chem.* **267,** 14998–15004.
16. Brett, J., Schmidt, A.-M., Zou, Y.-S., Yan, S.-D., Weidman, E., Pinsky, D., Neeper, M., Przysiecki, M., Shaw, A., Migheli, A., and Stern, D. (1993). Tissue distribution of the receptor for advanced glycation end products (RAGE): Expression in smooth muscle,

cardiac myocytes, and neural tissue in addition to the vasculature. *Am. J. Pathol.* **143**, 1699–1712.
17. Ritthaler, U., Deng, Y., Zhang, Y., Greten, J., Abel, M., Allenberg, J., Otto, G., Roth, H., Bierhaus, A., Ziegler, R., Schmidt, A. M., Waldherr, R., Wahl, P., Stern, D., and Nawroth, P. (1995). Expression of receptors for advanced glycation end products in peripheral occlusive vascular disease. *Am. J. Pathol.* **146**, 688–694.
18. Wautier, J.-L., Wautier, M. P., Schmidt, A. M., Anderson, G. M., Zoukourian, C., Capron, L., Chappey, O., Yan, S. D., Brett, J., Guillausseau, P. J., and Stern, D. (1994). Advanced glycation end products (AGEs) on the surface of diabetic red cells bind to the vessel wall via a specific receptor inducing oxidant stress in the vasculature: A link between surface-associated AGEs and diabetic complications. *Proc. Natl. Acad. Sci. U.S.A.* **91**, 7742–7746.
19. Wautier, J. L., Paton, C., Wautier, M. P., Pintigny, D., Abadie, E., Passa, P., and Caen, J. (1981). Increased adhesion of erythrocytes to endothelial cells in diabetes mellitus and its relation to vascular complications. *N. Engl. J. Med.* **305**, 237–242.
20. Schmidt, A.-M., Hasu, M., Popov, D., Zhang, J.-H., Yan, S.-D., Brett, J., Cao, R., Kuwabara, K., Costache, G., Simionescu, N., Simionescu, M., and Stern, D. (1994). The receptor for Advanced Glycation End products (AGEs) has a central role in vessel wall interactions and gene activation in response to AGEs in the intravascular space. *Proc. Natl. Acad. Sci. U.S.A.* **91**, 8807–8811.
21. Miyata, T., Hori, O., Zhang, J. H., Yan, S. D., Ferran, L., Iida, Y., and Schmidt, A. M. (1996). The Receptor for Advanced Glycation Endproducts (RAGE) mediates the interaction of AGE-β_2-Microglobulin with human mononuclear phagocytes via an oxidant-sensitive pathway: Implications for the pathogenesis of dialysis-related amyloidosis. *J. Clin. Invest.* **98**, 1088–1094.
22. Schmidt, A.-M., Yan, S. D., and Stern, D. (1995). The dark side of glucose (news and views). *Nat. Med.* **1**, 1002–1004.
23. Yan, S.-D., Schmidt, A.-M., Anderson, G., Zhang, J., Brett, J., Zou, Y.-S., Pinksy, D., and Stern, D. (1994). Enhanced cellular oxidant stress by the interaction of advanced glycation endproducts with their receptors/binding proteins. *J. Biol. Chem.* **269**, 9889–9897.
24. Wautier, J. L., Chappey, O., Wautier, M. P., Boval, B., Stern, D., and Schmidt, A. M. (1996). Interaction of diabetic erythrocytes bearing advanced glycation endproducts with the endothelial receptor RAGE induces generation of reactive oxygen intermediates and cellular dysfunction. *Circulation* **94**(8), 4139.
25. Li, H., Cybulsky, M., Gimbrone, M., and Libby, L. (1993). An atherogenic diet rapidly induces VCAM-1, a cytokine-regulatable mononuclear leukocyte adhesion molecule, in rabbit aortic endothelium. *Arterioscler. Thromb.* **13**, 197–204.
26. Richardson, M., Hadcock, S., DeReske, M., and Cybulsky, M. (1994). Increased expression in vivo of VCAM-1 and E-selectin by the aortic endothelium of normolipemic and hyperlipemic diabetic rabbits. *Arterioscler. Thromb.* **14**, 760–769.
27. Schmidt, A. M., Crandall, J., Cao, R., Hori, O., and Lakatta, E. (1996). Elevated plasma levels of Vascular Cell Adhesion Molecule-1 (VCAM-1) in diabetic patients with microalbuminuria: A marker of vascular dysfunction and progressive vascular disease. *Br. J. Haematol.* **92**, 747–750.
28. Sugaya, K., Fukagawa, T., Matsumoto, K.-I., Mita, K., Takahashi, E.-I., Ando, A., Inoko, H., and Ikemura, T. (1994). Three genes in the human MHC Class III region near the junction with the Class II: Gene for Receptor of Advanced Glycosylation End Products, PBX2 homeobox gene and a notch homolog, human counterpart of mouse mammary tumor gene int-3. *Genomics* **23**, 408–419.
29. Hori, O., Brett, J., Slattery, T., Cao, R., Zhang, J., Chen, J., Nagashima, M., Nitecki, D., Morser, J., Stern, D., and Schmidt, A. M. (1995). The Receptor for Advanced Glycation Endproducts (RAGE) is a cellular binding site for amphoterin: Mediation of neurite out-

growth and co-expression of RAGE and amphoterin in the developing nervous system. *J. Biol. Chem.* **270**, 25752–25761.
30. Rauvala, H., Merenmies, J., Pihlaskari, R., Korkolainen, M., Huhtala, M.-L., and Panula, P. (1987). The adhesive and neurite-promoting molecule p30: Analysis of the amino terminal sequence and production of antipeptide antibodies that detect p30 at the surface of neuroblastoma cells and of brain neurons. *J. Cell Biol.* **107**, 2293–2305.
31. Rauvala, H., and Pihlaskari, R. (1987). Isolation and some characteristics of an adhesive factor of brain that enhances neurite outgrowth in central neurons. *J. Biol. Chem.* **262**, 16625–16635.
32. Yan, S.-D., Chen, X., Schmidt, A. M., Brett, J., Godman, G., Scott, C. W., Caputo, C., Frappier, T., Yen, S. H., and Stern, D. (1994). The presence of glycated tau in Alzheimer's disease: A mechanism for induction of oxidant stress. *Proc. Natl. Acad. Sci. U.S.A.* **91**, 7787–7791.
33. Yan, S.-D., Yan, S. F., Chen, X., Fu, J., Chen, M., Kuppusamy, P., Yen, S. H., Smith, M., Perry, G., Nawroth, P., Godman, G., Zweier, J., and Stern, D. (1995). Nonenzymatically glycated tau in Alzheimer's disease induces neuronal oxidant stress resulting in cytokine gene expression and release of amyloid-β peptide. *Nat. Med.* **1**, 693–699.
34. Yan, S.-D., Chen, X., Fu, J., Chen, M., Zhu, H., Roher, A., Slattery, T., Nagashima, M., Morser, J., Migheli, A., Nawroth, P., Godman, G., Stern, D., and Schmidt, A. M. (1996). RAGE in Alzheimer's disease: A receptor mediating amyloid-β peptide-induced oxidant stress and neurotoxicity, and microglial activation. *Nature (London)* **382**, 685–691.
35. Yanker, B., Duffy, L., and Kirschner, D. (1990). Neurotrophic and neurotoxic effects of amyloid beta protein: Reversal by tachykinin neuropeptides. *Science* **250**, 279–282.

Alexander D. Macrae
Robert J. Lefkowitz[1]

Departments of Medicine (Cardiology) and Biochemistry
Howard Hughes Medical Institute
Duke University Medical Center
Durham, North Carolina 27710

The Function and Regulation of the G-Protein-Coupled Receptor Kinases

Signaling by G-protein-coupled receptors (GPCR) is finely controlled. One of the most important regulators of this process is through receptor phosphorylation and desensitization by the G-protein-coupled receptor kinases (GRKs), a family of serine/threonine kinases which rapidly phosphorylate agonist-bound receptors and uncouple them from their cognate G protein. A number of recent findings have brought exciting developments in this field. The GRKs have been shown to be regulated by their interaction with membrane-bound G proteins $\beta\gamma$ subunits and to be activated by membrane lipids. They have also been shown to interact with the intracellular second messenger, PKC. A number of lines of mice have been described that transgenically overexpress the GRKs or an inhibitory C terminal segment. In addition GRK2 has been deleted in mice where it results in cardiac abnormalities and embryonic death. Finally, there is an increasing awareness of the importance of the GRKs in the pathophysiology of disease, particularly ischemic and congestive heart disease, where changes in the levels and functions of the GRKs have been shown to occur in patients and models of cardiac disease.

[1] To whom correspondence should be addressed. Supported by NIH Grant HL16037.

I. Introduction

The regulation of cellular signaling is exquisitely controlled, and our understanding of the processes underlying cellular activation by G-protein-coupled receptors is rapidly increasing. This chapter aims to review the mechanisms controlling this process through the action of the group of enzymes known as the G-protein-coupled receptor kinases (GRKs). We will highlight some of the recent advances in the understanding of the regulation and specificity of the GRKs and describe some of the insights brought about through the use of animal models and the study of human disease states.

G-protein-coupled receptors (GPCR) mediate the action of a vast array of cellular activators, ranging from the senses of light, taste, and smell, to the circulating hormones and cytokines, to the complexity of neurotransmitters within the synapse. These receptors consist of a central seven-transmembrane section flanked by extracellular N-terminal and cytosolic C-terminal domains. It is believed that ligands generally interact with the transmembrane domains and cause a conformational change which activates the receptor. The resulting change is transmitted to heterotrimeric G proteins through interactions involving the third cytoplasmic loop, the adjacent parts of the transmembrane domains, and the cytoplasmic tail of the receptor.

The G proteins themselves are composed of G_α, G_β, and G_γ subunits. Activation by the receptor increases the affinity of the G_α subunit for GTP resulting in dissociation of the G_α subunit from the membrane associated $G_{\beta\gamma}$ subunits. Although it was initially thought that it was solely the G_α subunit that was responsible for cellular signaling through subtype-specific activation (or inhibition) of second messengers, more recent work has demonstrated the independent activation of effectors such as adenylyl cyclase, phospholipase C (PLC), and membrane channels by the $G_{\beta\gamma}$ subunits themselves.

II. The Regulation of Cellular Signaling

The process by which a ligand-activated receptor couples to a G protein to influence cellular function is regulated in a number of ways. The effect of loss of these controls is best seen in constitutively activated receptors where a mutated receptor is locked in an activated conformation and constantly activates cellular signaling producing a range of disease states.

There are four main methods of control of receptor signaling. The expression of each of the components can be controlled through changes in the levels of the mRNA either through alterations in the rate of transcription or by affecting mRNA stability in a process that that can take place over hours to days. Second, the number of receptors expressed on the cell surface can be altered. The process of sequestration removes the receptor from the cell surface into intracellular endocytotic vesicles. The receptor-containing vesicles can subsequently be degraded or dephosphorylated and recycled to the cell membrane. The final method of regulation, and the subject of this review, involves the uncoupling of the receptor from its G protein. Here the receptor undergoes phosphorylation either heterologously, in a nonagonist-dependent manner by the second-messenger-dependent protein kinases PKA and PKC, or homologously by the agonist dependent receptor specific G-protein-coupled receptor kinases. The availability of inhibitors with some specificity for each of these desensitization processes [PKI for PKA, heparin for the GRKs, and concanvalin for sequestration (Lohse et al., 1990a)] has allowed the dissection of their individual roles.

The GRKs phosphorylate agonist-occupied receptors (Fig. 1). Although the second-messenger-dependent kinases are more effective at phosphorylating agonist-occupied receptors, this preference is much less striking than that of the GRKs. Desensitization of the β-adrenoceptor by PKA occurs at an agonist concentration 100-fold less than that required by GRK2 (40–50% desensitization at 10 nM isoproterenol for PKA; 30–50% desensitization at 1 μM isoproterenol for GRK2) (Lohse et al., 1990a). GRK2 rapidly desensitizes agonist-occupied receptors ($t_{\frac{1}{2}} < 15$ sec). PKA desensitization occurs more slowly ($t_{\frac{1}{2}}$ 3.5 min) and sequestration is even slower, with only 30% of receptors being sequestered after 10 min (Roth et al., 1991). Therefore, it is likely that within the synapse (where GRK2 is localized (Arriza et al., 1992) the GRKs are ideally situated to respond to the rapid release of concentrated neurotransmitter, and that the action of PKA may be more important in maintaining a basal level of receptor phosphorylation in response to low levels of circulating agonist.

Phosphorylation of agonist-occupied receptor alone is insufficient for full receptor desensitization. During the original purification of βARK it was noted that crude protein preparations, were better able to desensitize receptors than were more purified preparations, suggesting the need for further components in the desensitization process. Within bovine retina, a highly expressed 48-kDa protein called arrestin was isolated and shown to enhance the desensitization of light-activated rhodopsin by rhodopsin kinase (Benovic et al., 1987). Addition of arrestin in vitro enhances GRK2 mediated desensitization of the β_2-adrenoreceptor (Lohse et al., 1990b). Several arrestin-like proteins have now been described. In the retina and pineal gland two forms are found, rod and cone arrestin (Craft et al., 1994; Murakami et al., 1993). Elsewhere it is β-arrestin (Lohse et al., 1990b) and β-arrestin 2 (or arrestin 3) which bind to phosphorylated receptors (Attramadal et al., 1992; Sterne-Marr et al., 1993). Each of the four arrestin subtypes has two splice variants, increasing the repertoire of arrestin proteins (Palczewski et al., 1994; Sterne-Marr et al., 1993).

Although the arrestins can bind to unphosphorylated receptor they bind much more avidly (10- to 30-fold higher affinity) to GRK-phosphorylated

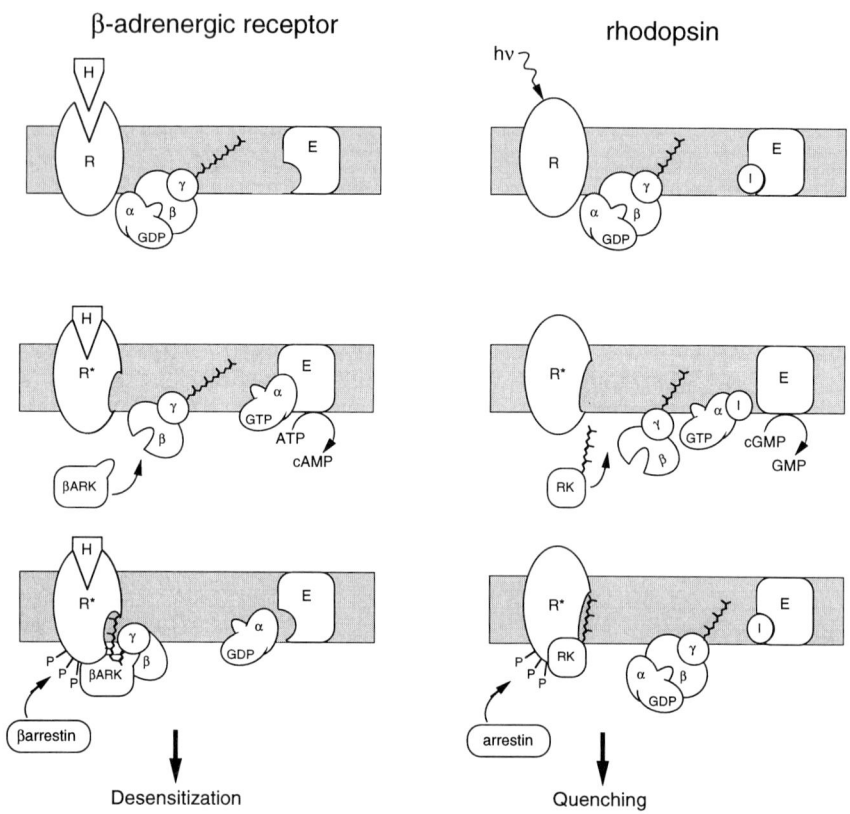

FIGURE I Schematic representation of signal transduction and desensitization for either the β-adrenoceptor binding hormone (H) or light activated rhodopsin. For the β-adrenoceptor agonist binding causes the $G_\alpha{}^s$ subunit to dissociate from the activated receptor (R*), and leads to $G_\alpha{}^s$ activation of the effector enzyme adenylyl cyclase (E). βARK (GRK2) translocates to the membrane through its interactions with Gβγ subunits and membrane lipid (PIP) and is able to bind to and phosphorylate the agonist occupied receptor. β-Arrestin binds to the phosphorylated receptor enhancing desensitization.

receptors and serve to increase the uncoupling of the receptor from its G protein (Lohse *et al.*, 1990b). The exact site of the receptor–arrestin coupling has not been determined; however, it is known that peptides derived from the first and third intracellular loops of rhodopsin inhibit arrestin binding, and that arrestin can bind to truncated receptors lacking a cytoplasmic tail (Ferguson *et al.*, 1996; Gurevich and Benovic, 1993; Gurevich *et al.*, 1995). This suggests that the phosphorylation and receptor binding sites on the receptor are distinct.

While it has been established that binding of arrestin to the receptor impedes G protein coupling and uncouples the receptor from the signaling pathway, it now appears that arrestin binding may provide a number of other important functions. Although not within the scope of this review, it appears that arrestin is involved in receptor sequestration into clathrin-coated vesicles and is one of the determinants regulating the resensitization of the phosphorylated receptor (Ferguson et al., 1996; Goodman et al., 1996; Zhang et al., 1996).

III. The G-Protein-Coupled Receptor Kinases

A. Cloning and Structure of the GRKs

The nomenclature for the GRKs was originally based on the first substrate identified for the kinase (rhodopsin kinase or β-adrenergic receptor kinase). With the cloning of additional β-adrenergic receptor kinase-like genes these enzymes have now been grouped as a family, the G-protein-coupled receptor kinases.

Six GRKs have been recognized in man (Table I). Light (agonist) phosphorylation of rhodopsin in rod outer segments was first described by Kuhn in 1972, and the kinase responsible, rhodopsin kinase (or GRK1) was purified by Kuhn in 1978 (Kuhn, 1978). The cDNA was isolated based on partially purified peptide sequence and shown to encode a protein of 450 aa (Lorenz et al., 1991). GRK2, or β-adrenergic receptor kinase (β ARK1), was first isolated through the purification of a protein fraction which phosphorylated agonist-occupied β-adrenoceptor (Benovic et al., 1986). Peptide sequencing and cDNA library screening allowed isolation of a 689-aa protein (Benovic et al., 1989). A second GRK2-like sequence, GRK3 (β ARK2), was isolated by screening of a bovine brain cDNA library with GRK2 (Benovic et al., 1991a).

TABLE I Sequence Homology and Chromosomal Localization of the Six Known GRKs

Kinase	Homology to GRK2: identity (similarity)	Homology to GRK4: identity (similarity)	Chromosomal localization
GRK1 Rhodopsin kinase	33 (58)	47 (68)	
GRK2 (βARK1)	100 (100)	37 (57)	11q13
GRK3 (βARK2)	84 (92)	37 (57)	22q11
GRK4 (IT11)	37 (57)	100 (100)	4p16.3
GRK5	37 (58)	69 (82)	10q24-qter
GRK6	39 (60)	68 (82)	5q35 13pter-q21

Note. Homology scores are based on Sterne Marr and Benovic (1996). Chromosomal localization is described in Benovic et al. (1991b), Calabrese et al. (1994), Ambrose et al. (1992), and Bullrich et al. (1995).

Three further kinases have been isolated. GRK4 (full length 578 aa) was isolated as IT11 during exon trapping within the Huntington disease locus (Ambrose et al., 1992), GRK5 (590 aa) was cloned by degenerate PCR from both a human heart (Kunapuli and Benovic, 1993; Kunapuli et al., 1994a) and bovine tongue cDNA library (Premont et al., 1994), and GRK6 (576 aa) was isolated through hybridization of a human heart cDNA library with GRK2 fragments (Benovic and Gomez, 1993).

The sequence of GRK4 is of particular interest. When the initial sequence, described by Ambrose et al. (1992), was compared with that of the known kinases a gap within the N terminal region was apparent. Later reports of alternate transcripts (Sallese et al., 1994), and of the genomic structure of GRK4 (Premont et al., 1996), added a 32-aa N-terminal and a 46-aa C-terminal exon. This completed the missing sequence which was then contiguous with that of the other kinases, GRK5 and GRK6. Subsequent analysis has shown that there are four splice variants of GRK4 formed by alternate splicing of the N- and C-terminal exons. Although the longest form, GRK4α, appears to be the most prevalent, the function of the shorter forms is as yet unknown, and all but GRK4γ seem to function equally well in inhibiting LH receptor signaling (Premont et al., 1996).

In addition two GRK like proteins have been isolated from *Drosophilla* (GPRK1 and GPRK2 representing GRK2- and 4-like proteins, respectively) (Cassill et al., 1991). Finally, two GRK2- and GRK4-like sequences have been unearthed during the *C. elegans* sequencing project.

B. Expression of the GRKs

Each of the GRKs has a distinct expression pattern. GRK1 is localized exclusively to the retina and the pineal gland (Lorenz et al., 1991). In contrast, the distribution of GRK2 is the most widespread: it appears to be expressed in all tissues thus far tested with the exception of sperm and olfactory cilia (Benovic et al., 1989). The highest levels of expression are seen in leukocytes and brain. GRK3 is expressed at levels approximately 10–20% that of GRK2. Again, like GRK2, the expression is widespread. There are very few regions where GRK3 predominates over GRK2, but these include the olfactory cilia (Dawson et al., 1993; Schleicher et al., 1993) and mature sperm (Walensky et al., 1995) (where GRK2 is absent), and to a lesser degree, within the pituitary and localized regions of the brain (olfactory tubercule, substantia nigra, granullar cell layer of the cerebellum) (Benovic et al., 1991a; Arriza et al., 1992). GRK5 expression is greatest in the heart, lung, and placenta and to lesser levels in the muscle and brain (Kunapuli and Benovic, 1993; Premont et al., 1994). GRK6 is widely expressed with highest levels in the brain and skeletal muscle (Benovic and Gomez, 1993). Finally GRK4 has a unique distribution (Ambrose et al., 1992; Sallese et al., 1994; Premont et al., 1996). Expression is almost exclusively confined to the testis, although RT-PCR has detected expression in the brain and peripheral nerve tissue. It must be emphasized, however, that all of these descriptions—with the exception of the studies of GRK2 and -3 in the brain localized (Arriza *et al.*, 1992)—have depended on total organ Northern blots and give no indication of any cellular distribution or preference.

Although the six kinases share many common features, they can be grouped into three subfamilies according to molecular and functional criteria—GRK1, GRK2-like, and GRK4-like. The sequence homologies are shown in Table I. It can be seen that GRK2 and -3 show a high degree of sequence homology and that GRK4, -5, and -6 form a distinct subgroup. This subclassification is further validated in the analyses of the genomic structure of these genes. Although only the genomic structures of GRK1 (Khani *et al.*, 1996), GRK2 (Jaber *et al.*, 1996), and GRK4 (Premont *et al.*, 1996) have been published we have recently cloned the mouse genomic homologues of GRK3 (K. Peppel, personal communication), GRK5, and GRK6. Again, the intron/exon boundaries within the subgroups are identical but differ completely from those of the other subgroup. The report of distinct GRK2-like and GRK4-like genes in lower species such as *Drosophilla* and *C. elegans* adds further support to this hypothesis and suggests a distinct evolution for these three subgroups.

IV. The Regulation of the G-Protein-Coupled Receptor Kinases

A. Regulation by Membrane Association

Each of the six kinases shares a similar structure, with a central (263–266 aa) catalytic domain, within which sequence homology is highest, and divergent N and C terminals (Fig. 2) (Inglese *et al.*, 1993). While initial characterization concentrated on the central catalytic domain, there is an increasing awareness of the importance of the N and C terminals in GRK function and regulation. Upon agonist activation of receptor, GRK1, -2, and -3 are seen to translocate from the cytosolic fraction to the membrane fraction of the cell (Chuang *et al.*, 1992; Inglese *et al.*, 1992a,b). This targeting is dependent on the interaction of the kinase with receptor and membrane components, and is different for each of the kinases.

GRK1 contains a "CAAX" motif (CVLS) within the C-terminal domain. Light-induced translocation from cytosol to membrane requires posttranslational (C15) farnesylation of the cysteine, removal of the terminal three amino acids, and carboxyl methylation. Without this modification GRK1 is unable to interact with the rod outer segment (ROS) membranes in a light-dependent manner and is less able to phosphorylate rhodopsin (Inglese

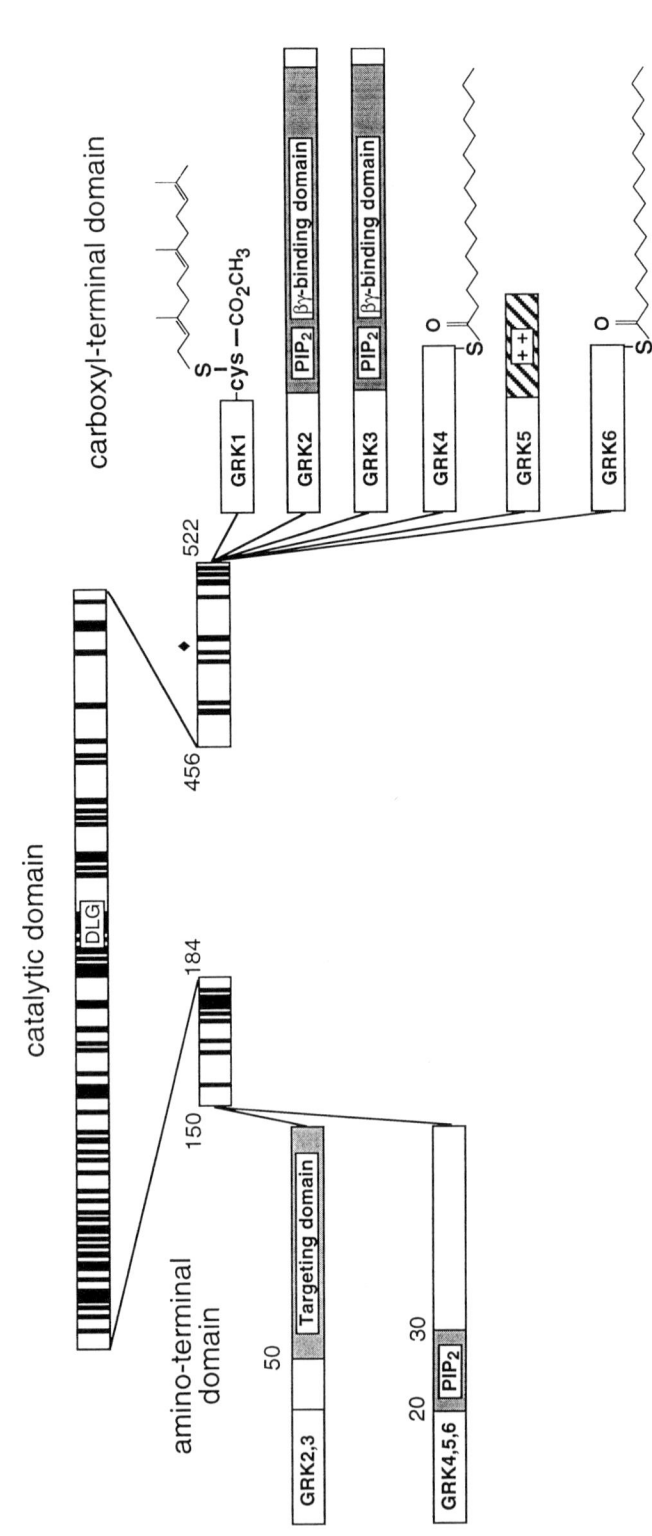

FIGURE 2 Structure of the GRKs. Each is composed of a conserved central catalytic domain and divergent amino- and carboxyl-terminal domains. Conserved amino acids are indicated by a vertical bar. The GRK2 and GRK3 N-terminal targeting domain (Murga et al., 1996) and the GRK4, 5, and 6 PIP_2 binding domains (Pitcher et al., 1996) are shown. Each of the proposed C-terminal membrane/lipid interacting schemes are represented diagramatically and discussed further in the text (Sections IV.A and IV.B). DLG represents the catalytic domain invariant sequence. The autophosphorylation sites for GRK1 and GRK5 are shown as a diamond.

et al., 1992a). A mutated form of the kinase with a C-terminal geranylgeranyl group is constitutively attached to the membrane and is thus able to associate with rhodopsin in a light-independent manner (Inglese *et al.*, 1992b). GRK2 and GRK3 are not isoprenylated; instead, they translocate to the membrane via an association with the G protein $G_{\beta\gamma}$ subunits (Haga and Haga, 1990, 1992; Pitcher *et al.*, 1992). Both kinase translocation and receptor phosphorylation are facilitated by the addition of exogenous $G_{\beta\gamma}$ subunits; indeed, GRK2 translocation can be inhibited by the addition of G protein G_{α} subunits.

In this case it is the G_{γ} subunit which anchors the GRK2/$\beta\gamma$ complex to the membrane via posttranslational modification of its CAAX box and addition of a (C20) geranylgeranyl group. When compared to GRK1, GRK2 and -3 have 124 and 125 extra amino acids at the C terminal, and the $G_{\beta\gamma}$ binding domain has been mapped to this region (Koch *et al.*, 1993, 1994). Subseqent studies have shown (Touhara *et al.*, 1994) that this region contains a Pleckstrin homology domain, a region of protein sequence homology found in a number of proteins including some involved in cellular signaling and known to interact with $G_{\beta\gamma}$ subunits. As with GRK1, removal of this C-terminal region significantly reduces the function of the kinase and replacement of this domain with a CAAX box restores this function.

$\beta\gamma$-Induced translocation of GRK2 and -3 brings a further level of control to the regulation of the GRKs: the $G_{\beta\gamma}$ subunits are available for translocation only following activation of the receptor and dissociation of the heterotrimeric G protein complex, thus ensuring that the kinase is available to phosphorylate activated receptors. In contrast to GRK1, -2, and -3, GRK4, -5, and -6 are membrane bound in the absence of ligand (Inglese *et al.*, 1993). It is thought that GRK5 associates with the negatively charged phospholipid membrane via a highly basic carboxyl-terminal region. In support of this, and in contrast to GRK2 and -3, GRK5 activity is not significantly influenced by the addition of exogenous $G_{\beta\gamma}$ subunits (Premont *et al.*, 1994), an observation which allows the experimental dissociation of the role of the two kinases.

In a further variation, GRK4 (Premont *et al.*, 1996) and GRK6 (Stoffel *et al.*, 1994) have been shown to be palmitoylated. For GRK6 this posttranslational modification has been localized to three cysteines (561, 562, and 565) (Stoffel *et al.*, 1994) within the C-terminal domain. As dynamic palmitoylation has been demonstrated for other components of the signaling pathway, the authors were tempted to speculate that acylation of GRK6 may serve a regulatory function for this kinase.

B. Regulation by Membrane Lipids

Two further lines of investigation have served to show that the interaction of the GRKs with the membrane and receptor is more complex (Fig.

2). It has become apparent that GRK-mediated receptor phosphorylation is affected by the lipid milieu surrounding their receptor substrate. This effect of lipids was first noted for GRK5 where autophosphorylation at residue Ser 484 is enhanced by the addition of phospholipids in the form of liposomes of phosphatidylcholine, serine, and inositol (Kunapuli et al., 1994b). In a mutant GRK5, without this phospholipid-induced autophosphorylation, GRK5 is 15- to 20-fold less able to phosphorylate the β-adrenoceptor. It does, however, still phosphorylate peptide substrates, suggesting that the role of the autophosphorylation is to enhance the association of the kinase with the membrane and consequently with the receptor.

More recently, Pitcher et al. (1995), using receptor reconstituted in purified phosphatidylcholine vesicles supplemented with varying amounts of phosphatidylinositol-4,5,-bisphosphate (PIP_2), have shown that PIP_2 and $G_{\beta\gamma}$ are able to synergistically enhance GRK2 translocation and phosphorylation. In this system, 100 nM $G_{\beta\gamma}$ and 3–10% PIP_2 increase the receptor phosphorylation by 25- to 65-fold, without affecting nonreceptor substrate phosphorylation. The effect of both $G_{\beta\gamma}$ and PIP_2 could be inhibited by a fusion protein containing the C-terminal portion of GRK2, leading the authors to suggest that the two substrates bound synergistically to the Pleckstrin homology domain and that full activity required the simultaneous presence of both ligands.

Onorato et al. (1995), using dodecyl maltoside-solubilized receptors, described inhibition of kinase activity by PIP_2, an action that was blocked by the addition of $G_{\beta\gamma}$ subunits. They suggested that the lipid promoted kinase binding to the vesicles, but also inhibited kinase activity. In support of this view, DebBurman et al. (1995a) showed increased membrane association of GRK2 in PIP_2-containing vesicles, but again reported a PIP_2-mediated inhibition of kinase function.

These apparently contradictory results are readily explained when the concentrations of PIP_2 utilized in the experiments are compared. Thus, both groups have found that low concentrations of PIP_2 (0.5–2.0 µg) stimulate receptor phosphorylation and that higher concentrations (10–50 µg) do indeed inhibit both receptor and peptide phosphorylation.

Similar experiments studying the effect of lipids on the activity of GRK5 have recently been reported (Pitcher et al., 1996). Again, using β_2-adrenoceptor reconstituted into lipid vesicles of defined lipid composition, 5% PIP_2 (0.5–2 µg) results in a dramatic enhancement of GRK5 activity toward receptor but not peptide substrates. It was proposed that this is due to an increased membrane association of the kinase. However, unlike previous reports of lipid dependence of GRK5, this effect does not involve the autophosphorylation site, as mutation of this amino acid did not alter the effect of PIP_2. In contrast to the interaction of GRK2 with PIP_2, mutations within the C terminus of GRK5 had no effect on PIP_2 enhancement of GRK5-mediated phosphorylation. For GRK5 it is mutations within the N-terminal domain of the kinase which fail to show PIP_2 enhancement of receptor phosphorylation. This is of some interest given the conservation of the N-terminal basic region among the GRK4-like kinases, suggesting that the other members of the family may also be regulated in this way by lipids.

C. Microsomal Membrane Association

A third role for the noncatalytic domains has been reported by Mayor and colleagues, who have shown that a significant amount of the cellular pool of GRK2 associates with intracellular microsomal membranes (Garcia-Higuera *et al.*, 1994; Murga *et al.*, 1996). This association is rapid, reversible, and saturable, and has a K_d of 20 nM, similar to the affinity of GRK2 for $G_{\beta\gamma}$ subunits. While bound, the GRK is in an inactive form. Although the nature of the binding site is unclear, the observation that proteases or heat treatment strongly inhibits this association and Na_2CO_3 has little effect suggests that the anchor is an integral membrane protein. The binding is not, however, due to an interaction with $G_{\beta\gamma}$ subunits in the microsomal membrane as it is not influenced by modulators of G proteins, or by a loss of G protein G_β subunits from the membrane. Furthermore, the interaction was not inhibited by a C-terminal fusion protein described above, which has been shown to inhibit $G_{\beta\gamma}$-mediated translocation, but was prevented by the addition of an N-terminal fusion protein (88–145), suggesting that it is the N-terminus that attaches GRK2 to microsomal membranes.

Although the GRK2 binding protein was not identified, this concept of an anchor protein for GRK2 would not be without precedent, as similar proteins have been reported for PKA and PKC.

D. Regulation by PKC

A further level of regulation of GRK2 has been described by Lohse and co-workers, who studied the effect of α_{1B}-adrenoceptor or PKC stimulation on GRK2 and found that stimulation of PKC resulted in an activation of GRK2 and a reduction in the fraction of GRK2 found in the cytosol (Winstel *et al.*, 1996). This activation was associated with a PKC-dependent phosphorylation of GRK2 (probably at the C terminus) and served to increase membrane translocation, rather than enhancing catalytic activity, as no increase was seen in the phosphorylation of a peptide substrate. The idea of cross talk between PKC and GRK2 has been expanded through the use of antisense technology (Shih and Malbon, 1994, 1996). Here, antisense oligonucleotides to either GRK2 or PKA inhibited desensitization in a cell-type-specific manner. In contrast, antisense constructs to PKC amplified desensitization. Furthermore, the recovery from this desensitization was prolonged, supporting a role for PKC in the regulation of resensitization.

V. Kinase Enzymology

The GRKs are serine threonine protein kinases (Inglese et al., 1993). A number of methods have been used to identify the GRK phosphorylation sites. Most work has concentrated on receptors such as the β_2-adrenoceptors, which have short cytoplasmic loops and long cytoplasmic C termini. Using truncated receptor mutants, proteolytic digestion, or selective mutation of the serine and threonine residues in the carboxyl tail, the phosphorylation sites for GRK2 and GRK5 for the β_2-adrenoceptor have been localized to the carboxyl tail (Bouvier et al., 1988; Dohlman et al., 1987). For the α_{2A}-adrenoceptor, which has a large third intracellular loop and a short carboxyl tail, sites conferring the capacity for desensitization are located in the third intracellular loop (Liggett et al., 1992).

GRKs are also able to phosphorylate peptide substrates, although at less than 1/1000 the efficiency of the receptor substrate. Here GRK2 shows a preference for residues with a neighboring N-terminal acidic amino acid (Onorato et al., 1991). In contrast, rhodopsin kinase prefers acidic residues on the carboxyl side of the phosphorylated amino acid (Palczewski et al., 1989). Although ligand-activated β-adrenoceptor or light-activated rhodopsin enhances peptide phosphorylation by GRK2 by up to 100-fold, this effect may not involve the C-terminal portion of the receptor (implicated in receptor phosphorylation) as truncated receptor mutants are still able to enhance phosphorylation without themselves being phosphorylated.

The interpretation of the results with mutated receptors or peptides is limited by the effect of the mutation on the total conformation of the receptor. To overcome this problem the phosphorylation sites have been mapped by proteolytic digestion and protein sequence and phosphamino analysis of the β_2-adrenoceptor phosphorylated by GRK2 or -5 (Fredericks et al., 1996). All the phosphorylation sites (only phosphoserine and phosphothreonine) were found in the carboxyl terminal 40 aa. Even at stoichiometries of 1.0 mol Pi/mol receptor GRK2 phosphorylated four and GRK5 phosphorylated six of the seven possible sites. No change was seen at higher stoichiometries. Although the phosphorylation sites for both kinases overlap, some differences were noted. GRK5 alone phosphorylated Thr 393 and was more effective at Ser 411, while GRK2 was more effective at Ser 407. It remains unclear whether these sites are phosphorylated sequentially or nonsequentially.

VI. Receptor—Kinase Specificity

To date, six kinases and a multitude of G-protein-coupled receptors have been cloned. Although GRK2 was purified and cloned as the β-adrenergic receptor kinase it quickly became apparent that GRK2 could phosphorylate and desensitize many more receptors. It is therefore of some interest to determine whether there is a receptor repertoire for each kinase or whether each kinase serves a distinct role across a range of receptors. The kinase repertoire may: be determined by receptor type—rhodopsin/adrenergic-like, glutamate-like, or secretin-like; be a function of a specific G protein coupling; or be controlled by tissue or cellular localization. A number of investigators have endeavored to examine these possibilities and their results are detailed below.

Initial attempts to describe a kinase—receptor relationship by describing receptor desensitization and the translocation of kinase from cytosol to membrane quickly revealed that the β_2-adrenoceptor (Strasser *et al.*, 1986), somatostatin (Mayor *et al.*, 1987), PGE_1 (Strasser *et al.*, 1986), and PAF receptors (Chuang *et al.*, 1992) all induced GRK translocation. As these receptors represent members of different classes, and couple to a range of G protein G_α subunits, it is difficult to suggest receptor subtype as the rationale for kinase specificity. Moreover, as this work preceded the cloning of the GRK4-like kinases (GRK4, -5, and -6) the assays at that time were unable to determine the GRK responsible for desensitization.

Other groups have tried to show preferential phosphorylation or desensitization of a receptor by a particular GRK. Freedman *et al.* (1995) used whole-cell phosphorylation and *in vitro* reconstitution to study the effect of GRK2, -3, and -5 on β_1-adrenoceptor phosphorylation. All three kinases increased receptor phosphorylation and decreased the agonist-induced rise in cellular cAMP. No preferential phosphorylation by one kinase was seen. In similar studies on the angiotensin II receptor which signals through phosphoinositide hydrolysis expression of either of the three kinases augmented desensitization and produced a 1.5- to 1.7-fold increase in receptor phosphorylation (Oppermann *et al.*, 1996a). Overexpression of a dominant negative mutant GRK2 (K220R) reduced endogenous desensitization by 40–50%, suggesting a functional role for endogenous kinases in desensitization.

The difficulties in purifying sufficient receptor to study receptor phosphorylation *in vitro* has limited the number of receptors that have been studied for their interactions with the GRKs. More recently, two alternative methods of obtaining larger quantities of membrane-bound receptor have been described either using baculovirus overexpression and sucrose gradient enrichment of membranes (Pei *et al.*, 1994), or stripping membranes with 4 M urea (DebBurman *et al.*, 1995b). Using these methods, Pei and colleagues (1994) found that the α_{2C}-adrenoceptor was a better substrate for GRK2 and -3 than for GRK5, while no difference was seen for the three kinases with the β_2-adrenoceptor. DebBurman, using urea-stripped membranes, showed that the m2 and m3 muscarinic receptors are better substrates for GRK2 and -3 than for GRK5 and -6 (DebBurman *et al.*, 1995b). However, there are definite examples of receptor-kinase specificity. For example, GRK1 is 2-fold better at phosphorylating rhodopsin than the β_2-adrenoceptor;

conversely, GRK2 is 20-fold better at phosphorylating the β_2-adrenoceptor than at phosphorylating rhodopsin.

Another example of this specificity is that of GRK3 and desensitization of the odor receptors. Stimulation of isolated nasal cilia with concentrated odors produces a rapid rise in cAMP followed by an equally rapid (100 msec) fall due to receptor uncoupling (Dawson et al., 1993; Schleicher et al., 1993). This signal termination is partially inhibited by heparin (indicating a role for the GRKs). Antibodies raised to GRK3 but not GRK2 were also able to abrogate the desensitization, suggesting that it is mediated by GRK3. This result can be explained at least partially by GRK expression. Immunohistochemistry shows that GRK3 and not GRK2 is expressed within the nasal cilia, in contrast to most other sites (Dawson et al., 1993). GRK3 is also the only GRK2-like kinase found in sperm, where it has been suggested that it plays a role in desensitizing the odor-like receptors which may be involved in sperm chemotaxis and motility (Walensky et al., 1995).

Although specificity of GRK3 for the odor receptor can be proposed based on tissue specificity for nasal cilia, the same is not obvious for the thrombin receptor and GRK3. Ishii and colleagues coexpressed the thrombin receptor with GRK1, -2, and -3 (Ishii et al., 1994). GRK3 was 10- to 25-fold more potent than GRK2 in inhibiting thrombin-induced Ca^{2+} flux. This action was blocked by mutations engineered in the catalytic domain of GRK3 or by removing the serines and threonines in the receptor's carboxyl tail. Coexpression of GRK3 with thrombin receptor in ^{32}P-labeled cos cells was more potent than GRK2 in increasing immunoprecipitated receptor phosphorylation (2- to 2.5-fold increase in phosphorylation for GRK3, 0.5- to 0.8-fold for GRK2).

A further, elegant example of GRK3 specificity is seen in the work of Diverse-Pierluissi et al. (1996). This group used embryonic chick dorsal root ganglion neurons as a single-cell assay system. Here, norepinephrine acting on α_2-adrenoceptors produces two effects: via $G_{\alpha o}$, there is a non-PKC-dependent slowing of the calcium current which slows with time at positive potentials (termed kinetic slowing or KS), and via $G_{\alpha i}$, there is a PKC-dependent sustained reduction in calcium current (termed steady state inhibition or SSI). Together, these pathways act to reduce synaptic transmitter release and both show evidence of desensitization to repeated stimulation.

Injection of recombinant GRKs into these neurons had surprising effects. GRK3 enhanced the rate of desensitization of both components of norepinephrine stimulation by twofold. No effect was seen for GRK1, -2, or -5 even when they were injected at significantly higher concentrations. To confirm these results, synthetic peptides from the $G_{\beta\gamma}$ binding site of GRK2 or GRK3 were injected. Only the peptide corresponding to the $G_{\beta\gamma}$ binding domain of GRK3 was able to inhibit desensitization. This degree of specificity is particularly unusual in view of the homology between GRK2 and -3. However, closer examination of the sequences of GRK2 and -3 shows that the $G_{\beta\gamma}$ binding domains of the two proteins are the most dissimilar, perhaps explaining the observed receptor-kinase selectivity. As a caveat it must be noted that the chick GRKs are not known. Western blotting of chick dorsal root ganglion with antibodies that detect both GRK2 and -3 produces a single band with a molecular weight similar to that of recombinant mammalian GRK3.

Little is known of the specificity of the GRK4-like genes. Most studies have looked solely at GRK2 and -3 or, more recently, included only GRK5. It is known that GRK4 is as able as GRK2 to desensitize signaling by the LH and FSH receptor (Premont *et al.*, 1996). In a study of desensitization of the D_{1A} receptor Tiberi found that although GRK2, -3, and -5 were equally able to phosphorylate D_{1A} receptors, the desensitization induced by GRK5 resulted in an increase in the EC_{50} and 40% reduction in the maximal activation by the receptor. In contrast, transfection of cells with GRK2 and GRK3 led to significant rightward shifts of the dose–response curve to dopamine without any change in the maximal response (Tiberi *et al.*, 1996).

One approach to determining the role of GRK2-like and GRK4-like gene families is through the use of subtype-specific monoclonal antibodies mABs (Oppermann *et al.*, 1996b). Anti GRK2 and -3 mABs inhibit isoproterenol-induced β-adrenoceptor phosphorylation in permeabilized myocytes by 77%, while GRK4, -5, and -6 mABs have no effect. Although this suggests that desensitization in this system is predominantly mediated by GRK2 and/or -3, it may also be that, as they are cytosolic enzymes, they are more susceptible to blockade by mABs.

VII. Animal Models

A. Animal Studies by Transgenic Gene Expression

In an attempt to understand the physiological significance of the GRKs we have taken advantage of the ability to manipulate the function of these enzymes in genetically modified mouse lines through transgenic overexpression or by gene deletion through homologous recombination.

Continuing work developed in the overexpression of β-adrenoceptors in the mouse heart, Koch *et al.* (1995) described the effect of transgenic overexpression of either GRK2 or the C-terminal portion of GRK2 which acts as an intracellular inhibitor of GRK2 translocation and receptor desensitization. These genes were cloned under the control of the α myosin heavy chain promoter which directs gene expression in mice to the atria in embryonic life and to the atria and ventricles in the adult mouse. Mouse lines transmitting and expressing the transgenes were identified and studied for their receptor binding, receptor phosphorylation, and cardiac function. Hearts from mice overexpressing GRK2 (TGβK12) had three times the kinase activity of control animals. In contrast those overexpressing the C-terminal transgene showed twofold less rhodopsin phosphorylation following the addition of $G_{\beta\gamma}$ subunits than did control mice. β_1-Adrenoceptor binding isotherms for the GRK2-overexpressing mice were shifted to the right, indicating an uncoupling of the receptor. In addition, GRK2-overexpressing mice had significantly lower basal and agonist-stimulated cyclase levels. No changes were seen in either of these parameters for the C-tail transgenic mice.

Cardiac catheterization was used to study the consequences of these changes on myocardial function. Although the basal measurements of cardiac function for the GRK2-overexpressing mice were unchanged, both the inotropic and chronotropic responses to isoproterenol were blunted. In contrast, basal measurements of LV dP/dt_{max}, LV dP/dt_{min}, and LV systolic pressure were all significantly increased in GRK2 C-terminal transgene mice, as were LV responses to isoproterenol. No change was seen in heart rate.

These findings have potentially interesting implications. They imply that at basal levels of receptor activation, GRK2 is active, reducing β-adrenoceptor-mediated contractility. The lack of effect of GRK2 overexpression on the basal cardiac parameters, in combination with the observation that the C-terminal transgene inhibits basal receptor uncoupling, suggests that it is the $G_{\beta\gamma}$ subunits, and not the GRK, which are the rate-limiting factors in the basal state. With further cardiac stimulation, overexpression of GRK2 attenuates cardiac function. This effect of GRK2 overexpression mimics the clinical situation in cardiac failure discussed below, and may suggest some pathophysiological role for the GRKs in that condition.

B. Animal Studies by Gene Deletion

As an alternative to transgenic overexpression, the physiological function of the GRKs can be examined through gene deletion. To this end a targeted disruption of GRK2 has been described (Jaber et al., 1996). For this, the mouse GRK2 gene was cloned and exons 5–8 (encompassing the catalytic domain of the kinase) were replaced with a neomycin selectable marker. Mice heterozygous for the exon deletion were generated by homologous recombination and blastocyst injection. Subsequent breeding of heterozygote mice failed to generate homozygous offspring, despite the screening of 623 pups. Genotyping of the offspring showed that although the appropriate ratios of homozygotes/heterozygotes/wild-type mice were present at Embryonic Days 9–15.5, following this date, no viable homozygote offspring were seen. Autopsy of the homozygote embryos showed them to be smaller and paler than their litter mates. Detailed cardiac examination demonstrated dilated atria, hypoplasia of both atria and ventricles, and dysplasia of the interventricular septum. The ventricular wall was thinned with randomly ordered trabeculae—reminiscent of the "thin myocardium" syndrome. The functional significance of these findings was confirmed by intracavital microscopy *in utero* which measured an LV ejection fraction in the homozygote embryos of 9–16% compared to an average of 56% in age-matched wild-type controls. This phenotype is similar to that described following inactivation of a number of transcription factor genes, including RXRa (Sucov *et al.*, 1994; Kastner *et al.*, 1994), WT1 (Kreidberg *et al.*, 1993), TEF-1 (Moens *et al.*, 1993), and N-myc (Charron *et al.*, 1992).

There are several possible interpretations for the GRK2 knock-out phenotype and its similarities to that of the transcription factor knock-out animals. It is possible that signaling by GPCRs is important in the formation of the heart in the mouse embryo. The loss of desensitization of these receptors through deletion of GRK2 impairs cardiac morphogenesis in a way that is fundamental enough to give a phenotype that is indistinguishable from the other deleted genes. Alternatively, it is possible that the function of the transcription factors is actually regulated by GPCRs, suggesting a convergence of the two pathways. Loss of desensitization of these as yet unidentified GPCRs affects the regulation of the transcription factors, producing a phenotype similar to that seen when the transcription factors themselves are deleted. Finally, it is possible that the GRKs fill some as yet undescribed role, and that deletion of GRK2 is embryonically lethal through a mechanism which does not necessarily involve desensitization of GPCRs.

VIII. Physiological Significance

The importance of regulation of receptor signaling is put into a clinical perspective by the changes in signaling that take place during disease processes. Much of this research relates to cardiac disease. Congestive heart failure (CHF) results in a decreased responsiveness to β-adrenergic receptor agonists. This is due both to down-regulation of the β_1-adrenoceptor and to an uncoupling of the receptors from their G proteins, at least in part due to GRK activity (Brodde, 1991).

Several studies have looked at the level of expression and function of GRK2 in models of heart failure. In explanted hearts of patients undergoing cardiac transplantation for congestive heart failure there is a 70% reduction in β-adrenoceptor responsiveness accompanied by a 50% reduction in β-adrenoceptor mRNA and approximately 50% decrease in β-adrenoceptor binding (Ungerer *et al.*, 1993). These changes are accompanied by a two- to three-fold increase in GRK2 mRNA and a concomitant increase in β-ARK activity . This work was later expanded to include the effect of CHF on both GRK2 and -3 and the β-arrestin isoforms (Ungerer *et al.*, 1994). Again using quantitative PCR the levels of GRK2 were threefold higher than those in control tissue. No changes were seen in either GRK3 or β-arrestin mRNA. GRK2 activity was doubled in failing hearts. The authors interpreted the increase in GRK2 but not β-arrestin levels as being the result of the relative difference in expression of the two proteins. Within the human heart the ratio of β-arrestin to GRK2 is 20:1. Thus in CHF it may be that the expression of GRK2 and not β-arrestin is the limiting factor in desensitization and it is the GRK2 levels that are modulated by receptor signaling.

It is assumed that the effect of CHF on GRK expression is mediated via increased catecholamine release acting through adrenergic receptors. Elevated catecholamine levels are also present during periods of myocardial ischemia. There is an increase in β-adrenoceptor expression and decrease in isoproterenol stimulated cAMP stimulation in pig hearts when subjected to periods of stop-flow cardiac perfusion used to model myocardial ischemia (Ungerer et al., 1996). Reduced cardiac flow also leads to an increase in GRK activity which reaches a maximum of twofold greater than control after 20 min and lasts at least 6 hr. These changes can be mimicked by perfusion of norepinephrine and inhibited by perfusion of desimipramine, which suppresses ischemia induced release of norepinephrine, inferring a catecholamine-induced effect. Hypoperfusion also results in a threefold increase in GRK2 mRNA which is maximal at 20 min and returns to normal within 40 min. No change in β-arrestin levels were seen.

To determine the effect of decreased receptor stimulation, and in an attempt to try to explain recent findings that β_2-adrenoceptor blockade increases survival in CHF, Ping and colleagues (1995) described a pig model in which the animals were given the β_1-adrenoceptor antagonist bisoprolol for 35 days. This produced an increase in β-adrenoceptor binding and a twofold increase in receptor-mediated cAMP generation. After bisoprolol treatment the majority of receptors were in a high-affinity state, implying increased coupling. There was a 25% decrease in soluble kinase activity and a 35% decrease in membrane-bound kinase activity which was found only in the left ventricle. Interestingly, no change was found in GRK2 protein levels by Western immunoblot, leading the authors to suggest that some of the change in GRK activity may be due to other kinases such as GRK5 which are expressed in the heart.

Collectively, these studies may have important implications for future therapy of cardiac disease states. Although it is most likely that the increase in GRK2 induced by increased catecholamine levels brought on by either CHF or ischemia serves to desensitize the β-adrenoceptor and protect the myocyte from excessive activation in the short term, in the longer term the decreased responsiveness to catecholamines may be detrimental. It will therefore be important to study the effect of GRK inhibitors in the treatment of these conditions. There are few non-cardiac sites where the physiological relevance of the GRKs has been described. Treatment of mononuclear leukocytes with either isoproteronol or PAF can induce GRK2 translocation to the membrane (De Blasi et al., 1995). PHA (acting through PKC induction) induces T lymphocyte activation and produces a heparin inhibited increase in cytosolic (but not membrane) kinase activity which reaches a maximum increase over basal of over 300% at between 48 and 72 hr. This increase is accompanied by a threefold increase in GRK2 mRNA and greater than twofold increase in GRK3 mRNA. No change was seen in the mRNA levels for GRK5 and GRK6.

Finally, the response of GRKs to opiate treatment and withdrawal has been described (Terwilliger *et al.,* 1994). Rats were treated with either chronic opiate treatment (75 mg morphine for 5 days, withdrawn for the 6th day prior to sacrifice) or acute treatment (30 mg/kg 45 min prior to sacrifice). Only the chronic treatment/withdrawal resulted in an increase in GRK2 immunoreactivity to 135% that of the control values. This change was found only in the locus coeruleus, which also showed a smaller but significant increase in levels of β-arrestin. These results are of particular interest as Nestler's group has previously shown that the process of adaptation after opiate withdrawal is accompanied by behavioral changes, increased neuronal firing within the locus coeruleus, and an increase in a number of components of the opiate signal transduction pathway including GRK2. It is therefore possible to postulate that the expression of increased levels of the GRK is induced during opiate administration, serving to reduce signaling by the $G_{\alpha i}$-coupled opiate receptors. Upon abruptly stopping the drug there is a sudden increase in neuronal firing (previously supressed by opiate administration) within the locus coeruleus. The elevated GRK levels continue to phosphorylate and desensitize the opiate receptor and act to to prolong the withdrawal process.

IX. Conclusions

Thus it can be seen that the study of receptor regulation by phosphorylation and desensitization has grown immensely in the past 10 years. The idea that each of the kinases can be regulated by membrane-bound proteins, lipids, or cytosolic second messengers adds a further layer of complexity to the study of the GRKs. Further, potentially exciting information will be obtained through the development of additional transgenic animals, and by gene deletion through conventional, tissue-specific, and inducible knock-out animals. Finally, the possibility that the GRKs may be involved in the pathogenesis of disease states, or that they are potential targets for therapeutic intervention, offers many possibilities for new avenues of research.

References

Ambrose, C., James, M., Barnes, G., Lin, C., Bates, G., Altherr, M., Duyao, M., Groot, N., Church, D., Wasmuth, J. J. *et al.* (1992). A novel G protein-coupled receptor kinase gene cloned from 4p16.3. *Hum. Mol. Genet.* **1,** 697–703.

Arriza, J. L., Dawson, T. M., Simerly, R. B., Martin, L. J., Caron, M. G., Snyder, S. H., and Lefkowitz, R. J. (1992). The G-protein-coupled receptor kinases βARK1 and βARK2 are widely distributed at synapses in rat brain. *J. Neurosci.* **12**, 4045–4055.

Attramadal, H., Arriza, J. L., Aoki, C., Dawson, T. M., Codina, J., Kwatra, M. M., Snyder, S. H., Caron, M. G., and Lefkowitz, R. J. (1992). β-arrestin2, a novel member of the arrestin/β-arrestin gene family. *J. Biol. Chem.* **267**, 17882–17890.

Benovic, J. L., and Gomez, J. (1993). Molecular cloning and expression of GRK6. A new member of the G protein-coupled receptor kinase family. *J. Biol. Chem.* **268**, 19521–19527.

Benovic, J. L., Strasser, R. H., Caron, M. G., and Lefkowitz, R. J. (1986). β-adrenergic receptor kinase: Identification of a novel protein kinase that phosphorylates the agonist-occupied form of the receptor. *Proc. Natl. Acad. Sci. U.S.A.* **83**, 2797–2801.

Benovic, J. L., Kuhn, H., Weyand, I., Codina, J., Caron, M. G., and Lefkowitz, R. J. (1987). Functional desensitization of the isolated β-adrenergic receptor by the β-adrenergic receptor kinase: Potential role of an analog of the retinal protein arrestin (48-kDa protein). *Proc. Natl. Acad. Sci. U.S.A.* **84**, 8879–8882.

Benovic, J. L., DeBlasi, A., Stone, W. C., Caron, M. G., and Lefkowitz, R. J. (1989). β-adrenergic receptor kinase: primary structure delineates a multigene family. *Science* **246**, 235–240.

Benovic, J. L., Onorato, J. J., Arriza, J. L., Stone, W. C., Lohse, M., Jenkins, N. A., Gilbert, D. J., Copeland, N. G., Caron, M. G., and Lefkowitz, R. J. (1991a). Cloning, expression, and chromosomal localization of β-adrenergic receptor kinase 2. A new member of the receptor kinase family. *J. Biol. Chem.* **266**, 14939–14946.

Benovic, J. L., Stone, W. C., Huebner, K., Croce, C., Caron, M. G., and Lefkowitz, R. J. (1991b). cDNA cloning and chromosomal localization of the human β-adrenergic receptor kinase. *FEBS Lett.* **283**, 122–126.

Bouvier, M., Hausdorff, W. P., DeBlasi, A., O'Dowd, B. F., Kobilka, B. K., Caron, M. G., and Lefkowitz, R. J. (1988). Removal of phosphorylation sites from the β_2-adrenergic receptor delays onset of agonist-promoted desensitization. *Nature (London)* **333**, 370–373.

Brodde, O. E. (1991). Beta 1- and beta 2-adrenoceptors in the human heart: properties, function, and alterations in chronic heart. *Pharmacol. Rev.* **43**, 203–242.

Bullrich, F., Druck, T., Kunapuli, P., Gomez, J., Gripp, K. W., Schelgelberger, B., Lasota, J., Aronson, M., Cannizzaro, L. A., Huebner, K., and Benovic, J. L. (1995). Chromosomal mapping of the genes encoding G protein-coupled receptor kinases GRK5 and GRK6. *Cytogenet. Cell Genet.* **70**, 250–254.

Calabrese, G., Sallese, M., Stornaiulolo, A., Morizio, E., Palka, G., and DeBlasi, A. (1994). Chromosome mapping of the human arrestin (SAG), β-arrestin 2 (ARRB2), and β-adrenergic receptor kinase 2 (ADRBK2) genes. *Genomics* **23**, 286–288.

Cassill, J. A., Whitney, M., Joazeiro, C. A., Becker, A., and Zuker, C. S. (1991). Isolation of Drosophila genes encoding G protein-coupled receptor kinases. *Proc. Natl. Acad. Sci. U.S.A.* **88**, 11067–11070.

Charron, J., Malynn, B. A., Fisher, P., Stewart, V., Jeannotte, L., Goff, S. P., Robertson, E. J., and Alt, F. W. (1992). Embryonic lethality in mice homozygous for a targeted disruption of the N-*myc* gene. *Genes Dev.* **6**, 2248–2257.

Chuang, T. T., Sallese, M., Ambrosini, G., Parruti, G., and DeBlasi, A. (1992). High expression of β-adrenergic receptor kinase in human peripheral blood leukocytes. Isoproterenol and platelet activating factor can induce kinase translocation. *J. Biol. Chem.* **267**, 6886–6892.

Craft, C. M., Whitmore, D. H., and Wiechmann, A. F. (1994). Cone arrestin identified by targeting expression of a functional family. *J. Biol. Chem.* **269**, 4613–4619.

Dawson, T. M., Arriza, J. L., Jaworsky, D. E., Borisy, F. F., Attramadal, H., Lefkowitz, R. J., and Ronnett, G. V. (1993). β-adrenergic receptor kinase-2 and β-arrestin-2 as mediators of odorant-induced desensitization. *Science* **259**, 825–829.

De Blasi, A., Parruti, G., and Sallese, M. (1995). Regulation of G protein-coupled receptor kinase subtypes in activated T lymphocytes. Selective increase of β-adrenergic receptor kinase 1 and 2. *J. Clin. Invest.* **95**, 203–210.

DebBurman, S. K., Ptasienski, J., Boetticher, E., Lomasney, J. W., Benovic, J. L., and Hosey, M. M. (1995a). Lipid-mediated regulation of G protein-coupled receptor kinases 2 and 3. *J. Biol. Chem.* **270**, 5742–5747.

DebBurman, S. K., Kunapuli, P., Benovic, J. L., and Hosey, M. M. (1995b). Agonist-dependent phosphorylation of human muscarinic receptors in *Spodoptera frugiperda* insect cell membranes by G protein-coupled receptor kinases. *Mol. Pharmacol.* **47**, 224–233.

Diverse-Pierluissi, M., Inglese, J., Stoffel, R. H., Lefkowitz, R. J., and Dunlap, K. (1996). G protein-coupled receptor kinase mediates desensitization of norepinephrine-induced Ca^{2+} channel inhibition. *Neuron* **16**, 579–585.

Dohlman, H. G., Bouvier, M., Benovic, J. L., Caron, M. G., and Lefkowitz, R. J. (1987). The multiple membrane spanning topography of the β_2-adrenergic receptor. Localization of the sites of binding, glycosylation, and regulatory phosphorylation by limited proteolysis. *J. Biol. Chem.* **262**, 14282–14288.

Ferguson, S. S., Downey, W. E. R., Colapietro, A. M., Barak, L. S., Menard, L., and Caron, M. G. (1996). Role of β-arrestin in mediating agonist-promoted G protein-coupled receptor internalization. *Science* **271**, 363–366.

Fredericks, Z. L., Pitcher, J. A., and Lefkowitz, R. J. (1996). Identification of the G protein-coupled receptor kinase phosphorylation sites in the human β_2-adrenergic receptor. *J. Biol. Chem.* **271**, 13796–13803.

Freedman, N. J., Liggett, S. B., Drachman, D. E., Pei, G., Caron, M. G., and Lefkowitz, R. J. (1995). Phosphorylation and desensitization of the human β_1-adrenergic receptor. Involvement of G protein-coupled receptor kinases and cAMP-dependent protein kinase. *J. Biol. Chem.* **270**, 17953–17961.

Garcia--Higuera, I., Penela, P., Murga, C., Egea, G., Bonay, P., Benovic, J. L., and Mayor, F., Jr. (1994). Association of the regulatory β-adrenergic receptor kinase with rat liver microsomal membranes. *J. Biol. Chem.* **269**, 1348–1355.

Goodman, O. B., Krupnick, J., Santini, F., Gurevich, V. V., Penn, R. B., Gagnon, A. W., Keen, J. H., and Benovic, J. L. (1996). β-arrestin acts as a clathrin adaptor in the endocytosis of the β_2-adrenergic receptor. *Nature (London)* **383**, 447–450.

Gurevich, V. V., and Benovic, J. L. (1993). Visual arrestin interaction with rhodopsin. Sequential multisite binding ensures strict selectivity toward light-activated phosphorylated rhodopsin. *J. Biol. Chem.* **268**, 11628–11638.

Gurevich, V. V., Dion, S. B., Onorato, J. J., Ptasienski, J., Kim, C. M., Sterne-Marr, R., Hosey, M. M., and Benovic, J. L. (1995). Arrestin interactions with G protein-coupled receptors. Direct binding studies of wild type and mutant arrestins with rhodopsin, β_2-adrenergic, and m2 muscarinic cholinergic receptors. *J. Biol. Chem.* **270**, 720–731.

Haga, K., and Haga, T. (1990). Dual regulation by G proteins of agonist-dependent phosphorylation of muscarinic acetylcholine receptors. *FEBS Lett.* **268**, 43–47.

Haga, K., and Haga, T. (1992). Activation by G protein beta gamma subunits of agonist- or light-dependent phosphorylation of muscarinic acetylcholine receptors and rhodopsin. *J. Biol. Chem.* **267**, 2222–2227.

Inglese, J., Glickman, J. F., Lorenz, W., Caron, M. G., and Lefkowitz, R. J. (1992a). Isoprenylation of a protein kinase. Requirement of farnesylation/alpha-carboxyl methylation for full enzymatic activity of rhodopsin kinase. *J. Biol. Chem.* **267**, 1422–1425.

Inglese, J., Koch, W. J., Caron, M. G., and Lefkowitz, R. J. (1992b). Isoprenylation in regulation of signal transduction by G-protein-coupled receptor kinases. *Nature (London)* **359**, 147–150.

Inglese, J., Freedman, N. J., Koch, W. J., and Lefkowitz, R. J. (1993). Structure and mechanism of the G protein-coupled receptor kinases. *J. Biol. Chem.* **268**, 23735–23738.

Ishii, K., Chen, J., Ishii, M., Koch, W. J., Freedman, N. J., Lefkowitz, R. J., and Coughlin, S. R. (1994). Inhibition of thrombin receptor signaling by a G-protein coupled receptor kinase. Functional specificity among G-protein coupled receptor kinases. *J. Biol. Chem.* **269**, 1125–1130.

Jaber, M., Koch, W. J., Rockman, H., Smith, B., Bond, R. A., Sulik, K. K., Ross, J., Lefkowitz, R. J., Caron, M. G., and Giros, B. (1996). Essential role of β-adrenergic receptor kinase in cardiac development and function. *Proc. Natl. Acad. Sci. U.S.A.* **93**, 12974–12979.

Kastner, P., Grondona, J. M., Mark, M., Gansmuller, A., Lemeur, M., Decimo, D., Vonesch, J.-L., Dolle, P., and Chambon, P. (1994). Genetic analysis of RXRα developmental function: Convergence of RXR and RAR signaling pathways in heart and eye morphogenesis. *Cell (Cambridge, Mass.)* **8**, 987–1003.

Khani, S. C., Abitbol, M., Yamamoto, S., Maravicmagovcevic, I., and Dryja, T. P. (1996). Characterization and chromosomal localization of the gene for human rhodopsin kinase. *Genomics* **35**, 571–576.

Koch, W. J., Inglese, J., Stone, W. C., and Lefkowitz, R. J. (1993). The binding site for the βγ subunits of heterotrimeric G proteins on the β-adrenergic receptor kinase. *J. Biol. Chem.* **268**, 8256–8260.

Koch, W. J., Hawes, B. E., Inglese, J., Luttrell, L. M., and Lefkowitz, R. J. (1994). Cellular expression of the carboxyl terminus of a G protein-coupled receptor kinase attenuates $G_{\beta\gamma}$-mediated signaling. *J. Biol. Chem.* **269**, 6193–6197.

Koch, W. J., Rockman, H. A., Samama, P., Hamilton, R. A., Bond, R. A., Milano, C. A., and Lefkowitz, R. J. (1995). Cardiac function in mice overexpressing the β-adrenergic receptor kinase or a βARK inhibitor. *Science* **268**, 1350–1353.

Kreidberg, J. A., Sariola, H., Loring, J. M., Maeda, M., Pelletier, J., Housman, D., and Jaenisch, R. (1993). WT-1 is required for early kidney development. *Cell (Cambridge, Mass.)* **74**, 679–691.

Kuhn, H. (1978). Light-regulated binding of rhodopsin kinase and other proteins to cattle photoreceptor membranes. *Biochemistry* **17**, 4389–4395.

Kunapuli, P., and Benovic, J. L. (1993). Cloning and expression of GRK5: A member of the G protein-coupled receptor kinase family. *Proc. Natl. Acad. Sci. U.S.A.* **90**, 5588–5592.

Kunapuli, P., Onorato, J. J., Hosey, M. M., and Benovic, J. L. (1994a). Expression, purification, and characterization of the G protein-coupled receptor kinase GRK5. *J. Biol. Chem.* **269**, 1099–1105.

Kunapuli, P., Gurevich, V. V., and Benovic, J. L. (1994b). Phospholipid-stimulated autophosphorylation activates the G protein-coupled receptor kinase GRK5. *J. Biol. Chem.* **269**, 10209–10212.

Liggett, S. B., Ostrowski, J., Chesnut, L. C., Kurose, H., Raymond, J. R., Caron, M. G., and Lefkowitz, R. J. (1992). Sites in the third intracellular loop of the α_{2A}-adrenergic receptor confer short term agonist-promoted desensitization. Evidence for a receptor kinase-mediated mechanism. *J. Biol. Chem.* **267**, 4740–4746.

Lohse, M. J., Benovic, J. L., Caron, M. G., and Lefkowitz, R. J. (1990a). Multiple pathways of rapid β2-adrenergic receptor desensitization. Delineation with specific inhibitors. *J. Biol. Chem.* **265**, 3202–3211.

Lohse, M. J., Benovic, J. L., Codina, J., Caron, M. G., and Lefkowitz, R. J. (1990b). β-Arrestin: A protein that regulates β-adrenergic receptor function. *Science* **248**, 1547–1550.

Lorenz, W., Inglese, J., Palczewski, K., Onorato, J. J., Caron, M. G., and Lefkowitz, R. J. (1991). The receptor kinase family: Primary structure of rhodopsin kinase reveals similarities to the β-adrenergic receptor kinase. *Proc. Natl. Acad. Sci. U.S.A.* **88**, 8715–8719.

Mayor, F., Jr., Benovic, J. L., Caron, M. G., and Lefkowitz, R. J. (1987). Somatostatin induces translocation of the β-adrenergic receptor kinase and desensitizes somatostatin receptors in S49 lymphoma cells. *J. Biol. Chem.* **262**, 6468–6471.

Moens, C. B., Stanton, B. R., Parada, L. F., and Rossant, J. (1993). Defects in heart and lung development in compound heterozygotes for two different targeted mutations in the N-myc locus. *Development (Cambridge, UK)* **119**, 485–499.

Murakami, A., Yajima, T., Sakuma, H., McLaren, M. J., and Inana, G. (1993). X-arrestin: A new retinal arrestin mapping to the X chromosome. *FEBS Lett.* **334**, 203–209.

Murga, C., Ruizgomez, A., Garciahiguera, I., Kim, C. M., Benovic, J. L., and Mayor, F. (1996). High affinity binding of β-adrenergic receptor kinase to microsomal membranes—Modulation of the activity of bound kinase by heterotrimeric G protein activation. *J. Biol. Chem.* **271**, 985–994.

Onorato, J. J., Palczewski, K., Regan, J. W., Caron, M. G., Lefkowitz, R. J., and Benovic, J. L. (1991). Role of acidic amino acids in peptide substrates of the β-adrenergic receptor kinase and rhodopsin kinase. *Biochemistry* **30**, 5118–5125.

Onorato, J. J., Gillis, M. E., Liu, Y., Benovic, J. L., and Ruoho, A. E. (1995). The β-adrenergic receptor kinase (GRK2) is regulated by phospholipids. *J. Biol. Chem.* **270**, 21346–21353.

Oppermann, M., Freedman, N. J., Alexander, R. W., and Lefkowitz, R. J. (1996a). Phosphorylation of the type 1A angiotensin II receptor by G protein-coupled receptor kinases and protein kinase C. *J. Biol. Chem.* **271**, 13266–13272.

Oppermann, M., Diverse-Pierluissi, H., Drazner, M. H., Dyer, S. L., Freedman, N. J., Peppel, K. C., and Lefkowitz, R. J. (1996b). Monoclonal antibodies reveal receptor specificity G-protein-coupled receptor kinases. *Proc. Natl. Acad. Sci. U.S.A.* **93**, 7649–7654.

Palczewski, K., Arendt, A., McDowell, J. H., and Hargrave, P. A. (1989). Substrate recognition determinants for rhodopsin kinase: studies with synthetic peptides, polyanions, and polycations. *Biochemistry* **28**, 8764–8770.

Palczewski, K., Buczylko, J., Ohguro, H., Annan, R. S., Carr, S. A., Crabb, J. W., Kaplan, M. W., Johnson, R. S., and Walsh, K. A. (1994). Characterization of a truncated form of arrestin isolated from bovine rod outer segments. *Protein Sci.* **3**, 314–324.

Pei, G., Tiberi, M., Caron, M. G., and Lefkowitz, R. J. (1994). An approach to the study of G-protein-coupled receptor kinases: An in vitro-purified membrane assay reveals differential receptor specificity and regulation by $G_{\beta\gamma}$ subunits. *Proc. Natl. Acad. Sci. U.S.A.* **91**, 3633–3636.

Ping, P., Gelzer-Bell, R., Roth, D. A., Kiel, D., Insel, P. A., and Hammond, H. K. (1995). Reduced β-adrenergic receptor activation decreases G-protein expression and β-adrenergic receptor kinase activity in porcine heart. *J. Clin. Invest.* **95**, 1271–1280.

Pitcher, J. A., Inglese, J., Higgins, J. B., Arriza, J. L., Casey, P. J., Kim, C., Benovic, J. L., Kwatra, M. M., Caron, M. G., and Lefkowitz, R. J. (1992). Role of beta gamma subunits of G proteins in targeting the β-adrenergic receptor kinase to membrane-bound receptors. *Science* **257**, 1264–1267.

Pitcher, J. A., Touhara, K., Payne, E. S., and Lefkowitz, R. J. (1995). Pleckstrin homology domain-mediated membrane association and activation of the β-adrenergic receptor kinase requires coordinate interaction with Gβγ subunits and lipid. *J. Biol. Chem.* **270**, 11707–11710.

Pitcher, J. A., Fredericks, Z. L., Stone, W. C., Premont, R. T., Stoffel, R. H., Koch, W. J., and Lefkowitz, R. J. (1996). Phosphatidylinositol 4,5-bisphosphate (PIP_2)-enhanced G protein-coupled receptor kinase (GRK) activity—location, structure, and regulation of the PIP_2 binding site distinguishes the GRK subfamilies. *J. Biol. Chem.* **71**, 24907–24913.

Premont, R. T., Koch, W. J., Inglese, J., and Lefkowitz, R J. (1994). Identification, purification, and characterization of GRK5, a member of the family of G protein-coupled receptor kinases. *J. Biol. Chem.* **269**, 6832–6841.

Premont, R. T., Macrae, A. D., Stoffel, R. H., Chung, N., Pitcher, J. A., Ambrose, C., Inglese, J., MacDonald, M. E., and Lefkowitz, R. J. (1996). Characterization of the G protein-coupled receptor kinase GRK4. Identification of four splice variants. *J. Biol. Chem.* **271**, 6403–6410.

Roth, N. S., Campbell, P. T., Caron, M. G., Lefkowitz, R. J., and Lohse, M. J. (1991). Comparative rates of desensitization of β-adrenergic receptors by the β-adrenergic receptor kinase and the cyclic AMP-dependent protein kinase. *Proc. Natl. Acad. Sci. U.S.A.* **88**, 6201–6204.

Sallese, M., Lombardi, M. S., and DeBlasi, A. (1994). Two isoforms of G protein-coupled receptor kinase 4 identified by molecular cloning. *Biochem. Biophys. Res. Commun.* **199**, 848–854.

Schleicher, S., Boekhoff, I., Arriza, J., Lefkowitz, R. J., and Breer, H. (1993). A β-adrenergic receptor kinase-like enzyme is involved in olfactory signal termination. *Proc. Natl. Acad. Sci. U.S.A.* **90**, 1420–1424.

Shih, M., and Malbon, C. C. (1994). Oligodeoxynucleotides antisense to mRNA encoding protein kinase A, protein kinase C, and β-adrenergic receptor kinase reveal distinctive cell-type-specific roles in agonist-induced desensitization. *Proc. Natl. Acad. Sci. U.S.A.* **91**, 12193–12197.

Shih, M., and Malbon, C. C. (1996). Protein kinase C deficiency blocks recovery from agonist-induced desensitization. *J. Biol. Chem.* **271**, 21478–21483.

Sterne-Marr, R., and Benovic, J. L. (1996). Regulation of G protein-coupled receptors by receptor kinases and arrestins. *Vitam. Horm. (N.Y.)* **51**, 193–234.

Sterne-Marr, R., Gurevich, V. V., Goldsmith, P., Bodine, R. C., Sanders, C., Donoso, L. A., and Benovic, J. L. (1993). Polypeptide variants of β-arrestin and arrestin3. *J. Biol. Chem.* **268**, 15640–15648.

Stoffel, R. H., Randall, R. R., Premont, R. T., Lefkowitz, R. J., and Inglese, J. (1994). Palmitoylation of G protein-coupled receptor kinase, GRK6. Lipid modification diversity in the GRK family. *J. Biol. Chem.* **269**, 27791–27794.

Strasser, R. H., Benovic, J. L., Caron, M. G., and Lefkowitz, R. J. (1986). β-agonist- and prostaglandin E1-induced translocation of the β-adrenergic receptor kinase: Evidence that the kinase may act on multiple adenylate cyclase-coupled receptors. *Proc. Natl. Acad. Sci. U.S.A.* **83**, 6362–6366.

Sucov, H. M., Dyson, E., Gumeringer, C. L., Price, J., Chien, K. R., and Evans, R. M. (1994). RXRα mutant mice establish a genetic basis for vitamin A signaling in heart morphogenesis. *Genes Dev.* **8**, 1007–1018.

Terwilliger, R. Z., Ortiz, J., Guitart, X., and Nestler, E. J. (1994). Chronic morphine administration increases β-adrenergic receptor kinase (βARK) levels in the rat locus coeruleus. *J. Neurochem.* **63**, 1983–1986.

Tiberi, M., Nash, S. R., Bertrand, L., Lefkowitz, R. J., and Caron, M. G. (1996). Differential regulation of dopamine D_{1A} receptor responsiveness by various G protein-coupled receptor kinases. *J. Biol. Chem.* **271**, 3771–3778.

Touhara, K., Inglese, J., Pitcher, J. A., Shaw, G., and Lefkowitz, R. J. (1994). Binding of G protein βγ-subunits to pleckstrin homology domains. *J. Biol. Chem.* **269**, 10217–10220.

Ungerer, M., Bohm, M., Elce, J. S., Erdmann, E., and Lohse, M. J. (1993). Altered expression of β-adrenergic receptor kinase and $β_1$-adrenergic receptors in the failing human heart. *Circulation* **87**, 454–463.

Ungerer, M., Parruti, G., Bohm, M., Puzicha, M., De Blasi, A., Erdmann, E., and Lohse, M. J. (1994). Expression of β-arrestins and β-adrenergic receptor kinases in the failing human heart. *Circ. Res.* **74**, 206–213.

Ungerer, M., Kessebohm, K., Kronsbein, K., Lohse, M. J., and Richardt, G. (1996). Activation of β-adrenergic receptor kinase during myocardial ischemia. *Circ. Res.* **79**, 455–460.

Walensky, L. D., Roskams, A. J., Lefkowitz, R. J., Snyder, S. H., and Ronnett, G. V. (1995). Odorant receptors and desensitization proteins colocalize in mammalian sperm. *Mol. Med.* **1**, 130–141.

Winstel, R., Freund, S., Krasel, C., Hoppe, E., and Lohse, M. J. (1996). Protein kinase crosstalk: Membrane targeting of the β-adrenergic receptor kinase by protein kinase C. *Proc. Natl. Acad. Sci. U.S.A.* **93**, 2105–2109.

Zhang, J., Ferguson, S. S. G., Barak, L. S., Menard, L., and Caron, M. G. (1996). Dynamin and β-arrestin reveal distinct mechanisms for G protein-coupled receptor internalization. *J. Biol. Chem.* **271**, 18302–18305.

George G. J. M. Kuiper
Stefan Nilsson*
Jan-Åke Gustafsson
Center for Biotechnology and Department of Medical Nutrition
Karolinska Institute
Huddinge, Sweden
*KaroBio AB
Huddinge, Sweden

Characteristics and Function of the Novel Estrogen Receptor β

We have cloned a novel member of the nuclear receptor superfamily: estrogen receptor β (ERβ). The cDNA of ERβ was isolated from a rat prostate cDNA library, and it encodes a protein of 485 amino acid residues with a calculated molecular weight of 54,200. The ERβ protein is highly homologous to the previously cloned ERα protein, particularly in the DNA-binding domain (95%) and in the C-terminal ligand binding domain (55%). Expression of ERβ in rat tissues was investigated by *in situ* hybridization and RT-PCR; moderate to high expression was found in prostate (secretory epithelial cells), ovary (granulosa cells), lung, bladder, brain, uterus, epidydimis, and testis. Saturation ligand-binding analysis of *in vitro*-synthesized rat ERβ protein revealed a single binding component for 16α-iodo-3,17β-estradiol with high affinity (K_d = 0.4 nM). In ligand-competition experiments the binding affinity decreased in the order dienestrol > 4-OH-tamoxifen > diethylstilbestrol > ICI-164384 > 17β-estradiol > estrone > estriol > tamoxifen. In cotransfection experiments of Chinese hamster ovary cells with an ERβ expression vector and an estrogen-regulated reporter gene, maximal stimulation of reporter gene activity was found during incubation with 1 nM 17β-estradiol. The detailed biological significance of the existence of two different ERs is at this moment unclear. Differences in the ligand-binding properties and/or transactivation function on certain target genes may exist.

I. Introduction

Estrogens influence the growth, differentiation, and functioning of many target tissues. These include tissues of the male and female reproductive systems such as mammary gland, uterus, ovary, testis, and prostate (Clark *et al.*, 1992). Estrogens also play an important role in bone maintenance and in the cardiovascular system, where estrogens have certain cardioprotective effects (Farhat *et al.*, 1996; Turner *et al.*, 1994). Estrogens are mainly produced in the ovaries and testis. They diffuse into and out of cells, but are retained with high affinity and specificity in target cells by an intranuclear binding protein, termed the estrogen receptor (ER). Once bound by estrogens, the ER undergoes a conformational change allowing the receptor to bind with high affinity to chromatin and to modulate transcription of target genes (Murdoch and Gorski, 1991). Important examples of genes regulated by ER-mediated mechanisms are the progesterone receptor, epidermal growth factor receptor, certain growth factors (IGF-I, TGF-α, TGF-β), cathepsin D, certain proto-oncogenes (c-*fos*, c-*myc*, and c-*jun*), and several heat-shock proteins (Ciocca and Vargas-Roig, 1995, and references therein). The estrogen receptor-encoding cDNAs were cloned in the mid-1980s from several species (Green *et al.*, 1986; Greene *et al.*, 1986; Koike *et al.*, 1987; Krust *et al.*, 1986; White *et al.*, 1987). Estrogen receptors were subsequently found to consist of a hypervariable N-terminal domain that contributes to the transactivation function, a highly conserved central domain responsible for specific DNA-binding, dimerization, and nuclear localization, and a C-terminal domain involved in ligand binding and ligand-dependent transactivation function (Beato *et al.*, 1995, and references therein).

In addition to these receptors for known ligands numerous so-called orphan receptors, which are putative receptors interacting with unknown ligands, have been found. Today, the nuclear receptor superfamily contains more than 100 different members, found in a large diversity of animal species from worm to insect to human (Mangelsdorf *et al.*, 1995). In an effort to clone and characterize novel nuclear receptors or unknown isoforms of existing (steroid) receptors, we designed degenerate primers based upon conserved regions within the DNA- and ligand-binding domains of nuclear receptors (Laudet *et al.*, 1992). These were used to PCR-amplify cDNA from several rat tissues. Prostate was selected as an organ of interest given the high incidence of prostate cancer and benign prostatic hyperplasia. There is firm evidence that many biological processes of prostate epithelial and stromal cells are controlled by androgens and estrogens (Gleave and Chung, 1995; Prins and Birch, 1995). Reports on the presence of ERs in the prostate have been conflicting. Although ERs could readily be detected in ligand-binding studies in rat and human prostate tissue (Jung-Testas *et al.*, 1981; van Beurden-Lamers *et al.*, 1974; Ekman *et al.*, 1983), it was difficult if not impossible to detect ERs with the existing ER antibodies (Prins and Birch, 1995; Brolin *et al.*, 1992), pointing to the possible existence in the prostate of an unknown isoform of the ER. Various naturally occuring ER isoforms have been described in breast tumors, meningiomas, and rat brain (Castles and Fuqua, 1996; Skipper *et al.*, 1993; Koehorst *et al.*, 1993).

II. Cloning of ERβ cDNA

The RT-PCR method, used before to identify new members of the nuclear receptor family (Enmark *et al.*, 1994; Pettersson *et al.*, 1996), was used to screen rat prostatic tissue. This resulted in the cloning of DNA sequences identical to several known members of the nuclear receptor family and one putative novel nuclear receptor partial cDNA clone (Kuiper *et al.*, 1996). This novel clone was found to be highly homologous to the cDNA of the rat ER (65%), which was previously cloned from rat uterus (Koike *et al.*, 1987). The amino acid residues predicted by this novel clone suggested that this DNA fragment encoded part of the DNA-binding domain, hinge region, and the beginning of the ligand-binding domain of a nuclear receptor protein. Two PCR primers were made (Fig. 1) to generate a probe consisting of the hinge region of the putative novel receptor. This probe was used to screen a rat prostate cDNA library, resulting in a strongly positive clone of about 2.5 kb, which was sequenced completely (Fig. 1).

Two in-frame ATG codons are located at nucleotides 424 and 448, preceded by an in-frame stop codon at nucleotide 319, which suggests that they are possible start codons. The open reading frame encodes a protein of 485 amino acid residues with a calculated molecular weight of 54,200. Analysis of the proteins synthesized by *in vitro* translation of the cDNA shown in Fig. 1 in rabbit reticulocyte lysate revealed a doublet protein band with an apparent molecular weight of 61,000 during SDS-PAGE (data not shown), confirming the open reading frame. The doublet protein band is probably caused by the use of both ATG codons for initiation of protein synthesis. It remains to be seen if the same protein heterogeneity also exists *in vivo*. Protein sequence comparison (Figs. 2–4) showed that the novel nuclear receptor clone is most related to the rat ER, cloned from rat uterus (Koike *et al.*, 1987), with 95% identity in the DNA-binding domain and 55% identity in the putative ligand-binding domain. It was therefore decided to tentatively name the novel nuclear receptor cDNA clone rat ERβ, and consequently the previously cloned receptor rat ERα.

A number of functional characteristics have been identified within the DNA-binding domain of nuclear receptors (Zilliacus *et al.*, 1994). The ERβ

```
ggaattcCGGGGGAGCTGGCCCAGGGGGAGCGGCTGGTGCTGCCACTGGCATCCCTAGGC   60
ACCCAGGTCTGCAATAAAGTCTGGCAGCCACTGCATGGCTGAGCGACAACCAGTGGCTGG  120
GAGTCCGGCTCTGTGGCTGAGGAAAGCACCTGTCTGCATTTAGAGAATGCAAAATAGAGA  180
ATGTTTACCTGCCAGTCATTACATCTGAGTCCCATGAGTCTCTGAGAACATAATGTCCAT  240
CTGTACCTCTTCTCACAAGGAGTTTTCTCAGCTGCGACCCTCTGAAGACATGGAGATCAA  300
AAACTCACCGTCGAGCCTTAGTTCCCTGCTTCCTATAACTGTAGCCAGTCCATCCTACCC  360
CTGGAGCACGGCCCCATCTACATCCCTTCCTCCTACGTAGACAACCGCCATGAGTATTCA  420
GCTATGACATTCTACAGTCCTGCTGTGATGAACTACAGTGTTCCCGGCAGCACCAGTAAC  480
         M  T  F  Y  S  P  A  V  M  N  Y  S  V  P  G  S  T  S  N
CTGGACGGTGGGCCTGTCCGACTGAGCACAAGCCCAAATGTGCTATGGCCAACTTCTGGG  540
 L  D  G  G  P  V  R  L  S  T  S  P  N  V  L  W  P  T  S  G
CACCTGTCTCCTTTAGCGACCCATTGCCAATCATCGCTCCTCTATGCAGAACCTCAAAAG  600
 H  L  S  P  L  A  T  H  C  Q  S  S  L  L  Y  A  E  P  Q  K
AGTCCTTGGTGTGAAGCAAGATCACTAGAGCACACCTTACCTGTAAACAGAGAGACACTG  660
 S  P  W  C  E  A  R  S  L  E  H  T  L  P  V  N  R  E  T  L
AAGAGGAAGCTTAGTGGGAGCAGTTGTGCCAGCCCTGTTACTAGTCCAAACGCAAAGAGG  720
 K  R  K  L  S  G  S  S  C  A  S  P  V  T  S  P  N  A  K  R
GATGCTCACTTCTGCCCCGTCTGCAGCGATTATGCATCTGGGTATCATTACGGCGTTTGG  780
 D  A  H  F  C  P  V  C  S  D  Y  A  S  G  Y  H  Y  G  V  W
TCATGTGAAGGATGTAAGGCCTTTTTTAAAGAAGCATTCAAGGACATAATGATTATATC  840
 S  C  E  G  C  K  A  F  F  K  R  S  I  Q  G  H  N  D  Y  I
TGTCCAGCCACGAATCAGTGTACCATAGACAAGAACCGGCGTAAAAGCTGCCAGGCCTGC  900
 C  P  A  T  N  Q  C  T  I  D  K  N  R  R  K  S  C  Q  A  C
CGACTTCGCAAGTGTTATGAAGTAGGAATGGTCAAGTGTGGATCCAGGAGAGAACGGTGT  960
 R  L  R  K  C  Y  E  V  G  M  V  K  C  G  S  R  R  E  R  C
GGGTACCGTATAGTGCGGAGGCAGAGAAGTTCTAGCGAGCAGGTACACTGCCTGAGCAAA 1020
 G  Y  R  I  V  R  R  Q  R  S  S  E  Q  V  H  C  L  S  K
GCCAAGAGAAACGGTGGGCATGCACCCCGGGTGAAGGAGCTACTGCTGAGCACCTTGAGT 1080
 A  K  R  N  G  G  H  A  P  R  V  K  E  L  L  L  S  T  L  S
CCAGAGCAACTGGTGCTCACCCTCCTGGAAGCTGAACCACCCAATGTGCTGGTGAGCCGT 1140
 P  E  Q  L  V  L  T  L  L  E  A  P  P  N  V  L  V  S  R
CCCAGCATGCCCTTCACCGAGGCCTCCATGATGATGTCCCTCACTAAGCTGGCGGACAAG 1200
 P  S  M  P  F  T  E  A  S  M  M  M  S  L  T  K  L  A  D  K
GAACTGGTGCACATGATTGGCTGGGCCAAGAAAATCCCTGGCTTTGTGGAGCTCAGCCTG 1260
 E  L  V  H  M  I  G  W  A  K  K  I  P  G  F  V  E  L  S  L
TTGGACCAAGTCCGGCTCTTAGAAAGCTGCTGGATGGAGGTGCTAATGGTGGGACTGATG 1320
 L  D  Q  V  R  L  L  E  S  C  W  M  E  V  L  M  V  G  L  M
TGGCGCTCCATCGACCACCCCGGCAAGCTCATTTTCGCTCCCGACCTCGTTCTGGACAGG 1380
 W  R  S  I  D  H  P  G  K  L  I  F  A  P  D  L  V  L  D  R
GATGAGGGGAAGTGCGTAGAAGGGATTCTGGAAATCTTTGACATGCTCCTGGCGACGACG 1440
 D  E  G  K  C  V  E  G  I  L  E  I  F  D  M  L  L  A  T  T
TCAAGGTTCCGTGAGTTAAAACTCCAGCACAAGGAGTATCTCTGTGTGAAGGCCATGATC 1500
 S  R  F  R  E  L  K  L  Q  H  K  E  Y  L  C  V  K  A  M  I
CTCCTCAACTCCAGTATGTACCCCTTGGCTTCTGCAAACCAGGAGGCAGAAAGTAGCCGG 1560
 L  L  N  S  S  M  Y  P  L  A  S  A  N  Q  E  A  E  S  S  R
AAGCTGACACACCTACTGAACGCGGTGACAGATGCCCTGGTCTGGGTGATTGCGAAGAGT 1620
 K  L  T  H  L  L  N  A  V  T  D  A  L  V  W  V  I  A  K  S
GGTATCTCCTCCCAGCAGCAGTCAGTCCGACTGGCCAACCTCCTGATGCTTCTTTCTCAC 1680
 G  I  S  S  Q  Q  Q  S  V  R  L  A  N  L  L  M  L  L  S  H
GTCAGGCACATCAGTAACAAGGGCATGGAACATCTGCTCAGCATGAAGTGCAAAAATGTG 1740
 V  R  H  I  S  N  K  G  M  E  H  L  L  S  M  K  C  K  N  V
GTCCCGGTGTATGACCTGCTGCTGGAGATGCTGAATGCTCACACGCTTCGAGGGTACAAG 1800
 V  P  V  Y  D  L  L  L  E  M  L  N  A  H  T  L  R  G  Y  K
TCCTCAATCTCGGGGTCTGAGTGCAGCTCAACAGAGGACAGTAAGAACAAAGAGAGCTCC 1860
 S  S  I  S  G  S  E  C  S  S  T  E  D  S  K  N  K  E  S
CAGAACCTACAGTCTCAGTGATGGCCAGGCCTGAGGCGGACAGACTACAGAGATGGTCAA 1920
 Q  N  L  Q  S  Q  *
AAGTGGAACATGTACCCTAGCATCTGGGGGTTCCTCTTAGGGCTGCCTTGGTTACGCACC 1980
CCTTACCCACACTGCACTTCCCAGGAGTCAGGGTGGTTGTGTGGCGGTGTTCCTCATACC 2040
AGGATGTACCACCGAATGCCAAGTTCTAACTTGTATAGCCTTGAAGGCTCTCGGTGTACT 2100
TACTTTCTGTCTCCTTGCCCACTTGGAAACATCTGAAAGGTTCTGGAACTAAAGGTCAAA 2160
GTCTGATTTGGAAGGATTGTCCTTAGTCAGGAAAAGGAAATATGGCATGTGACACAGCTAT 2220
AAGAAATGGACTGTAGGACTGTGTGGCCATAAAATCAACCTTTGGATGGCGTCTTCTAGA 2280
CCACTTGATTGTAGGATTGAAAACCACATTGACAATCAGCTCATTTCGCATTCCTGCCTC 2340
ACGGGTCTGTGAGGACTCATTAATGTCATGGGTTATTCTATCAAAGACCAGAAAGATAGT 2400
GCAAGCTTAGATGTACCTTGTTCCTCCTCCCAGACCCTTGGGTTACATCCTTAGAGCCTG 2460
CTTATTTGGTCTGTCTGAATGTGGTCATTGTCATGGGTTAAGATTTAAATCTCTTTGTAA 2520
TATTGGCTTCCTTGAAGCTATGTCATCTTTCTCTCTCCCGgaattc 2568
```

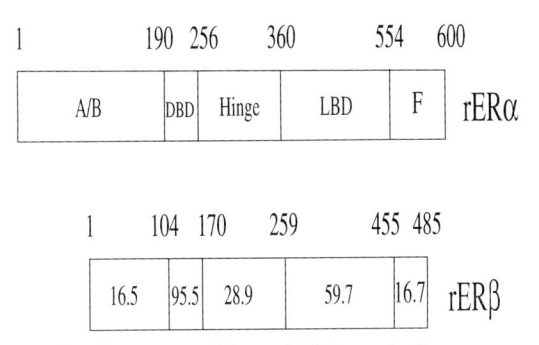

```
1        190  256      360          554  600
┌─────────┬────┬──────┬────────────┬────┐
│   A/B   │DBD │ Hinge│    LBD     │ F  │  rERα
└─────────┴────┴──────┴────────────┴────┘

1     104  170      259          455  485
┌─────┬────┬────────┬────────────┬────┐
│16.5 │95.5│  28.9  │    59.7    │16.7│  rERβ
└─────┴────┴────────┴────────────┴────┘
```

FIGURE 2 Comparison between rat ERα and ERβ protein. Percentage amino acid identity in the domains A/B (N terminus), DBD (DNA-binding), hinge, and LBD/F (ligand binding, dimerization, and ligand-dependent transactivation) are depicted.

protein P-box and D-box sequences of EGCKA and PATNQ, respectively, are identical to the corresponding boxes in the ERα protein (Koike et al., 1987), thus predicting that ERβ protein binds to estrogen response element (ERE) sequences. The putative ligand-binding domain of ERβ protein shows closest homology to the ligand-binding domain of ERα protein (Figs. 2 and 3), whereas the homology with the estrogen receptor-related receptors ERR1 and ERR2 (Giguere et al., 1988; Pettersson et al., 1996) is considerably less. The ERR1 and ERR2 nuclear orphan receptors do not bind estradiol. Several amino acid residues described to be close to or part of the ligand-binding pocket of the human ERα protein (Cys 530, Asp 426, and Gly 521) are conserved in the putative ligand-binding domain of rat ERβ protein (Cys 436, Asp 333, and Gly 427) and in the ligand-binding domain of ERs from various species (Harlow et al., 1989, Fawell et al., 1990). An alignment of rat ERα and rat ERβ protein ligand-binding domains (Fig. 4) reveals several completely conserved stretches as well as a central stretch which is essentially nonconserved (boxed in Fig. 4). Comparative studies of the ligand-binding domain of ERα and ERβ protein by peptide mapping using mass spectrometric techniques (Hegy et al., 1996; Seielstad et al., 1995) and crystal structure determination should provide more detailed information on the structural requirements for ligand binding.

The core domain of the ligand-dependent transactivation function TAF-2, identified in the ERα protein (Danielian et al., 1992), is almost completely conserved in ERβ protein (Fig. 4; amino acid residues 441–457). Steroid hormone receptors are phosphoproteins (Kuiper and Brinkmann, 1994, and

---

FIGURE 1 Sequence of rat ERβ cDNA and predicted amino acid sequence of ERβ protein. Two potential translation start sites are indicated in bold. The predicted DNA-binding domain is double underlined and the PCR primers, used for generation of the probe for screening of a rat prostate cDNA library, are single underlined.

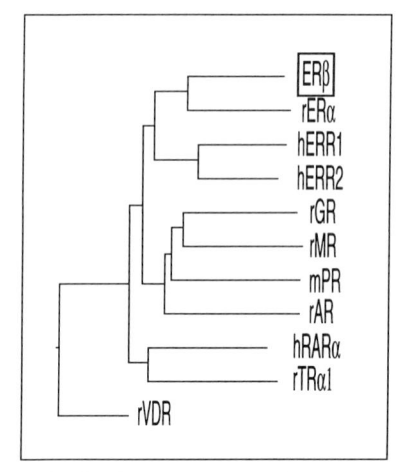

Domain	A/B	C	D	E/F
rERα	16.5	95.5	28.9	53.5
hERR1	16.5	69.7	15.6	34.2
hERR2	13.7	72.7	19.8	31.0
rAR	16.5	59.1	12.8	19.1
rGR	19.4	59.1	14.6	18.8
rMR	17.5	57.6	17.8	19.1
mPR	15.5	56.1	14.4	19.6
rTRα1	15.4	50.0	13.3	17.7
rRARα	26.1	59.1	14.8	17.3
rVDR	21.7	47.0	15.6	18.8

FIGURE 3 Comparison of ERβ protein with several representative members of the nuclear receptor family. Percentage amino acid identity in the domains A/B (N terminus), C (DNA-binding), D (hinge region), and E/F (ligand binding, dimerization, and ligand-dependent transactivation function) are depicted. For the alignment and phylogenetic tree, Clustal analysis of the full-length receptor sequences using the MEGALIGN/DNASTAR software was used.

references therein) and ER phosphorylation has been implicated in ligand-dependent and ligand-independent transactivation functions of the ER (Ali *et al.*, 1993; Kato *et al.*, 1995). Several phosphorylation sites identified in the N-terminal domain and ligand-binding domain of the human ERα protein (Ali *et al.*, 1993; Arnold *et al.*, 1995) are present at similar positions in the rat ERβ protein (Ser 30, Ser 42, Ser 94, and Tyr 443). The functional significance of these putative phosphorylation sites within the ERβ protein remains to be established. The rat ERβ protein consists of 485 amino acid

```
223 LVLTLLEAEPPNVLVS-RPSMPFTEASMMSLTKLADKELVHMIGWAKKI    rat ERβ
320 M.SA..D....LIYSEYD..R..S.....GL..N...R......N...RV   rat ERα

272 PGFVELSLLDQVRLLESCWMEVLMVGLMARSIDHPGKLIFAPDLVLDRDE    rat ERβ
370 ...GD.N.H...H...CA.L.I...I..V...ME....L...N.L...NQ   rat ERα

322 GKCVEGILEIFDMLLATTSRFRELKLQHKEYLCVKAMILLNSSMYP-LAS    rat ERβ
420 ......MV.........S....MMN..GE.FV.L.SI.....GV.TF.S.   rat ERα

371 ANQEAESSRKLTHLLNAVIDALVWVIAKSGISSQQQSVRLANLLMLLSHV    rat ERβ
470 ILKSL.EKDHIHRV.DKIN.T.IHLM..A.LTL...HR...Q..LI...I   rat ERα

421 RHISNKGMEHLLSMKCKNVVPVYDLLLEMLNAHTLRG 457            rat ERβ
520 ..M........YN........L........D..R.HAPA 558         rat ERα
```

FIGURE 4 Alignment of the amino acid sequences of rat ERα protein (GenBank database Y00102) ligand-binding domain (amino acid residues 320–558), and rat ERβ protein (GenBank database U57439) ligand-binding domain (amino acid residues 223–457).

residues, whereas the ERα protein from human, mouse, and rat consists of 590–600 amino acid residues. The main difference is a much shorter N-terminal domain in ERβ protein, i.e., 103 amino acid residues as compared to 185–190 amino acid residues in the ERα protein. Also, the nonconserved, so-called F-domain at the C-terminal end of ERβ protein is 15 amino acid residues shorter than in ERα proteins. Recently, the mouse and human homologues of rat ERβ were cloned in our laboratory (E. Enmark, G. Bertilsson, K. Grandien, and K. Pettersson, unpublished observations) and the human ERβ cDNA also by others (Mosselman *et al.*, 1996).

III. Ligand-Binding Characteristics of ERβ Protein

The ER protein can be isolated from the cytosol of target cell extracts as a large, nontransformed (i.e., non-DNA binding), 7–8S oligomeric complex, which contains hsp90 and hsp70 (Murdoch and Gorski, 1991, and references therein). It is believed that heat shock proteins function to help fold the ER protein properly and to protect the hydrophobic hormone binding domain from inappropiate interactions (Murdoch and Gorski, 1991; Sabbah *et al.*, 1996). Rabbit reticulocyte lysates contain large amounts of several heat shock proteins as hsp90 and hsp70, and have been used extensively for the study of ER complex formation with hsps as well as for the study of requirements for steroid binding and interactions with DNA (Beekman *et al.*, 1993; Sabbah *et al.*, 1996). It was therefore decided to use human ERα and rat ERβ protein synthesized in reticulocyte lysates for the ligand binding experiments. When ERβ protein was labeled with a saturating dose of [^{3}H] 17β-estradiol and analyzed on sucrose density gradients, a single peak of specifically bound radioactivity was observed (Fig. 5). The sedimentation coefficient of this complex was about 7S, and it shifted to 4S in the presence of 0.4 M NaCl (Fig. 5).

In order to obtain optimal conditions for the determination of equilibrium dissociation constants (K_d) and relative binding affinities (RBA) of various ligands, the ER protein concentration in the binding assays was lowered to 10–20 pM. At these low ER protein concentrations radioligand and/or competitor depletion can be excluded, while maintaining high ER protein recovery during separation of bound and unbound ligand by the use of a gel filtration assay instead of the more traditional charcoal adsorption assay (Salomonsson *et al.*, 1994). The low ER protein concentration made it necessary to use radioiodinated estradiol as a probe since the specific radioactivity of tritiated estradiol was too low to maintain sufficient accuracy. Radioiodinated estradiol ([^{125}I]16α-iodoestradiol) binds to the ER protein with high affinity and specificity as shown by its use in dry mount autoradiographic techniques and in various ligand binding assays (Berns *et al.*, 1985; Hochberg and Rosner, 1980). In Fig. 6 the result of a saturation

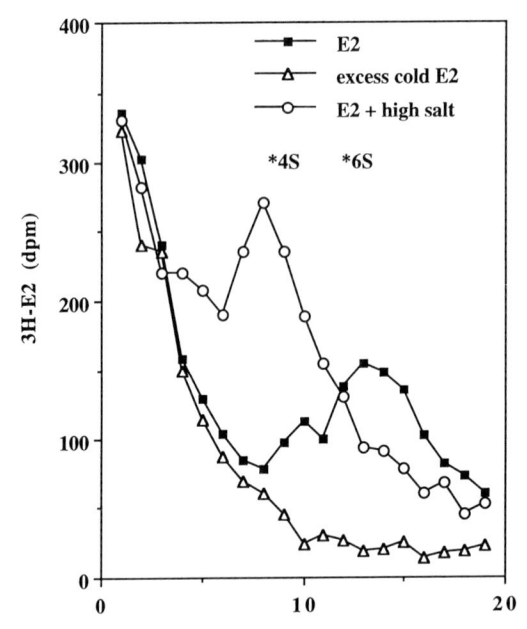

FIGURE 5 Sucrose density gradient analysis of rat ERβ protein synthesized *in vitro*. The ERβ protein was labeled with [³H] estradiol in the presence or absence of 200-fold unlabeled estradiol for 4 hr at 8°C. Samples were run on 5–20% sucrose density gradients (with or without 0.4 M NaCl) for 20 hr at 48,000 rpm in a SW-60 rotor at 8°C.

ligand binding assay with [^{125}I]16α-iodoestradiol is shown. The nonspecific binding was ≤8% of total binding over the whole radioligand concentration range used. The K_d values calculated from the saturation curves (Fig. 6) are 0.06 nM for ERα protein and 0.24 nM for ERβ protein. Linear transformation of saturation data (Scatchard plots in Fig. 6) revealed a single population of binding sites for 16α-iodoestradiol with a K_d of 0.1 nM ($n = 2$) for the ERα protein and 0.4 nM ($n = 2$) for the ERβ protein. Although the ERβ protein has four times lower affinity for 16α-iodoestradiol in this system compared to the ERα protein, both K_d values are within the range (0.1—1 nM) generally reported for estradiol binding to ER in various systems (Clark *et al.*, 1992, and references therein).

Measurements of the equilibrium binding of the radioligand in the presence of different concentrations of unlabeled competitors provide readily interpretable information about the affinities of the latter, provided that radioligand and/or competitor depletion are avoided. Competition experiments (Table I) are indicative of an ER in that only estrogens and antiestrogens competed efficiently with 16α-iodoestradiol for binding. For the ERα as well as ERβ protein the estradiol binding was stereospecific because 17α-estradiol showed, respectively, 2 and 10 times lower affinity (Table 1) compared to 17β-estradiol, which is in agreement with previous findings on stereospecific binding of estradiol by the ER (Noteboom and Gorski,

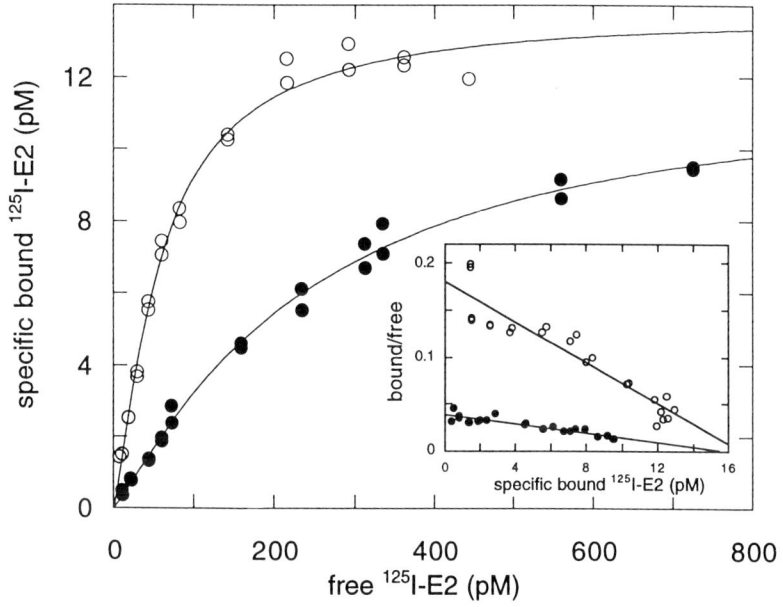

FIGURE 6 Binding of [^{125}I]16α-iodoestradiol to *in vitro*-synthesized ERα and ERβ protein in the presence or absence of a 300-fold excess of diethylstilbestrol for 16 hr at 4°C. Unbound radioactivity was removed as described (Salomonsson *et al.*, 1994) and specific counts (ERα, ○; ERβ, ●) were calculated by subtracting nonspecific bound counts from total bound counts. Inset shows Scatchard analysis of specific binding giving a K_d of 0.1 nM for ERα protein and a K_d of 0.4 nM for ERβ protein.

1965). Several differences in the relative binding affinities are present between receptor subtypes. For instance, 4OH-tamoxifen and ICI-164384 have 2 times higher affinity for the ERβ protein. It should be kept in mind that most if not all ER ligand binding studies performed in the past 30 years actually involved mixtures of ERα and ERβ protein. This is certainly the case for many studies in which rat uterus cytosol was used (see below). Therefore, caution is needed when comparing relative binding affinities for the individual receptors with the relative binding affinities measured involving mixtures of both ER subtypes.

It has been known for a long time that a number of compounds classified as androgens (C19 steroids) can evoke estrogen-like effects in the female genital tract and in mammary glands (Huggins *et al.*, 1954). Of several androgens tested only those with a hydroxyl group at C3 and C17 have significant affinity for both ER subtypes (Table I). The binding affinity of 5α-androstane-3β,17β-diol and 5-androstene-3β,17β-diol for both ER subtypes is in agreement with previous studies showing specific binding to the rat uterus ER and estrogenic responses in rat uterus and mammary tumors for both steroids (Garcia and Rochefort, 1979; van Doorn *et al.*, 1981). Recently, homologous recombination in mouse embryonic stem cells

TABLE I Relative Binding Affinity of Various Compounds for Estrogen Receptor α and β

Compound	ERα	ERβ
17β-Estradiol	100	100
17α-Estradiol	58	11
Diethylstilbestrol	468	295
Dienestrol	223	404
Hexestrol	302	234
Estrone	60	37
Estriol	14	21
Moxestrol	43	5
Tamoxifen	7	6
4-OH-Tamoxifen	178	339
ICI-164384	85	166
Nafoxidine	44	16
Clomifene	25	12
5-Androstenediol	6	17
3β-Androstanediol	3	7
3α-Androstanediol	0.07	0.3
5α-DHT	0.05	0.17
Progesterone	<0.001	<0.001
Testosterone	<0.01	<0.01
Corticosterone	<0.001	<0.001

Note. The relative binding affinity of each competitor was calculated as the ratio of concentration of 17β-estradiol and competitor required to reduce the specific radioligand binding by 50%. The relative binding affinity of 17β-estradiol was arbitrarily set at 100. 3β-Androstanediol = 5α-androstane-3β,17β-diol. 3α-Androstanediol = 5α-androstane-3α,17β-diol. 5-Androstenediol = 5-androstene-3β,17β-diol. 5α-DHT = 5α-dihydrotestosterone.

was used to produce male and female mice with a disruption (null allele) in the 5α-reductase type I isozyme gene (Mahendroo *et al.*, 1996). Female mice exhibited a parturition defect that is maternal in origin, and which could be reversed by administration of 5α-androstane-3α,17β-diol. Enzymes that synthesize 5α-androstane-3α,17β-diol are induced in wild-type mouse uterus during late gestation. Although ERα and ERβ protein are expressed in uterus, it seems unlikely that 5α-androstane-3α,17β-diol acts via ERα or ERβ protein during late gestation given the very low binding affinity (Table I).

A question of considerable interest is why, despite the numerous ligand binding assays carried out with the ER protein, an indication of the existence of two ER subtypes has never been reported. Distinguishing between a mixed population of receptor subtypes and a homogeneous receptor population by saturation or homologous/heterologous competition is generally difficult. This is only possible with certainty when the two subtypes differ sufficiently in affinity (10- to 100-fold) and the range of ligand concentrations examined is wide. Furthermore, the proportion of the two subtypes must be appropiate (Swillens *et al.*, 1995). Of all the radioligands used in ER assays the difference in affinity for moxestrol between both ER subtypes is the greatest (Table I). Most ER ligand-binding assays have been performed with uterus extracts and breast tumor extracts or cell lines, and it could be that the right conditions are not fulfilled in these experiments (Hähnel *et al.*, 1973; Notides, 1970, Clark *et al.*, 1992, and references therein). It remains to be seen if the differences in relative binding affinity of some ligands for ERα and ERβ protein are also reflected in transactivation assay systems using different cellular backgrounds.

IV. Transactivation Function of ERβ Protein

In order to investigate the transcriptional regulatory properties of ERβ protein, cotransfection experiments in which CHO (Chinese hamster ovary) cells were transfected with an ERβ protein expression vector and an estrogen responsive reporter gene construct were performed. The pERE-ALP reporter construct contains a secreted form of the placental alkaline phosphatase gene (Berger *et al.*, 1988) and the mouse mammary tumor virus long-terminal repeat in which the glucocorticoid response elements were replaced by a single consensus vitellogenin promoter estrogen response element (Martinez *et al.*, 1987). In the absence of exogenously added estradiol, the ERβ protein showed considerable transcriptional activity, which could be further increased by the addition of estradiol (Fig. 7). Simultaneous addition of a 10-fold excess of tamoxifen partially supressed the estradiol stimulated activity (Fig. 7). The constitutive transcriptional activity of ERβ protein could be suppressed by tamoxifen or ICI-164384 (not shown). Although we have done everything possible to exclude contaminating estrogenic compounds by using phenol red-free serum replacement medium for the transfection experiments, we cannot exclude the presence of very low amounts of estrogenic compounds. Ligand-independent ERα protein-mediated transcriptional activation by dopamine and growth factors has been described (Smith *et al.* 1995; Ignar-Trowbridge *et al.* 1996), and might also be the cause of the ERβ protein-mediated constitutive activity. In dose–response experiments, ERβ protein began to respond at 0.1 nM estradiol (Fig. 8) and maximal stimulation was observed at 1 nM estradiol. The maximal stimulation factor in the CHO cells was 2.6 ± 0.5-fold (mean ± SD, $n = 4$) as compared to incubation in the absence of estradiol. Dexamethasone, testosterone, progesterone, thyroid hormone, or all-*trans*-retinoic acid did not stimulate transcriptional activity of ERβ protein, even at the highest

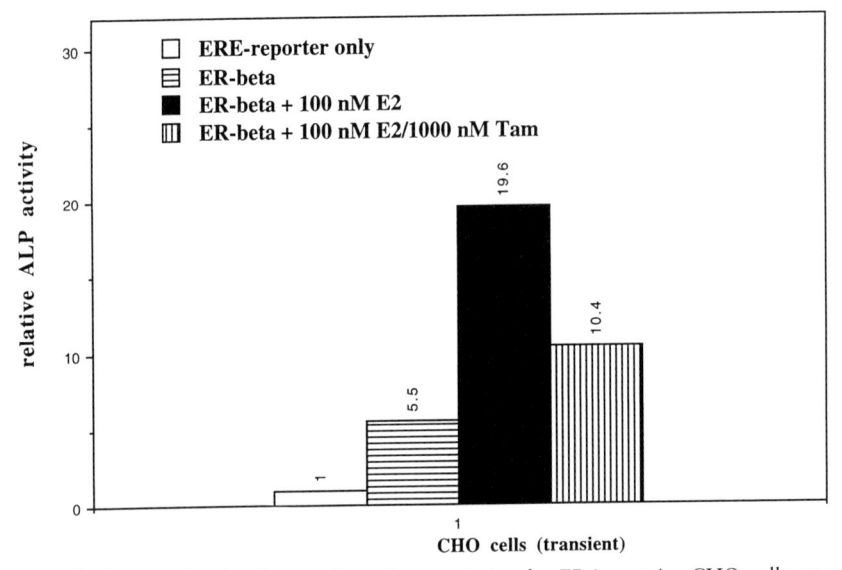

FIGURE 7 E2-stimulated activation of transcription by ERβ protein. CHO cells were transiently transfected with ERE-reporter plasmid alone (ERE-reporter) or together with an ERβ protein expression plasmid. Cells were incubated with estradiol alone or estradiol and tamoxifen as indicated.

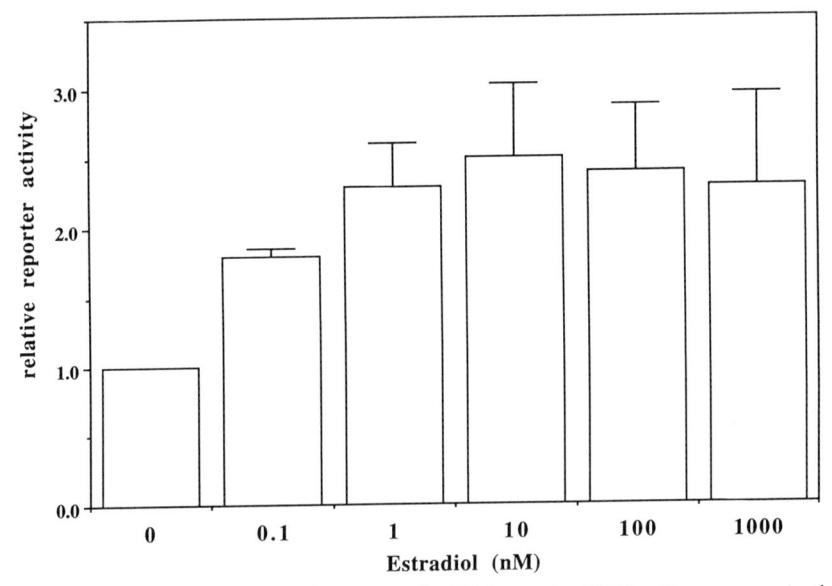

FIGURE 8 Estrogen-stimulated activation by ERβ protein. CHO cells were transiently transfected with the ERE-reporter plasmid and the ERβ protein expression plasmid. Cells were incubated with estradiol as indicated.

concentration (1000 nM) tested (not shown). In control experiments, the wild-type human ERα protein also showed transcriptional activity in the absence of estradiol which could be stimulated to a similar extent as for ERβ protein by estradiol.

In addition to forming homodimers, ERα and ERβ protein form heterodimers when bound to DNA (Pettersson *et al.*, 1977). Heterodimerization between other steroid hormone receptors, for instance the mineralocorticoid and glucocorticoid receptors has been described (Liu *et al.*, 1995). The transcriptional activity of the ER heterodimers as compared to the respective ER homodimers on various target gene promoters remains to be investigated. The ERα protein contains two independent transcriptional activation domains: AF-1, which is located in the N-terminal A/B domain, and AF-2, located in the ligand-binding domain (Lees *et al.*, 1989; Tora *et al.*, 1989). The transcriptional activities of AF-1 and AF-2 are promoter- and cell-type-specific, and synergism between AF-1 and AF-2 is required for full transcriptional activity. Estradiol has been shown to promote association of the amino- and carboxy-terminal regions of ERα, leading to transcriptional synergism between AF-1 and AF-2 (Kraus *et al.*, 1995). The N-terminal A/B domain is poorly conserved between ERα and ERβ protein (Fig. 2) and it remains to be determined if the N-terminal domain of ERβ protein has any transcriptional activation function. The transcription activation function AF-1 is necessary for the activation of ERα protein by growth factors as EGF and IGF-I (Ignar-Trowbridge *et al.*, 1996), and in more detailed studies it was shown that phosphorylation of human ERα protein at serine residue 118 is required for full activity of AF-1 (Kato *et al.*, 1995; Bunone *et al.*, 1996). Serine residue 118 of human ERα protein is phosphorylated by MAP kinase (MAPK) *in vitro* and in intact cells incubated with EGF and IGF-I (Kato *et al.*, 1995). Several potential phosphorylation sites for MAPK are present in the N-terminus of rat ERβ protein and it would be interesting to see if they are involved in the constitutive transcriptional activity of ERβ (Fig. 7).

V. Expression of ERβ mRNA

In situ hybridization with oligonucleotide probes revealed high expression of ERβ in male and female rat reproductive tissues, whereas in all other tissues the expression was much lower or below the level of detection by *in situ* hybridization with oligonucleotide probes (Kuiper *et al.*, 1996). In the prostate ERβ is expressed in the epithelial cells of the secretory alveoli, whereas in the ovary the granulosa cells in primary, secondary, and mature follicles showed expression of ERβ. Examination of ERα expression at the cellular level, by *in situ* hybridization, showed that ERα mRNA is expressed at a low level throughout the rat ovary with no particular cellular localization

(Byers et al., 1997). In rat prostate and ovary it has been difficult to demonstrate the presence of ER protein by immunostaining with the available ER protein antibodies, although specific binding of estradiol could be measured in the prostate (van Beurden-Lamers et al., 1974; Jung-Testas et al., 1981) and ovary (Clark et al., 1992, and references therein; Richards, 1975). In human and rat prostate weak staining in stromal cells was detected by immunohistochemistry, while the glandular epithelial cells were negative (Prins and Birch, 1995; Brolin et al., 1992; Ehara et al., 1995), which is in contrast with our *in situ* hybridization results (Kuiper et al., 1996). In the rat ovary specific binding of estradiol was found in intact follicles and granulosa cells (Kudolo et al., 1984; Richards, 1975; Kawashima and Greenwald, 1993), but it has been difficult to detect ER protein with the available ER protein antibodies (Hild-Petito et al., 1988). These discrepancies can be explained by the fact that the most frequently used ER protein antibodies, H-222 and H-226 (Greene and Jensen, 1982), do not crossreact with rat ERβ protein on immunoblots (G. Kuiper, unpublished observations). The above findings and our results indicate that the ERβ protein is the predominant ER subtype in rat prostate and ovary.

In order to determine the relative distribution of ERα and ERβ mRNA, total RNA was isolated from rat tissues and used for RT-PCR experiments, using primers which are specific for each ER subtype (Fig. 9). Highest expression of ERβ mRNA was found in the ovary and prostate, which is in agreement with our previous *in situ* hybridization experiments using male and female rats (Kuiper et al., 1996). In addition testis, uterus, bladder, and lung revealed moderate expression, while pituitary, epididymis, thymus, various brain sections (thalamus/hypothalamus, cerebellum, olfactory lobe, brain stem), and spinal cord reveal low expression of ERβ mRNA. The ERα mRNA is highly expressed in epididymis, testis, pituitary, uterus, kidney, and adrenal. Apart from weak expression in thalamus/hypothalamus the

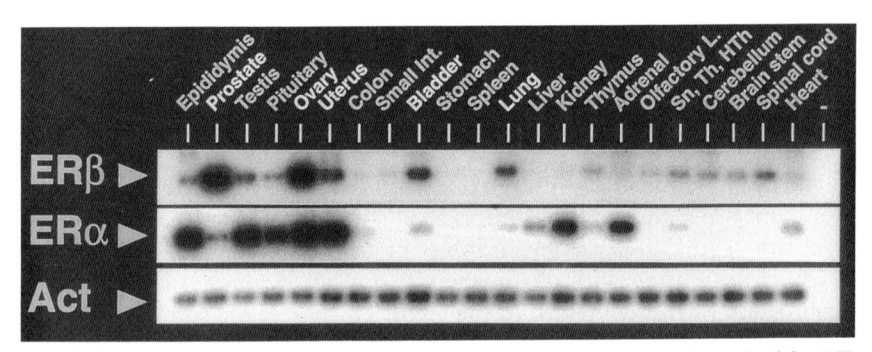

FIGURE 9 Rat tissue distribution of ERα-mRNA and ERβ-mRNA determined by RT-PCR. Shown are autoradiograms of blots after hybridization with oligonucleotide probes specific for ERβ (top), ERα (middle), and actin (bottom). Sn, substantia nigra, preopticus; Th, thalamus; Hth, hypothalamus; Olfactory L., olfactory lobes; Small Int., small intestine.

brain sections tested were negative for ERα mRNA. Ovary and uterus, which are known to contain high amounts of ER protein, clearly express both ER subtypes. All organs from male rats previously described to display specific binding of estradiol to an 8S cytosolic protein, i.e., lung, adrenal, pituitary, prostate, epididymis, and testis, showed clear expression of either or both ER subtype mRNA (van Beurden-Lamers et al., 1974; Morishige and Uetake, 1978).

The widespread use of tamoxifen for the treatment of metastatic breast cancer made it possible to evaluate the chronic toxicology of this drug. Among others, a significant increase in the development of endometrial cancers was demonstrated in women treated with tamoxifen for longer periods (Grainger and Metcalfe, 1996, and references therein), most likely as a consequence of the agonist-like activity of tamoxifen in uterine cells. Furthermore, tamoxifen is a strong estrogen-like agonist for bone density maintenance in rats and women, yet at the same time tamoxifen is a full antagonist of estrogen-stimulated breast cancer cell proliferation (Katzenellenbogen et al., 1996, and references therein). The coexpression of ERα and ERβ mRNA in various tissues and cells (Fig. 9, and G. Kuiper and K. Grandien, unpublished observations) is of particular interest in view of these as yet poorly understood tissue-specific agonist/antagonist activities of tamoxifen and other anti-estrogens. In general, different target tissues may respond differently to the same hormonal stimulus because they have a different composition of receptors. This could include variations in the ratio and/or concentration of receptor subtypes. Different ratios and concentrations of the ERα and ERβ protein in tissues and tumor cells could constitute a hitherto unrecognized mechanism involved in the tissue- and cell-specific effects of ER ligands. More specifically, different patterns of ER protein homo- and heterodimerization on different target gene response elements could lead to differential responses. Investigations on the interactions of ERα and ERβ protein with different target gene promoters in different (tumor) cell lines could be a first step in the evaluation of the described hypothesis.

VI. Physiological Role of ERβ Protein

A. ERβ in the ERα Knock-Out Mouse

Since the ERβ protein has been discovered very recently no definite statements regarding its unique function(s) can be made; in this section, some interesting observations and speculations will be discussed. Estrogens synergize with follicle-stimulating hormone (FSH) in ovarian weight augmentation, which is associated with a pronounced proliferation of granulosa cells, and the growth of small- and medium-sized follicles (Richards, 1994).

A mutant mouse line without a functional ERα protein was created and assessed for estrogen responsiveness (Lubahn et al., 1993). Female mice were infertile and showed hypoplastic uteri and hyperemic ovaries with no detectable corpora lutea. The fact that disruption of the ERα gene did not eliminate the ability of small follicles to grow, as was evident from the presence of secondary and antral follicles in the knock-out mouse ovary, pointed to the possible existence of alternative estrogen receptors mediating the intraovarian effects of estradiol. Indeed, in some tissues from the ERα knock-out mice residual binding of estradiol with a K_d of 0.7 nM could be measured (Lubahn et al., 1993; Couse et al., 1995). The authors ascribed this to a possible "splicing over" event (Couse et al., 1995), resulting in the production of a smaller mutant ERα protein that could be the source of the residual estradiol binding. In retrospect it seems more likely that the remaining estradiol binding is caused by the presence of ERβ protein. The mouse and human homologues of rat ERβ have recently been cloned in our laboratory (E. Enmark, G. Bertilsson, K. Grandien, and K. Pettersson, unpublished observations).

We (Byers et al., 1997) have investigated the expression and regulation of the two ER subtypes in the rat ovary. Examination of ERβ mRNA expression at the cellular level, by in situ hybridization, showed that ERβ mRNA is expressed preferentially in granulosa cells of small, growing, and preovulatory follicles. In contrast, similar studies for ERα mRNA revealed expression at a low level throughout the ovary with no particular cellular localization. Thus, the ovarian responsiveness to estrogen in the ERα knock-out mice probably results from the expression of ERβ protein. In the rat pituitary high levels of ERα mRNA and very low levels of ERβ mRNA are expressed (Fig. 9). Therefore, the fact that in the ERα knock-out mouse the ovarian follicles do not develop beyond the FSH-independent stage could be explained by a dysregulation in the release of FSH by the pituitary, and not by the absence of ERα protein in the ovary. Estrogens sensitize the pituitary to gonadotropin-releasing hormone (GnRH) input, and the sexually dimorphic transcriptional responses to GnRH of gonadotroph cells require chronic exposure to estradiol (Colin et al., 1996). The ERβ knock-out mice will probably show no follicular growth at all, although this remains to be proven.

The presence of estrogen receptors (ERα) in preimplantation mouse embryos (Hou and Gorski, 1993; Hou et al., 1996) and the absence of reported human ER mutations have been interpreted as indications for an essential role of estrogens during embryonic development. This view seemed to be challenged by the survival of the ERα knock-out mice (Lubahn et al., 1993) and the existence of a male with (partial) estrogen resistance caused by a mutation in the ERα gene (Smith et al., 1994). With the identification of a second ER protein, the unexpected viability of the ERα knock-out mouse and human could also be explained by complementation through

ERβ protein during embryo development. On the other hand, the existence of a karyotypically female patient with pseudohermaphroditism caused by a null mutation in the aromatase cytochrome P450 gene (Ito et al., 1993) raises further questions about the validity of the lethality hypothesis.

Although no data on ERβ mRNA expression or protein concentration in tissues of the ERα knock-out mouse are available, the possible presence of ERβ protein should be kept in mind when interpreting experiments using the ERα knock-out mouse (Merchenthaler et al., 1996). In the brain of the ERα knock-out mouse specific binding of estradiol and modulation of progesterone receptor gene expression by estradiol was observed (Merchenthaler et al., 1996). Again, this is most likely caused by the presence of ERβ protein, since the ERβ mRNA is broadly expressed in the rat brain (Fig. 9) and probably also in mouse brain.

B. ERβ and the Prostate

Estrogen receptors are present in human and rat prostate, as evidenced by ligand-binding studies (Ekman et al., 1983; van Beurden-Lamers et al., 1974; Jung-Testas et al., 1981). Estrogens, in addition to androgens, are implicated in the growth of the prostate (Griffiths et al., 1991), and estrogens have been implicated in the pathogenesis of benign prostatic hyperplasia (Habenicht et al., 1993). Estrogen treatment of neonatal rats was shown to down-regulate androgen receptor expression in the epithelial cells of all the three prostate lobes, thus leading to an overall prostate growth retardation (Prins and Birch, 1995). Various means of androgen deprivation are in use for the treatment of prostate cancer (Furr et al., 1996). Diethylstilbestrol (DES) treatment leads to long-term suppression of serum testosterone due to inhibition of LH release by the pituitary, although direct effects of DES on the rat and human prostate are not excluded (Yamashita et al., 1996; Cox and Crawford, 1995). In cotransfection experiments ERα was capable of inhibiting transcriptional activation of MMTV-LTR-CAT reporter constructs by the androgen receptor in an estrogen-dependent manner (Kumar et al., 1994). The coexpression of androgen receptor and ERβ protein in the secretory epithelial cells of the rat prostate is of interest in this regard, and the putative inhibitory effect of ERβ should be investigated in more detail in normal prostate cells as well as in prostate tumor cells.

So far, only one male with (partial) estrogen resistance due to a mutation in the ERα gene has been described (Smith et al., 1994). This patient had normal male genitalia with bilateral descended testis and a normally sized prostate gland, indicating that the absence of ERα protein in the testis and prostate could be compensated for by the presence of ERβ or that estrogens are not essential for normal prostate and testis development. On the other hand, this patient had severe osteoporosis, unfused epiphyses, and continuing linear growth in adulthood, demonstrating that estrogens acting through ERα play a crucial role in bone development and maintenance in the male.

C. ERβ and Environmental Endocrine Disruptors

Abnormal sexual development in reptiles in Florida as well as the increasing incidence of certain human reproductive tract abnormalities (hypospadias, cryptorchidism), testicular cancer, and declining male reproductive health has been associated with increased exposure to and body burden of so-called estrogenic environmental chemicals (Kelce *et al.*, 1995; Arnold *et al.*, 1996; Jensen *et al.*, 1995). These effects from environmental chemicals, such as the pesticides methoxychlor and chlordecone, the plastics ingredient bisphenol A, and the detergent component alkylphenol, are postulated to be mediated via the ER since these compounds have estrogenic effects (increase of uterine weight) in female rats (Bicknell *et al.*, 1995; Hammond *et al.*, 1979). The most critical period for exposure to these estrogenic environmental chemicals seems to be during the development of the reproductive tissues. Between 1945 and 1971, treatment of several million pregnant women with the synthetic estrogen DES led to an increase in the incidence of cryptorchidism and hypospadias as well as decreased sperm counts in the sons of these women (Jensen *et al.*, 1995, and references therein). The similarity between the effects of DES treatment and the reported adverse changes in male reproductive development and function is regarded as further evidence that increased exposure to environmental estrogenic chemicals during early fetal development is the underlying cause for the malformations and defects mentioned. The high expression of ERβ in male reproductive tissues is of particular interest in this regard. Bisphenol A and methoxychlor were able to inhibit the binding of estradiol by ERα and ERβ protein, and the interaction seemed to be stronger for the ERβ protein (Kuiper *et al.*, 1997).

Studies are underway to determine the cellular localization of ERβ in the developing rat testis, and during mouse embryogenesis. In addition the interaction of ERα and ERβ with a large group of suspected environmental estrogenic chemicals is under active investigation.

VII. Concluding Remarks

The ERβ protein displays high-affinity binding of estrogens and in a transactivation assay system activation of an ERE-containing reporter gene construct was measured. The detailed biological significance of the existence of two ER subtypes is at this moment unclear. Perhaps the existence of two ER subtypes provides, at least in part, an explanation for the selective actions of estrogens in different target tissues. The expression patterns of both subtypes are quite different, and this could be regarded as support for this hypothesis. The possible presence of ERβ protein should be kept in mind when interpreting experiments using the ERα knock-out mouse. The production of the ERβ knock-out mouse and hopefully also the ERα/ERβ double knock-out mouse will undoubtedly provide more detailed insight in the physiological role of each subtype in different tissues.

Acknowledgments

G.G.J.M.K. was supported, in part, by grants from the Netherlands Organization for Scientific Research (NWO) and from the Karolinska Institute. J.-Å.G. was supported by a grant from the Swedish Cancer Society.

References

Ali, S., Metzger, D., Bornert, J.-M., and Chambon, P. (1993). Modulation of transcriptional activation by ligand dependent phosphorylation of the human oestrogen receptor A/B region. *EMBO J.* **12,** 1153–1160.

Arnold, S. F., Obourn, J. D., Jaffe, H., and Notides, A. C. (1995). Phosphorylation of the human estrogen receptor on tyrosine 537 in vivo and by src family tyrosine kinases in vitro. *Mol. Endocrinol.* **9,** 24–33.

Arnold, S. F., Klotz, D. M., Collins, B. M., Vonier, P. M., Guilette, L. J., and McLachlan, J. A. (1996). Synergistic activation of estrogen receptor with combinations of environmental chemicals. *Science* **272,** 1489–1492.

Beato, M., Herrlich, P., and Schutz, G. (1995). Steroid hormone receptors: Many actors in search of a plot. *Cell (Cambridge, Mass.)* **83,** 851–857.

Beekman, J. M., Allan, G. F., Tsai, S. Y., Tsai, M.-J., and O'Malley, B. W. (1993). Transcriptional activation by the estrogen receptor requires a conformational change in the ligand binding domain. *Mol. Endocrinol.* **7,** 1266–1272.

Berger, J., Hauber, J., Hauber, R., Geiger, R., and Cullen, B. R. (1988). Secreted placental alkaline phosphatase: A powerful new quantitative indicator of gene expression in eukaryotic cells. *Gene* **66,** 1–10.

Berns, E. M. J. J., Rommerts, F. G. G., and Mulder, E. M. (1985). Rapid and sensitive detection of oestrogen receptors in cells and tissue sections by autoradiography with [^{125}I]-oestradiol. *Histochem. J.* **17,** 1185–1196.

Bicknell, R. J., Herbison, A. E., and Sumpter, J. P. (1995). Oestrogenic activity of an environmentally persistent alkylphenol in the reproductive tract but not the brain of rodents. *J. Steroid Biochem. Mol. Biol.* **54,** 7–9.

Brolin, J., Skoog, L., and Ekman, P. (1992). Immunohistochemistry and biochemistry in detection of androgen, progesterone, and estrogen receptors in benign and malignant human prostatic tissue. *Prostate* **20,** 281–295.

Bunone, G., Briand, P.-A., Miksicek, J., and Picard, D. (1996). Activation of the unliganded estrogen receptor by EGF involves the MAP kinase pathway and direct phosphorylation. *EMBO J.* **15,** 2174–2183.

Byers, M., Kuiper, G. G. J. M., Gustafsson, J.-Å., and Park-Sarge, O.-K. (1997). Estrogen receptor-β mRNA expression in the rat ovary: Down regulation by gonadotropins. *Mol. Endocrinol.* **11,** 172–182.

Castles, C. G., and Fuqua, S. A. W. (1996). Alterations within the estrogen receptor in breast cancer. *In* "Hormone-dependent cancer" (J. R. Pasqualini and B. S. Katzenellenbogen, eds.), pp. 81–105. Dekker, New York.

Ciocca, D. R., and Vargas-Roig, L. M. (1995). Estrogen receptors in human nontarget tissues: Biological and clinical implications. *Endocr. Rev.*, **16**, 35–62.

Clark, J. H., Schrader, W. T., and O'Malley, B. W. (1992). Mechanisms of action of steroid hormones. *In* "Textbook of Endocrinology" (J. D. Wilson and D. W. Foster, eds.), pp. 35–90. Saunders, Philadelphia.

Colin, I. M., Bauer-Dantoin, A. C., Sundaresan, S., Kopp, P., and Jameson, J. L. (1996). Sexually dimorphic transcriptional responses to gonadotropin-releasing hormone require chronic *in vivo* exposure to estradiol. *Endocrinology (Baltimore)* **137**, 2300–2307.

Couse, J. F., Curtis, S. W., Washburn, T. F., Lindzey, J., Golding, T. S., and Lubahn, D. B. (1995). Analysis of transcription and estrogen insensitivity in the female mouse after targeted disruption of the estrogen receptor gene. *Mol. Endocrinol.* **9**, 1441–1454.

Cox, R. L., and Crawford, E. D. (1995). Estrogens in the treatment of prostate cancer. *J. Urol.* **154**, 1991–1998.

Danielian, P. S., White, R., Lees, J. A., and Parker, M. (1992). Identification of a conserved region required for hormone dependent transcriptional activation by steroid hormone receptors. *EMBO J.* **11**, 1025–1031.

Ehara, H., Koji, T., Deguchi, T., Yoshii, A., Nakano, M., Nakane, P. K., and Kawada, Y. (1995). Expression of estrogen receptor in diseased human prostate assessed by non-radioactive in situ hybridization and immunohistochemistry. *Prostate* **27**, 304–313.

Ekman, P., Barrack, E. R., Greene, G., Jensen, E. V., and Walsh, P. C. (1983). Estrogen receptors in human prostate: Evidence for multiple binding sites. *J. Clin. Endocrinol. Metab.* **57**, 166–176.

Enmark, E., Kainu, T., Pelto-Huikko, M., and Gustafsson, J.-Å. (1994). Identification of a novel member of the nuclear receptor superfamily which is closely related to rev-erbA. *Biochem. Biophys. Res. Commun.* **204**, 49–56.

Farhat, M. Y., Lavigne, M. C., and Ramwell, P. W. (1996). The vascular protective effects of estrogen. *FASEB J.* **10**, 615–624.

Fawell, S. E., Lees, J. A., White, R., and Parker, M. G. (1990). Characterization and colocalization of steroid binding and dimerization activities in the mouse estrogen receptor. *Cell (Cambridge, Mass.)* **60**, 953–962.

Furr, B. J. A., Blackledge, G. R. P., and Cockshott, I. D. (1996). Casodex: Preclinical and clinical studies. *In* "Hormone-dependent Cancer" (J. R. Pasqualini, and B. S. Katzenellenbogen, eds.), pp. 397–424. Dekker, New York.

Garcia, M., and Rochefort, H. (1979). Evidence and characterization of the binding of two ^{3}H-labeled androgens to the estrogen receptor. *Endocrinology (Baltimore)* **104**, 1797–1804.

Giguere, V. Yang, N., Segui, P., and Evans, R. M. (1988). Identification of a new class of steroid hormone receptors. *Nature (London)* **331**, 91–94.

Gleave, M. E., and Chung, L. W. K. (1995). Stromal-epithelial interaction affecting prostate tumour growth and hormonal responsiveness. *Endocr. Relat. Cancer* **2**, 243–265.

Grainger, D. J. and Metcalfe, J. C. (1996). Tamoxifen: Teaching an old drug new tricks? *Nat. Med.* **2**, 381–385.

Green, S., Walter, P., Kumar, V., Krust, A., Bornert, J.-M., Argos, P., and Chambon, P. (1986). Human oestrogen receptor cDNA: Sequence, expression and homology to v-erb-A. *Nature (London)* **320**, 134–139.

Greene, G. L., and Jensen, E. V. (1982). Monoclonal antibodies as probes for estrogen receptor detection and characterization. *J. Steroid Biochem.* **16**, 353–359.

Greene, G. L., Gilna, P., Waterfield, M., Baker, A., Hort, Y., and Shine, J. (1986). Sequence and expression of human estrogen receptor complementary cDNA. *Science* **231**, 1150–1154.

Griffiths, K., Davies, P., Eaton, C. L., Harper, M. E., Turkes, A., and Peeling, W. B. (1991). Endocrine factors in the initiation, diagnosis, and treatment of prostatic cancer. *In* "Endocrine-dependent Tumors" (K. D. Voigt and C. Knabbe, eds.) pp. 83–125.

Habenicht, U.-F., Tunn, U. W., Senge, T., Schröder, F. H., Schweikert, H. U., Bartsch, G., and El Etreby, M. F. (1993). Management of benign prostatic hyperplasia with particular emphasis on aromatase inhibitors. *J. Steroid Biochem. Mol. Biol.* **44,** 557–563.

Hähnel, R., Twaddle, E., and Ratajczak, T. (1973). The specificity of the estrogen receptor of human uterus. *J. Steroid Biochem.* **4,** 21–31.

Hammond, B., Katzenellenbogen, B. S., Krauthammer, N., and McConnell, J. (1979). Estrogenic activity of the insecticide chlordecone and interaction with uterine estrogen receptors. *Proc. Natl. Acad. Sci. U.S.A.* **76,** 6641–6645.

Harlow, K. W., Smith, D. N., Katzenellenbogen, J. A., Greene, G. L., and Katzenellenbogen, B. S. (1989). Identification of cysteine 530 as the covalent attachment site of an affinity-labeling estrogen and antiestrogen in the human estrogen receptor. *J. Biol. Chem.* **264,** 17476–17485.

Hegy, G. B., Shackleton, C. H. L., Carlquist, M., Bonn, T., Engström, O., Sjöholm, P., and Witkowska, H. E. (1996). Carboxymethylation of the human estrogen receptor ligand-binding domain-estradiol complex: HPLC/ESMS peptide mapping shows that cysteine 447 does not react with iodoacetic acid. *Steroids* **61,** 367–373.

Hild-Petito, S., Stouffer, R. L., and Brenner, R. M. (1988). Immunocytochemical localization of estradiol and progesterone receptors in the monkey ovary throughout the menstrual cycle. *Endocrinology (Baltimore)* **123,** 2896–2904.

Hochberg, R. B., and Rosner, W. (1980). Interaction of 16α-[^{125}I]iodo-estradiol with estrogen receptor and other steroid-binding proteins. *Proc. Natl. Acad. Sci. U.S.A.* **77,** 328–332.

Hou, Q., and Gorski, J. (1993). Estrogen receptor and progesterone receptor genes are expressed differentially in mouse embryos during preimplantation development. *Proc. Natl. Acad. Sci. U.S.A.* **90,** 9460–9464.

Hou, Q., Paria, B. C., Mui, C., Dey, S. K., and Gorski, J. (1996). Immunolocalization of estrogen receptor protein in the mouse blastocyst during normal and delayed implantation. *Proc. Natl. Acad. Sci. U.S.A.* **93,** 2376–2381.

Huggins, C. V., Jensen, E. V., and Cleveland, A. S. (1954). Chemical structure of steroids in relation to promotion of growth of the vagina and uterus of the hypophysectomized rat. *J. Exp. Med.* **100,** 225–236.

Ignar-Trowbridge, D. M., Pimentel, M., Parker, M. G., McLachlan, J. A., and Korach, K. S. (1996). Peptide growth factor cross-talk with the estrogen receptor requires the A/B domain and occurs independently of protein kinase C or estradiol. *Endocrinology (Baltimore)* **137,** 1735–1744.

Ito, Y., Fisher, C., Conte, F. A., Grumbach, M. M., and Simpson, E. R. (1993). Molecular basis of aromatase deficiency in an adult female with sexual infantilism and polycystic ovaries. *Proc. Natl. Acad. Sci. U.S.A.* **90,** 11673–11677.

Jensen, T. K., Toppari, J., Keiding, N., and Skakkebaek, N. E. (1995). Do environmental estrogens contribute to the decline in male reproductive health? *Clin. Chem. (Winston-Salem, N. C.)* **41,** 1896–1901.

Jung-Testas, I., Groyer, M.-T., Bruner-Lorand, J., Hechter, O., Baulieu, E.-E., and Robel, P. (1981). Androgen and estrogen receptors in rat ventral prostate epithelium and stroma. *Endocrinology (Baltimore)* **109,** 1287–1289.

Kato, S., Endoh, H., Masuhiro, Y., Kitamoto, T., Uchiyama, S., Sasaki, H. *et al.* (1995). Activation of the estrogen receptor through phosphorylation by mitogen-activated protein kinase. *Science* **270,** 1491–1494.

Katzenellenbogen, J. A., O'Malley, B. W., and Katzenellenbogen, B. S. (1996). Tripartite steroid hormone receptor pharmacology: Interaction with multiple effector sites as a basis for the cell- and promoter- specific action of these hormones. *Mol. Endocrinol.* **10,** 119–131.

Kawashima, M., and Greenwald, G. S. (1993). Comparison of follicular estrogen receptors in rat, hamster and pig. *Biol. Reprod.* **48,** 172–179.

Kelce, W. R., Stone, C. R., Laws, S. C., Gray, L. E., Kemppainen, J. A., and Wilson, E. M. (1995). Persistent DDT metabolite p,p'-DDE is a potent androgen receptor antagonist. *Nature (London)* **375,** 581–585.

Koehorst, S. G. A., Jacobs, H. M., Thijssen, J. H. H., and Blankenstein, M. A. (1993). Wild type and alternatively spliced estrogen receptor messenger RNA in human meningioma tissue and MCF-7 breast cancer cells. *J. Steroid Biochem. Mol. Biol.* **45,** 227–233.

Koike, S., Sakai, M., and Muramatsu, M. (1987). Molecular cloning and characterization of rat estrogen receptor cDNA. *Nucleic Acids Res.* **15,** 2499–2513.

Kraus, W. L., McInerney, E. M., and Katzenellenbogen, B. S. (1995). Ligand-dependent, transcriptional productive association of the amino- and carboxyl-terminal regions of a steroid hormone nuclear receptor. *Proc. Natl. Acad. Sci. U.S.A.* **92,** 12314–12318.

Krust, A., Green, S., Argos, P., Kumar, V., Walter, P., Bornert, J., and Chambon, P. (1986). The chicken oestrogen receptor sequence: Homology with v-erb A and the human oestrogen and glucocorticoid receptors. *EMBO J.* **5,** 891–897.

Kudolo, G. B., Elder, M. G., and Myatt, L. (1984). A novel oestrogen-binding species in rat granulosa cells. *J. Endocrinol.* **102,** 83–91.

Kuiper, G. G. J. M., and Brinkmann, A. O. (1994). Steroid hormone receptor phosphorylation: Is there a physiological role? *Mol. Cell. Endocrinol.* **100,** 103–107.

Kuiper, G. G. J. M., Enmark, E., Pelto-Huikko, M., Nilsson, S. and Gustafsson, J.-Å. (1996). Cloning of a novel estrogen receptor expressed in rat prostate and ovary. *Proc. Natl. Acad. Sci. U.S.A.* **93,** 5925–5930.

Kuiper, G. G. J. M., Carlsson, B., Grandien, K., Enmark, E., Häggblad, J., Nilsson, S. and Gustafsson, J.-Å. (1997). Comparison of the ligand binding specificity and transcript tissue distribution of estrogen receptors α and β. *Endocrinology (Baltimore)* **3,** 863–870.

Kumar, M. V., Leo, M. E. and Tindall, D. J. (1994). Modulation of androgen receptor transcriptional activity by the estrogen receptor. *J. Androl.* **15,** 534–542.

Laudet, V., Hänni, C., Coll, J., Catzeflis, F., and Stehelin, D. (1992). Evolution of the nuclear receptor gene superfamily. *EMBO J.* **11,** 1003–1013.

Lees, J. A., Fawell, S. E., and Parker, M. G. (1989). Identification of two transactivation domains in the mouse oestogen receptor. *Nucleic Acids Res.* **17,** 5477–5487.

Liu, W., Wang, J., Sauter, N. K., and Pearce, D. (1995). Steroid receptor heterodimerization demonstated in vitro and in vivo. *Proc. Natl. Acad. Sci. U.S.A.* **92,** 12480–12484.

Lubahn, D. B., Moyer, J. S., Golding, T. S., Couse, J. F., Korach, K. S., and Smithies, O. (1993). Alteration of reproductive function but not prenatal sexual development after insertional disruption of the mouse estrogen receptor gene. *Proc. Natl. Acad. Sci. U.S.A.* **90,** 11162–11166.

Mahendroo, M. S., Cala, K. M., and Russell, D. W. (1996). 5α-reduced androgens play a key role in murine parturition. *Mol. Endocrinol.* **10,** 380–392.

Mangelsdorf, D. J., Thummel, C., Beato, M., Herrlich, P., Schutz, G., Umesono, K. Blumberg, B., Kastner, P., Mark, M., Chambon, P., and Evans, R. M. (1995). The nuclear receptor superfamily: The second decade. *Cell (Cambridge, Mass.)* **83,** 835–839.

Martinez, E., Givel, F. and Wahli, W. (1987). The estrogen responsive element as an inducible enhancer: DNA sequence requirements and conversion to a glucocorticoid-reponsive element. *EMBO J.* **6,** 3719–3727.

Merchenthaler, I., Shughrue, P. J., Lubahn, D. B., Negro-Vilar, A., and Korach, K. S. (1996). Estrogen responses in estrogen receptor-disrupted mice: An in vivo autoradiographic and in situ hybridisation study. *Program Int. Congr. Endocrinol. 10th,* San Francisco, p. 744 (abstr.).

Morishige, W. K., and Uetake, C.-A. (1978). Receptors for androgen and estrogen in the rat lung. *Endocrinology (Baltimore)* **102,** 1827–1836.

Mosselman, S., Polman, J. and Dijkema, R. (1996). ERβ: Identification and characterization of a novel human estrogen receptor. *FEBS Lett.* **392,** 49–53.

Murdoch, F. E., and Gorski, J. (1991). The role of ligand in estrogen receptor regulation of gene expression. *Mol. Cell. Endocrinol.* **78,** C103–C108.
Noteboom, W. D., and Gorski, J. (1965). Stereospecific binding of estrogens in the rat uterus. *Arch. Biochem. Biophys.* **111,** 559–568.
Notides, A. C. (1970). Binding affinity and specificity of the estrogen receptor of the rat uterus and anterior pituitary. *Endocrinology (Baltimore)* **87,** 987–992.
Pettersson, K., Svensson, K., Mattsson, R., Carlsson, B., Ohlsson, R., and Berkenstam, A. (1996). Expression of a novel member of estrogen response element-binding nuclear receptors is restricted to the early stages of chorion formation during mouse embryogenesis. *Mech. Dev.* **54,** 211–223.
Pettersson, K., Grandien, K., Kuiper, G. G. J. M., and Gustafsson, J.-Å. (1997). Mouse estrogen receptor β forms estrogen response element binding heterodimers with estrogen receptor α. *Mol. Endocrinol.,* in press.
Prins, G. S., and Birch, L. (1995). The developmental pattern of androgen receptor expression in rat prostate lobes is altered after neonatal exposure to estrogen. *Endocrinology (Baltimore)* **136,** 1303–1314.
Richards, J. A. (1994). Hormonal control of gene expression in the ovary. *Endocr. Rev.* **15,** 725–751.
Richards, J. S. (1975). Estrogen receptor content in rat granulosa cells during follicular development: Modification by estradiol and gonadotropins. *Endocrinology (Baltimore)* **97,** 1174–1184.
Sabbah, M., Radanyi, C., Redeuilh, G., and Baulieu, E.-E. (1996). The 90 kDa heat-shock protein (hsp90) modulates the binding of the oestrogen receptor to its cognate DNA. *Biochem. J.* **314,** 205–213.
Salomonsson, M., Carlsson, B., and Häggblad, J. (1994). Equilibrium hormone binding to human estrogen receptors in highly diluted cell extracts is non-cooperative and has a Kd of approximately 10 pM. *J. Steroid Biochem. Mol. Biol.* **50,** 313–318.
Seielstad, D. A., Carlson, K. E., Kushner, P. J., Greene, G. L., and Katzenellenbogen, J. A. (1995). Analysis of the structural core of the human estrogen receptor ligand binding domain by selective proteolysis/mass spectrometric analysis. *Biochemistry* **34,** 12605–12615.
Skipper, J., Young, L. J., Bergeron, J. M., Tetzlaff, M. T., Osborn, C. T., and Crews, D. (1993). Identification of an isoform of the estrogen receptor messenger RNA lacking exon four and present in the brain. *Proc. Natl. Acad. Sci. U.S.A.* **90,** 7172–7175.
Smith, C. L., Conneely, O. M., and O'Malley, B. W. (1995). Oestrogen receptor activation in the absence of ligand. *Biochem. Soc. Trans.* **23,** 935–939.
Smith, E. P., Boyd, J., Frank, G. R., Takahashi, H., Cohen, R. M., Specker, B., Williams, T. C., Lubahn, D. B., and Korach, K. S. (1994). Estrogen resistance caused by a mutation in the estrogen-receptor gene in a man. *N. Engl. J. Med.* **331,** 1056–1061.
Swillens, S., Waelbroeck, M., and Champeil, P. (1995). Does a radiolabelled ligand bind to a homogeneous population of non-interacting receptor sites? *Trends Pharm. Sci.* **16,** 151–155.
Tora, L., White, J., Brou, C., Tasset, D., Webster, N., Scheen, E., and Chambon, P. (1989). The human estrogen receptor has two independent non-acidic transcriptional activation functions. *Cell (Cambridge, Mass.)* **59,** 477–487.
Turner, R. T., Riggs, B. L., and Spelsberg, T. C. (1994). Skeletal effects of estrogen. *Endocr. Rev.* **15,** 275–296.
van Beurden-Lamers, W. M. O., Brinkmann, A. O., Mulder, E., and van der Molen, H. J. (1974). High-affinity binding of oestradiol by cytosols from testis interstitial tissue, pituitary, adrenal, liver and accessory sex glands of the male rat. *Biochem. J.* **140,** 495–502.
van Doorn, L. G., Poortman, J., Thijssen, J. H. H., and Schwarz, F. (1981). Actions and interactions of 5-androstene-3β,17β-diol and 17β-estradiol in the immature rat uterus. *Endocrinolgy (Baltimore)* **108,** 1587–1593.

White, R., Lees, J. A., Needham, M., Ham, J., and Parker, M. (1987). Structural organization and expression of the mouse oestrogen receptor gene. *Mol. Endocrinol.* **1**, 735–744.

Yamashita, A., Hayashi, N., Sugimura, Y., Cunha, G. R., and Kawamura, J. (1996). Influence of diethylstilbestrol, leuprolin, finasteride and castration on the lobar subdivisions of the rat prostate. *Prostate* **29**, 1–14.

Zilliacus, J., Carlstedt-Duke, J., Gustafsson, J.-Å., and Wright, A. P. H. (1994). Evolution of distinct DNA-binding specificities within the nuclear receptor family of transcription factors. *Proc. Natl. Acad. Sci. U.S.A.* **91**, 4175–4179.

Careen K. Tang
Marc E. Lippman[1]

Lombardi Cancer Center
Georgetown University Medical Center
Washington, DC 20007

EGF Family Receptors and Their Ligands in Human Cancer

Abnormalities in the expression, structure, or activity of proto-oncogene products contribute to the development and maintenance of malignant phenotypes. Amplification or overexpression of epidermal growth factor (EGF) family receptors is frequently implicated in human cancer. In addition, the EGF family ligands are thought to play significant roles in the genesis or progression of a number of human malignancies. Understanding the function, biology, and interactions of these growth factor receptors and their ligands will have important implications for the detection and treatment of human cancer. This review will summarize current knowledge of the involvement of EGF family receptors and their ligands in human neoplasia. It will also provide information on the clinical applications that could result from selective targeting of these receptors and ligands. A combination of conventional therapies and molecular gene therapies, such as those covered here, could eventually lead to a new dimension in cancer therapy.

[1] To whom correspondence should be addressed at E501 New Research Building, Lombardi Cancer Center, Georgetown University Medical Center, 3970 Reservoir Road NW, Washington, DC 20007. Fax: (202) 687-6402.

I. Introduction

In the last decade, numerous studies have indicated that growth factors and their receptors play an important role in cancer biology. Of the receptors, the epidermal growth factor receptor (EGFR) family is the most frequently implicated in human cancers. This class I subfamily is composed of four members identified to date: EGFR/p170^{erbB-1} (Savage et al., 1972; Wrann and Fox, 1979; Lin et al., 1984; Ullrich et al., 1984; Haley et al., 1987), HER2/p185^{erbB-2}/neu (Schechter et al., 1984, 1985; Coussens et al., 1985; Yamamoto et al., 1986; Bargmann et al., 1986a), HER3/p160^{erbB}-3 (Kraus et al., 1989; Plowman et al., 1990a), and HER4/p180^{erbB-4} (Plowman et al., 1993a). All EGF-like receptors are transmembrane glycoproteins with intrinsic tyrosine kinase activity (Stern and Kamps, 1988; Connelly and Stern, 1990; Stern et al., 1986). They are activated by binding of their respective ligands, and are implicated in the autocrine/paracrine growth of normal and malignant epithelial cells (Adamon, 1987).

There are at least 15 different agonists for ErbB family receptors, all of which exert their function by binding to their respective receptor at the cell surface. These EGF-related peptides can be divided into two groups, depending on their binding specificities. One group of ligands binds predominantly to the EGFR. It includes EGF, transforming growth factor α (TGF-α), amphiregulin (AR), betacellulin (BTC), and heparin-binding EGF-like growth factor (HB-EGF). It also includes a series of DNA pox virus-derived peptides such as vaccinia virus growth factor (VGF), shope fibroma growth factor (SFGF), and myxoma virus growth factor (MGF) (Salomon et al., 1995). Each of these peptides competes with EGF for receptor binding. A second group of EGF-related peptides, the product of a single gene, is composed of the variously named neu differentiation factors: (NDF)/heregulin (HRG), gp30, acetylcholine-receptor inducing activity (ARIA), and glial growth factor (GGF). These ligands have been found to bind directly to ErbB-4 with high affinity. They also bind to ErbB-3, with lower affinity.

This review will describe our current knowledge of the members of the c-erbB family of receptors, their ligands, and their involvement in human neoplasia. It will also present some information about the targeting of these receptors and ligands for potential antitumor therapy.

II. EGF Family Receptors

A. Overview

Epidermal growth factor was first identified over 30 years ago from extracts of mouse submaxillary glands (Savage et al., 1972). This small peptide was subsequently shown to have growth-promoting effects (Savage *et al.*, 1972). Isolation of EGF led to the purification of the epidermal growth factor receptor (EGFR/c-erbB-1) from the squamous carcinoma cell line A431 (Wrann and Fox, 1979). In 1984, the EGFRs cDNA sequence (Lin *et al.*,1984; Ullrich *et al.*, 1984) and genomic structure (Haley *et al.*, 1987) were determined.

Following the EGFR discovery, another related putative receptor, c-erbB-2/neu/HER2 proto-oncogene, was identified (Schechter *et al.*, 1984, 1985; Coussens *et al.*, 1985; Yamamoto *et al.*, 1986; Bargmann *et al.*, 1986b). More recently, c-erbB-3/HER3 and c-erbB-4/HER4 were discovered by low-stringency probing of human cDNA libraries with sequences from the EGFR gene or its avian viral homologue v-erbB (Kraus *et al.*,1989; Plowman *et al.*, 1990b, 1993b).

Each of the receptors in the EGFR family consists of an extracellular ligand-binding domain, a single amphipathic transmembrane domain, a short juxtamembrane portion followed by the protein tyrosine kinase domain, and a C-terminal autophosphorylation domain. Each receptor also presents several potential sites for glycosylation. The overall homology between all four receptors is 40 to 50% (Bargmann *et al.*, 1986a; Earp *et al.*, 1995). Figure 1 illustrates the percentage of homology between EGF receptor

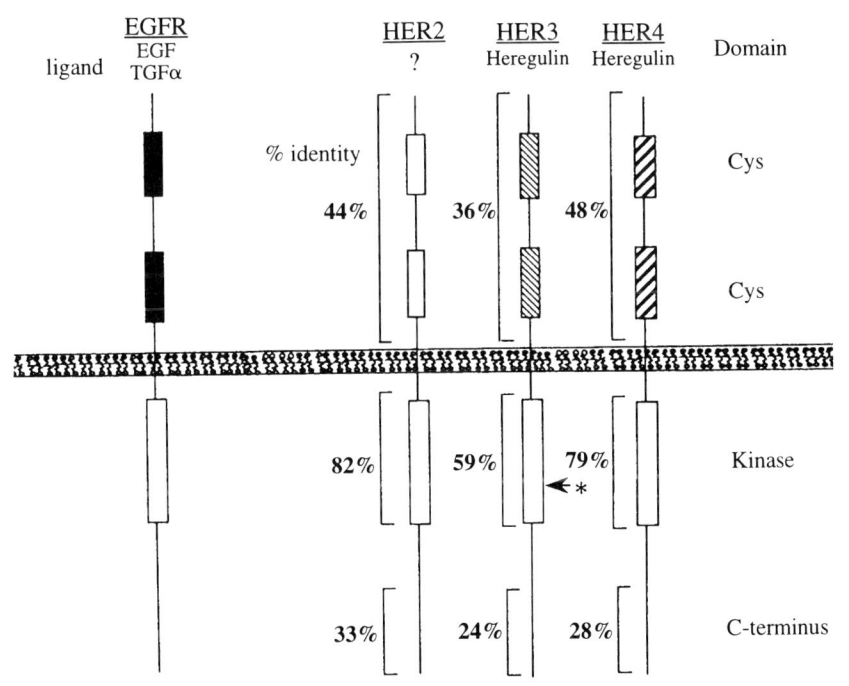

FIGURE 1 Diagrammatic representation of the structure and sequence homology in the different domains of EGF family receptors. (Reproduced by permission from *Breast Cancer Res. Treatment* **35**, 119, 1995.)

family members. Note that EGFR, ErbB-2, and ErbB-4 are homologous with the highest degree of consensus in the intracellular (290 amino acid) tyrosine kinase domain. This region exhibits >80% similarity between each of the members of the EGFR gene family (Hanks et al., 1988; Earp et al., 1995). The sequence dissimilarity in the carboxyl-terminal region implies potential divergence in downstream signaling. The following section will describe the structure and expression of each of the members of the EGFR family.

B. Structure, Expression, and Transforming Potential of EGF Family Receptors

1. Epidermal Growth Factor Receptor (EGFR)

The EGF receptor is encoded by the proto-oncogene *c-erb*B (Downward et al., 1977) and is the cellular homolog of the v-erbB oncogene. The v-erbB oncogene was originally identified as a transforming protein in avian erythroblastosis viruses (Downward et al., 1984; Salomon et al., 1995; Groenen et al., 1994). EGFR is a single-chain polypeptide (M_r 170,000) composed of three major domains: an extracellular ligand-binding (LB) domain, a transmembrane (TM) domain, and a cytoplasmic domain which contains the catalytic protein tyrosine kinase (PTK) and C-terminal regulatory domains. The external LB region of the EGFR can be divided into four subdomains based on amino acid homology. Subdomains I and III appear to be primarily responsible for ligand binding specificity (Lax et al., 1988), whereas subdomains II and IV are cysteine-rich domains containing 21 conserved cysteine residues. These domains may perform a role in the protein's tertiary structure. The ligand binding pocket is located near the junction of domains II and III. The TM region is distinguished by its hydrophobicity, while the PTK region is defined by its structural homology with other kinases (Hanks et al., 1988). The mature receptor is glycosylated, and thus is transported through the Golgi apparatus before reaching plasma membrane.

EGFR is the first receptor in which ligand-dependent tyrosine kinase activation was demonstrated (Carpenter and Cohen, 1990; Carpenter and Wahl, 1990). Mutational analysis suggests that the C-terminus contains the three major important sites for receptor autophosphorylation and maximum biological activity: Y-1068, Y-1148, and Y-1173 (Downward et al., 1984; Velu et al., 1989; Helin et al., 1991). The C-terminal tail may also be involved in negative regulation and internalization (Chen et al., 1989; Wells et al., 1990). In most situations, the activated receptor/ligand complex is endocytosed and degraded within the lysosomes. There are some exceptions, such as hepatocytes, where the receptor is recycled into the cell membrane. Nevertheless, internalization is believed to be an essential process in the control of normal mitogenic signaling, because a truncated EGFR mutant with normal kinase activity possesses elevated transforming activity and is not internalized (Wells et al., 1990).

EGFR is expressed in a wide variety of adult rodent and human tissues. Exceptions are parietal endoderm, mature skeletal muscle, and hemopoietic tissues (Partanen, 1990). The normal ovary expresses low levels of EGFR mRNA.

Overexpression of EGFR has been implicated in many human cancers. In addition, structural alterations and rearrangements have been observed in cancer cells. Rearrangements in both the extracellular and intracellular domains of EGFR have been reported in the A431 vulalcarcinoma cell line. Rearrangement of EGFR has often been found in glioblastomas (Ekstrand et al., 1991; Di Carlo et al., 1992; Agosti et al., 1992; Chaffanet et al., 1992; Wong et al., 1987). These rearrangements often result in abnormal size—usually truncated—protein products (Wong et al., 1987). Truncation in turn results in ligand-independent tyrosine kinase activity, altered subcellular location, increased stability (Ekstrand et al., 1994, 1995), and enhanced tumorigenicity (Nishikawa et al., 1994). The behavior of overexpressed, structurally altered, and rearranged EGFR conforms closely to that described for several viral ErbB (v-ErbB) oncogenes (Downward et al., 1984).

The transforming capacity of EGFR has been analyzed in several different cell lines, including primary chicken fibroblasts, erythroblasts, NIH 3T3 cells, and murine hematopoietic cell lines. In chicken embryo fibroblasts, overexpression of normal human EGFR results in a ligand-dependent transformed phenotype. Overexpression of EGFR is capable of transforming NIH3T3 cells in the presence of EGF (Pierce, 1990). Expression of EGFR in 32D hematopoietic cells also confers the ability to utilize EGF for transduction of a mitogenic signal. Overexpression of either ligand or receptor in the absence of the other does not usually result in full neoplastic transformation in vitro (Pierce, 1990), but high expression of both components together can lead to transformation of a variety of cell types (Rosenthal et al., 1986; Di Marco et al., 1989; Watanabe et al., 1987; McGeady et al., 1984; Shankar et al., 1989). Interestingly, however, transgenic animal experiments suggest that overexpression of the EGF receptor does not transform cells in vivo even when their ligand is available (Merlino, 1990).

2. ErbB-2/Neu

The *neu* oncogene was initially identified in rat neuroglioblastomas (Bargmann et al., 1986b; Hung et al., 1986; Schechter et al., 1984, 1985). Subsequently, Bargmann and Weinberg discovered that a point mutation in the transmembrane region, generating a single amino acid substitution (Val-664—Glu), activates the c-ErbB-2 gene (Bargmann et al., 1986b). This alteration results in constitutive activity of its intrinsic kinase and in malignant transformation of the cells (Bargmann and Weinberg, 1988), possibly because the mutation stabilizes dimeric forms of ErbB-2 (Weiner et al.,

1989). The human counterpart of the *neu* gene was named HER-2 or *c-erb*B-2 (King *et al.*, 1985; Coussens *et al.*, 1985; Yamamoto *et al.*, 1986). The *neu* proto-oncogene (HER-2 or *c-erb*B-2) encodes a 185-kDa receptor tyrosine kinase.

Neu/ErbB-2 is highly homologous with, but distinct from, EGFR (Yamamoto *et al.*, 1986; Schechter *et al.*, 1984) and EGF does not bind to ErbB-2. The ErbB-2 transmembrane glycoprotein is present in many epithelial (lung, salivary gland, breast, pancreas, ovary, gastrointestinal tract, and skin) and neural tissues (Gullick *et al.*, 1987; Quirke *et al.*, 1989; Natali *et al.*, 1990; Press *et al.*, 1990). Although adult tissues generally exhibit a lower level of p185^{erbB-2} expression than the corresponding fetal tissues, p185^{erbB-2} expression levels are frequently elevated in certain human neoplasms and associated with poor prognosis. The mutation observed in the mouse *neu* gene has not been reported in human malignancies (Slamon *et al.*, 1989a). However, ectopic overexpression of ErbB-2 in rodent fibroblasts (NIH 3T3 cells) causes phenotypic transformation and tumorigenicity, even in the absence of ligands (Hudziak *et al.*, 1987; Di Fiore *et al.*, 1987). This contrasts with the behavior of EGFR, which is dependent on the presence of ligand(s) for transformation of NIH3T3 cells, and suggests that the transformation potential of ErbB-2 is greater than that of EGFR. This intrinsic characteristic of ErbB-2 is apparently relevant to human adenocarcinomas, which often display remarkable amplification and/or overexpression of ErbB-2 (Slamon *et al.*, 1989b). In addition, transgenic mice carrying either the normal or the mutated ErbB-2/neu gene develop tumors in a variety of tissues, including mammary epithelium (Muller *et al.*, 1988; Suda *et al.*, 1990). A transgenic mouse study has suggested that ErbB-2 gene expression induces long latency, metastatic breast tumors (Guy *et al.*, 1992). Recently, it has been demonstrated that mice homozygous for disruptions in the *neu* gene develop defects in the heart and nervous system, and die in utero at 10.5 days (Lee *et al.*, 1995).

Several reports have characterized a soluble ErbB-2 ECD protein found in the conditioned medium of SkBr-3 and BT-474 cells, which overexpress ErbB-2 (Alper *et al.*, 1990; Lin and Clinton, 1991; Zabrecky *et al.*, 1991). In 1993, Benz and colleagues isolated cDNA encoding a 100-kDa truncated ECD form of ErbB-2 from the above cell lines. This 2.3-kb-truncated transcript appears to be produced by an alterative RNA processing mechanism. Expression of this transcript in COS-1 cells produces both secreted and cytosolic forms of ErbB-2 ECD. Clinical analysis has detected ErbB-2 ECD protein in the blood of patients with ErbB-2-overexpressing tumors (Leitzel *et al.*, 1991).

Several lines of evidence shown that p185^{erbB-2} may have unusually complex activation pathways due to its homomeric and heteromeric associations within the EGFR family of receptors. In a later section, we will describe the details of these possible mechanisms.

3. ErbB-3

Recently, Aaronson's and Todaro's groups have cloned a third member of the ErbB-receptor tyrosine kinase family (Kraus et al., 1989; Plowman et al., 1990b). The predicted structure of the ErbB-3 protein is similar to that of the EGF receptor and ErbB-2. The mature protein is glycosylated extensively and has a molecular weight of 160 kDa. The genomic locus of ErbB-3 has been assigned to 12q13 (Kraus et al., 1993). Several other cancer-related genes have been mapped in this region of chromosome 12, in close proximity to ErbB-3. These include the melanoma associated antigen, ME491 (Hotta et al., 1988), several histone genes (Tripputi et al., 1986), the gene for lactalbumin (Davies et al., 1987), and two proto-oncogenes, INT1 (Turc-Carel et al., 1987) and GLI (Kinzler et al., 1987).

ErbB-3 is a unique type I receptor tyrosine kinase because its catalytic sequence contains several unusual amino acids. A sequence analysis of the amino acids encoded by the ErbB-3 gene reveals an overall 40–60% homology with other members of the ErbB family (Plowman et al., 1990a, 1993a), with the least similarity found in the predicted c-terminal domain of the protein. Strikingly, this divergent area is the kinase domain of the human ErbB-3 protein. It has deviations at three positions thought to be invariantly conserved in all known protein tyrosine kinases. Cys-721, His-740, and Asn-815 in ErbB-3 correspond to Ala, Glu, and Asp, in all other known protein tyrosine kinases (Plowman et al., 1990b; Hanks and Quinn, 1991). Analysis of the X-ray crystallographic structure of other protein kinases, such as protein kinase A, indicates that the conserved Glu residue plays a key role in recognition of the phosphates of MgATP, and that the conserved Asp residue functions as a catalytic base (Knighton et al., 1991a,b). Furthermore, the substitution of Asn for the conserved Asp residue in the Kit and Fps proteins abolishes their kinase activities (Tan et al., 1990; Moran et al., 1988). This variation in sequence suggests that ErbB-3 is an impaired kinase receptor. A second important difference in ErbB-3 is the presence of multiple repeats of a specific carboxyl-terminal tyrosine autophosphorylation site. This YXXM sequence is found seven times in ErbB-3 and is missing from other EGFR family members (Soltoff et al., 1994; Kim et al., 1994; Sun et al., 1991). The multiple presence of the YXXM motif strongly suggests that the ErbB-3 gene product may act as an efficient recruiter of PI-3 kinase, an enzyme which has been implicated in cellular transformation and mitogenesis (Ling et al., 1992; Escobedo and Williams, 1988). The repeated YXXM sequence may thus confer on ErbB-3 a specific signaling ability which is not found in other EGFR family members.

The cellular transformation potential of ErbB-3 has been demonstrated in the NIH 3T3 and Ba/F3 or 32D cell systems. Coexpression of either EGFR or ErbB-2 with ErbB-3 in NIH 3T3 cells can induce foci formation in the presence of HRG (Zhang et al., 1996). Recent studies have also shown that coexpression of ErbB-2 and ErbB-3 can stimulate mitogenesis in 32D cells in the presence of HRG. However, transformation by ErbB-3 relies on the coexpression of either EGFR or ErbB-2 (Riese II et al., 1996; Pinkas-Kramarski et al., 1996).

ErbB-3 is normally expressed in cells of epithelial and neuroectodermal origin, such as placenta, stomach, lung, kidney, and brain cells. However, it is not detectable in skin fibroblasts, skeletal muscle cells, or lymphoid cells (Kraus et al., 1989).

4. ErbB-4

In 1993, ErbB-4 cDNA was cloned from the breast cancer cell line MDA-MB-453 by using degenerate oligonucleotide primers designed on the basis of conserved amino acids in the tyrosine kinase domains of EGFR, ErbB-2, ErbB-3, and xmrk. ErbB-4 cDNA contains a single open reading frame that encodes 1308 amino acids, yielding a 180-kDa protein. ErbB-4 has all the structural features of the EGFR family of RTKs (Hudziak et al., 1987). The extracellular domain of ErbB-4 is most similar to that of ErbB-3. ErbB-4 also conserves all 50 cysteines present in the extracellular portion of other members of the EGFR family. The cytoplasmic juxtamembrane region of 37 amino acids shares the highest degree of homology with EGFR (73% amino acid identity). ErbB-4 lacks a site analogous to Thr-654 in EGFR, which is a major site for protein kinase C-induced phosphorylation. However, the EGFR's major EGF-stimulated mitogenesis-activating protein kinase phosphorylation site—Thr-669—is conserved. It is located at Thr-699 in ErbB-4 (Takishima et al., 1991).

Overexpression of ErbB-4 in NIH 3T3 cells enables the formation of foci in the absence of ligand. This transforming activity is further stimulated by the addition of HRG-β2 (Cohen et al., 1996). Gene knockout experiments have shown that homozygous mice exhibit defects in the development of the heart and nervous system similar to those observed in ErbB-2 and HRG gene knockout mice. As with the ErbB-2 and HRG knockouts, the mutant embryos died *in utero* at 10.5 days (Gassmann et al. 1995). This suggests that ErbB-2, ErbB-4, and HRG are essential for development (Meyer and Birchmeier, 1995).

High levels of ErbB-4 transcripts have been found in brain, heart, kidney, parathyroid, cerebellum, pituitary, spleen, testis, and breast tissues. Lower levels have been found in thymus, lung, salivary gland, and pancreatic tissues, and low or undetectable expression has been found in liver, prostate, ovary, adrenal, colon, duodenum, epidermis, and bone marrow tissues. Elevated ErbB-4 levels have found in some breast cancer cells (Plowman et al., 1990b).

Table I below summarizes chromosome, transcript, and protein information on the EGF family receptors.

TABLE I Chromosome, Transcript, and Protein Information on the EGF Family Receptors

Receptor	Chromosome localization	Transcript size (kb)	Protein size (kDa)
EGFR	7q12-13	5.6	170
ErbB-2	17q12-21.3	4.8	185
ErbB-3	12q11-13	6.2	160
ErbB-4	Not available	6.0	180

III. EGF-like Growth Factors

A. Overview

Growth factors have been implicated in differentiation and morphogenic processes. All the biological actions of growth factors are mediated by specific cell surface receptors that transduce the biochemical signal through stimulatory associations with cytoplasmic proteins. The fundamental function of the receptors for growth factors is to share catalytic function which results in the phosphorylation of tyrosine residues (Savage et al., 1973; Aaronson, 1991).

B. Common Structure of EGF-like Growth Factor Family Members

EGF growth factor family members (Table II), including EGF, TGF-α, heparin-binding EGF, AR, and BTC, share several common features. EGF-related growth factors are synthesized as large precursors with N-terminal

TABLE II Transcripts, Precursors, Processed Peptides, and Percent Homologies of These EGF-like Growth Factors

Ligand	Size of mRNA (kb)	No. of aa and M_r (in kDa) of processed peptide	Chromosome localization	Precursor
EGF	4.8	53 (6)	4q25	1217 aa (M_r ~170 kDa)
TGF-α	4.8	50 (5.6)	2q11-13	M_r ~22 kDa
AR	1.4	78–84 (22 to 26)	4q13-4q21	252 aa (M_r ~34, 36 kDa)
HB-EGF	2.5	86 (23) kDa	5q23	208 aa (M_r ~23kDa)
HRGs	1.8, 2.6, 6.8	228–241 (26)	8p12-q21	640, 645, 637, 231 aa (M_r ~45 kDa)
BTC	3.0	80 (32)	n/a	178 aa
CR-1	2.2	188 (36, 28)	3p21.3	188 aa

Note. aa, amino acid; M_r, molecular weight; n/a, data not available.

signal peptide sequences, and a six-cysteine EGF-like domain (Savage et al., 1973), with the cysteines characteristically spaced over a sequence of 35–40 amino acids (Carpenter and Cohen, 1979; Engel, 1989; Davis, 1990). These cysteines pair to form three disulfide bonds. This EGF-like domain is believed to have a role in mediating protein–protein interactions, and it is presumed that its structure is conserved in all other EGF-like repeats. Following the cysteine/EGF-like domain is a hydrophobic transmembrane domain, and a cytoplasmic domain of unknown function. NDF/HRG also belongs to the EGF family and shares similar molecular architecture and membrane topology (Ullrich and Schlessinger, 1990).

An important feature of all of the above peptides is that they are synthesized via large, membrane-bound, glycosylated precursors which have been shown to possess biological activity. The extracellular domains of these precursors are proteolytically processed to release the biologically active mature growth factors. Interestingly, the general structure of EGF-like precursors is conserved from lower organisms to mammalian cells.

C. Function of EGF-like Growth Factors

In general, tumor cells exhibit a reduction in their requirement for exogenously supplied growth factors to maintain their proliferation. This relaxation in growth factor dependency as compared to nontransformed cells may be due in part to the ability of tumor cells to synthesize and respond to endogenously produced growth factors. Tumor-derived growth factors may function via intracrine, juxtacrine (membrane-bound forms of growth factors activate the receptor on adjacent cells), autocrine (ligands secreted by the same cells), or paracrine (ligands secreted by other cells) mechanisms on cells that express the proper cognate receptor (Aaronson, 1991; Sporn and Roberts, 1992; Logan, 1990). The major known biological role of all these proteins, or their soluble extracellular portions, is to guide cell growth and differentiation through specific protein–protein interactions. Membrane-bound forms may interact with receptors on cell surfaces through a juxtacrine pathway, thereby serving as cell–cell adhesion molecules, as well as via cell–cell stimulation. Growth factors have been shown to be involved in regulating normal cellular proliferation. They have also been implicated in the initiation and/or maintenance of cellular transformation (Massague, 1983a,b, 1990).

The conserved (EGF-like) regions of EGF-like growth factors contain an essential motif for protein–protein interactions. The EGF-like domain is responsible for receptor recognition. In general, secreted EGF-like peptides bind to their target receptor and stimulate the intrinsic protein tyrosine kinase activity and autophosphorylation of the receptor. Receptor autophosphorylation then mediates the recruitment of specific intracellular proteins and triggers cascades of events that propagate the signal in the nucleus, culminating in a biological response (Bishop, 1991).

The activity or expression of EGF-like ligands can be controlled by various oncogenes and tumor suppressor genes that are involved in the ligands' intracellular signal transduction pathways. Reciprocally, EGF-like ligands can regulate the activity or expression of these oncogenes and tumor suppressor genes. The following section will describe the biological role of each of the growth factors in the EGF family.

D. Expression and Biological Role of EGF-like Growth Factors

1. Epidermal Growth Factor

EGF is a 53-amino-acid peptide (M_r 6 kDa) that is encoded by a 4.8-kb mRNA transcript from a gene that is 110 kb in length, contains 24 exons, and is located on human chromosome 4q25. EGF contains three disulfide bridges which form a triple-loop structure (the loops are labeled A, B, and C in Fig. 2). This triple-loop domain is involved in both the receptor-binding and biological stimulation activities of the molecule. EGF mRNA and protein are expressed in a number of adult tissues, especially in epithelial cells of the gastrointestinal tract (Kajikawa *et al.*, 1991). EGF has been found to stimulate the growth of both normal and transformed human mammary epithelial cells.

EGF binds to the EGF receptor and is a mitogen for a number of cell types. It has been reported that EGF carries out a variety of functions *in vitro* and *in vivo*, including stimulation of metabolite transport, activation

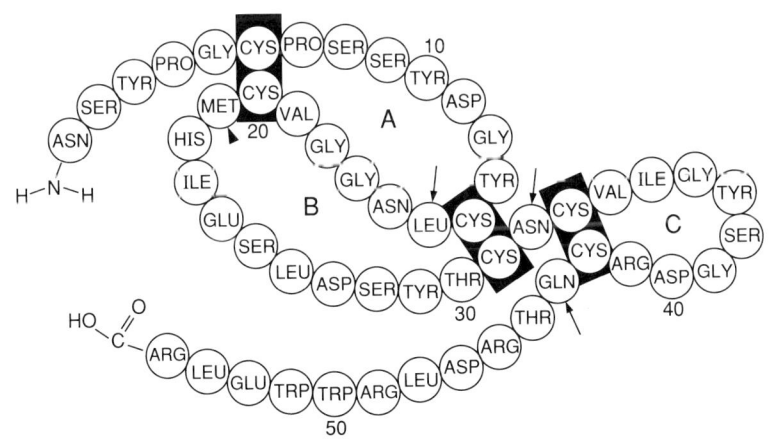

FIGURE 2 Illustration of the structure of mouse EGF. Three disulfide bridges form a triple-loop structure. The loops are labeled A, B, and C. (Reproduced by permission from *Proc. Natl. Acad. Sci. U.S.A.* **83**, 6367, 1986.)

of glycolysis, stimulation of production of RNA, protein, and DNA, enhancement of cell proliferation, alteration of cell morphology, and inhibition of gastric acid secretion.

Both EGF and TGF-α act as autocrine regulators of human breast cancer cell growth *in vitro* and *in vivo* (Madsen *et al.*, 1992; Clarke *et al.*, 1989). EGF mRNA has been detected in a majority of human breast cancer cell lines, as well as in 83% of breast cancer biopsies (L. C. Murphy *et al.*, 1990; Dotzlaw *et al.*, 1990). It has been reported that an inverse correlation between ER status and EGF protein expression is correlated with poor prognosis in breast carcinomas (Mizukami *et al.*, 1991). In poorly differentiated gastric carcinomas, EGF expression is detected at a frequency of 42% (Yasui *et al.*, 1990). Several reports also indicate that elevated EGF expression is observed in other human cancers, for example, in 86% of pleomorphic adenomas of the salivary glands (Yamahara *et al.*, 1988), 12% of pancreatic carcinomas (Barton *et al.*, 1991), 38% of prostatic carcinomas (Fowler *et al.*, 1988), and 30% of ovarian, endometrial, and cervical carcinomas (Bauknecht *et al.*, 1989b).

2. Transforming Growth Factor α

TGF-α was first found in the culture fluids of various oncogenically transformed cells (Todaro *et al.*, 1980; DeLarco and Todaro, 1978). TGF-α is related to EGF both structurally and functionally (Anzano *et al.*, 1983; Marquardt *et al.*, 1984; Massague, 1990). TGF-α is a 50-amino-acid peptide, has an M_r of 5.6 kDa, and exhibits a 30–40% sequence homology with EGF. The secondary structure is identical to that of EGF, containing the same disulfide-bound triple-loop motif. Figure 3 illustrates the structure of EGF-like domain of human TGF-α. TGF-α is believed to bind to, and function exclusively through, the EGF receptor. The different biological activities of TGF-α and EGF may be due to their ability to bind to different regions of the EGF receptor, and to differences in receptor internalization and degradation. TGF-α binds with a 100-fold higher affinity than EGF to the chicken EGF receptor (Lax *et al.*, 1988). In addition, a monoclonal antibody against the EGF receptor can block the binding of TGF-α but not EGF, suggesting that the two growth factors interact with different sites in the extracellular domain of the receptor, or cause different conformational changes in the receptor (Winkler *et al.*, 1989).

TGF-α has a potency similar to that of EGF as a mitogen for fibroblasts and as an inducer of epithelial development *in vivo* (Massague, 1990; Schreiber *et al.*, 1986; Tam *et al.*, 1985; Smith *et al.*, 1985). However, TGF-α is more potent than EGF as an angiogenic factor *in vivo* (R29, 58). For example, in epidermal keratinocyte cultures, TGF-α has a greater activity than EGF in promoting epidermal regeneration *in vivo* after topical application to burn wounds (Schultz *et al.*, 1987). TGF-α stimulates a variety of biological responses in cell culture and animal models, including mitogenesis

EGF Family Receptors and Their Ligands 125

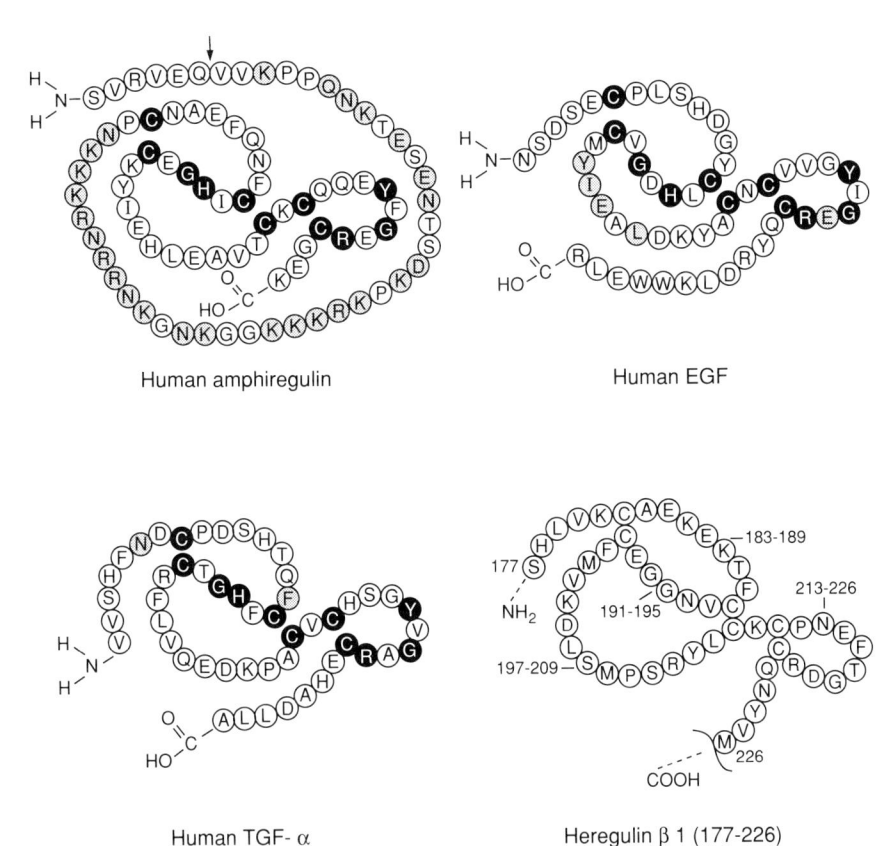

FIGURE 3 Schematic representation of four members of the EGF family.

(Anzano et al., 1983), tumor formation (Rosenthal et al., 1986), angiogenesis (Schreiber et al., 1986), bone resorption (Ibbotson et al., 1985), and wound healing (Schultz et al., 1987).

An increase in TGF-α synthesis and secretion occurs in several types of human carcinoma cell lines, in primary human tumors, and in fibroblasts and epithelial cell lines that have been transformed with a number of different oncogenes, such as point mutated c-Ha- or cKi-ras genes. However, TGF-α is also expressed during normal embryogenesis and is produced by a number of normal adult tissues, especially in regenerating or stem cell populations of epithelial cells.

In vivo transgenic mice overexpressing TGF-α developed benign skin lesions, and liver and mammary cancers (Jhappan et al., 1990; Sandgren et al., 1990).

Several clinical and experimental studies have demonstrated that TGF-α is an important modulator of the malignant progression of mammary epithelial cells in breast cancer (Normanno et al., 1994). In addition, overexpression was found in 60% of lung carcinomas (Kuniyasu et al., 1991; Sandgren et al., 1990; Liu et al., 1990; Liu and Tsao, 1993; Tateishi et al., 1991).

3. Betacellulin (BTC)

Betacellulin was originally isolated from the conditioned medium of a mouse pancreatic beta tumor cell line (Shing et al., 1993) derived from a transgenic mouse expressing the SV40 large T antigen gene under the control of the insulin promoter (Hanahan, 1985; Folkman, 1989). The cDNA encoding human BTC has been cloned from a cDNA library prepared from the MCF-7 human breast adenocarcinoma cell line (Sasada et al., 1993; Seno et al., 1996). The released form of BTC is composed of 80 amino acid residues exhibiting an apparent molecular size of about 32 kDa, with extensive glycosylation. BTC has been shown to promote the proliferation of epithelial and vascular smooth muscle cells, but not endothelial cells (Shing et al., 1993). Mouse BTC is also expressed in normal tissues, such as those of the lung, uterus, and kidney (Sasada et al., 1993). The expression of the human BTC gene in some tumor cells has also been described. The physiological and biochemical mechanism of action of BTC is still unclear.

4. Amphiregulin (AR)

AR was isolated from the conditioned medium of MCF-7 cells treated with phorbol ester (Shoyab et al., 1988, 1989). Both the structure and the function of AR are related to those of EGF and TGF-α. However, AR has several features which distinguish it from TGF-α. One of them is that AR has a hydrophilic 43-amino-acid extension rich in lysine and arginine residues at its N-terminus. This motif is usually associated with nuclear localization and DNA binding (Plowman et al., 1990b). Precursor sequences flanking mature AR lack homology with TGF-α cleavage sites, so processing may be mediated by different enzymes and thus subject to different regulatory factors. High expression of AR has been detected in normal placenta, testis, and ovary tissue. Pancreas, colon, and breast tissue also reveal significant levels of AR expression (L. D. Murphy et al., 1990).

AR binds to and activates the EGFR with low affinity when compared to either EGF or TGF-α (Shoyab et al., 1989; Johnson et al., 1991). AR has been described as a bifunctional growth modulator, with its functionality depending on the cell type and concentration of AR (Shoyab et al., 1988; Kenney et al., 1993). It can inhibit the growth of several human carcinoma cells in vitro, yet it induces the proliferation of human fibroblast, ovarian, mammary epithelial, and certain other tumor cell lines (Shoyab et al., 1988; Johnson et al., 1991, 1993; Cook et al., 1988). AR protein is expressed in 80% of primary breast carcinomas (Qi et al., 1994), 50% of colon tumors, and 64% of adenomas (Kitadai *et al.*, 1991). In addition, a strong correlation exists between AR expression and ER expression.

5. Heparin-Binding EGF-like Growth Factor (HB-EGF)

HB-EGF was derived from the conditioned medium of the macrophage-like U-937 cell. It is a 22-kDa secreted polypeptide, whose amino acid sequence (predicted from complementary cDNA clones) indicates that it is structurally a member of the EGF family. HB-EGF has the ability to bind heparin. It also binds to the EGF receptor on A-431 epidermal carcinoma cells and smooth muscle cells. Purified HB-EGF is mitogenic for BALB-3T3 cells and smooth muscle cells in a dose-dependent manner. It is also a mitogen for keratinocytes (Higashiyama *et al.*, 1991). However, it is not mitogenic for endothelial cells (Higashiyama *et al.*, 1991). HB-EGF exhibits a higher affinity for EGFR on smooth muscle cells than does EGF. Moreover, it requires only 100 pg/ml of HB-EGF to stimulate smooth muscle cell proliferation to the same extent as does EGF at 4 ng/ml. Thus, HB-EGF is a more potent mitogen for smooth muscle cells than either EGF or TGF-α (Higashiyama *et al.*, 1991).

The HB-EGF gene is expressed not only by macrophage-like U-937 cells, but by cultured human macrophages as well. Since macrophages appear to mediate fibroblast migration and proliferation in wound healing, and smooth muscle cell hyperplasia in atherosclerosis (Aqel *et al.*, 1985; Ross, 1986), HB-EGF may play an important role in these processes, perhaps binding to heparin-like sites on cell surfaces and in the extracellular matrix (Higashiyama *et al.*, 1991).

6. Cripto (CR-1)

Cripto was originally identified and cloned from human embryonal carcinoma cells (Ciccodocola *et al.*, 1989). CR-1 is an EGF-related gene expressed in a majority of human colorectal tumors. The cripto gene encodes a 188-amino-acid protein. It contains a 37-amino-acid region which shares the cysteine-rich motif common in other members of EGF supergene family. This region could potentially form the three intramolecular disulfide bond region which is one of the conserved features in other members of the EGF/TGF-α family. However, in the CR-1 protein, there is no A-loop, and the B-loop is truncated. In addition, unlike other members of the EGF family, human CR-1 lacks a conventional hydrophobic signal peptide and a transmembrane domain (Derynck, 1988; Todaro *et al.*, 1990; Salomon *et al.*, 1990; Massague, 1990; Dono *et al.*, 1993; Ciardiello *et al.*, 1991a,b; Qi *et al.*, 1994; Kuniyasu *et al.*, 1991). Although CR-1 contains a cysteine-rich EGF-like domain and has been considered a candidate for membership in the EGF family of ligands (Ciccodocola *et al.*, 1989), a recent report clarifies that CR-1 does not bind to any EGF receptor family members (Brandt *et al.*, 1994). Thus Cripto does not belong to the EGF-related peptide family.

CR-1 can function as a dominantly acting oncogene in 60–70% of human primary and metastatic colorectal tumors (Saeki et al., 1992; Ciardiello et al., 1991b). CR-1 may also be involved in the pathogenesis of human breast cancer, though its mechanism of action remains unknown.

7. Neu Differentiation Factor (NDF)/Heregulin (HRG)

HRGs are a large group of secreted and membrane-attached growth factors, expressed as alternative RNA spliced isoforms from a single gene mapped to human chromosome 8p22-11 (Lee and Wood, 1993; Orr-Urtreger et al., 1993). These peptides include the 44-kDa glycoprotein heregulin (Holmes et al., 1992), Neu differentiation factor (NDF) (Wen et al., 1992); a 25-kDa NEL/GF purified from bovine kidney (Huang et al., 1992), gp30 (Lupu et al., 1992), acetylcholine receptor inducing activity (ARIA) (Falls et al., 1993), glial growth factor (GGF) (Marchionni et al., 1993), and sensory and motor neuron-derived factor (Ho et al., 1995). Each protein performs distinct tissue-specific functions (Wen et al., 1994; Marchionni et al., 1993), and is involved in diverse biological activities. NDF was originally isolated from Ha-ras-transformed EJRat-1 fibroblasts, and HRG was isolated from human MDA-MB-231 breast cancer cells. At least 10 isoforms of NDF and 4 isoforms of HRG exist. They fall into two groups, α and β, that differ in the EGF-like domains and in receptor binding affinity (Wen et al., 1994). NDF and HRG were originally purified as putative ligands for ErbB-2, based on their ability to stimulate the phosphorylation of ErbB-2 in a number of breast carcinoma cell lines (Holmes et al., 1992; Peles et al., 1992; Wen et al., 1992). NDF can induce synthesis of milk components (casein and lipids) in certain breast carcinoma cell lines (Wen et al., 1992), while HRG can activate ErbB-2 phosphorylation in breast cancer cell lines (Holmes et al., 1992). However, neither NDF nor HRG binds to ErbB-2-expressing ovarian and fibroblastic cell lines (Peles et al., 1993). In addition, none of 15 ErbB-2 monoclonal antibodies inhibited cellular binding of NDF/HRG. Moreover, ectopic expression of ErbB-2 does not enhance NDF/HRG binding. In 1994, Yarden and Carraway's groups demonstrated that ErbB-3 and ErbB-4 are the receptors for NDF/HRG (Tzahar et al., 1994) and not ErbB-2, as thought at first. Of the two, ErbB-3 displays lower ligand binding affinity than ErbB-4 (Tzahar et al., 1994). Coexpression of ErbB-2 and ErbB-3 proteins constitutes a high-affinity receptor for NDF/HRG (Sliwkowski et al., 1994). Both ErbB-3 and ErbB-4 receptors bind preferentially to the β-isoforms of HRG, over the α class. (Tzahar et al., 1994). Very recently, it has been reported that NDF can partially inhibit EGF binding in a subset of human mammary carcinoma cells. This effect appears to be influenced by the relative expression levels of ErbB proteins in the cell (Karunagaran et al., 1995).

The HRG cDNA sequence predicts a transmembrane glycoprotein precursor (pro-NDF). The basic structure of NDF/HRG includes an N-terminal region, an immunoglobulin homology unit, a glycosylation-rich space motif, an EGF-like domain, a hydrophobic transmembrane domain, and a variable-length cytoplasmic domain. The isomers of NDF/HRG are probably generated through alternative splicing of the EGF-like domain, juxtaposed to the transmembrane domain and the cytoplasmic tail. In spite of the low sequence similarity (45, 27, and 32% homology to HB-EGF, EGF, and TGF-α, respectively), the EGF-like domain of HRG is similar to HB-EGF, EGF, and TGF-α (Nagata et al., 1994; Kline et al., 1990; Harvey et al., 1991; Hommel et al., 1991; Kohda and Inagaki, 1992a,b; Montelione et al., 1992; Moy et al., 1993). Nine residues, including the six cysteine residues, Gly-194, Gly-218, and Arg-220, are thoroughly conserved in the EGF-like domain of HRG. In EGF, Arg220 is essential for EGFR recognition (R90, Engler et al., 1990; Hommel et al., 1991). Thus Arg220 in HRG-α is probably required functionally to confer to the molecule an affinity for p180^{erbB-4} (Nagata et al., 1994). Despite these similarities, the EGF-like domains of HRG and EGF family growth factors are biologically distinct. The EGF-like domains of HRG bind specifically to p180^{erbB-4} but not p170^{erbB-1}(EGFR), whereas EGF, TGF-α, HB-EGF, AR, and BTC all bind to p170^{erbB-1} (Holmes et al., 1992; Wen et al., 1992; Plowman et al., 1993b). The EGF-like domain of HRG-β1 is illustrated in Fig. 3.

The possible role of HRG in regulating proliferation and differentiation of human breast cancer cells has been addressed. Several studies have demonstrated that HRG-β1 can stimulate the anchorage-dependent, serum-free growth of nontransformed human MCF-10A mammary epithelial cells. Unlike EGF, TGF-α, or AR, HRG-β1 (but not the α-form) produces a significant dose-dependent three- to fourfold stimulation in the agar growth of nontransformed MCF-10A cells. In addition, HRG-β1 is also able to stimulate the anchorage-independent growth of c-Ha-ras- or c-erbB-2-transformed MCF-10A or mouse NOG-8 mammary epithelial cells (Mincione et al., 1996). MCF-10A cells exhibit very high endogenous expression of various HRG mRNA transcripts and protein isoforms. It has been suggested that endogenous heregulin might function as an autocrine growth factor for Ha-ras- or ErbB-2-transfected mammary epithelial cells. NDF/HRG-induced morphologic alterations in mammary cells correlate with up-regulation of intracellular adhesion molecule 1 (ICAM-1) (Bacus et al., 1994). Marikovsky et al. (1995) have shown that different isoforms of NDF-β (but not NDF-α) can replace EGF as mitogens, to stimulate anchorage-dependent serum-restricted proliferation in EGF-dependent mouse Balb/MK keratinocytes which express ErbB-3. Furthermore, this proliferative response correlates with the binding affinities of the different NDF isoforms to the ErbB-3 protein. However, different NDF-β isoforms were generally two- to threefold less potent than EGF in stimulating the Balb/MK keratinocytes (Marikovsky et al., 1995). HRG also can stimulate mitogenesis in NIH 3T3 cells that express ErbB-3 or ErbB-4. However, HRG-dependent transformation via

ErbB-3 and ErbB-4 relies on the coexpression of either EGFR or ErbB-2 (Zhang et al., 1996). Transfection of HRG-β into ER-positive MCF-7 cells induces ErbB-2, ErbB-3, and ErbB-4 phosphorylation. Clones expressing high levels of HRG (β2) are estrogen-independent and resistant to antiestrogen agents *in vitro* (Tang et al., 1996). Furthermore, these HRG-transfected MCF-7 cells can form tumors in nude mice in the absence of estradiol or in the presence of tamoxifen (Pietras et al., 1995; Tang et al., 1996).

The physiological relevance of transmodulation is supported by gene-targeting experiments in transgenic mice. Transgenic mice lacking NDF (the mouse homologue of HRG) exhibit defects in heart development identical to those observed in the ErbB-2 and ErbB-4 gene knock-out experiments mentioned above. The mutant embryos die *in utero* at Day 10.5. However, the defects in nervous system development overlap with, but are not identical to, those observed in ErbB-2 and ErbB-4 knock-out mice (Meyer and Birchmeier, 1995).

IV. Activation of EGF Family Receptors

A. Interaction between the EGF Family Receptors and Their Ligands

Cell growth and differentiation are regulated in part by polypeptide-mediated extracellular signals (Aaronson, 1991). The interaction between growth factors and EGF family receptors initiates a biochemical cascade culminating in nuclear events that regulate gene expression and DNA replication (Ullrich and Schlessinger, 1990). Deregulation of this process may lead to oncogenic transformation, which can be induced by constitutive production of growth and regulatory factors, or by altered forms of the latter's cognate receptors (Yarden and Ullrich, 1988).

EGFR binds many different ligands, including EGF, BTC, HB-EGF, TGF-α, and AR, whereas ErbB-3 and ErbB-4 bind to more than a dozen isoforms of HRG (Groenen et al., 1994). In addition, a recent report has demonstrated that BTC can activate ErbB-4 with mitogenic effect in Ba/F3 cell system (Riese II et al., 1995). However, no fully characterized ligand binds directly to the closely related ErbB-2 protein (Dougall et al., 1994).

Binding of ligands induces receptor dimerization and increases intrinsic tyrosine kinase activity, which leads to receptor autophosphorylation and tyrosine phosphorylation of various cellular substrates (Stern and Kamps, 1988). In this section, we will describe the mechanisms involved in receptor activation, and the heterodimerization within EGF receptor family members.

B. The Mechanisms of Receptor Activation

In general, the biological actions of EGF-like ligands are transmitted solely by their receptor tyrosine kinases. Following ligand binding, the kinase is activated and phosphorylates the receptor's carboxyl terminus on tyrosine residues. In addition, C-terminal autophosphorylation domains of growth factor receptor protein tyrosine kinases play an important role in the first step of the transduction of external signals to the nucleus. Tyrosine residues in the autophosphorylation domains of these receptors become rapidly phosphorylated upon binding of their cognate ligands. The phosphorylated tyrosine residues function as high-affinity binding sites for SH2 (src homology 2) domain-containing intracellular proteins (Sadowski *et al.*, 1986; Songyang *et al.* 1993; Koch *et al.* 1991). After receptor autophosphorylation, a series of substrates becomes phosphorylated by the activated receptor. These in turn stimulate a series of downstream signal pathways. Depending on the cell type, within 5–90 min ligand-dependent tyrosine phosphorylation gradually decreases and becomes undetectable. Receptors are then internalized along with the ligand by receptor-mediated endocytosis (McCune *et al.*, 1993). In this section, we describe the possible mechanisms of signal transduction used by each of the EGF family receptor members.

I. EGFR

EGFR was the first receptor for which ligand-dependent tyrosine kinase activity was demonstrated. Several lines of evidence have shown that ligand binding to the extracellular ligand binding domain of EGFR activates the cytoplasmic kinase domain, which undergoes self-phosphorylation, subsequently phosphorylates various cellular substrates, and initiates a cascade of signaling events (Ushiro and Cohen, 1980; Cohen *et al.*, 1982; Ullrich and Schlessinger, 1990; Fantl *et al.*, 1993). Point mutations in the kinase domain of EGFR that prevent ATP binding abolish ligand-dependent kinase activity of EGFR and abrogate EGF/TGF-α-induced mitogenesis (Carpenter and Cohen, 1990). Activation of EGFR by ligand-induced conformational alteration leads to homo- or heterodimerization (the latter with other EGFR family members), which in turn results in the activation of a variety of downstream substrates and the autophosphorylation of the receptor itself (Heldon, 1995).

In many tumors, amplification of the EGFR gene is accompanied by gene rearrangements (Libermann *et al.*, 1985; Wong *et al.*, 1987; Yamazaki *et al.*, 1988; Malden *et al.*, 1988). These rearrangements result in a truncated EGFR gene, a hybrid mRNA, and ultimately a mutant EGFR with no extracytoplasmic domain. The predicted amino acid sequence of the EGFR protein extracted from these tumors is remarkably similar to that described for several v-ErbB oncogenes. The end result is a protein that is unable to bind EGF but can still phosphorylate other substrates, and is constitutively activated (Kris *et al.*, 1985; Yarden and Ullrich, 1988). It is possible that the rearrangement of EGFR causes the receptor to be constitutively activated in the absence of any stimulant, as has been shown to occur with the v-ErbB protein tyrosine kinase (Kris *et al.*, 1985; Shu *et al.*, 1991). These experiments indicate that the tyrosine kinase activity of EGFR is essential for signal transduction and receptor function (Downward *et al.*, 1984; Weiss, 1993; Slamon *et al.*, 1989a; Fry *et al.*, 1993; Birchmeier *et al.*, 1993; Lax *et al.*, 1989).

2. ErbB-2

Multiple genetic and biochemical mechanisms may be involved in ErbB-2 oncogenic potential. Oncogenically activated forms of *neu* are permanently active as tyrosine kinases, and are therefore constitutively coupled to their effector pathways. The constitutive activation of the ErbB-2 protein kinase has been detected in NIH 3T3 cells that overexpress the protein, and also in some human cancer cell lines (Leonardo *et al.*, 1990; Peles *et al.*, 1993; Di Fiore *et al.*, 1990). These results indicate that the constitutive activation of the catalytic activity of these proteins may be involved in cellular transformation and in the pathogenesis of cancer.

Possible mechanisms for neu/ErbB-2 oncogenic transformation include the following: (1) Certain single amino acid substitutions at the transmembrane domain (Bargmann and Weinberg, 1988) may result in the formation of dimers between wild-type *neu* monomers which lead to constitutive activation of ErbB-2 (Sternberg and Gullick, 1990; Brandt-Rauf *et al.*, 1990). (2) Overexpression of wild-type protein at the cell surface may lead to elevated basal tyrosine kinase activity that could exceed a threshold needed for cell stimulation (Di Fiore *et al.*, 1987; Hudziak *et al.*, 1987; Segatto *et al.*, 1988). (3) Truncation of noncatalytic portions of ErbB-2 may affect transduction of the mitogenic signal via constitutive activation of the receptor (Di Fiore *et al.*, 1987; Bargmann and Weinberg, 1988). (4) ErbB-2 can form heterodimers with the EGF-receptor or ErbB-3, and perhaps ErbB-4 receptors (see below), where only the non-ErbB-2 receptor is ligand-occupied (Goldman *et al.*, 1990; Wada *et al.*, 1990). The difference between ErbB-2 and EGFR heterodimers is that the dimerized ErbB-2 is not ligand-bound, while the dimerized EGFR is. Thus, activation of ErbB-2 can result in transphosphorylation through the non-ErbB-2 member of the heterodimer. These heterodimers elevate tyrosine kinase activity and induce mitogenic signaling.

All of these different mechanisms could converge on a single cell activation pathway. Alternatively, each oncogenic form could utilize a distinct signaling pathway (Fig. 4).

3. ErbB-3

Activation of the ErbB-3 receptor appears to be different from that of other receptors within the family. The cytoplasmic domain of ErbB-3 displays a multiple YXXM motif, which suggests that the ErbB-3 gene product may act as an efficient recruiter of phosphatidylinositol-3" (PI-3) kinase. In order to constitute a functional signal transduction pathway involving PI

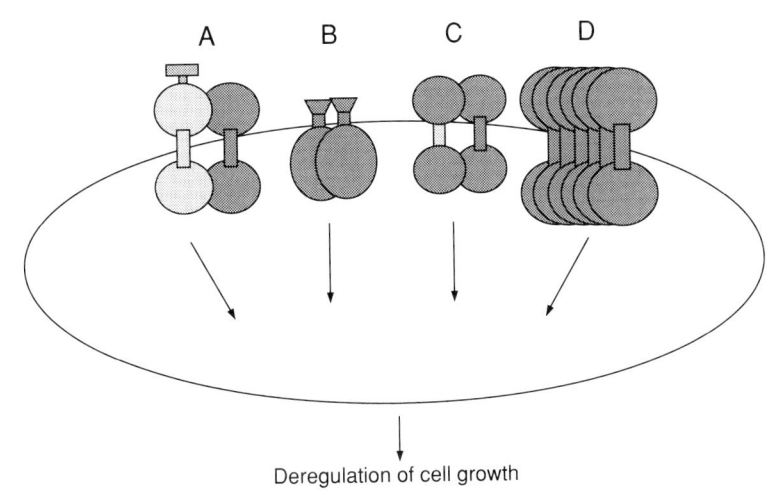

FIGURE 4 Schematic representation of possible mechanisms for neu/ErbB-2 oncogenic transformation. (Redrawn from *J. Steroid Biochem. Mol. Biol.* Vol. 43, 98, 1992, with kind permission from Elsevier Science Ltd., The Boulevard, Langford Lane, Kidlington OX5 1GB, United Kingdom.)

3-kinase, the ErbB-3 protein would need to be phosphorylated on tyrosine residues. This prerequisite could be fulfilled by either autophosphorylation, or cross-phosphorylation mediated by another cellular protein tyrosine kinase. But it is probable that autophosphorylation cannot occur, since ErbB-3 has impaired kinase activity. Thus, cross-phosphorylation is likely mediated by heterodimerization between ErbB-3 and other receptors in the family. Recently, it has been reported that EGFR/ErbB-3 heterodimers appear to enable recruitment of PI-3 kinase by the EGF-receptor (Kim *et al.*, 1994; Soltoff *et al.*, 1994). In addition, Kraus and colleagues have demonstrated that cooperation between ErbB-3 and ErbB-2 involves heterodimerization and increased ErbB-3 tyrosine phosphorylation by HRG, resulting in increased PI-3-kinase recruitment. A cross-phosphorylation mechanism involving ErbB-3 may be particularly relevant in breast cancer cells in which the concurrent overexpression of ErbB-2 and ErbB-3 is common. Simultaneous overexpression of the EGFR and ErbB-3 protein is observed in the human breast cancer cell line MDA-MB-468 (Kraus *et al.*, 1987), and high levels of both ErbB-2 and ErbB-3 have been detected in several breast cancer cell lines, including MDA-MB-453, MDA-MB-361, SKBr-3, and BT-474 (Peles *et al.*, 1993; Kraus *et al.*, 1987, 1993). Therefore, it appears that interreceptor interactions allow ErbB-3 to amplify and diversify its signaling pathway.

4. ErbB-4

ErbB-4 can be activated by its ligand, NDF/HRG, and form homodimers. ErbB-4 can also form heterodimers with other receptors in the family. When

ErbB-4 heterodimerizes with ErbB-2, NDF/HRG activates tyrosine phosphorylation of both receptors. This scenario is analogous to the EGF-dependent EGFR activation of ErbB-2 (Qian et al., 1994). Activation of ErbB-2 is thus possible through transphosphorylation by either ErbB-4 or EGFR. In the Ba/F3 cell system, EGF is able to induce ErbB-4/EGFR heterodimerization and activate both EGFR and ErbB-4 signaling pathways (Riese et al., 1995)

Despite the different activation pathways described above, the tyrosine kinase activity of the EGF family receptors is ultimately regulated by ligand binding. Ligand binding is thus likely to be the major signaling mechanism of these receptors.

C. Heterodimerization, Transphosphorylation

Data from numerous laboratories suggest that EGFR family members may play a complex and ultimately more flexible role in signaling by forming heterodimers. Protein-tyrosine kinase receptors are activated by ligand-induced dimerization or oligomerization. It appears that ligand-induced dimerization is a generally applicable mechanism for regulation of EGF family receptor signal transduction. The interaction between the intracellular domains of the receptors in the dimer may promote a conformational change that leads to increased kinase activity.

Very recently, the Stern and Yarden groups have utilized a Ba/F3 or 32D cell system, in which no endogenous expression of ErbB-receptor family receptors has been detected, to demonstrate that binary receptor interaction enables the tuning and amplification of growth factor signaling (Riese et al., 1995; Pinkas-Kramarski et al., 1996). These studies conclude that receptor interactions are selective rather than random. An example of receptor hierarchy is that ErbB-2/ErbB-3 heterodimers have been found to be more active than EGFR/ErbB-3 heterodimers. This is despite the fact that EGFR homodimer signaling displays dominance over EGFR/ErbB-3 heterodimers when these two receptors are coexpressed.

Hierarchy of receptor cross talk is reflected biochemically in two ways: (1) coexpression of EGF family receptors enhances ligand binding affinity. (2) Heterodimers tend to be more potent than homodimers due to elevation of receptor tyrosine phosphorylation and mediation of different biological responses. Combinatorial receptor interaction diversifies signal transduction and confers double regulation. The critical role of receptor dimerization and tyrosine phosphorylation in the EGF receptor family will be discussed below.

1. EGFR/ErbB-2 Heterodimerization

The EGF receptor was the first protein-tyrosine kinase receptor to be shown to dimerize after ligand binding (Yarden and Schlessinger, 1987). EGFR and ErbB-2 heterodimerization was observed when ErbB-2 tyrosine phosphorylation was induced by the binding of EGF to EGFR (King et al.,

1988; Stern and Kamps, 1988). Subsequently, heterodimers of EGFR and ErbB-2 were detected in rodent fibroblast NR6 cells, providing direct evidence of their physical interaction (Wada et al., 1990; Qian et al., 1992). Heterodimer association of EGFR and ErbB-2 has also been found in the human breast cancer cell line SkBr-3 (Goldman et al., 1990) and in NIH 3T3 cells cotransfected with EGFR and ErbB-2 (Spivale, 1992). Studies have shown that expression of ErbB-2 alone at a moderately high level (10^5 receptors/cell) in NIH3T3 cells or NR6 cells does not cause transformation (Hung et al., 1989; Kokai et al., 1989), unless EGFR is coexpressed at an equivalent level (Kokai et al., 1989). Down-regulation of either EGFR or ErbB-2 from the cell surface by anti-receptor antibody treatment reverses the transformed phenotype (Wada et al., 1990). These results suggest that two distinct, moderately overexpressed tyrosine kinases can interact synergistically, leading to cellular transformation. Furthermore, there is evidence to suggest that the predominance of the heterodimer of EGFR and ErbB-2 reveals preference for heterodimerization over either form of homodimerization (Qian et al., 1994).

2. EGFR/ErbB-3 Heterodimerization

Homodimerization of ErbB-3 exhibits extremely low tyrosine kinase activity due to ErbB-3's impaired kinase domain. It is insufficient for any detectable biological response. However, ErbB-3 can interact with either EGFR or ErbB-2 and reconstitute its biological activity. It has also been reported that the EGFR/ErbB-3 dimer stimulates PI-3 kinase activity through the EGFR with EGF stimulation in SkBr-3 mammary cells. This activation of an ErbB-3-associated PI-3 kinase via the EGFR-associated ligand EGF provides evidence of the formation of the EGFR/ErbB-3 heterodimer (Soltoff et al., 1994). HRG can also activate the EGFR/ErbB-3 heterodimer. However, only the HRG-β form, not the HRG-α form, can activate the heterodimer and abrogate the IL-3-dependent pathway in 32D cotransfected EGFR and ErbB-3 cells (Yarden, 1996).

3. ErbB-2/ErbB-3 Heterodimerization

HRG has been found to induce heterodimeric complexes between ErbB-2 and ErbB-3 or ErbB-2 and ErbB-4 (Peles et al., 1993; Plowman et al., 1993a; Sliwkowski et al., 1994). The presence of ErbB-3 or ErbB-4 is necessary for high-affinity binding of HRG, and for signal transduction through ErbB-2. Since ErbB-3 is an impaired kinase receptor, the major function of ErbB-3 in the ErbB-2+3 heterodimer is to act as a substrate for the ErbB-2 kinase and provide a docking site for downstream SH2 domain-containing signal transduction molecules (Carraway and Cantley, 1994).

4. EGFR/ErbB-4 and ErbB-2/ErbB-4 Heterodimerizations

EGFR/ErbB-4 and ErbB-2/ErbB-4 heterodimerization has been observed in NIH 3T3 cells and Ba/F3 cells (Cohen et al., 1996; Riese et al., 1995). This expression of dimerized receptors diversifies NDF/HRG and EGF signaling.

V. Clinical Significance of EGF Family Receptors

Abnormalities in the expression, structure, or activity of proto-oncogene products contribute to the development and maintenance of a malignant phenotype. Overexpression and amplification of EGF family receptors is frequently implicated in human cancer. Increasing evidence indicates that aberrant activation of EGF family receptors may be pathogenically significant and may contribute to tumorigenesis or progression. In this section, we will discuss the clinical significance of these receptors.

A. EGFR

Gene amplification is one of the most common genetic alterations occurring in the oncogenic transformation and malignant progression of cells. The first demonstration of elevated levels of EGFR expression was by Hendler and Ozanne (1984). Amplification and rearrangement of the EGFR locus has been identified in a variety of human cancers (Libermann *et al.*, 1984, 1985; Merlino *et al.*, 1985; Humphery *et al.*, 1988, 1990), most commonly in squamous carcinomas of various sites and less commonly in adenocarcinomas (particularly in pancreatic and gastric cancers) (Lemoine *et al.*, 1991a,b). A common property of many tumor cells is coexpression of EGFR and its TGF-α ligand, to establish an autocrine loop (Watanabe *et al.*, 1987; McGeady *et al.*, 1984; Shankar *et al.*, 1989). High levels of EGFR in breast tumors correlate strongly with a poor prognosis, independent of ER status (Fitzpatrick *et al.*, 1984; Sainsbury *et al.*, 1985; Klijn *et al.*, 1992).

Overexpression of EGFR occurs in a wide variety of human tumor types, including breast carcinomas (47%) (Klijn *et al.*, 1992), primary ovarian carcinomas (35–70%) (Morishige *et al.*, 1991; Bauknecht *et al.*, 1989a,b, 1990, 1993; Kohler *et al.*, 1989; Johnson *et al.*, 1991; Berns *et al.*, 1992; Battaglia *et al.*, 1989; Henzen-Longmans *et al.*, 1992a,b; Cambia *et al.*, 1992; Owens *et al.*, 1992), glioblastomas (40–50%) (Ekstrand *et al.*, 1991; Yung *et al.*, 1990; Di Carlo *et al.*, 1992; Agosti *et al.*, 1992; Chaffanet *et al.*, 1992; Jones *et al.*, 1990; Tuzi *et al.*, 1991; Wong *et al.*, 1987), liver carcinomas (32%) (Nonomura *et al.*, 1988), pancreatic carcinomas (30–35%) (Yamanaka *et al.*, 1990, 1993; Barton *et al.*, 1991; Korc *et al.*, 1992), and endometrial and cervical carcinomas (91%) (Sato *et al.*, 1991). Overexpression is sometimes associated with gene amplification; 40–80% of lung carcinomas show amplification of the EGFR gene, involving large amplicons of DNA spanning up to 1000 kb. Overexpression or amplification of EGFR may be associated with poor prognosis in breast cancer, lung cancer, and bladder cancer. Recently, the most frequently identified receptor mutant, EGFRvIII (lost amino acids: 6–273), had been detected in up to 57% of high-grade and 86% of low-grade glial tumors, 78% of breast cancers, 73%

of ovarian cancers, and 16% of non-small-cell lung cancers, but not in any normal tissues examined to date (Garcia de Palazzo *et al.*, 1993; Wikstrand *et al.*, 1995; Moscatello *et al.*, 1995).

B. ErbB-2

Overexpression and amplification of ErbB-2 are associated with a variety of human cancers. Overexpression of ErbB-2 has been detected frequently in human adenocarcinomas from several tissues (Slamon *et al.*, 1987; van de Vijver *et al.*, 1987; Slamon *et al.*, 1989b; Allred *et al.*, 1990; Venter *et al.*, 1987; Yokota *et al.*, 1986), including 30% of lung and stomach adenocarcinomas. Amplification is detected more frequently in metastatic tumors than in primary tumors (Tsujino *et al.*, 1990; Mizutani *et al.*, 1993). Overexpression and amplification is also associated with 25% of carcinomas of the breast, stomach, and ovary (Aaronson, 1991). ErbB-2 plays a more important role in the initiation rather than the progression of ductal carcinomas. No relationship has been detected between ErbB-2 expression and age of patients, or size and histological type of tumor (van de Vijver *et al.*, 1988; Tandon *et al.*, 1989; Tikannen *et al.*, 1992; Gusterson *et al.*, 1992; Wright *et al.*, 1989; Paik *et al.*, 1990; Noguchi *et al.*, 1992; Gasperini *et al.*, 1992; Schonborn *et al.*, 1994; Schroeter *et al.*, 1992; Lonn *et al.*, 1994; Henry *et al.*, 1993; Ro *et al.*, 1989; Allred *et al.*, 1992; Bianchi *et al.*, 1993; Clark and McGuire, 1991; Nagai *et al.*, 1993). Over 90% of comedo-type ductal carcinoma *in situ* (DCIS) mammary tumors have been found to overexpress ErbB-2 (van de Vijver *et al.*, 1988). Overexpression of ErbB-2 in DCIS tumors appears to associated with a greater invasive potential (Barnes *et al.*, 1992). In invasive cancers, overexpression of ErbB-2 is essentially confined to the more inflammatory ductal carcinomas, rather than noninflammatory ones (Guerin *et al.*, 1989; Garcia *et al.*, 1989). Epidemiological and clinical findings suggest that steroid hormones (most notably estrogen) are one of the major factors in the stimulation of breast cancer growth (Antoniotti *et al.*, 1992; Russell and Hung, 1992; Dati *et al.*, 1990). Expression of either ErbB-2 or ER in human breast cancer provides important prognostic information (Slamon *et al.*, 1987, 1989b; Nicholson *et al.*, 1990; Benz *et al.*, 1992; Wright *et al.*, 1992; Borg *et al.*, 1994; Elledge *et al.*, 1994). It is clear that cross-regulation occurs between the estrogen receptor and ErbB-2 signaling pathways. *In vitro* experiments have demonstrated that down-regulation of ErbB-2 mRNA and protein expression occurs when estrogen receptor (ER)-positive cells are treated with estradiol (E2). Meanwhile, tamoxifen and other anti-estrogenic agents can up-regulate ErbB-2 in ER-positive cells, and inhibit cell growth (Dati *et al.*, 1990; Read *et al.*, 1990). Various sources show an inverse correlation between ErbB-2 overexpression and ER expression. Compared with ER-positive, ErbB-2-negative (ER+/ErbB-2-) patients, patients with ER+/ErbB-2+ tumors show a significantly decreased response to endocrine therapy. These findings have led to the speculation of anti-estrogen resistance in ER+/ErbB-2+ coexpressing patients (Wright et al., 1992). Ovarian cancer patients overexpressing ErbB-2 show resistance to cisplatinum and a fivefold lower complete response rate at second-look laparotomy, compared with ErbB-2 negative patients (Berchuch et al., 1990).

The relationship between overexpression of ErbB-2 and response to conventional chemotherapy in breast cancer patients has also been studied. There is evidence that overexpression of ErbB-2 may be a marker not only for increasing tumor aggressiveness but also for resistance to chemotherapy. The Intergroup 110 study notes that node-negative/ErbB-2 overexpressors did not benefit from adjuvant chemotherapy. The survival rates are the same whether or not these patients received chemotherapy with CMF (cyclophosphamide methotrexate 5-fluorouracil) (Allred et al., 1992). Data from CALGB study 8541 indicate a dose-responsive effect to doxorubicin in patients with overexpression of ErbB-2, which suggests that overexpression of ErbB-2 correlate with resistance to alkylating agents (Muss et al., 1994). Overexpression and amplification of ErbB-2 also correlate with depth of tumor invasion and poor prognosis (Liotta, 1984; Goldolphin et al., 1981). National Surgical Adjuvant Breast and Bowel Project (NSABP) studies on women indicate that overexpression of ErbB-2 results in significantly worse overall survival, with twice the mortality rate relative to women without detectable ErbB-2 expression (Paik et al., 1990). In addition, the shed extracellular domain (ECD) of ErbB-2 could represent a new marker of human cancer (Langton et al., 1991). Elevated levels of soluble ErbB-2 can be detected in the serum and effusions of about 25% patients with locally advanced or metastatic breast cancer (Leitzel et al., 1992).

Since ErbB-2 protein can heterodimerize with other EGFR family members, coexpression of all family members must be taken into account as future clinical studies proceed.

C. ErbB-3

ErbB-3 is activated in some breast tumors. In addition, the ErbB-3 transcript has been observed in a wide range of human carcinomas including those of the colon, lung, kidney, pancreas, and skin. Immunohistochemically detected membrane staining of the ErbB-3 receptor is rare in primary carcinomas. Most staining in advanced breast cancer specimens is cytoplasmic (Lemoine et al., 1992). Elevated ErbB-3 expression has been detected in 53% of primary colorectal tumors (Ciardiello et al., 1991b; Cook et al., 1992; Saeki et al., 1992; Kuniyasu et al., 1991), in 56% of liver metastases (Cicardiello et al., 1991b), and in 90% of pancreatic cancers (Lemoine et al., 1992).

D. ErbB-4

ErbB-4 has been found to be elevated in some human breast cancer cells (Kraus *et al.*, 1989; Plowman *et al.*, 1993b). There are no clinical data available at the present time. However, one report indicates that 75% of infiltrating ductal carcinomas express ErbB-4, and that a positive correlation exists between ErbB-4 expression and estrogen and progesterone receptor expression (Bacus *et al.*, 1994).

VI. Potential Clinical Application by Targeting of EGF Family Receptor Members and Ligands

Our present knowledge of the role of EGF receptors and ligands in cancer offers possibilities for improvements in diagnosis and prognosis, and opportunities for therapeutic intervention (Fig. 5). Overexpression or gene amplification in EGF receptor family members is frequently implicated in human cancer. Overexpression and amplification of ErbB-2 is correlated with poor prognosis in breast cancer, although for other cancers this relationship is less well defined. In addition, because of heterodimerization within the EGFR family, other receptors such as ErbB-3 or ErbB-4 may also play a role in tumorigenicity. Identification of multiple copies of receptor genes by quantitative PCR could provide a diagnostic indicator. Furthermore, it would be appropriate to screen for these receptors and their expression levels in tumor biopsies, and most importantly to identify high-risk patients for more aggressive therapy.

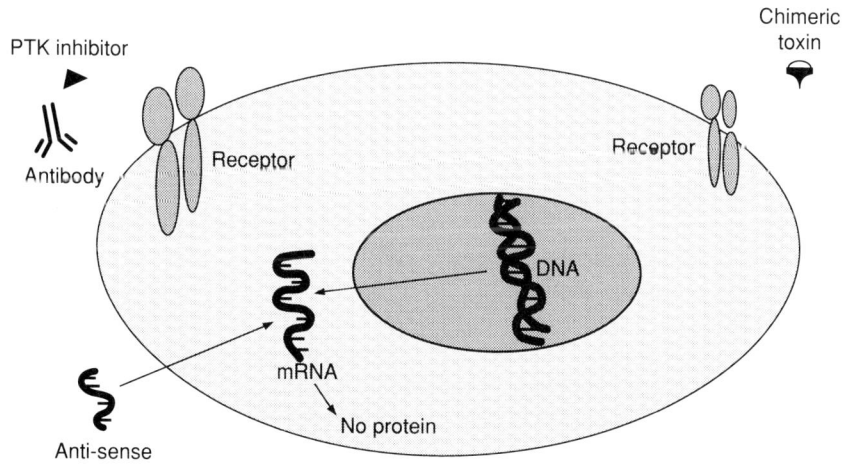

FIGURE 5 Schematic representation of different potential approaches for targeting receptors and ligands.

Theoretically, intervention could be achieved by inhibiting ligand binding, receptor dimerization, tyrosine kinase activation, or protein expression of ligands or receptors. Specific approaches include (A) antibody immunotherapy, (B) coupling of receptor antibodies or ligands to toxin molecules, (C) antisense strategies, and (D) receptor tyrosine kinase inhibitors.

A. Immuno (Antibody) Therapy

Immunotherapy has long been considered a promising approach for the treatment of cancer (Hellström and Hellström, 1985; Mellstedt, 1990). Monoclonal antibodies have been raised against both ligands and receptors for the EGF receptor system and against the extracellular domain of ErbB-2.

1. Receptor Antagonists

a. Monoclonal Antibodies   Mendelsohn and co-workers have generated monoclonal antibodies against the EGF receptor (Sato *et al.*, 1983; Kawamoto *et al.*, 1983). Mendelsohn's antibodies 225 and 528 bind to EGFR with an affinity very similar to that of EGF, and compete with EGF for receptor binding (Sato *et al.*, 1983; Kawamoto *et al.*, 1983). These antibodies inhibit EGF-dependent tyrosine kinase activation, inhibit the proliferation of cultured A431 cells (Gill *et al.*, 1984; Kawamoto *et al.*, 1984), and inhibit tumor formation *in vivo*. Another group of monoclonal antibodies against the EGF receptor has been produced by Schlessinger *et al.* (108.4, 96, and 42). Monoclonal antibody 108.4 has been shown to inhibit proliferation of human KB epidermoid carcinoma cell lines *in vitro* and *in vivo* (Aboud-Pirak *et al.*, 1988).

Many monoclonal antibodies have been developed for ErbB-2. The most success has been achieved with an antibody known as 4D5, which reacts with an extracellular epitope on ErbB-2 (Hudziak *et al.*, 1989). This antibody has been shown to exhibit strong antiproliferative activity on human breast cancer cell lines expressing this oncogene product, and to increase the sensitivity of cancer cells to tumor necrosis factor α (Hudziak *et al.*, 1989). Furthermore, 4D5 is able to enhance the cytotoxic effect—*in vitro* and *in vivo*—of diammedichloroplatinum in breast cancer cell lines expressing high levels of ErbB-2. It is currently being used in clinical trials for tumor localization and possible efficacy, in combination with CDDP (Shepard *et al.*, 1991). The 4D5 antibody has recently been humanized, and bispecific antibodies containing Fv fragments of both 4D5 and anti-CD3 antibody have been engineered. It is hoped that this combination will promote T-cell recruitment to the tumor site (Shalaby *et al.*, 1992). The following section discusses the details of bispecific antibodies.

b. Bispecific Monoclonal Antibodies   Many monoclonal antibodies (mAbs) have been developed against human tumors (Schlom, 1986; Dillman, 1994).

However, mAbs have had little direct therapeutic effect due to lack of specificity in binding to tumors. This lack of efficacy is compounded by the inability of many murine mAbs to activate immune-effector pathways. Although humanized mAbs are able to activate immune-effector pathways, they are often impaired by high concentrations of nonspecific immunoglobulins (Igs). Moreover, large amounts of immunologically active mAbs may be directed to Fc receptors on cells that are not cytotoxic to tumor cells. Therefore, bispecific mAbs (BsAbs) are one approach for increasing the immunologic effectiveness of immunotherapy with mAbs. BsAbs are hybrid antibodies constructed from two parent mAbs: one specific to the tumor target cell and the other specific to an immune-effector cell (Fanger *et al.*, 1991, 1993). BsAbs can direct the cytotoxic activity of monocytes, (Fanger *et al.*, 1991, 1993), monocyte-derived macrophages, T cells (Perez *et al.*, 1985; van Dijk *et al.*, 1989; Weiner *et al.*, 1994), natural-killer (NK) cells (Weiner *et al.*, 1993a,b), and neutrophils (Valerius *et al.*, 1993), so as to kill and /or ingest tumor target cells *in vitro* and *in vivo*. BsAb MDX-210 is constructed from mAb 520C9 and mAb 22 (Valone *et al.*, 1995). mAb 520C9 recognizes the extracellular domain of oncogene ErbB-2 (Ring *et al.*, 1991), while mAB 22 recognizes FcγRI (Guyre *et al.*, 1989; Shen *et al.*, 1986). FcγRI refers to high-affinity type I Fc receptors on immune-effector cells. The BsAb MDX-210 phase Ia/Ib trial in patients with advanced ErbB-2 overexpressing breast or ovarian cancer has demonstrated that MDX210 is immunologically active at well-tolerated doses (Valone *et al.*, 1995).

2. Ligand Antagonists

In addition to anti-receptor-blocking antibodies, anti-growth factor neutralizing antibodies such as EGF- and TGF-α neutralizing monoclonal antibodies have been generated. It has been shown that these neutralizing monoclonal antibodies are effective *in vitro*, but they have not been used successfully *in vivo* (Gullick, 1990). Combination therapy with different antibodies has proven to be advantageous.

B. Coupling of Receptor Antibodies or Ligands to Toxin Molecules

The use of targeted toxins is a promising approach for the therapy of cancer and autoimmune diseases, as well as other disorders (Pastan *et al.*, 1986, 1992; Pastan and FitzGerald, 1989; Vitetta *et al.*, 1987). One of the toxins that has proven versatile in producing chimeric toxins is *Pseudomonas* exotoxin A (PE) (Pastan *et al.*, 1986, 1992; Pastan and FitzGerald, 1989). The PE molecule possess three structural domains: The N-terminal domain (I) is responsible for cell recognition and binding, domain II facilitates translocation of toxin across the membrane, and the C-terminal domain (III) catalyzes the adenine phosphate-ribosylation of elongation factor-2.

Thus domain III plays a role in the inhibition of protein synthesis, leading to cell death. *Pseudomonas* exotoxin A can be chemically coupled to antibodies. It can also be coupled to growth factors by using recombinant DNA to construct chimeric toxins from genes encoding growth factors, or to single-chain antibody genes fused to toxin genes in order to kill target cells with differential surface properties (Pastan *et al.*, 1992; Vitetta *et al.*, 1987). Recombinant toxins possess the potential advantage of extreme potency, small molecular size, and ease of manufacture.

1. Chimeric Toxins

Chimeric toxins have been generated which express the genes of growth factor such as TGF-α or HB-EGF, fused to the II and III domains of *Pseudomonas* exotoxin A. TGF-α-PE40 *Pseudomonas* exotoxin has been shown to exert specific cytotoxic effects on a series of human cancer cell lines expressing the EGF receptor, including ovary, liver, breast, kidney, and colon cell lines (Siegall *et al.*, 1989). Pastan and colleagues have demonstrated that the continuous infusion of TGF-α-PE40 via miniosmotic pump placed in the peritoneal cavity of nude mice has antitumor effects on A431 human epidermoid carcinoma cells and DU-145 prostate carcinoma cells (Siegall *et al.*, 1989)

Recently, a chimeric toxin combining the EGF-like domain of HRG with PE38KDEL, a truncated recombinant form of *Pseudomonas* exotoxin, has been shown to interact with ErbB-3 and ErbB-4, which are heregulin receptors. The cytotoxic activity HRG-PE38KDEL targets ErbB-4 and ErbB-2 + 3 coexpressing cells, *in vitro* and *in vivo* (Kihara and Pastan, 1995; Siegall *et al.*, 1995).

2. Immunotoxins

Pseudomonas exotoxin has also used in conjugation with monoclonal antibodies. The King and Pastan groups have developed recombinant anti-erbB-2 immunotoxins which directly target the p185^{erbB-2}-expressing tumor cell. A specific ErbB-2 single-chain antibody (e23Fv) is coupled with a portion of *Pseudomonas* exotoxin (PE38KDEL) (Chaudhary *et al.*, 1989, 1990). This recombinant molecule OLX-209 [e23(Fv) PE38KDEL] has been found to kill cancer cells *in vitro* and also to have antitumor activities in mice bearing human tumor xenografts. Preclinical testing of OLX-209 has focused on tumors having very high levels of overexpression caused by gene amplification (Peiter *et al.*, 1996; Batra *et al.*, 1992; Kasprzyk *et al.*, 1996). However, gene amplification does not account for overexpression in many cancers. In most cases of lung cancer, overexpression of ErbB-2 is not due to gene amplification. Nevertheless, recent studies have demonstrated that OLX-209 antitumor efficacy is observed in a variety of human lung cancer cell lines with varying levels of ErbB-2 expression, even in the absence of gene amplification. This implies that patients with moderate levels of p185^{erbB-2} expression in lung cancer could be candidates for OLX-209 therapy.

C. Antisense Strategies

The notion that oligonucleotides can modulate gene-specific expression was established more than a decade ago. Triplex DNA, antisense DNA/RNA, and ribozymes have been used for suppressing activated oncogenes (Yokoyama and Imamoto, 1987; McManaway *et al.*, 1990; Goodchild *et al.*, 1988).

1. Triple Helix

Triplex DNA has numerous potential applications as a molecular biological tool. The advantages of using triplex DNA over other strategies to inhibit gene expression is that the triplex-forming oligonucleotide targets the gene directly at the transcription level rather than its mRNA at the translation level. It has been reported that inhibition of ErbB-2 mRNA (42%) and ErbB-2 protein level (59%) with a 28-nt phosphodiester triple helix-forming oligonucleotide targeted to the promoter region of the human ErbB-2 oncogene was observed in MCF-7 cells (Porumb *et al.*, 1996). However, no effects on cellular proliferation were reported.

2. Antisense Oligonucleotides

Antisense oligonucleotide technology uses single-stranded RNA or DNA to modulate gene expression by altering intermediate metabolism of mRNA. Numerous modifications have been attempted to increase oligonucleotide stability. Antisense oligonucleotides have shown effectiveness both *in vitro* and *in vivo* as modulators of gene expression. Examples include the targeting of the ras gene in melanoma (Kashani-Sabet *et al.*, 1994) and bladder (Kashani-Sabet *et al.*, 1992; Tone *et al.*, 1993) and lung (Mukhopadhyay *et al.*, 1991; Georges *et al.*, 1993) carcinomas, and myc (Yokoyama and Imamoto, 1987), myb (Hijiya *et al.*, 1994), BCR-ABL (Szczylik *et al.*, 1991), and BCL-2 (Reed *et al.*, 1993) genes for leukemia. Antisense oligonucleotides have also been designed to target ErbB-2. Liposome-mediated ErbB-2 antisense phosphorothioate oligonucleotide has been shown to efficiently inhibit expression of ErbB-2 mRNA and protein in the SkBr-3 human breast cancer cell line. The cell-cycle profile of anti-sense-treated cells exhibits an increased time of arrest in the G1 phase.

EGFR anti-sense RNA complementary to the entire coding region, or to parts of the EGFR mRNA, has been shown to effectively block translation of EGFR mRNA. In addition, upon microinjection into KB cells, the antisense RNAs were able to transiently inhibit the synthesis of EGFR. This inhibition was concentration-dependent, both *in vitro* and *in vivo*. Expression of antisense EGFR RNA in the KB human epidermoid carcinoma cell line results in a suppression of the transformed phenotype of KB cells and restores serum and anchorage-dependent growth (Moroni et al., 1992). In addition, the degree of inhibition in the transformed phenotype is proportionate to the decrease in expression of EGFR (Moroni et al., 1992).

Chemically modified phosphorothioate or methyl-phosphonate antisense oligonucleotides against growth factors also have been studied. It has been reported that anti-sense TGF-α oligonucleotides effectively inhibit autocrine-stimulated proliferation of a colon carcinoma cell line (Sizeland and Burgess, 1992).

Despite some of the potential problems that still exist in optimizing anti-sense oligonucleotide-mediated inhibition of gene expression, clinical trials using this strategy have been approved and it may soon begin to fulfill its promise as an important tool in gene therapy.

3. Ribozymes

Ribozymes can also be used to target various oncogenes (e.g. ras, c-fos, BCR-ABL), and can be used to help study gene expression, as well as to determine the malignancy of a phenotype. The advantages of ribozyme strategies are site-specific cleavage activity and catalytic potential. We have designed and generated three specific hammerhead ribozymes (Rz) targeted to ErbB-4 mRNA. ErbB-4 ribozyme, stably transfected into T47D human breast cancer cells, has been shown to down-regulate expression of the ErbB-4 receptor in T47D cells. We have observed that down-regulation of the ErbB-4 receptor in T47D cells results in a reduction in colony formation, as well as a reduction in transfection efficiency, compared with mock (vector) transfection. The low efficiency of selection of Rz-expressing clones suggests that ErbB-4 expression and mitogenic signaling may be essential for T47D cell survival. These preliminary findings suggest that down-regulation of ErbB-4 expression diminishes ErbB-4-mediated intracellular signaling. Because of heterodimerization between EGFR family receptors, down-regulation of ErbB-4 receptor may also indirectly interrupt other family receptor signaling pathways. This could result in a phenotype of diminished pathogenicity in T47D cells. These preliminary results also suggest that ErbB-4 may play a role in human breast cancer; and thus support the potential use of ribozymes as therapeutic agents for human breast cancers (Tang et al., 1997).

Although ribozyme technology is still in its infancy, the broad and potentially powerful uses of ribozymes have placed it among the prospective tools for gene therapy.

D. Receptor Tyrosine Kinase Inhibitors (EGFR PTK Inhibitors)

Proliferation of normal cells is dependent on more than one growth factor, and one growth factor can activate multiple intracellular signaling pathways. Abnormalities in the EGFR-signaling pathway have been associated with the development of many human cancers. EGFR is therefore used as a potential target for chemoprevention. Overexpression of EGFR or binding of EGFR with its ligands leads to constitutive activation of the EGFR tyrosine kinase signaling pathway. This kinase activation is a crucial event in mitogenic signaling. Because of the redundancy of growth factor networks, all of which lead to a single tyrosine kinase signaling pathway, blocking of EGFR tyrosine kinase activity could result in the inhibition of cellular proliferation. A series of specific EGFR tyrosine kinase inhibitors have been synthesized (Yaish *et al.*, 1988), including benzylidene malononitriles, dianilinophthalimides, quinazolines, [(alkylamino) methl]acrylophenones, enollactones, dihydroxybenzylaminosalicylates, 2-thiondoles, aminoflavones, and tyrosine analogue-containing peptides. Recently Parke-Davis Pharmaceutical Research has synthesized a series of compounds for evaluation as tyrosine kinase inhibitors. One of these small molecules, PD 153035, inhibits the EGF-receptor tyrosine kinase at the picomolar range and might also be competitive with ATP. PD 153035 selectively blocks EGF-mediated cellular processes including mitogenesis, early gene expression, and oncogenic transformation. This compound also demonstrates an increase in potency of four to five orders of magnitude over other tyrosine kinase inhibitors (Fry *et al.*, 1994).

VII. Conclusion

Clearly the EGF receptor family and their ligands play important roles in human neoplasia. These receptors and ligands are potentially useful targets for anticancer therapy. An improved fundamental understanding of the biochemical processes involved in normal receptor function and the transcription factors responsible for overexpression of these receptors, as well as the mechanism of action of EGF-like ligands, could provide opportunities for intervention. In addition, a knowledge of the three-dimensional structure of these receptors will assist in the design of peptides or other molecules capable of inhibiting dimerization. Finally, a combination of conventional therapies and molecular gene therapies could lead to a new dimension in cancer therapy.

Acknowledgments

We gratefully thank Dr. C. Richter King for his helpful discussions and suggestions. We also thank Kevin Brennan and Xiao-Zheng Alice Wu for assistance in the preparation of the manuscript.

References

Aaronson, S. A. (1991) Growth factors and cancer. *Science* **254**, 1146–1153.

Aboud-Pirak, E., Hurwitz, E., Pirak, M., et al. (1988). Efficacy of antibodies of epidermal growth factor receptor against KB carcinoma in vitro and in nude mice. *J. Natl. Cancer Inst.* **80**, 1605–1611.

Adamon, E. D. (1987). Oncogenes in development. *Development (Cambridge, UK)* **99**, 449–471.

Agosti, R. M., Leuthold, M., Gullick, W. J., Yasargil, M. G., and Wiestler, O. D. (1992). Expression of the epidermal growth factor receptor in astrocytic tumors is specifically associate with glioblastoma multiform. *Virchows Arch. A: Pathol. Anat. Histopathol.* **420**, 321–325.

Allred, D., Clark, G., Tandon, A., et al. (1990). HER-2/neu expression identified a group of node-negative breast cancer patients at high risk for recurrence. Presented at the annual meeting of the American Society of Clinical Oncology, Washington, DC, May.

Allred, D. C., Clark, G. M., Tandon, A. K. et al. (1992). HER-2/neu in node negative breast cancer: Prognostic significance of overexpression influenced by the presence of in situ carcinoma. *J. Clin. Oncol.* **10**, 599–605.

Alper, O., Yamaguchi, J., Hitomi, J., Honda, S., Matsushima, T., and Abe, K. (1990). The presence of c-erbB-2 gene product-related protein in culture medium conditioned by breast cancer cell line SK-Br-3. *Cell Growth Differ.* **1**, 591–599.

Antoniotti, S., Maggiora, P., Dati, C. et al. (1992). Tamoxifen up-regulates c-*erb*B-2 expression in oestrogen-responsive breast cancer cells *in vitro*. *Eur. J. Cancer* **28**, 318–321.

Anzano, M. A., Roberts, A. B., Smith, J. M., Sporn, M. B., and DeLarco, J. E. (1983). Sarcoma growth factor from conditioned medium of virally transformed cells is composed of both type alpha and type beta transforming growth factors. *Proc. Natl. Acad. Sci. U.S.A.* **80**, 6264–6268.

Aqel, N. M., Ball, R. Y., Walsmann, H., and Mitchinson, M. J. (1985). Identification of macrophages and smooth muscle cells in human atherosclerosis using monoclonal antibodies. *J. Pathol.* **146**, 197.

Bacus, S. S., Zelnizk, C. R., Plowman, G., and Yarden, Y. (1994). Expression of the erbB-2 family of growth factor receptors and their ligands in breast cancers. *Am. J. Pathol.* **102**, Suppl. 1, S13–S24.

Bargmann, C. I., and Weinberg, R. A. (1988). Oncogenic activation of the neu-encoded receptor protein by point-mutation and deletion. *EMBO J.* **7**, 2043–2052.

Bargmann, C. I., Hung, M. C., and Weinberg, R. A. (1986a). The *neu* oncogene encodes an epidermal growth factor receptor-related protein. *Nature (London)* **319**, 226–230.

Bargmann, C. I., Hung, M.-C., and Weinberg, R. A. (1986b). Multiple independent activations of neu oncogene by a point mutation altering the transmembrane domain of p185. *Cell (Cambridge, Mass.)* **45**, 649–657.

Barton, C. M., Hall, P. A., Hughes, C. M., Gullick, W. J., and Lemoine, N. R. (1991). Transforming growth factor alpha and epidermal growth factor in human pancreatic cancer. *J. Pathol.* **163**, 111–116.

Batra, J. K., Kasprzyk, P. G., Bird, R. E., Pastan, I., and King, C. R. (1992). Recombinant anti-erbB-2 immunotoxins containing *Pseudomonas* exotoxin. *Proc. Natl. Acad. Sci. U.S.A.* **89**, 5867–5871.

Battaglia, F., Scambia, G., and Benedetti Panici, P. (1989). Epidermal growth factor receptor expression in gynecological malignancies. *Gynecol. Obstet. Invest.* **27**, 42–44.

Bauknecht, T., Janz, I., Kohler, M., and Pfleiderer, A. (1989a). Human ovarian carcinomas: Correlation of malignancy and survival with the expression of epidermal growth factor receptors (EGF-R) and E9F-like factors (EGF-F). *Med. Oncol. Tumor. Pharmacother* **6**, 121–127.

Bauknecht, T., Kohler, M., Janz, I., and Pfleiderer, A. (1989b). The occurrence of epidermal growth factor receptors and the characterization of EGF-like factors in human ovarian, endometrial, cervical and breast cancer. EGF receptors and factors in gynecological carcinomas. *J. Cancer Res. Clin. Oncol.* 115, 193–199.

Bauknecht, T., Birmelin, G., and Kommoss, F. (1990). Clinical significance of oncogenes and growth factors in ovarian carcinomas. *J. Steroid Biochem. Mol. Biol.* 37, 855–862.

Bauknecht, T., Angel, P., Kohler, M. *et al.* (1993). Gene structure and expression analysis of the epidermal growth factor receptor, transforming growth factor-alpha, *myc, jun,* and metallothionein in human ovarian carcinomas: Classification of malignant phenotypes. *Cancer (Philadelphia)* 71, 419–429.

Benz, C. C., Scott, G. K., Robles, R., Parks, J. W., Montgomery, P. A., Daniel, J., Holmes, W. E., *et al.* (1993). A truncated intracellular HER2/neu receptor produced by alternative RNA processing affects growth of human carcinoma cells. *Mol. Cell. Biol.* 13(4), 2247–2257.

Benz, C. C., Scott, G. K., Sarup, J. C., Johnson, R. M., Tripathy, D., Coronado, E., Shepard, H. M., and Osborne, C. K. (1992). Estrogen-dependent, tamoxefin-resistant tumorigenic growth of MCF-7 cells transfected with HER2/neu. *Breast Cancer Rese. Treat.* 24, 85–95.

Berchuch, A., Kamel, A., Whitaker, R., *et al.* (1990). Overexpression of HER-2/neu is associated with poor survival in advanced epithelial ovarian cancer. *Cancer Res.* 50, 4087–4091.

Berns, E. M. J. J., Klijn, J. G. M., Henzen-Logmans, S. C., Rodenburg, C. J., Van Der Burg, M. E. L., and Foekens, J. A. (1992). Receptors for hormones and growth factors and (onco)-gene amplification in human ovarian cancer. *Int. J. Cancer* 52, 218–224.

Bianchi, S., Paglierani, M., Zampi, G. *et al.* (1993). Prognostic significance of c-erbB-2 expression in node negative breast cancer. *Br. J. Cancer* 67, 625–629.

Birchmeier, C., Sonnenberg, E., Weidner, K. M., and Walter, B. (1993). Tyrosine kinase receptors, in the control of epithelial growth and morphegenesis during development. *Bio. Essays* 15, 185–190.

Bishop, J. M. (1991). Molecular themes in oncogenesis. *Cell (Cambridge, Mass.)* 64, 235–248.

Borg, A., Baldetorp, B., Ferno, M., Killander, D., Olsson, H., Ryden, S., and Sigurdsson, H. (1994). ERBB2 amplification is associated with tamoxifen resistance in steriod-receptor positive breast cancer. *Cancer Lett.* 81, 137–144.

Brandt, R., Normanns, N., Gullick, W. J., Lin, J.-H., Harkins, R., Schneider, M., Jones, B. W., Ciardiello, F., Persico, M. G., Armenante, F., Kim, N., and Salomon, D. S. (1994). Identification and biological characterization of an epidermal growth factor-related protein: Cripto-1. *J. Biol. Chem.* 269, 17320–17328.

Brandt-Rauf, P. W., Rackovsky, S., and Pincus, M. R. (1990). Correlation of the neu oncogene-encoded p185 protein with its function. *Proc. Natl. Acad. Sci. U.S.A.* 87, 8660–8664.

Cambia, G., Benedetti Panici, P., Battaglia, F. *et al.* (1992). Significance of epidermal growth factor receptor in advanced ovarian cancer. *J. Clin. Oncol.* 10, 529–535.

Cantley, L. C., Auger, K. R., Carpenter, C. *et al.* (1991). Oncogenes and signal transduction. *Cell (Cambridge, Mass.)* 64, 281–302.

Carpenter, G., and Cohen, S. (1979). Epidermal growth factor. *Annu. Rev. Biochem.* 48, 193–216.

Carpenter, G., and Cohen, S. (1990). Epidermal growth factor. *J. Biol. Chem.* 265, 7709–7712.

Carpenter, G., and Wahl, M. I. (1990). The epidermal growth factor family. *Pept. Growth Factors Their Recept.* 1, 69–171.

Carraway, K. L., III, and Cantley, L. C. (1994). A neu acquaintance for erbB-3 and erbB-4: A role for receptor heterodimerization in growth signaling. *Cell (Cambridge, Mass.)* 78, 5–8.

Chaffanet, M., Chauvin, C., Laine, M. *et al.* (1992). EGF receptor amplification and expression in human brain tumors. *Eur. J. Cancer* 28, 11–17.

Chaudhary, V. K., Queen, C., Junghans, R. P., Waldmann, T. A., FitzGerald, D. J., and Pastan, I. (1989). A recombinant immunotoxin consisting of two antibody variable domains fused to pseudomonas exotoxin. *Nature (London)* 339, 394–397.

Chaudhary, V. K., Batra, J. K., Gallo, M., Willingham, M. C., FitzGerald, D. J., and Pastan, I. (1990). A rapid method of cloning functional variable-region antibody genes in *Esherichia coli* as single-chain immunotoxins *Proc. Natl. Acad. Sci. U.S.A.* **87**, 1066–1070; published erratum appears in *Proc. Natl. Acad. Sci. U.S.A.* **87**(8), 3253 (1990).

Chen, W. S., Lazar, C. S., Lund, K. A., Welsh, J. B., Chang, C.-P., Walton, G. M., Der, C. J., Wiley, H. S., Gill, G. N., and Rosenfeld, M. G. (1989). Functional independence of epidermal growth factor receptor from a domain required for ligand-induced internalization and calcium regulation. *Cell (Cambridge, Mass.)* **59**, 33–43.

Ciardiello, F., Dono, R., Kim, N., Persico, M. G., and Salomon, D. S. (1991a). Expression of cripto, a novel gene of the epidermal growth factor gene family, leads to in vitro transformation of a normal moousemammary epithelial cell line. *Cancer Res.* **51**, 1051–1054.

Ciardiello, F., Kim, N., Saeki, T., Dono, R., Persico, M. G., Plowman, G. D., Garrigues, J., Radke, S., Todaro, G. J., and Salomon, D. S. (**1991b**). Differential expression of epidermal growth factor-related proteins in human colorectal tumors. *Proc. Natl. Acad. Sci. U.S.A.* **88**, 7792–7796.

Ciccodocola, A., Dono, R., Obisci, S., Simeone, A., Zollo, M., and Persico, M. (1989). Molecular characterization of a gene of the 'EGF family' expressed in undifferentiated human NTERA2 teratocarcinoma cells. *EMBO J.* 1987–1991.

Clark, G. M., and McGuire, W. L. (1991). Follow-up study of HER-2/neu amplification in primary breast cancer. *Cancer Res.* **51**, 944–948.

Clarke, R., Brunnere, N., Katz, D., Glanz, P., Dickson, R. B., Lippman, M. E., and Kern, F. G. (1989). The effects of a constituitive expression of transforming growth factor- on the growth of MCF-7 human breast cancer cell *in vitro* and *in vivo*. *Mol. Endocrinol.* **3**, 372–380.

Cohen, B. D., Green, J. M., Foy, L., and Fell, H. P. (1996). HER4-mediated biological and biochemical properties in NIH 3T3 cells. *J. Biol. Chem.* **271**(9), 4813–4818.

Cohen, S., Ushiro, H., Stostcheck, C., and Chinkers, M. (1982). A native 170,000 epidermal growth factor receptor-kinase complex from shed plasma membrane vesicles. *J. Biol. Chem.* **257**, 1523–1531.

Connelly, P. A., and Stern, D. F. (1990). The epidermal growth factor and the product of the neu protooncogene are members of a receptor tyrosine phosphorylation cascade. *Proc. Natl. Acad. Sci. U.S.A.* **87**(16), 6054–6057.

Cook, P. W., Mattox, P. A., Keeble, W. W. *et al.* (1991). A heparin sulfate-regulated human keratinocyte autocrine factor is similar or identical to amphiregulin. *Mol. Cell. Biol.* **11**, 2547–2557.

Cook, P. W., Pittelkow, M. R., Keeble, W. W., Graves-Deal, R., Coffey, R. J., Jr., and Shipley, G. D. (1992). Amphiregulin messenger RNA is elevated in psoriatic epidermis and gastrointestinal carcinomas. *Cancer Res.* **52**, 3224–3227.

Coussens, L., Yang-Feng, T. L., Liao, Y. C. Chen, E., Gray, A., McGrath, J., Seeburg, P. H., Libermann, T. A., Schlesinger, J., Francke, U., and Toyoshima, K. (1985). Tyrosine kinase receptor with extensive homology to EGF receptor shares chromosomal location with *neu* oncognene. *Science* **230**, 1132–1139.

Dati, C., Antoniiotto, S., Taverna, D. *et al.* (1990). Inhibition of c-*erb*B-2 oncogene expression by estrogens in human breast cancer cell. *Oncogene* **5**, 1001–1006.

Davies, M. S., West, L. F., Davies, M. B., Povey, S., and Craig, R. K. (1987). The gene for human alpha-lactalbumin is assigned to chromosome 12q13. *Ann. Hum. Genet.* **51**, 183–188.

Davis, C. G. (1990). The many faces of epidermal growth factor repeats. *New Biol.* **2**, 410–419.

DeLarco, J. E., and Todaro, G. J. (1978). Growth factors from murine sarcoma virus-transformed cells. *Proc. Natl. Acad. Sci. U.S.A.* **75**, 4001–4005.

Derynck R. (1988). Transforming growth factor. *Cell (Cambridge, Mass.)* **54**, 593–595.

Di Carlo, A., Mariano, A., Macchia, P. E., Moroni, M. C., Beguinot, L., and Macchia, V. (1992). Epidermal growth factor receptor in human brain tumors. *J. Endocrinol. Invest.* **15**, 31–37.

Di Fiore, P. P., Pierce, J. H., Kraus, M. H. *et al.* (1987). ErbB-2 is a potent oncogene when overexpressed in NIH/3T3 cells. *Science* **237**, 178–182.

Di Fiore, P. P., Lonardo, F., Di Marco, E., King, C. R., Pierce, J. H., Segatto, O., and Aaronson, S. A. (1990). The mormal erB-2 product is an atypical receptor-like tyrosine kinase with constitutive activity in the absence of ligand. *New Biol.* **2**(11), 992–1003.

Dillman, R. O. (1994). Antibodies as cytotoxic therapy. *J. Clin. Oncol.* **12**, 1497–1515.

Di Marco, E., Pierce, J. H., Fleming, T. P. *et al.* (1989). Autocrine interaction between TGF alpha and the EGF-receptor: quantitative requirements for induction of the malignant phenotype. *Oncogene* **4**, 831–838.

Dono, R., Scalera, L., Pacifico, F., Acampora, D., Persico, M. G. and Simeone, A. (1993). The murine cripto gene: Expression during mesoderm induction and early heart morphogenesis. *Development (Cambridge, Mass.)* **118**, 1157–1168.

Dotzlaw, H., Miller, T., Karvelas, J., and Murphy, L. C. (1990). Epidermal growth factor gene expression in human breast cancer biopsy samples: Relationship to estrogen and progesterone receptor gene expression. *Cancer Res.* **50**, 4204–4208.

Dougall, W. C., Quian, X., Peterson, N. C., Miller, M. J., Samanta, A., and Greene, M. I. (1994). The new-oncogene: Signal transduction pathways, trasformation mechanisms and evolving therapies. *Oncogene* **9**, 2109–2123.

Downward, J., Yarden, Y., Mayer, E. *et al.* (1977). *Proc. Natl. Acad. Sci. U.S.A.* **74**, 565–569.

Downward, J., Yarden, Y., Mayes, E., Scrace, G., Totty, N., Stockwell, P., Ullrich, A., Schlessinger, J., and Waterfield, M. D. (1984). Close similarity of epidermal growth factor receptor and v-erbB oncogene protein sequences. *Nature (London)* **307**, 521–527.

Earp, S. H., Dawson, T. L., Li, X. and Yu, H. (1995). Heterodimerization and functional interaction between EGF receptor family member: A new signaling paradigm with implications for breast cancer research. *Breast Cancer Res. and Treat.* **35**, 115–132.

Ekstrand, A. J., Longo, N., Hamid, M. L., Olson, J. L., Liu, L., Collins, V. P., and James, C. D. (1994). Functional characterization of an EGF receptor with a truncated extracellular domain expressed in glioblastomas with EGFR gene amplification. *Oncogene* **9**, 2312–2320.

Ekstrand, A. J., Liu. L., He, J., Hamid, M. L., Longo, N., Collins, V. P., and James, C. D. (1995). Altered subcellular location of an activated and tumor-associated epidermal growth factor receptor. *Oncogene* **10**, 1455–1460.

Ekstrand, E. J., James, C. D., Cavenee, W. K., Seliger, B., Pettersson, R. F., and Collins, V. P. (1991). Genes for epidermal growth factor receptor, transforming growth gactor alpha and epidermal growth factor and their expression in human gliomas in vivo. *Cancer Res.* **51**, 2164–2172.

Elledge, R. M., Clark, G. M., Chamness, G. C., and Osborne, C. K. (1994). Tumor biological factors and breast cancer prognosis among white, Hispanic, and black women in the United States. *J. Natl. Cancer Inst.* **86**, 705–711.

Engel, J. (1989). EGF-like domains in extracellular matrix proteins: localized signals for growth and differentiation? *FEBS Lett.* **251**, 1–7.

Engler, D. A., Montelione, G. T., and Niyogi, S. K. (1990). Human epidermal growth factor. Distinct roles of tyrosine 37 and arginine 41 in receptor binding as determined by site-directed mutagenesis and nuclear magnetic resonance spectroscopy. *FEBS Lett.* **271**, 47–50, published erratum appears in *FEBS Lett.* **29**, 273(1–2), 261 (1990).

Escobedo, J. A., and Williams, L. T. (1988). A PDGF receptor domain essential for mutagenesis but not for many other responses to PDGF. *Nature (London)* **335**, 85–87.

Falls, D. L., Rosen, K. M., Corfas, G., Lane, W. S., and Fischbach, G. D. (1993). AIRA, a protein that stimulates acetylcholine receptor synthesis, is a member of the neu ligand family. *Cell (Cambridge, Mass.)* **72**, 801–815.

Fanger, M. W., Segal, D. M., and Romet-Lemonne, J. L. (1991). Bispecific antibodies and targeted cellular cytotoxicity. *Immunol. Today* **12**, 51–54.

Fanger, M. W., Morganelli, P. M., and Guyre, P. M. (1993). Use of bispecific antibodies in the therapy of tumors. *Cancer Treat. Res.* **68**, 181–194.

Fantal, W. J., Johnson, D. E., and William, L. T. (1993). Signaling by receptor tyrosine kinases. *Annu. Rev. Biochem.* **62**, 453–481.

Fitzpatrick, S. L., Brightwell, J., Wittliff, J., Barrows, G. H., and Schultz, G. S. (1984). Epidermal growth factor binding by brest tumor biopsies and relationship to estrogen and progestin receptor levels. *Cancer Res.* **44**, 3448–3453.

Fowler, J. E., Lau, J. L. T., Ghash, L., Mills, S. E., and Mounzer, A. (1988). Epidermal growth factor and prostatic carcinoma: an immunohistochemical study. *J. Urol.* **139**, 857–861.

Fry, D. W., Kraker, A. J., McMichael, A., Ambroso, L. A., Nelson, J. M., Leopold, W. R., Connors, R. W., and Bridges, A. J. (1994). A specific inhibitor of the epidermal growth factor receptor tyrosine kinase. *Science* **265**, 1093–1095.

Fry, M. J., Panayotou, G., Booker, G. W., and Waterfield, M. D. (1993). New insights into protein-tyrosine kinase receptor siganling complexes. *Protein Sci.* **2**, 1785–1797.

Garcia de Palazzo, I. E., Adams, G. P., Sundareshan, P., Wong, A. J., Testa, J. R., Bigner, D. D., and Weiner, L. M. (1993). Expression of mutated epidermal growth factor by non-small cell lung carcinomas. *Cancer Res.* **53**, 3217–3220.

Gasperini, G., Gullick, W. J., Bevilacqua, P. *et al.* (1992). Human breast cancer: Prognostic significance of the erbB-2 oncoprotein compared with Epidermal Growth Factor receptor, DNA ploidy, and conventional pathological features. *J. Clin. Oncol.* **10**, 686–695.

Georges, R. N., Mukhopadhyay, T., Zhang, Y., Yen, N., and Roth, J. A. (1993). Prevention of orthotopic human lung cancer growth by intratracheal instillation of a retroviral antisense K-*ras* construct. *Cancer Res.* **53**, 1743–1746.

Gill, G. N., Kawamoto, Y., Cochet, C., *et al.* (1984). Monoclonal anti-epidermal growth factor receptor antibodies which are inhibitors of epidermal growth factor binding and antagonist of epidermal growth factor-stimulated tyrosine protein kinase activity. *J. Biol. Chem.* **259**, 7755–7760.

Godolphin, W., Elwood, J. M., and Spinelli, J. J. (1981). Estrogen receptor quantitation and staging as complimentary prognostic inducators in breast cancer. *Int. J. Cancer* **28**, 677–683.

Goldman, R., Ben-Levy, R., Peles, E. *et al.* (1990). Heterodimerization of the erbB-1 and erbB-1 receptors in human breast carcinoma cells: A mechanism for receptor transregulation. *Biochemistry* **29**, 11024–11028.

Goodchild, J., Sun, D., Klotman, M., Agrawal, S., Zamecnik, P., and Gallo, R. (1988). Inhibition of human immunodeficiency virus type 1 replication by anti-sense oligonucleotides: An in vitro model for treatment. *Proc. Natl. Acad. Sci. U.S.A.* **89**, 11209–11213.

Groenen, L. C., Nice, E. C., and Burgess, A. W. (1994). Structure-function relationships for the EGF/TGF-alpha family of mitogens. *Growth Factors* **11**, 235–257.

Gullick, W. J. (1990). Inhibitors of growth factor receptors. In "Genes and Cancer" (D. Carney and K. Sikora eds.) pp. 263–273. Wiley, Chichester.

Gullick, W. J., Barger, M. S., Bennett, P. L., Rothbard, J. B., and Waterfield, M. D. (1987). Expression of the c-erbB-2 protein in normal and transformed cells. *Int. J. Cancer.* **40**, 246–254.

Gusterson, B. A., Machin, L. G., Gibbs, N. M., *et al.* (1988a). c-cerbB-2 expression in benign and malignant breast disease. *Br. J. Cancer* **58**, 453–547.

Gusterson, B. A., Machin, L. G., Gibbs, N. M., *et al.* (1988b). Immunochemical distrubtion of c-erbB-2 in infilrating and in situ breast cancer. *int. J. Cancer* **42**, 842–845.

Gusterson, B. A., Gelber, R. D., Goldhirsch, K. N., *et al.* (1992). Prognostic importance of c-erbB-2 expression in breast cancer. *J. Clin. Oncol.* **10**, 1049–1056.

Guy, C., Schuller, M., Parsons, T., Cardiff, R. D., and Muller, W. J. (1992). Induction in mammary tumors in transgenic mice expressing the unactivated c-neu oncogene. *J. Cell. Biochem. Suppl.* **16D**, 100 (abstr).

Guyre, P. M., Graziano, R. F., Vance, B. A., et al. (1989). Monoclonal antibodies that bind to distinct epitopes on Fc-gamma-RI are able to trigger receptor function. *J. Immunol.* **143**, 1650–1655.

Haley, J., Whittle, N., Bennett, P., Kinchington, D., Ullrich, A., and Waterfield, M. D. (1987). The human EGF receptor gene: Structure of the 110kb locus and identification of sequences regulating its transcription. *Oncogene Res.* **1**, 375–396.

Hanahan, D. (1985). Heritable formation of pancreatic beta-cell tumors in transgenic mice expressing recombinant insulin/simian virus 40 oncogenes. *Nature* **315**, 115–122.

Hanks, S. K., and Quinn, A. M. (1991). Protein kinase catalytic domain sequence database: Identification of conserved features of primary structure and classification of family members. *In* Methods in Enzymology (T. Hunter and B. M. Sefton, eds.) vol. 200, pp. 38–62. Academic Press, San Diego, CA.

Hanks, S. K., Quinn, A. M., and Hunter, T. (1988). The protein kinase family: conserved features and deduced phologeny of the catalytic domain. *Science* **241**, 48–52.

Harris, D. T., and Mastrangelo, M. J. (1989). Serotherapy of cancer. *Semin. Oncol.* **16**, 180–198.

Harvey, T. S., Wilkinson, A. J., Tappin, M. J., Cooke, R. M., and Campbell, I. D. (1991). The solution structure of human transforming growth factor alpha. *Eur. J. Biochem.*, **198**, 555–562.

Heldon, C.-H. (1995). Dimerization of cell surface receptors in signal transduction. *Cell (Cambridge, Mass.)* **80**, 213–223.

Helin, K., Velu, T., Martin, P., Vass, W. C., Allevato, G., Lowy, D. R., Beguinot, L. (1991). The biological activity of the human epidermal growth factor receptor is positively regulated by its C-terminal tyrosine. *Oncogene* **6**, 825–832.

Hellstrom, K. E., and Hellström, I. (1985). Oncogene-associated tumor antigens as target for immunotherapy. *FASEB J.* **3**, 1715–1722.

Hendler, F. J., Ozanne, B. W. (1984). Squamous cell cancers express increased EGF receptors. *J. Clin. Invest.* **74**, 647–665.

Henry, J. A., Hennessy, C., Levett, D. L. et al. (1993). Int-2 amplification in breast cancer: Association with decreased survival and relationship to amplification of erbB-2 and c-myc. *Int. J. Cancer* **53**, 774–780

Henzen-Longmans, S. C., Berns, E. M. J. J., Klijn, J. G. M., Van Der Burg, M. E. L., and Foekens, J. A. (1992a). Epidermal growth factor receptor in ovarian tumours: Correlation of immunohistochemistry with liyand binding assay. *Br. J. Cancer* **66**, 1015–1021.

Henzen-Longmans, S. C., Van Der Burg, M. E. L., Foekens, J. A. et al. (1992b). Occurrence of epidermal growth factor receptors in benign and malignant ovarian tumors and normal ovarian tissues: An immunohistochemical study. *J. Cancer Res. Clin. Oncol.* **118**, 303–307.

Higashiyama, S., Abraham, J. A., Miller, J., Fiddes, J. C., and Klagsbrun, M. (1991). A heparin-binding growth factor secreted by macrophage-like cells that is related to EGF. *Science* **251**, 936–939.

Hijiya, N., Zhang, J., Ratajczak, M. Z., Kant, J. A., DeRiel, K., Herlyn, M., Zon, G., and Gewirtz, A. (1994). Biologic and therapeutic significance of MYB expression in human melanoma. *Proc. Natl. Acad. Sci. U.S.A.* **91**, 4499–4503.

Holmes, W. E., Sliwkowski, M. X., Akita, R. W., Henzel, W. J., Lee, J., Park, J. W., Yansura, D., Abadi, N., Reeb, H., and Lewis, G. D. (1992). Identification of Heregulin, a specific activator of p185erbB2. *Science* **256**, 1206–1210.

Hommel, U., Dudgeon, T. J., Fallon, A., Edwards, R. M., and Campbell, I. D. (1991). Structure-function relationships in human epidermal growth factor studied by site-directed mutagenesis and 1H NMR. *Biochemistry* **30**, 8891–8898.

Hotta, H., Ross, A. H., Huebner, K., Isobe, M., Wendeborn, S., Chao, M. V., Ricciardi, R. P., Tsujimoto, Y., Croce, C. M., and Koprowski, H. (1988). Molecular cloning and characterization of an antigen associated with early stages of melanoma tumor progression. *Cancer Res.* **48**, 2955-2962.

Huang, S. S., and Huang, J. S. (1992). Purification and characterization of the neu/erbB2 ligand growth factor from bovine kidney. *J. Biol. Chem.* **267**, 11508-11512.

Hudziak, R. M., Schlessingeer, J., and Ullrich, A. (1987). Increased expression of the putative growth factor receptor p185^{HER2} causes transformation and tumorigenesis of NIH-3t3 cells. *Proc. Natl. Acad. Sci. U.S.A.* **84**, 7159-7163.

Hudziak, R. M., Lewis, G. D., Winget, M., Fendly, B. M., Shepard, H. M., and Ullrich, A. (1989). Monoclonal antibody has antiproliferative effects in vitro and sensitizes human breast tumor cells to tumor necrosis factor. *Mol. Cell. Biol.* **9**, 1165-1172.

Humphery, P. A., Wong, A. J., Vogelstein, B., Friedman, H. S., Werner, M. H., Bigner, D. D., and Bigner, S. H.. (1988). Amplification and expression of the epidermal growth factor receptor gene in human glioma xenografts. *Cancer Res.* **48**, 2231-2238.

Humphery, P. A., Wong, A. J., Vogelstein, B., Zalutsky, M. R., Fuller, G. N., Archer, G. E., Friedman, H. S., Kwatra, M. M., Bigner, S. H., and Bigner, D. D. (1990). Anti-synthetic peptide antibody reacting at the fusion junction of deletion-mutant epidermal growth factor receptors in human globlastomas. *Proc. Natl. Acad. Sci. U.S.A.* **87**, 4207-4211.

Hung, M.-C., Schechter, A. L., Chevary, P.-Y., Stern, D. F., and Weinberg, R. G. (1986). Molecular cloning of the neu gene: Absence of gross structural alteration in oncogenic alleles. *Proc. Natl. Acad. Sci. U.S.A.* **83**, 261-264.

Hung, M.-C., Yan, D.-H., and Zhao, X. (1989). Amplifications of the proto-neu oncogene facilitates oncogenic activation by a single point mutation. *Proc. Natl. Acad. Sci. U.S.A.* **86**, 2545-2548.

Ibbotson, K. J., Twardzik, D. R., D'Souzea, M. S., Hargreaves, W. R., Todaro, G. J., and Mundy, G. R. (1985). Stimulation of bone resoption *in vivo* by synthetic transforming growth factor-alpha. *Science* **228**, 1007-1009.

Jardines, L., Weiss, M., Fowble, B., and Greene, M. (1993). neu (c-erbB-2/HER2) and the epidermal growth factor receptor in breast cancer. *Pathobiology* **61**, 268-282.

Jhappan, C., Stahle, C., Harkins, R. N., Fausto, N., Smith, G. H., and Merlino, G. T. (1990). TGF- alpha overexpression in transgenic mice induces liver neoplasia and abnormal development of the mammary gland and pancreas. *Cell (Cambridge, Mass.)* **61**, 1137-1146.

Johnson, G. R., Saeki, T., Auersperg, N. *et al.* (1991). Response to and expression of amphiregulin by ovarian carcinoma and normal ovarian surface epithelial cells: Nuclear localization of endogenous amphiregulin. *Biochem. Biophys. Res. Commun.* **180**, 481-488.

Johnson, G. R., Kannan, B., Shoyab, M., and Stromberg, K. (1993). Amphiregulin induces tyrosine phosphorylation of the epidermal growth factor receptor and pl 85erbB2: Evidence that amphiregulin acts exclusively through the epidermal growth factor receptor at the surface of human epithelial cells. *J. Biol. Chem.* **268**, 2924-2931.

Jones, N. R., Rossi, M. L., Gregoriou, M., and Hughes, J. T. (1990). Investigation of the expression of epidermal growth factor receptors and blood group A antigen in 110 human gliomas. *Neuropathol. Appl. Neurobiol.* **16**, 185-192.

Kajikawa, K., Yasui, W., Sumiyoshi, H. *et al.* (1991). Expression of epidermal growth factor in human tissues: Immunohistochemical and biochemical analysis. *Virchows Arch. A: Pathol. Anat. Histopathol.* **418**, 27-32.

Karunagaran, D., Tzahar, E., Liu, N., Wen, D., and Yarden, Y. (1995). Neu differentiation factor inhibits EGF binding, a model for trans-regulation within the erbB family of receptor tyrosine kinases. *J. Biol. Chem.* **270**(17), 9982-9990.

Kashani-Sabet, M., Funato, T., Tone, T. *et al.* (1992). Reversal of the malignant phenotype by an anti-*ras* ribozyme. *Antisense Res. Dev.* **2**, 3-15.

Kashani-Sabet, M., Funator, T., Florenes, V. A. *et al.* (1994). Sppression of the neoplastic phenotype in vivo by an anti-*ras* ribozyme. *Cancer Res.* **54**, 900-902.

Kasprzyk, P. G., Sullivan, T. L., Hunt, J. D., Gubish, C. T., Scoppa, C. A., Oelkuct, M., Bird, R., Fischer, P. H., Siegfried, J. M., and King, C. R. (1996). Activity of anti-erbB-2 recombinant toxin OLX-209 on lung cancer cell lines in the absence of erbB-2 gene amplification. *Clin. Cancer Res.* **2**, 75–80.

Kawamoto, T., Sato, J. D., Le, A. D., *et al.* (1983). Growth stimulation of A431 cells by epidermal growth factor: Identification of high-affinity receptors for epidermal growth factor by an anti-receptor monoclonal antibody. *Proc. Natl. Acad. Sci. U.S.A.* **80**, 1337–1341.

Kawamoto, T., Medelson, J., Le, A. D., *et al.* (1984). Relation of epidermal growth factor receptor concentration of growth of human epidermoid carcinoma A341 cells. *J. Biol. Chem.* **259**, 7761–7766.

Kellof, G. J., Fay, J. R., Steele, V. E., Lubet, R. A., Boone, C. W., Crowell, J. A., and Sigman, C. C. (1996). Epidermal growth factor receptor tyrosine kinase inhibitors as potential cancer chemopreventives. *Cancer Epidemiol., Biomarkers Prev.* **5**, 657–666.

Kenney, N., Johnson, G., Selvam, M. P. *et al.* (1993). Transforming growth factor a (TGFalpha) and amphiregulin (AR) as autocrine growth factors in nontransformed immortalized 184A1N4 human mammary epithelial cells. *Mol. Cell Differ.* **1**, 163–184.

Kihara, A., and Pastan, I. (1995). Cytotoxic activity of chimeric toxins containing the epidermal growth factor-like domain of heregulin fused to PE38KDEL, a truncated recombinant form of pseudomonas exotoxin. *Cancer Res.* **55**, 71–77.

King, C. R., Kraus, M. H., and Aaronson, S. A. (1985). Amplification of a novel v-erbB-2 related gene in a human mammary carcinoma. *Science* **229**, 974–976.

King, C. R., Borrello, I., Bellot, F., Comoglio, P., and Schlessinger, J. (1988). EGF binding to its receptor triggers a rapid tyrosine phosphorylation of erbB-2 protein in the mammary tumor cell line SKBR-3. *EMBO J.* **7**, 1647–1651.

Kinzler, K. W., Bigner, S. H., Bigner, D. D., Trent, J. M., Law, M. L., O'Brien, S. J., Wong, A. J., and Volgelstein, B. (1987). Identification of an amplified, highly expressed gene in a human glioma. *Science* **236**, 70–73.

Kitadai, Y., Yasui, W., Yokozaki, H. *et al.* (1991). Expression of amphiregulin, a novel gene of .the epidermal growth factor family, in human gastric carcinomas. *Jpn. J. Cancer Res.* **82**, 969–973.

Klagsbrun, M., and Baird, A. (1991). A dual receptor system is required for basic fibroblast growth factor activity. *Cell (Cambridge, Mass.)* **67**, 229–231.

Klijn, J. G. M., Berns, P. M. J. J., Schmitz, P. I. M., Foekens, J. A. (1992). The clinical significance of epidermal growth factor receptor (EGF-R) in human breast cancer: A review of 5232 patients. *Endocr. Rev.* **13**, 3–17.

Knighton, D. R., Zheng, J., Eyck, L. F. T., Ashford, V. A., Xuong, N., Talor, S. S., and Sowadski, J. M. (1991a). Crystal structure of the catalytic subunit of cyclic adenosine monophosphate-dependent protein kinase [see comments]. *Science* **253**, 407–414.

Knighton, D. R., Zheng, J., Eyck, L. F. T., Xuong, N., Talor, S. S., and Sowadski, J. M. (1991b). Structure of a peptide inhibitor bound to the catalytic subunit of cyclic adenosine monophosphate-dependent protein kinase [see comments]. *Science* **253**, 414–420.

Koch, C. A., Anderson, D., Moran, M. F. *et al.* (1991). SH2 and SH3 domains: Elements that control interactions of cytoplasmatic signaling proteins. *Science* **252**, 668–674.

Kohda, D., and Inagaki, F. (1992a). Structure of epidermal growth factor bound to perdeuterated dodecylphosphocholine micelles determined by two-dimensional NMR and stimulated annealing calculations. *Biochemistry* **31**, 677–685.

Kohda, D., and Inagaki, F. (1992b). Three-dimentional nuclear magnetic resonance structures of mouse epidermal growth factor in acdic and physiological PH solutions. *Biochemistry* **31**, 11928–11939.

Kohler, M., Janz, I., Wintzer, H.-O., Wagner, E., and Bauknecht, T. (1989). The expression of EGF receptors, EGF-like factors, and c-myc in ovarian and cervical carcinomas and their potential clinical significance. *Anticancer Res.* **9**, 1537–1548.

Kokai, Y., Myers, J. N., Wada, T., Brown, V. I., Levea, C. M., Davis, J. G., Dobashi, K., and Greene, M. I. (1989). Synergistic interaction of p185c-neu and the EGF receptor leads to transformation of rodent fibroblasts. *Cell (Cambridge, Mass.)* **58**, 287–292.

Korc, M., Chandrasekar, B., Yamanaka, Y., Friess, H., Buchler, M., and Beger, H. G. (1992). Overexpression of the epidermal growth factor receptor in human pancreatic cancer is associated with concomitant increases in the levels of epidermal growth factor and transforming growth factor alpha. *J. Clin. Invest.* **90**, 1352–1360.

Kraus, M. H., Popescu, N. C., Amsbaugh, S. C., and King, C. R. (1987). Overexpression of teh EGF receptor-related proto-oncogene erbB-2 in human mammary tumor cell lines by different molecular mechanisms. *EMBO J.* **6**, 605–610.

Kraus, M. H., Issing, W., Miki, T., Popeson, N. C., and Aaronson, S. A. (1989). Isolation and characterization of ERBB3, a third member of the ERBB/epidermal growth factor receptor family: Evidence for overexpression in an subset of human mammary tumors. *Proc. Natl. Acad. Sci. U.S.A.* **86**, 9193–9197.

Kraus, M. H., Fedi, P., Starks, V., Muraro, R., and Aaronson, S. A. (1993). Demonstration of ligand-dependent signaling by the erbB-3 tyrosine kinase and its constitutive activation in human breast tumor cells. *Proc. Natl. Acad. Sci. U.S.A.* **90**, 2900–2904.

Kris, R. M., Lax, I., Gullick, W., Waterfield, M. D., Ullrich, A., Fridkin, M., and Schlessinger, J. (1985). Antibodies against a synthetic peptide as a probe for thekinase activity of the avian EGF receptor and v-erbB protein. *Cell (Cambridge, Mass.)* **40**, 619–625.

Kuniyasu, H., Yoshida, K., Yokozaki, H., Yasui, W., Ito, H., Toge, T., Ciardiello, F., Persico, M. G., Saeki, T., Salomon, D. S. and Tahara, E. (1991). Expression of cripto, a novel gene of the epidermal growth factor family, in human gastrointestinal carcinomas. *Jpn. J. Cancer Res.* **82**, 969–973.

Langton, B. C., Crenshaw, M. C., Chao, L. A., Stuart, S. G., Akita, R. W., Jackson, J. E. (1991). An antigen immunologically related to the external domain of gpl 85 is shed from nude mouse tumours overexpressing the c-erbB-2 *(Her2/neu)* oncogene. *Cancer Res.* **51**, 2593–2598.

Lax, I. A., Johnson, A., Howk, R., Sap, J., Bellot, F., Winkler, M., Ullrich, A., Vennstrom, B., Schlessginer, J., and Givol, D. (1988). Chicken epidermal growth factor (EGF) receptor: cDNA cloning, expression in mouse cells, and differential binding of EGF and transforming growth factor alpha. *Mol. Cell. Biol.* **8**, 1970–1978.

Lax, I., Bellot, F., Howk, R., Ullrich, A., Givol, D., and Schlessinger, J. (1989). Functional analysis of the ligand binding site of EGF-receptor utilizing chimeric chicken/human receptor molecules. *EMBO J.* **8**, 421–427.

Lee, J., and Wood, W. I. (1993). Assignment of heregulin (HGL) to human chromosome 8p22-p11 by PCR analysis of somatic cell hybrid DNA. *Genomics* **16**, 790–791.

Lee, K. F., Simon, H., Chen, H., Bates, B., Hung, M. C., and Hauser, C. (1995). Requirement for neuregulin receptor erbB2 in neural and cardiac development. *Nature* **378**, 394–398.

Leitzel, K., Teramoto, Y., Sampson, E. L., Wallingford, G. A., Weaver, S., Dcmcro, L., Harvey, H., and Lipton, A. (1991). Elevated c-erbB-2 levels in the serum and tumor extracts of breast cancer patients. *Proc. Am. Assoc. Cancer Res.* **32**, 997.

Leitzel, K., Teramoto, Y., Sampson, E. et al. (1992). Elevated soluble c-erbB-2 antigen levels in the serum and effusions of a proportion of breast cancer patients. *J. Clin. Oncol.* **10**, 1436–1443.

Lemoine, N. R., Hughes, C. M., Gullick, W. J., Brown, C. L., and Wynford-Thomas, D. (1991a). Abnormalities of the receptor system in human thyroid neoplasia. *Int. J. Cancer* **49**, 1–4.

Lemoine, N. R., Jain, S., Silvestre, F., Lopes, C., Hughes, C. M., McLelland, E., Gullick, W. J., and Filipe, M. I. (1991b). Amplification and overexpression of EGF receptor and c-erbB-2 proto-oncogenes in human stomach cancer. *Br. J. Cancer* **64**, 79–83.

Lemoine, N. R., Lobresco, M., Leung, H. et al. (1992). The erbB-3 gene in human pancreatic cancer. *J. Pathol.* **168**, 269–273.

Leonardo, F., Di Marco, E., King, C. R. et al. (1990). The normal erbB-2 product is an atypical receptor-like tyrosine kinase with constitutive activity in the absence of ligand. *New Biol.* **2**, 992–1003.

Libermann, T. A., Razon, N., Battal, A. D., Yarden, Y., Schlessinger, J., Soreq, H. (1984). Expression of epidermal growth factor receptors in human brain tumors. *Cancer Res.* **44**, 753–760.

Libermann, T. A., Nusbaum, H. R., Razon, N., Kris, R., Lax, I., Soreq, H., Whittle, N., Waterfeld M. D., Ullrich, A., and Schlessinger, J. (1985). Amplification, enhanced expression and possible rearrangement of EGF receptor gene in primary brain tumors of gial origin. *Nature (London)* **313**, 144–147.

Lin, C. R., Chen, W. S., Kruiger, W., Stolarsky, L. S., Weber, W., Evans, R. M., Verma, I. M., Gill, G. N., and Rosenfeld, M. D. (1984). Expression cloning of human EGF receptor complementary DNA: Gene amplification and three related messenger RNA products in A431 cells. *Nature (London)* **224**, 843–848.

Lin, Y. J., and Clinton, G. M. (1991). A soluble protein related to the HER-2 proto-oncogene product is released from human breat carcinoma cells. *Oncogene* **6**, 639–643.

Ling, L. E., Drucker, B. J., Cantley, L. C., and Roberts, T. M. (1992). Transformation-defective mutants of polymavirus middle T antigen associate with phosphatidylinositol 3-kinase (PI 3-kinase) but are unable to maintain wild-type levels of PI 3-kinase products in intact cells. *J. Virol.* **66**, 1702–1708.

Liotta, L. A. (1984). Tumor invasion and metastasis: Role of the basement membrane. *Am. J. Pathol.* **117**, 339–348.

Liu, C., and Tsao, M.-S. (1993). In vitro and in vivo expressions of transforming growth factor-alpha and tyrosine kinase receptors in human non-small-cell lung carcinomas. *Am. J. Pathol.* **142**, 1155–162.

Liu, C., Woo, A., and Tsao, M.-S. (1990). Expression of transforming growth factor-alpha in primary human colon and lung carcinomas. *Br. J. Cancer* **62**, 425–429.

Logan, A. (1990). Intracrine regulation at the nucleus—a further mechanism of growth factor activity. *J. Endocrinol.* **125**, 339–343.

Lonn, U., Lonn, S., Nilsson, B., and Stevinkvist, B. (1994). Breast cancer: Prognostic significance of c-erbB-2 and int-2 amplificatopn compared with DNA ploidy, S-phase fraction, and conventional clinicopathological factors. *Breast Cancer Res. Treat.* **29**, 237–245.

Lupu, R., Colomer, R., Kannan, B. et al. (1992). Characterization of a growth factor that binds exclusively to the erbB-2 receptor and induces cellular response. *Proc. Natl. Acad. Sci. U.S.A.* **89**, 2287–2291.

Madsen, M. W., Lykkefeldt, A. E., Lausen, I., Nielson, K. V., and Briand, P. (1992). Altered gene expression of c-myc, epidermal growth factor receptor, transforming growth factor- and c-erb-B2 in an in an immortalized human breast epithilial cell line, HMT-3522, is associated with decreased growth factor requirements. *Cancer Res.* **52**, 1210–1217.

Malden, L. T., Novak, U., Kaye, A. H., and Burgess, A. W. (1988). Selective amplification of the cytoplasmic domain of the epidermal growth factor receptor gene in glioblastoma multiforme. *Cancer Res.* **48**, 2711–2714.

Marchionni, M. A., Gooderal, A. D. J., Chen, M. S., Bermingham-McDonogh, O., Kirk, C., Hendricks, M., Denehy, F., Misumi, D., Sudhalter, J., Kobayashi, K., Wrobleski, D., Lynch, C., Baldassare, M., Hiles, I., Davis, J. B., Hsuan, J. J., Totty, N. F., Otsu, M., McBurry, R. N., Waterfield, M. D., Stroobant, P., and Gwynne, D. (1993). Glial growth factors are alternatively spliced erbB2 ligands expressed in the nervous system [see comments] *Nature (London)* **362**, 312–318.

Marikovsky, M., Lavi, S., Pinkas-Kramarski, R., Karunagaran, D., Liu, N., Wen, D., and Yarden, Y. (1995). ErbB-3 mediates differential mitogenic effects of NDF/heregulin isoforms on mouse keratinocyte. *Oncogene* **10**(7), 1403–1411.

Marquardt, H., Hunkapiller, H. W., Hood, L. E., and Todaro, G. J. (1984). Rat transforming growth factor type 1: Structure and relation to epidermal growth factor. *Science* **233**, 1079–1082.

Massague, J. (1983a). Epidermal growth factor-like transforming growth factor. I. Isolation, chemical characterization, and potentiation by other transforming factors from feline sarcoma virus-transformed rat cells. *J. Biol. Chem.* **258**, 13606–13613.

Massague, J. (1983b). Epidermal growth factor-like transforming growth factor. II. Interaction with epidermal growth factor receptors in human placenta membranes and A431 cells. *J. Biol. Chem.* **258**, 13614–13620.

Massague, J. (1990). Transforming growth factor-alpha: A model for membrane-anchored growth factors. *J. Biol. Chem.* **256**, 21393–21396.

McCune, B. K., Hickle, W. R., and Earp, H. S. (1993). The role of endocytosis in epidermal growth factor signaling complexes. *Adv. Cell Mol. Biol. Membs.* **1**, 269–305.

McGeady, M. L., Kerby, S., Shankar, V., Ciardiello, F., Salomon, D., and Seidman, M. (1984). Infection with a TGF-alpha retroviral vector transforms normal mouse mammaryy epithelial cells but not normal rat fibroblasts. *Oncogene* **4**, 1375–1382.

McManaway, M. E., Neckers, L. M., Loke, S. L., Al-Nasser, A. A., Redner, R. L., Shiramizu, B. T., Goldschmidts, W. L., Huber, B. E., Bhatia, K., and Magrath, I. T. (1990). Tumor-specific inhibition of lymphoma growth by an anti-sense oligodeoxynucleotides. *Lancet* **335**, 808–811.

Mellstedt, H. (1990). Monoclonal antibodies in cancer therapy. *Curr. Opin. Immunol.* **2**, 708–713.

Merlino, G. T. (1990). Epidermal growth factor receptor regulation and function. *Semin. Cancer Biol.* **1**, 277–284.

Merlino, G. T., Ishii, S., Whang-Peng, J., Knutsen, T., Xu, Y. H., Clark, A. J., Stratton, R. H., Wilson, R. K., Ma, D. P., Roe, B. A., Hunts, J. H., Shimizu, N., and Pastan, I. (1985). Structure and localization of genes encoding aberrant and normal epidermal growth factor receptor RNAs from A431 human carcinoma cells. *Mol. Cell. Biol.* **5**, 1722–1734.

Meyer, D., and Birchmeier, C. (1995). Multiple essential functons of neuregulin in development [see comments]. *Nature (London)* **378**, 390–394.

Mincione, M., Bianco, C., Kannan, S., Colletta, G., Ciardiello, F., Sliwkowski, M., Yarden, Y., Normanno, N., Pramaggiore, A., Kim, N., and Salomon, D. S. (1996). Enhanced expression of heregulin in c-erbB-2 and c-Ha-ras transfored mouse and human mammary epithelial cells. *J. Cell. Biochem.* **60**, 437–446.

Mizukami, Y., Nonomura, A., Noguchi, M. *et al.* (1991). Immunohistochemical study of oncogene product Ras p21, c-myc and growth factor EGF in breast carcinomas. *Anticancer Res.* **11**, 1485–1494.

Mizutani, T., Onda, M., Tokunaga, A., Yamanaka, N., and Sugisaki, Y. (1993). Relationship of C-erbB-2 protein expression and gene amplification to invasion and metastasis in human gastric cancer. *Cancer (Philadelphia)* **72**, 2083–2088.

Moran, M. F., Koch, C. A., Sadowski, I., and Pawson, T. (1988). Mutational analysis of a phosphatransfer motif essential for v-fps tyrosine kinase activity. *Oncogene* **3**, 665–672.

Morishige, K., Kurachi, H., Amemiya, K. *et al.* (1991). Evidence for the involvement of transforming growth factor alpha and epidermal growth factor receptor autocrine growth mechanism in primary human ovarian cancers in vitro. *Cancer Res.* **51**, 5322–5328.

Moroni, M. C., Willingham, M. C., and Beguinot, L. (1992). EGF-R antisense RNA blocks expression of the epidermal growth factor receptor and suppresses the transforming phenotype of a human carcinoma cell line. *J. Biol. Chem.* **267**, 2714–2722.

Moscatello, D. K., Holgado-Madruga, N. M., Godwin, A. K., Ramirez, G., Gunn, G., Zoltick, P. W., Biegel, J. A., Hayes, R. L., and Wong, A. J. (1995). Frequent expression of a mutant epidermal growth factor receptor in multipele human tumors. *Cancer Res.* **55**, 5536–5539.

Moy, F. J., Rauenbuehler, P., Winkler, M. E., Scheraga, H. A., and Montelione, G. T. (1993). Solution structure of human type-alpha transforming factor determined by heteronuclear

NMR spectroscopy and refined by energy minimization with restraints. *Biochemistry,* **32,** 7334–7353.
Mukhopadhyay, T., Tainsky, M., Cavender, A. C., and Roth, J. A. (1991). Specific inhibition of K-*ras* expression and tumorigenicity of lung cancer cells by antisense RNA. *Cancer Res.* **51,** 1744–1748.
Muller, W. J., Sinn, E., Pattengale, P. K., Wallace, R., and Leder, P. (1988). Induction of mammary adenocarcinoma in transgenic mice bearing the actived c-neu oncogene. *Cell (Cambridge, Mass.)* **54,** 105–115.
Murphy, L. C., Dotzlaw, H., Wong, M. S. J. *et al.* (1990). Epidermal growth factor: Receptor and ligand expression in human breast cancer. *Semin. Cancer Biol.* **1,** 305–315.
Murphy, L. D., Valverius, E. M., Tsokos, M., Mickiey, L. A., Rosen, N., and Bates, S. E. (1990). Modulation of EGF receptor expression by differentiating agents in human colon carcinoma cell lines. *Cancer Commun.* **2,** 345–355.
Muss, H. B., Thor, A., Berry, D. A., Kute, T. *et al.* (1994). c-erbB-2 expression and response to adjuvant therapy in women with node-positive early breast cancer. *N. Engl. J. Med.* **5,** 1260–1266.
Nagai, M. A., Marques, L. A., Torloni, H., and Brentani, M. M. (1993). Genetic alterations in c-erbB-2 protooncogene as prognostic markers in human primary breast tumors. *Oncology* **50,** 412–417.
Nagata, K., Kohda, D., Hatanaka, H., Ichikawa, S., Matsuda, S., Yamamoto, T., Suzuki, A., and Inagaki, F. (1994). Solution structure of the epidermal growth factor-like domain of heregulin-α, a ligand for p180^{erbB-4}. *EMBO J.* **13**(15), 3517–3522.
Natali, P. G., Nicota, M. R., Bigotti, A., Venturo, I., Slamon, D. J., Fendly, B. M., and Ullrich, A. (1990). Expression of the p185 encoded by HER-2 oncogene in normal and transformed human tissues. *Int. J. Cancer* **45,** 457–461.
Nicholson, S., Wright, C., Sainsbury, J. R., Halcrow, O., *et al.* (1990). Epidermal growth factor receptor (EGFr) as a marker for poor prognosis in node-negative breast cancer patients: neu and tamoxifen failure. *J. Steroid Biochem. Mol. Biol.* **37,** 811–818.
Nishikawa, R., Ji, X. D., Harmon, R. C., Lazar, C. S., Gill, G. N., Cavenee, W. K., and Huang, H. J. (1994). A mutant epidermal growth factor receptor common in human glioma confers enhanced tumorigenicity. *Proc. Natl. Acad. Sci. U.S.A.* **91**(16), 7727–7731.
Noguchi, M., Koyasaki, N., Ohta, N. *et al.* (1992). c-erbB-2 oncoprotein expression versus internal mammmary lymph node metastasis as additional prognostic factors in patients with axillary lymph node-positive breast cancer. *Cancer (Philadelphia)* **69,** 2953–2960.
Nonomura, A., Ohta, G., Nakanuma, Y. *et al.* (1988). Simultaneous detection of epidermal growth factor receptor (EGF-R), epidermal growth factor (EGF) and *ras* p21 in cholangiocarcinoma by an immunocytochemical method. *Liver* **8,** 157–166.
Normanno, N., Ciardello, F., Brandt, R., and Salomon, D. S. (1994). Epidermal growth factor-related peptides in the pathogenesis of human breast cancer. *Breast Cancer Res. Treat.* **29,** 11–27.
Orr-Urtreger, A., Trakhtenbrot, L., Ben-Levy, R., Wen, D., Rechavi, G., Lonai, P., and Yarden, Y. (1993). Neural expression and chromosomal mapping of Neu differentiation factor to 8p12-p21. *Proc. Natl. Acad. Sci. U.S.A.* **90,** 1867–1871.
Owens, O. J., Stewart, C., Leake, R. E., and McNicol, A. M. (1992). A comparison of biochemical and immunohistochemical assessment of EGFR expression in ovarian cancer. *Anticancer Res.* **12,** 1455–1458.
Paik, S., Hazan, R., Fisher, E. R., Sass, R. E., Fisher, B., Redmond, C., Schlessinger, J., Lippman, M. E., and King, C. R. (1990). Pathologic findings from the national surgical adjuvant breast and bowel project: Prognostic significance of erbB-2 protein overexpression in primary breast cancer. *J. Clin. Oncol.* **8,** 103–112.
Partanen, A. M. (1990). Epidermal growth factor and transforming growth factor-alpha in the development of epithelial-mesenchymal organs of the mouse. *Curr. Top. Dev. Biol.* **24,** 33–55.

Pastan, I., and FitzGerald, D. (1989). Pseudomonas exotoxin: Chimeric toxins. *J. Biol. Chem.* **264**, 15157-15160.

Pastan, I., Willingham, M. C., and FitzGerald, D. (1986). Immunotoxins. *Cell (Cambridge, Mass.)* **47**, 641-648.

Pastan, I., Chaudhary, V., and FitzGerald, D. (1992). Recombinant toxins as novel therapeutic agents. *Annu. Rev. Biochem.* **61**, 331-354.

Peles, E., Bacus, S. S., Koski, R. A. *et al.* (1992). Isolation of the Neu/HER-2 stimulatory ligand: A 44 kd glycoprotein that induces differentiation of mammary tumor cells. *Cell (Cambridge, Mass.)* **69**, 205-216.

Peles, E., Ben, L. R., Tzahar, E., Liu, N., Wen, D., and Yarden, Y. (1993). Cell-type specific interaction of neu differenciation factor (NDF/heregulin) with neu/HER-2 suggests complex ligand-receptor relationships. *EMBO J.* **12**, 961-971

Perez, P., Hofffman, R. W., Shaw, S., *et al.* (1985). Specific targeting of cytotoxic T cells by anti-T3 linked to anti-target cell antibody. *Nature (London)* **316**, 354-356.

Pierce, J. H. (1990). Signal transduction through foreign growth factor receptors. *Adv. Regul. Cell Growth* **2**.

Pietras, R. J., Arboleda, J., Reese, D. M., Wongvipat, N., Pegram, M., Ramos, L., Gorman, C. M., Parker, M. G., Slikowski, M. X., and Slamon, D. J. (1995). HER-2 tyrosine kinase pathway targets estrogen receptor and promotes hormone-independent growth in human breast cancer cells. *Oncogene* **10**(12), 2435-2446.

Pinkas-Kramarski, R., Soussan, L., Waterman, H., Levkowitz, G., Alroy, I., Klapper, L., Lavi, S., Seger, R., Ratzkin, B. J., Sela, M., and Yarden, Y. (1996). Diversification of Neu differentiation factor and epidermal growth factor signaling by combinatorial receptor interactions. *EMBO J.* **15**(10), 2452-2467.

Plowman, G. D., Whitney, G. S., Neubauer, M. G., Green, J. M., McDonald, V. L., Todaro, G. J., and Shoyab, M. (1990a). Molecular cloning and expression of an additional epidermal growth factor receptor related gene. *Proc. Natl. Acad. Sci. U.S.A.* **87**, 4905-4909.

Plowman, G. D., Green, J. M., McDonald, V. L., Neubauer, M. G., Disteche, C. M., Todaro, G. J., and Shoyab, M. (1990b). The amphiregulin gene encodes a novel epidermal growth factor-related protein with tumor-inhibitory activity. *Mol. Cell. Biol.* **10**, 1969-1981.

Plowman, G. D., Colouscou, J. M., Whitney, G. S., Green, J. M., Carton, G. W., Foy, L., Newbaner, M. G., and Shoyab, M. (1993a). Ligand-specific activation of HER4/p180erbB4, a fourth member of the epidermal growth factor receptor family. *Proc. Natl. Acad. Sci. U.S.A.* **90**, 1746-1750.

Plowman, G. D., Grenn, J. M., Culouscou, J. M., Carlton, G. W., Rothwell, V. M., and Buckley, S. (1993b). Heregulin induces tyrosine phosphorylation of HER4/p180erbB-4. *Nature (London)* **366**, 473-475.

Porumb, H., Gousset, H., Letellier, R., Salle, V., Briane, D., Vassy, J., Amor-Gueret, M., Israel, L., and Taillandier, E. (1996). Temporary ex vivo inhibition of the expression of the human oncogene HER2 (NEU) by a triple helix-forming oligonucleotide. *Cancer Res.* **56**, 515-522.

Press, M. F., Cordon-Cardo, C., and Slamon, D. J. (1990). Expression of the HER-2/neu proto-oncogene in normal human adult and fetal tissues. *Oncogene* **5**, 953-962.

Qi, C., Liscia, D. S., Normanno, N., Merlo, G., Johnson, G. R., Gullick, W. J., Ciardiello, F., Brandt, R., Kim, N., Kenney, N., and Salomon, D. S. (1994). Expression of transforming growth factor alpha, amphiregulin and cripto-1 in human breast carcinomas. *Br. J. Cancer* **69**(5), 903-910.

Qian, X., LeVea, C. M., Freeman, J. K., Dougall, W. C., and Greene, M. I. (1994). Heterodimerization of epidermal growth factor receptor and wild-type or kinase-deficient neu: A mechanism of interreceptor kinase activation and transphosphorylation. *Proc. Natl. Acad. Sci. U.S.A.* **91**, 1500-1504.

Qian, X. L., Decker, S. J., and Greene, M. I. (1992). p185c-neu and epidermal growth factor receptor associate into a structure composed of activated kinases. *Proc. Natl. Acad. Sci. U.S.A.* **89**, 1330-1334.

Quirke, P., Pickles, A., Tuzi, N. L., Mohamdee, O., and Gullick, W. J. (1989). Pattern of expression of c-erbb-2 oncoprotein in human fetuses. *Br. J. Cancer* **60**, 64–69.

Read, L. D., Keith, D., Slamon, D., and Katzenellenbogen, B. S. (1990). Hormonal m odulation of HER-2/*neu* protooncogene messenger ribonucleic acid and p185 protein expression in human breast cancer cell lines. *Cancer Res.* **50**, 3947–3951.

Reed, J. C., Cuddy, M., Haldar, S. *et al.* (1993). BCL2-mediated tumorigenicity of a human T-lymphoid cell line: Synergy with MYC and inhibition by BCL2 antisense. *Proc Natl Acad Sci U.S.A.* **90**, 6340–6344.

Reiter, Y., Brinkman, U., Jung, S.-H., Lee, B., Kasprzyk, P. G., King, C. R., and Pastan, I. (1994). Improved binding and antitumor activity of a recombinant anti-erbB-2 immunotoxin by disulfied stabilization of the Fv fragment. *J. Biol. Chem.* **269**, 18327–18331.

Riese, D. J., II, Raaij, T. M. V., Ploman, G. D., Andrews, G. C., and Stern, D. F. (1995). The cellular response to neuregulins is governed by complex interactions of the erbB-receptor family. *Mol. Cell. Biol.* **15**, 5770–5776.

Riese, D. J., II, Bermingham, Y., Raaij, T. M. V., Buckley, S., Ploman, G. D., and Stern, D. F. (1996). Betacellulin activates the epidermal growth factor receptor and erbB-4 and induces cellular response patterns distinct from those stimulated by epidermal growth factor or neuregulin-β. *Oncogene* **12**, 345–353.

Ring, D. B., Clark, R., and Saxena, A. (1991). Identify of BCA200 and c-erbB-2 indicated by activity of monoclonal antibodies with recombinant c-erbB-2. *Mol. Immunol.* **28**, 915–917.

Ro, J., El-Naggar, A. E., Ro, J. Y. *et al.* (1989). c-erbB-2 amplification in node-negative human breast cancer. *Cancer Res.* **49**, 6941–6944.

Rosenthal, A., Lindquist, P. B., Bringman, T. B., Goeddel, D. V., and Derynck, R. (1986). Expression in rat fibroblats of a human transforming growth factor-alpha cDNA results in transformation. *Cell (Cambridge, Mass.)* **46**, 301–309.

Ross, R. (1986). The pathogenesis of atherosclerosis-an update. *N. Engl. J. Med.* **314**, 488.

Russell, K. S., and Hung, M. C. (1992). Transcriptional repression of the *neu* protooncogene by estrogen stimulated estrogen receptor. *Cancer Res.* **52**, 6624–6629.

Sadowski, I., Stone, J. C., and Pawson, T. (1986). A noncatalytic domain conserved among cytoplasmic protein-tyrosine kinases modifies the kinase function and transforming activity of Fujinami sarcoma virus P130gag-fps. *Mol. Cell. Biol.* **6**, 4396–4408.

Saeki, T., Stromberg, K., Qi, C. *et al.* (1992). Differential immunohistochemical detection of amphiregulin and cripto in human normal colon and colorectal tumors. *Cancer Res.* **52**, 3467–3473.

Sainsbury, J. R. C., Farndon, J. R., Sherbert, G. V., and Harris, A. L. (1985). Epidermal growth factor receptors and oestrogen receptors in human breast cancers. *Lancet* **1**, 364–366.

Salomon, D. S., Kim, N., Saeki, T., and Ciardello, F. (1990). Transforming growth factor-alpha: An oncodevelopmental growth factor. *Cancer Cells* **2**, 389–397.

Salomom, D. S., Brandt, R., Ciardiello, F., and Normanno, N. (1995). Epidermal growth factor-related peptides and their receptors in human malignancies. *Crit. Rev. Oncol.-Hematol.* **19**, 183–232.

Sandgren, E. P., Luetteke, N. C., Palmiter, R. D., Brinster, R. L., and Lee, D. C. (1990). Overexpression of TGF-alpha in transgenic mice: Induction of epithelial hyperplasia, pancreatic metaplasia, and carcinoma of the breast. *Cell (Cambridge, Mass.)* **61**, 1121–1135.

Sasada, R., Ono, Y., Taniyama, Y., Shing, Y., Folkman, J., and Igarashi, K. (1993). Cloning and expression of cDNA encoding human betacellulin, a new member of the EGF family. *Biochem. Biophys. Res. Commun.* **190**(3), 1173–1179.

Sato, J. D., Kawamoto, T., Le, A. D., *et al.* (1983). Biological effects in vitro of monoclonal antibodies to human epidermal growth factor receptors. *Mol. Biol. Med.* **1**, 511–529.

Sato, S., Ito, K., Ozawa, N., Yajima, A., and Sasano, H. (1991). Expression of c-myc, epidermal growth factor receptor and e-erbB-2 in human endomerial carcinoma and cervical adeno-carcinoma. *Tohoku J. Exp. Med.* **165**, 137–145.

Savage, C. R., Jr., Inagami, T., and Cohen, S. (1972). The primary structure of epidermal growth factor. *J. Biol. Chem.* **247,** 7612–7621.
Savage, C. R., Jr., Hash, J. H., and Cohen, S. (1973). Epidermal growth factor. *J. Biol. Chem.* **248,** 7669–7672.
Schechter, A. L., Stern, D. F., Vaidyanathan, L., Decker, S. J., Drebin, J. A., Greene, M. I., and Weinberg, R. A. (1984). The neu oncogene: An erbB-related gene encoding a 185,000-Mr tumor antigen. *Nature (London)* **312,** 513–516.
Schechter, A. L., Hung, M. C., Vaidyanathan, L., Weinberg, R. A., Yang-Feng, T. L., Francke, U., Ullrich, A., and Coussens, L. (1985). The *neu* gene: An erbB-homologous gene distinct from and unlinked to the gene encoding the EGF receptor. *Science* **229,** 976–987.
Schlom, J. (1986). Basic principles and applications of monoclonal antibodies in the management of carcinomas. *Cancer Res.* **46,** 3225–3238.
Schonborn, I., Zschiesche, W., Spitzer, E. *et al.* (1994). c-erbB-2 overexpression in primary breast cancer: Independent prognostic factor in patients at high risk. *Breast Cancer Res. Treat.* **29,** 287–295.
Schreiber, A. B., Winkler, M. E., and Derynck, R. (1986). Transforming growth factor-alpha: A more potent angiogenic mediator than epidermal growth factor. *Science* **232,** 1250–1253.
Schroeter, C. A., De Potter, C. R., Rathsmann, K. *et al.* (1992). c-erbB-2 positive breast tumors behave more aggresively in the first years after diagnosis. *Br. J. Cancer* **66,** 728–734.
Schultz, G. S., White, M., Mitchell, R., Brown, G., Lynch, J., Twardzik, D. R., and Todaro, G. J. (1987). Epithelial wound healing enhanced by transforming growth factor-alpha and vaccinia growth factor. *Science* **235,** 350–352.
Segatto, O., King, C. R., Pierce, J. H. *et al.* (1988). Different structural alterations upregulate in vitro tyrosine kinase activity and transforming potency of the erbB-2 gene. *Mol. Cell. Biol.* **8,** 5570–5574.
Seno, M., Tada, H., Kosada, M., Sasada, R., Igarashi, K., Shing, Y., Folkman, J., Ueda, M., and Yamada, H. (1996). Human Betacellulin, a member of the EGF family dominantly expressed in pancreas and small intestine, is fully active in a monomeric form. *Growth Factor* **13,** 1–11.
Shalaby, M. R., Shepard, H. M., Presta, L., Rodrigues, M. L., Beverly, P. C. L., Feldmann, M., and Carter, P. (1992). Development of humanized bispecific antibodies reactive with cytotoxic lymphocytes and tumor cells overexpressing the HER2 protooncogene. *J. Exp. Med.* **175,** 217–225.
Shankar, V., Ciardiello, F., Kim, N., Derynck, R., Liscia, D. S., Merlo, G., Langton, B. C., Sheer, D., Callahan, R., Bassin, R. H., Lippman, M. E., Hynes, N., and Salomon, D. S. (1989). Transformation of an established mouse mammary epithelial cell line following transfection with a human transforming growth factor alpha cDNA. *Mol. Carcinog.* **2,** 1–11.
Shen, L., Guyre, P. M., Anderson, C. L., *et al.* (1986). Heteroantibody-mediated cytotoxicity: Antibody to the high-affininy Fc receptor for IgG mediates cytotoxicity by human monocytes which is enhanced by interferon-gamma and is not blocked by human IgG. *J. Immunol.* **137,** 3378–3382.
Shepard, H. M., Lewis, G. D., Sarup, J. C., Fendly, B. M., Maneval, D., Mordenti, J., Figari, I., Kotts, C. E., Palladino, M. A., Jr., Ullrich, A., and Slamon, D. (1991). Monoclonal antibody therapy of human cancer: Taking the HER2 proto-oncogene to clinic. *J. Clin. Immunol.* **11,** 117–127.
Shing, Y., Christofori, G., Hanahan, D., Ono, Y., Sasada, R., Igarashi, K., and Folkman, J. (1993). *Science* **259,** 1604–1607.
Shoyab, M., McDonald, V. L., Bradley, J. F., and Todaro, G. J. (1988). Amphiregulin: A bifunctional growth-modulating glycoprotein produced by the phorbol 12-myristate 13-acetate-treated human breast adenocarcinoma cell line MCF-7. *Proc. Natl. Acad. Sci. U.S.A.* **85,** 6528–6532.

Shoyab, M., Plowman, G. D., McDonald, V. L., Bradley, J. G., and Todaro, G. J. (1989). Structure and function of human amphiregulin: A member of the epidermal growth factor family. *Science* **243**, 1074–1076.

Shu, H. K., Pelley, R. J., and Kung, H. J. (1991). Dissecting the activating mutations in v-erbB of avian erythroblastosis virus strain R. *J. Virol.* **65**, 6173–6180.

Siegall, C., Bacus, S. S., Cohcn, B. D., Plowman, G. D., Mixan, B., Chace, D., Chin, D. M., Goetze, A., Green, J. M., Hellström, I., Hellström, K. E., and Fell, P. H. (1995). HER4 expression correlates with cytotoxicity directed by a heregulin-toxin fusion protein. *J. Biol. Chem.* **270**(3), 7625–7630.

Siegall, C. B., Xu, Y.-H., Chaudhary, V. K., *et al.* (1989). Cytotoxic activities of a fusion protein comprised of TGF-alpha and *Pseudomonas* exotoxin. *FASEB J.* **3**, 2547–2653.

Sizeland, A. M., and Burgess, A. W. (1992). Anti-sense transforming growth factor alpha oligonucleotides inhibit autocrine stimulated proliferation of a colon carcinoma cell line. *Mol. Biol. Cell.* **3**, 1235–1243.

Skorski, T., Nieborowska-Skorska, M., Nicolaides, N. C., Szczylik, C., Iversen, P., Iozzo, R. V., Zon, G., and Calabretta, M. (1994). Suppression of Philadelphia leukemia cell growth in mice by BCR-ABL antisense oligodeoxynucleotide. *Proc. Natl. Acad. Sci. U.S.A.* **91**, 4504–4508.

Slamon, D. J., Clark, G. M., Wong, S. G. *et al.* (1987). Human breast cancer: Correlation of relapse and survival with amplification of the HER-2/neu oncogene. *Science* **235**, 177–182.

Slamon, D. J., Press, M. F., Godolphin, W., Ramos, L., Haran, P., Shek, L., Stuart, S.G., and Ullrich, A. (1989a). Studies of the HER2/neu protooncongene in human breast cancer. *In* "Cancer Cells 7-Molecular Diagnostics of Human Cancer" (M. Furth and M. Greaves, eds.), pp. 371–384. Cold Spring Harbor Lab. Press, Cold Spring Harbor, NY.

Slamon, D. J., Godolphin, W., Jones, L. A. *et al.* (1989b). Studies of the HER-2/neu protooncogene in human breast and ovarian cancer. *Science* **244**, 707–712.

Sliwkowski, M. X., Schaefer, G., Akita, R. W. *et al.* (1994). Coexpression of erbB-2 and erbB-3 proteins reconstitute a high affinity receptor for heregulin. *J. Biol. Chem.* **269**, 14661–14665.

Soltoff, S. P., Carraway, K. L., III, and Prigent, S. A. (1994). ErbB3 is involved in activation of phosphatidylinositol 3'-kinase by epidermal growth factor. *Mol. Cell. Biol.* **14**, 3550–3558.

Songyang, Z., Shoelson, S. E., Chandhuri, M. *et al.* (1993). SH2 domains recognize specific phosphopeptide sequences. *Cell (Cambridge, Mass.)* **72**, 767–778.

Spiva-Kroizman, T., Rotin, D., Pinchasi, D., Ullrich, A., Schlessinger, J., and Lax, I. (1992). Heterodimerization of c-erbB2 with different epidermal growth factor receptor mutants elicits stimulatory or inhibitory responses. *J. Biol. Chem.* **267**, 8056–8063.

Sporn, M. B., and Roberts, A. B. (1992). Autocrine secretion—10 years later. *Ann. Intern. Med.* **117**, 408–414.

Stern, D. F., and Kamps, M. P. (1988). EGF-stimulated tyrosine phosphorylation of p185^{neu}: A potential model for receptor interactions. *EMBO J.* **7**, 995–1001.

Stern, D. F., Heffernan, P. A., and Weinberg, R. A. (1986). p185, a product of the *neu* protooncogene, is a receptor like protein associated with tyrosine kinase activity. *Mol. Cell Biol.* **6**, 1729–1740.

Sternberg, M. J. E., and Gullick, W. J. (1990). A sequence motif in the transmembrane region of growth factor receptors with tyrosine kinase activity mediates dimerization. *Protein Eng.* **3**, 245–248.

Suda, Y., Aizawa, S., Furuta, Y., Yagi, T., Ikawa, Y., Saitoh, K., Yamada, Y., Toyoshima, K., and Yamamoto, T. (1990). Induction of variety of tumors by c-erbB-2 and clonal nature of lymphomas even with the mutated gene (Val^{659} => Glu^{659}). *EMBO J.* **9**, 181–190.

Sun, X. J., Rothenberg, P., Kahn, C., Backer, J. M., Araki, E., Wilden, P. A., Cahil, D. A., Goldstein, B. J., and White, M. F. (1991). Structure of the insulin receptor substrate IRS-1 defines a unique signal transduction protein. *Nature (London)* **352**, 73–77.

Szczylik, C., Skorski, T., Nicoliades, N. C., Manzella, L., Malaguarnera, L., Venturelli, D., Gewirtz, A. M., and Calabretta, B. (1991). Selective inhibition of leukemia cell proliferation by BCR-ABL antisense oligodeoxynucleotides. *Science* **253**, 562–565.

Takishima, K., Griswold-Prenner, I., Ingebritsen, T., and Rosneer, M. R. (1991). Epidermal growth factor (EGF) receptor T669 peptide kinase from 3T3-L1 cells is an EGF-stimulated "MAP" kinase. *Proc. Natl. Acad. Sci. U.S.A.* **88**, 2521–2524.

Tan, J. C., Nocka, K., Ray, P., Traktman, P., and Besmer, P. (1990). The domain W42 spotting phenotype results from a missense mutation in the c-kit receptor kinase. *Science* **247**, 209–212.

Tandon, A. K., Clark, G. M., Chamness, G. C. et al. (1989). HER-2/neu oncogene protein and prognosis in breast cancer. *J. Clin. Oncol.* **7**, 1120–1128.

Tang, C. K., Perez, C., Grunt, T., Cho, C., Waibel, C., and Lupu, R. (1996). Involvement of Heregulin-β2 in the acquisition of the hormone-independent phenotype of breast cancer cells. *Cancer Res.* **56**, 3350–3358.

Tang, C. K., Goldstein, D. J., Payne J., Czubayko, F., Alimandi, M., Wang, L. M., Pierce, J. H., and Lippman, M. E. (1997). ErbB-4 ribozymes abolish heregulin-induced mitogenesis *J. Biol. Chem.* submitted.

Tateishi, M., Ishida, T., Mitsudomi, T., and Sugimachi, K. (1991). Prognostic implication of transforming growth factor alpha in adenocarcinoma of the lung—an immunohistochemical study. *Br. J. Cancer* **63**, 130–133.

Tikannen, S., Helin, H., Isola, J., and Joensuu, H. (1992). Prognostic significance of HER-2 oncoprotein expression in breast cancer: A 30 year follow-up. *J. Clin. Oncol.* **10**, 1044–1048.

Todaro, G. J., Fryling, C., and DeLarco, J. E. (1980). Transforming growth factors produced by certain human tumor cells: Polypeptides that interact with epidermal growth factor receptors. *Proc. Natl. Acad. Sci. U.S.A.* **77**, 5258–5262.

Todaro, G. J., Rose, T. M., Spooner, C. E., Shoyab, M., and Plowman, G. D. (1990). Cellular and viral ligands that interact with the EGF receptor. *Semin. Cancer Biol.* **1**, 257–263.

Tone, T., Kashani-Sabet, M., Funato, T. et al. (1993). Suppression of EJ cells tumorigenicity. *In Vivo* **7**, 471–476.

Tripputi, P., Emanuel, B. S., Croce, C. M., Green, L. G., Stein, G. S., and Stein, J. L. (1986). Human histone genes map to multiple chromosomes. *Proc. Natl. Acad. Sci. U.S.A.* **83**, 3185–3188.

Tsujino, T., Yoshida, K., Nakayama, H. et al. (1990). Alerations of oncogenes in metastatic tumors of gas~ric carcinomas. *Br. J. Cancer* **62**, 226–230.

Turc-Carel, C., Pietrzak, E., Kakati, S., Kinniburgh, A. J., and Sandberg, A. A. (1987). The human int-1 gene is located at chromosome region 12q12-12q13 and is not rearranged in myxoid liposarcoma with t(12; 16) (q13; p11). *Oncogene Res.* **1**, 397–405.

Tuzi, N. L., Venter, D. J., Kumar, S., Staddon, S. L., Lemoine, N. R., and Gullick, W. J. (1991). Expression of growth factor receptors in human brain tumors. *Br. J. Cancer* **63**, 227–233.

Tzahar, E., Levkowitz, G., Karunagaran, D., Yi, L., Peles, E., Lavi, S., Chang, D., Liu, N., Yayon, A., Wen, D., and Yarden, Y. (1994). ErbB-3 and ErbB-4 function as the respective low and high affinity receptors of all neu differentiation factor/heregulin isoforms. *J. Biol. Chem.* **269**,(40) 25226–25233.

Ullrich, A., and Schlessinger, J. (1990). Signal transduction by receptors with tyrosine kinase activity. *Cell (Cambridge, Mass.)* **61**, 203–212.

Ullrich, A., Coussens, L., Hayflick, J. S., Dull, T. J., Gray, A., Tam, A. W., Lee, J., Yarden, Y., Libermann, T. A., Schlessinger, J., Downward, J., Mayes, E. L. V., Whittle, N., Waterfield, M. D., and Seeburg, P. H. (1984). Human epidermal growth factor receptor cDNA sequence and aberrant expression of the amplified gene in A431 epidermoid carcinoma cells. *Nature (London)* **309**, 418–425.

Ushiro, H., and Cohen, S. (1980). Identification of phosphotyrosine as a product of epidermal growth factor-activated protein kinase in A-431 cell membranes. *J. Biol. Chem.* **255**, 8363–8365.

Valerius, T., Repp, R., de Wit, T. P., *et al.* (1993). Involvement of the high-affinity receptor for Ig-G (Fc gamma RI; CD64) in enhanced tumor cell cytotoxicity of neutrophils during granulocyte colony-stimulating factor therapy. *Blood* **82**, 931–939.

Valone, F. H., Kaufman, P. A., Guyre, P. M., Lewis, L. D., Memoli, V., Deo, Y., Graziano, R., Fisher, J. L., Meyer, L., Mrozek-Orlowski, M., Wardwell, K., Guyre, V., Morley, T. L., Arvizu, C., and Fanger, M. W. (1995). Phase I1/Ib trial of bispecific antibody MDX-210 in patients with advanced breast or ovarian cancer that overexpresses the proto-oncogene HER-2/neu. *J. Clin. Oncol.* **13**, 2281–2292.

van de Vijver, M., van de Bersselaar, R., Devilee, P., *et al.* (1987). Amplification of the neu (c-erbB-2) oncogene in human mammary tumors is relatively frequent and is often accompanied by amplification of the linked c-erbA oncogene. *Mol. Cell. Biol.* **7**, 2019–2023.

van de Vijver, M. J., Johannes, L. P., Wolter, J. M. *et al.* (1988). Neu-protein overexpression in breast cancer. Association with comedo-type ductal carcinoma in situ and limited prognostic value in stage II breast cancer. *N. Engl. J. Med.* **319**, 1239–1245.

van Dijk, J., Warnaar, S. O., van Eendenburg, J. D. H., *et al.* (1989). Induction of tumor-cell lysis by bs-mAb recognizing renal-cell carcinoma and CD3 antigen. *Int. J. Cancer* **43**, 344–349.

Velu, T. J., Vass, W. C., Lowy, D. R., and Beguinot, L. (1989). Functional heterogeneity of proto-oncogene tyrosine kinases: The C terminus of the human epidermal growth factor receptor facilitates cell proliferation. *Mol. Cell. Biol.* **9**, 1772–1778.

Venter, D. J., Tsui, N. L., Kummar, S., and Gullick, W. J. (1987). Overexpression of the e-erbB-2 oncoprotein in human breast carcinomas: Immunohistological assessment correlated with gene amplification. *Lancet* **2**, 67–72.

Vitetta, E. S., Fulton, R. J., May, R. D., Till, M., and Uhr, J. W. (1987). Redesigning nature's poisons to create anti-tumor reagents. *Science* **238**, 1098–1104.

Wada, T., Qian, X., and Greene, M. (1990). Intermolecular association of the p185^{neu} protein and EGF receptor modulates EGF receptor function. *Cell (Cambridge, Mass.)* **61**, 1339–1347.

Watanabe, S., lazar, E., and Sporn, M. B. (1987). Tranformation of normal rat kidney (NRK) cells by an infectious retrovirus carrying a synthetic rat type alpha transforming growth factor gene. *Proc. Natl. Acad. Sci. U.S.A.* **84**, 1258–1262.

Weiner, D. B., Liu, J., Cohen, J. A., Williams, W. V., and Greene, M. I. (1989). A point mutation in the neu oncogene mimics ligand induction of receptor aggregation. *Nature (London)* **339**, 230–231.

Weiner, G. J., Kostelny, S. A., Hillstro¢m, J. R., *et al.* (1994). The role of T cell activation in anti-CD3 X antitumor bispecific antibody therapy. *J. Immunol.* **152**, 2385–2392.

Weiner, L. M., Holmes, M., Adams, G. P., *et al.* (1993a). A human tumor xenograft model of therapy with a specific monoclonal antibody targeting c-erbB-2 and CD 16. *Cancer Res.* **53**, 94–100.

Weiner, L. M., Holmes, M., Richeson, A., *et al.* (1993b). Binding and cytotoxicity characteristics of bispecific murine monoclonal antibody 2B1. *J. Immunol.* **151**, 2877–2886.

Weiss, A. (1993). T cell antigen receptor signal transduction: A tale of tails and cytoplasmic protein-tyrosine kinases. *Cell (Cambridge, Mass.)* **73**, 209–212.

Wells, A., Welsh, J. B., Lazar, C. S., Wiley, H. S., Gill, G. N., and Rosenfeld, M. G. (1990). Ligand-induced transformation by a noninternalizing epidermal growth factor receptor. *Science* **247**, 962–964.

Wen, D., Peles, E., Cupples, R., Suggs, S. V., Bacus, S. S., Luo, Y., Trail, G., Hu, S., Silbiger, S. M., Benlevy, R., Koski, R. A., Lu, H. S., and Yarden, Y. (1992). Neu differentiation

factor: A transmembrane glycoprotein containing an EGF domain and an immunoglobulin homology unit. *Cell (Cambridge, Mass.)* **69**, 559-572.
Wen, D., Suggs, S. V., Karanagaran, D., Liu, N., Cupples, R. L., Luo, Y., Jansen, A. M., Ben-Baruch, N., Trollinger, D. B., Jacobson, V. L., Meng, T., Lu, H. S., Hu, S., Chang, D., Yanigahara, D., Koski, R. A., and Yarden, Y. (1994). Structural and functional aspects of the multiplicity of Neu differentiation factors. *Mol. Cell. Biol.* **14**, 1909-1919.
Wikstrand, C. J., Hale, L. P., Batra, S. K., Hill, M. L., Humphrey, P. A., Kurpad, S. N., McLendon, R. E., Mostacello, D., Pegram, C. N., Reist, C. J., Traweek, S. T., Wong, A. J., Zalustsky, M. R., and Bigner, D. D. (1995). Monoclonal antibodies against EGFRvIII are tumor specific and react with breast and lung carcinomas and malignant gliomas. *Cancer Res.* **55**, 3140-3148.
Winkler, M. E., O'Connor, L., Winget, M., and Fendly, B. (1989). Epidermal growth factor and transforming growth factor alpha bind differently to the epidermal growth factor receptor. *Biochemistry* **28**, 6373-6378.
Wong, A. J., Bigner, S. H., Bigner, D. D., Kinzler, K. W., Hamilton, S. R., and Vogelstein, B. (1987). Increased expression of the epidermal growth factor receptor gene in malignant gliomas is invariably associated with gene amplification. *Proc. Natl. Acad. Sci. U.S.A.* **84**, 6899-6903.
Wong, A. J., Ruppert, J. M., Bigner, S. H., Grzeschik, C. H., Humphrey, P. A., Bigner, D. S., and Vogelstein, B. (1992). Structural alterations of the epidermal growth factor receptor gene in human gliomas. *Proc. Natl. Acad. Sci. U.S.A.* **89**, 2965-2969.
Wrann, M. M., and Fox, C. F. (1979). Identification of epidermal growth factor receptors in a hyperproducing epidermal carcinoma cell line. *J. Biol. Chem.* **254**, 8083-8086.
Wright, C., Angus, B., Nicholson, S. *et al.* (1989). Expression of c-erbB-2 oncoprotein: A prognostic indicator in human breast cancer. *Cancer Res.* **49**, 2087-2090.
Wright, C., Nicholson, S., Angus, B. *et al.* (1992). Relationship between c-*erb*B-2 protein product expression and response to endocrine therapy in advanced breast cancer. *Br. J. Cancer* **65**, 118-121.
Yaish, P., Gazit, A., Gilon, C., *et al.* (1988). Blocking of EGF-dependent cell proliferation by EGF receptor kinase inhibitor. *Science* **242**, 933-935.
Yamahara, M., Fujito, T., Ishikawa, T., Shimosato, T., Yokozaki, H., and Yasui, W. (1988). Phenotypic expression of human growth factor in foetal submandibular gland and pleomorphic adenoma of salivary gland. *Virchows Arch. A: Pathol. Anat. Histol.* **412**, 301-306.
Yamamoto, T., Ikawa, S., Akiyama, T., Semba, K., Numura, N., Miyajima, N., and Saito, T. (1986). Similarity of protein encoded by the human c-erbB-2 gene to epidermal growth factor receptor. *Nature (London)* **319**, 230-234.
Yamanaka, Y., Onda, M., Uchida, E. *et al.* (1990). Immunohistochemical study on epidermal growth factor and its receptor in human pancreatic carcinoma. *Nippon Shokakibyo Gakkai Zasshi* **87**, 1544-1550.
Yamanaka, Y., Friess, H., Kobrin, M. S., Buchler, M., Beger, H. G., and Korc, M. (1993). Coexpression of epidermal growth factor receptor and ligands in human pancreatic cancer is associated with enhanced tumor aggressiveness. *Anticancer Res.* **13**, 565-570.
Yamazaki, H., Fukui, Y., Ueyama, Y., Tamaoki, N., Kawamoto, T., Taniguchi, S., and Shibuya, M. (1988). Amplification of the structurally and functionally altered epidermal growth factor receptor gene (c-erbB) in human brain tumors. *Mol. Cell. Biol.* **8**, 1816-1820.
Yarden, Y., and Schlessinger, J. (1987). Epidermal growth factor induces rapid, reversible aggregation of the purified epidermal growth factor receptor. *Biochemistry* **26**, 1443-1451.
Yarden, Y., and Ullrich, A. (1988). Growth factor receptor tyrosine kinase. *Annu. Rev. Biochem.* **57**, 443-448.
Yarden, Y., Pinkas-Kramarski, R., Shelly, M., Glathe, S., and Ratzkin, B. J. (1996). Neu differenciation factor/neuregulin isoforms activate distint receptor combinations. *J. Biol. Chem.* **271**(32), 19029-19032.

Yasui, W., Sumiyoshi, H., Hata, J. *et al.* (1990). Expression of epidermal growth factor receptor in human gastric and colonic carcinomas. *Br. J. Cancer* **62**, 226–230.

Yokota, J., Yamamoto, T., Toyoshima, K., *et al.* (1986). Amplification of c-erbB-2 oncogene in human adenocarcinomas in vivo. *Lancet* **1**, 765–767.

Yokoyama, K., and Imamoto, F. (1987). Transcriptional control of the endogenous MYC proto-oncogene by antisense RNA. *Proc. Natl. Acad. Sci. U.S.A.* **84**, 7363–7367.

Yung, W. K. A., Zhang, X., Steck, P. A., and Hung, M.-C. (1990). Differential amplification of the TGF-alpha gene in human gliomas. *Cancer Commun.* **2**, 201–205.

Zabrecky, J. R., Lam, T., McKenzie, S. J., and Carney, W. (1991). The extracellular domain of p185/neu is released from the surface of human breast carcinoma cells, SK-Br-3. *J. Biol. Chem.* **266**, 1716–1720.

Zhang, K., Sun, J., Liu, N., Wen, D., Chang, D., Thomason, A., and Yoshinaga, S. K. (1996). Transformation of NIH 3T3 cells by HER3 or HER4 receptors requires the presence of HER1 or HER2. *J. Biol. Chem.* **217**(7), 3884–3890.

Timothy A. Quill
David L. Garbers

Howard Hughes Medical Institute and Department of Pharmacology
University of Texas Southwestern Medical Center
Dallas, Texas 75235

Fertilization: Common Molecular Signaling Pathways Across the Species

The union of spermatozoa and eggs has adapted to a striking variety of environments during the course of evolution, and concomitantly, a dramatic diversity of sperm cell morphologies and propulsion mechanisms have developed across the species. In each case, the primary functions of the spermatozoon including motility initiation, motility modulation by female factors, and adhesion to and fusion with the egg have remained constant features. Species-specificity of these sperm responses appears to arise from the uniqueness of both the factors that initiate the response as well as the detector modules exposed on the sperm surface. The intracellular signaling pathways that elicit the common sperm functional responses, however, appear relatively conserved across the species. Furthermore, although gametes may express unique isoforms of the various signaling molecules, these signaling pathways appear, in principle, similar to those of somatic cells.

I. Introduction

Fertilization encompasses a series of sequential, coordinated events ultimately achieving the union of the spermatozoon and the egg to initiate the development of a new individual. The fundamental importance of productive fertilization implies the presence of significant regulatory mechanisms to ensure success. In this review, we focus on the molecular aspects of the regulation of sperm function by the egg and egg-associated structures, including the acellular and cellular matrices around the egg, the female reproductive tract in the case of internal fertilizing organisms, and factors released from these structures.

From the perspective of the spermatozoon, gamete interaction potentially includes selective transport (internal fertilization), motility stimulation, attraction to the egg, interaction with cells surrounding the egg, adhesion to the egg extracellular matrix, acrosomal exocytosis, egg extracellular matrix penetration, egg plasma membrane adhesion, and fusion with the egg. In general, these events are conserved across the species, although the relative contribution of each event to successful fertilization may vary appreciably between species. Considerable experimental evidence indicates that each of the stages of gamete interaction exhibits some degree of species specificity. As discussed in detail below, species specificity is suggested to result principally from the binding of specific effectors to complementary receptors on the spermatozoon, as opposed to differential intracellular signaling pathways. One can liken this, as an example, to the large family of different odorant receptors (many hundreds), but apparently common downstream signaling pathways for each individual odor (Buck and Axel, 1991). Some of the downstream components, of course, may be sperm cell-specific isoforms of signal transducing proteins, and again using olfaction as a model, the olfactory neurons appear to contain specialized variants of a Gs-like protein and of a cyclic nucleotide-gated ion channel found in other tissues (Dhallan *et al.*, 1990; Jones and Reed, 1989). The variability in the degree of species-specificity (partial to complete) therefore most likely reflects, in most cases, specificity of interactions between the ligand/receptor or complementary adhesion molecules. Thus, our current model of gamete interaction is that a diversity of ligands and receptor binding domains exist across the species, and these are then coupled to relatively conserved signaling pathways present in all cells. Inherent in this model is the coevolution of both the effector and receptor binding domain in order to maintain productive fertilization.

Recent relevant reviews include Suzuki and Yoshino (1992), Eisenbach and Ralt (1992), Hardy and Garbers (1993), Ward and Kopf (1993), Wassarman (1995), and Snell and White (1996).

II. Oviductal Transport

In species where reproduction involves internal fertilization, spermatozoa are transported to the egg through the female reproductive tract by a combination of passive (reproductive tract muscle contractions and epithelial cell ciliary motion) and active (sperm flagellar motion) mechanisms. Along the length of the female reproductive tract, a substantial reduction in sperm cell number is observed as one approaches the site of fertilization. Whether or not this reflects an active selection of a specific subpopulation of spermatozoa by the female reproductive tract has not been resolved. Upon reaching the lower region of the oviduct, spermatozoa attach to the oviductal epithelium and become quiescent (Yanagimachi, 1994). Coincident with ovulation, a few sperm are released from the oviductal epithelium and migrate to the site of fertilization. The mechanism responsible for release is unknown, but could involve increased oviductal contractions, accelerated epithelial ciliary activity, and/or the release of chemotactic factors (Garbers, 1989). Recently, it was reported that the *in vitro* binding of spermatozoa to cow oviductal epithelium was independent of the stage of the estrous cycle (Lefebvre *et al.*, 1995; Lefebvre and Suarez, 1996). Therefore, the release may be regulated by an intrinsic property of the spermatozoa themselves (e.g., capacitation state).

During residence in the female reproductive tract, spermatozoa undergo a poorly understood process called capacitation. Capacitation has been defined as the ability of ejaculated spermatozoa to undergo an acrosome reaction in response to the physiological agonist, and hence gain the ability to fertilize an egg. This "operational" description is worthwhile, but failure to observe an acrosome reaction may not necessarily indicate that the biochemical events that define capacitation have not occurred. Only when we understand capacitation at the molecular level, in fact, will we be in a position to decide whether the ability to undergo an acrosome reaction adequately describes the molecular pathway defined as capacitation. The ability to undergo an apparent ligand-induced acrosome reaction can also develop during *in vitro* incubation in a chemically defined medium. Several modifications of the sperm cell occur during *in vitro* capacitation including removal and redistribution of lipids and peripheral membrane proteins, activation of ion channels, and protein phosphorylation (Harrison, 1996; Yanagimachi, 1994; Zeng *et al.*, 1995, 1996). One such modification is the efflux of cholesterol which appears to be mediated by albumin present in *in vitro* capacitation medium (Parks and Ehrenwald, 1990). Serum albumin and lipoproteins (e.g., HDL) also are present in follicular fluid and oviductal fluid where they may participate in sperm cell capacitation (Desnoyers and Manjunath, 1992). Tyrosine phosphorylation of several sperm proteins that temporally correlates with capacitation has recently been described in a variety of species (Emiliozzi and Fenichel, 1997; Galantino-Homer *et al.*, 1997; Leclerc *et al.*, 1996; Visconti *et al.*, 1995a). In every case, phosphorylation as well as capacitation was dependent on cAMP and PKA activity (Uguz *et al.*, 1994; Visconti *et al.*, 1995b). The identity of the phosphotyrosine-containing proteins is unknown except for a sperm-specific hexokinase in mouse spermatozoa (Kalab *et al.*, 1994). In human sperm cells, these proteins are localized in the fibrous sheath of the flagellum suggesting a role in motility regulation (Leclerc *et al.*, 1996). However, in each case, it is unclear if these changes are causal or only correlated with *in vitro* capacitation. The mechanisms that result in *in vivo* capacitation also remain unknown. While the cellular mechanisms leading to the ability to undergo the ligand-induced acrosome reaction *in vivo* and *in vitro* may be similar, it is also possible that the mechanisms are unique or perhaps only partially overlap. Reports of accelerated sperm cell capacitation rates when insemination occurs nearer the time of ovulation suggest that the oviduct environment regulates this process through mechanisms different from those of *in vitro* capacitation, although again the final steps of the process may be similar (Smith and Yanagimachi, 1989).

III. Motility Modulation

A. Motility Stimulation (Chemokinesis)

Sperm motility stimulation involves an increased forward velocity, increased percentage of motile cells, modified flagellar waveform, or any combination thereof. Numerous reports on widely divergent phyla indicate that soluble factors released from the egg or egg-associated structures can stimulate sperm motility. In eutherian mammals, the stimulatory factors have been associated with the genital tract and follicular fluid, although the nature of the factors (specific effectors or general metabolites) is unclear. There are also reports suggesting that cumulus oophorous cells can elaborate motility stimulatory factors (Bradley and Garbers, 1983; Tesarik *et al.*, 1990). Recent studies examining the migration of human spermatozoa from a Teflon well into a polyethylene capillary tube with follicular fluid in both compartments found increased accumulations of spermatozoa within the tube compared to cell medium alone, consistent with flagellar stimulation (Ralt *et al.*, 1994). Follicular fluid also induced "hyperactivated" motility (i.e., an exaggerated whip-like motion of the flagellum), but the active components have not been identified. Among external fertilizing species, it appears that sperm stimulatory factors originate from the egg or its associated extracellular matrix (Miller, 1985). An example in which the molecular basis of sperm motility activation by the egg is partially understood is the sea urchin.

Sea urchin sperm motility and respiration are stimulated by peptides released from sea urchin eggs. The first two peptides isolated were speract (GFDLNGGGVG) from *Strongylocentrotus purpuratus* and *Hemicentrotus pulcherrimus,* and resact (CVTGAPGCVGGGRL-NH$_2$) from *Arbacia punctulata* (Hansbrough and Garbers, 1981a; Suzuki and Garbers, 1984; Suzuki *et al.*, 1981). Each of these peptides was confirmed as the bioactive factor by chemical synthesis; the synthetic and isolated peptides stimulated conspecific sperm cell respiration and motility at equivalent concentrations (subnanomolar) (Garbers et al., 1982a; Shimomura and Garbers, 1986). Additionally, these studies defined the carboxy-terminal half as the critical determinant of activity. In the case of resact, this activity was potentiated by the amino-terminal half of the peptide. Both speract and resact were species-specific, stimulating respiration and motility solely in sea urchins in which the peptide was produced (Suzuki and Garbers, 1984).

More than 70 peptides have now been reported from nearly 20 species in the Class Echinoidea (Suzuki, 1995). These peptides can be sorted, based on structural similarity, into groups that have been designated Sperm Activating Peptide I–V (Table I). This grouping correlates with the taxonomic order or suborder classifications of the echinoderms (Suzuki and Yoshino, 1992). As might be expected, peptides from one species can stimulate respiration and motility of spermatozoa only from species within the same order or suborder.

Since more peptides have been identified than species examined, it is clear that some species produce multiple sperm activating peptides. The molecular cloning of the cDNA encoding speract from *S. purpuratus* ovarian libraries has provided an explanation for at least some of the intraspecies diversity. Two transcripts of 2.3 and 1.2 kb were identified that encoded precursor proteins for speract (Ramarao et al., 1990). The two predicted speract precursor proteins each contained four copies of speract. In addition, the larger transcript encoded six additional potential peptides, and the smaller transcript seven additional potential peptides. Some of these additional peptides are probably biologically active based on their structural similarity to speract and the finding that the speract-related peptides from *H. pulcherrimus* are biologically active (Suzuki and Yoshino, 1992). The

TABLE I Sperm Activating Peptides

Peptide	Structure[a]	Species
Speract (SAP I)	GFDLNGGGVG	*S. purpuratus* (Order Echinoida)
Resact (SAP IIA)[b,c]	CVTGAPGCVGGGRL-NH$_2$	*A. punctulata* (Order Arbacioida, Suborder Arbacina)
SAP IIB[b,c]	KLCPGGNCV	*G. crenularis* (Order Arbacioida, Suborder Phymosomatina)
SAP III	DSDAQNLIG	*C. japonicus* (Order Clypeasteroida)
SAP IV[b]	GCPWGGAVC	*D. setosum* (Order Diadematoida)
SAP V[b]	GCEGLFHGMGNC	*B. agassizii* (Order Spatangoida)

[a] Original peptide identified from each group.
[b] The cysteines in resact and SAP IIB, IV, and V form intramolecular disulfide bonds.
[c] Resact and SAP IIB demonstrate marginal activity across the suborders within the Order Arbacioida at high concentrations (~500 μM).

peptides were separated by a single lysine residue, suggesting proteolytic processing by a trypsin-like protease followed by a carboxypeptidase. The identity of the proteases and the site of proteolytic processing remain unknown. In contrast to the speract mRNA, only a single copy of resact, and no other similar peptide, is predicted from the corresponding cDNA (Burks, 1990). This is consistent with the purification data in that unlike speract and other sperm activating peptides, no peptide variant of resact has been found. It is unclear why multiple forms of speract exist when a single resact peptide suffices in *A. punctulata*. In addition, whether or not other portions of the propeptide have biological functions has not been determined. Other than a signal peptide encoded by the mRNA of the speract and resact precursor proteins, however, there are no regions of significant identity, suggesting that if additional portions do contain biological activity, the effects would also be species-specific.

A model depicting the molecular events initiated by the binding of the egg peptides to spermatozoa is shown in Fig. 1. The initial effects include a transient increase of cGMP and an associated membrane hyperpolarization (Cook and Babcock, 1993a). The membrane hyperpolarization activates a Na^+/H^+ exchanger producing an influx of Na^+ and an efflux of H^+, thus raising the pH_i (Hansbrough and Garbers, 1981b; Lee and Garbers, 1986). Subsequently, both sperm adenylyl cyclase and a Ca^{2+} channel are transiently activated, leading to increases of cellular cAMP and Ca^{2+} (Cook and Bab-

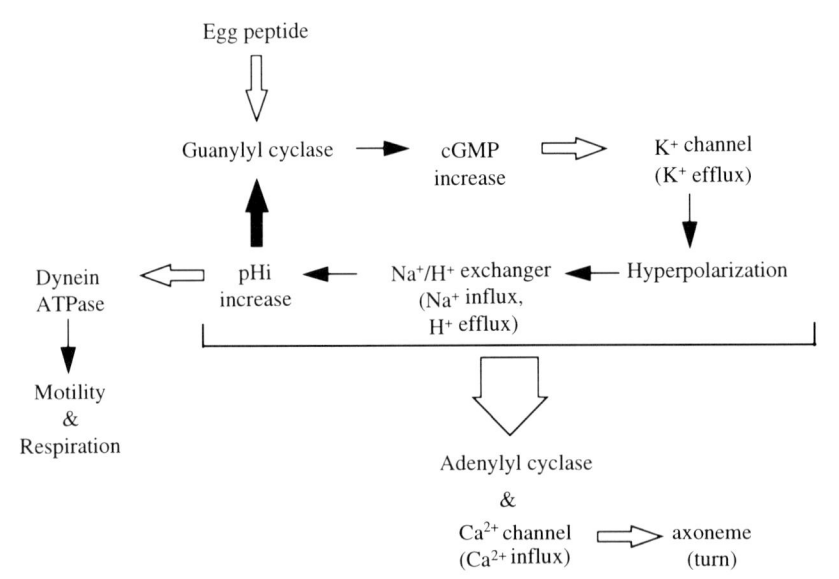

FIGURE 1 Egg peptide signal transduction in sea urchins. Solid arrows indicate direct activation and hollow arrows indicate activation through undefined mechanisms; the bold, solid arrow indicates inhibition through undefined mechanisms.

cock, 1993b; Schackmann and Chock, 1986). Other methods for increasing sperm pH$_i$ such as addition of NH$_4^+$ or monensin, an ionophore that mediates electroneutral Na$^+$/H$^+$ exchange, also produce a stimulation of sperm respiration and motility (Hansbrough and Garbers, 1981b; Repaske and Garbers, 1983). The elevated pH$_i$ appears to activate the flagellar dynein ATPase which is tightly coupled to mitochondrial respiration, thus accounting for both the increased respiration and flagellar activity (Christen et al., 1983). Therefore, alkalinization appears to be a primary mediator of motility and respiratory stimulation.

The sperm membrane hyperpolarization which activates the Na$^+$/H$^+$ exchanger results from the opening of a K$^+$ channel (Lee and Garbers, 1986). The mechanism by which egg peptides activate this K$^+$ channel remains to be demonstrated. However, inhibition of K$^+$ efflux prevents the physiological effects initiated by the peptides, except for increased cGMP (Harumi et al., 1992b). Thus, cGMP is a candidate regulatory factor for the K$^+$ channel as supported by the ability of 8-Br-cGMP to stimulate sea urchin sperm respiration and motility (Kopf et al., 1979). Two observations suggest that cellular cGMP elevation may not regulate the K$^+$ channel. First, both K$^+$ channel activation and half-maximal respiratory and motility stimulation occur at 10- to 100-fold lower peptide concentrations (10^{-11}–10^{-10} M) than required for half-maximal cGMP increases (Babcock et al., 1992; Hansbrough and Garbers, 1981a). Second, resact analogs have been synthesized that are only 2.5- to 5-fold less potent in stimulating respiration, but 100- to 1000-fold less potent in stimulating cGMP accumulation (Shimomura and Garbers, 1986). Similar disparity between the concentration of ligand that stimulates a physiological effect and half-maximal increases of cGMP, however, has been reported in other cGMP-mediated signaling systems (Drewett and Garbers, 1994). Several explanations for these observations have been proposed, including: (1) multiple receptors for a single ligand, (2) activation of alternative intracellular signal transduction pathways by a single receptor, and (3) small localized elevations of cGMP below the assay detection limits. None of these explanations has been eliminated with regard to the effects of the egg peptides on sea urchin spermatozoa. Recent data from Cook and Babcock (1993a) suggest that the sea urchin sperm K$^+$ channel, in fact, is regulated by cGMP, since a temporal and quantitative correlation is found between the open state of the channel and the accumulation of cellular cGMP obtained in the presence of IBMX. A cGMP-regulated K$^+$ channel with properties expected for the sea urchin sperm channel has been cloned from a rabbit genomic library (Yao et al., 1995). This K$^+$ channel appears related to both voltage- and cyclic nucleotide-gated ion channels, selectively responded to cGMP over cAMP, and was insensitive to tetraethylammonium ion inhibition, which reproduces the characteristics of the egg peptide-activated K$^+$ channel. Alternatively, the sperm K$^+$ channel could be G-protein regulated since GTP and GTPγS have been found to potentiate the speract-stimulated potassium transport in isolated flagellar membranes, while cGMP has no detectable effect (Lee, 1988). Either of these models is consistent with the presence of a diffusible second messenger detected in patch clamp studies (Babcock et al., 1992).

The consequences of the elevations of cAMP and Ca^{2+} in sea urchins are not well understood, although Ca^{2+} is required for chemotaxis (Section III.C). Both cAMP-dependent protein kinase and calmodulin are abundant cellular proteins which undoubtedly regulate sperm physiology via phosphorylation and dephosphorylation events (Garbers, 1981). However, no specific targets of the activated protein kinases and phosphatases have been reported. In the case of cAMP, a temporal and quantitative correlation found between the production of cAMP and the opening of a Ca^{2+} channel suggests that ion channels may also be targets for cAMP action (Cook and Babcock, 1993b).

Mammalian sperm motility also is regulated by cAMP, Ca^{2+}, and pH_i (Tash and Bracho, 1994; Tash et al., 1988; Tash and Means, 1988). A number of potential target proteins are present at high levels in spermatozoa including cAMP-dependent protein kinase, calmodulin, calmodulin-dependent protein kinase, and calmodulin-dependent protein phosphatase, as well as an AKAP (A Kinase Anchor Protein), which has been localized to the fibrous sheath surrounding the principle piece of the flagellum (Carrera et al., 1994). All of these components appear to be involved in the control of motility via modulation of the phosphorylation state of flagellar proteins (Tash, 1989).

The sperm adenylyl cyclase found in animals as diverse as echinoderms and mammals appears to be unique compared to other known adenylyl cyclase isoforms. All of the cloned mammalian somatic cell adenylyl cyclases are predicted as integral membrane proteins with 12 transmembrane segments and two intracellular consensus catalytic domains (one lies between the 6th and 7th transmembrane segment and the second at the carboxy-terminus of the enzyme) (Sunahara et al., 1996). Each of the somatic isoforms is regulated by a G protein and is stimulated by forskolin. In contrast, sperm adenylyl cyclase is not sensitive to either G proteins or forskolin (Hildebrandt et al., 1985). Rather, sperm adenylyl cyclase appears to be regulated by a variety of alternative factors, including calcium. Both sea urchin and mammalian sperm adenylyl cyclase appear to bind calmodulin and are inhibited by calmodulin antagonists (Bookbinder et al., 1990; Gross et al., 1987). Abalone sperm cell cAMP accumulation is increased 100-fold by a combination of Ca^{2+} and IBMX while IBMX alone elevates cellular cAMP only 3-fold (Kopf et al., 1983). Furthermore, following removal of endogenous calmodulin, abalone sperm adenylyl cyclase is activated by exogenous calmodulin (Kopf and Vacquier, 1984). In mammalian sperm cells, bicarbonate anion enhances the calcium-dependent enzyme stimulation (Garbers et al., 1982b). In fact, sperm adenylyl cyclase activity is inhibited by the competitive anion channel blockers SITS and DIDS, and binds to a SITS-affinity matrix, suggesting that bicarbonate directly interacts with the enzyme (Okamura et al., 1985, 1991). The bicarbonate effect appears to be independent of changes in energy metabolism and pH_i. In sea urchins, recent observations suggest that increased pH_i produced with NH_4^+ may also regulate sperm adenylyl cyclase activity (Cook and Babcock, 1993b). Beltran et al. (1996) also presented evidence that the adenylyl cyclase in sea urchin spermatozoa can be regulated by membrane potential without detectable changes in Ca^{2+}_i or pH_i. Efforts to isolate sperm adenylyl cyclase in order to directly characterize regulatory factors have so far proven unsuccessful, possibly because of the labile nature of the enzyme. However, two recently described eukaryotic adenylyl cyclases which are also unresponsive to G protein regulation may prove useful for comparison to the sperm adenylyl cyclase. One adenylyl cyclase, cloned from Dictyostelium, encodes a novel predicted structure containing an extracellular domain of approximately 320 amino acids, a single transmembrane segment, and an intracellular domain of approximately 500 amino acids (Pitt et al., 1992). The intracellular domain contains a single region showing homology to the mammalian adenylyl cyclase catalytic domains. The function of the extracellular region is unknown, but could represent a ligand binding domain. The other adenylyl cyclase was isolated from Paramecium (Schultz et al., 1992). As reported for the sea urchin sperm adenylyl cyclase, this enzyme was stimulated by membrane hyperpolarization. Of particular interest was the finding that the isolated enzyme, with a specific activity comparable to that of the mammalian somatic cell adenylyl cyclases, could be reconstituted into artificial lipid bilayers producing a cation channel. This channel activity was dependent on an enzymatically active adenylyl cyclase. The apparent existence of a voltage-regulated adenylyl cyclase that also functions as an ion channel is not completely unexpected as the proposed structure of the cloned mammalian somatic cell adenylyl cyclases described above is similar to ion channels/transporters (Krupinski et al., 1989).

B. Egg Peptide Receptors

The observation that the egg peptides produced similar sperm cell behavioral and physiological responses among sea urchin species led to receptor models in which the detector region varied, but the signal transducing component was highly conserved. Consequently, it was predicted that the identification of the receptor could lead to DNA probes that would be useful for identifying related receptors, possibly even in evolutionarily distant species.

The approach taken to identify the egg peptide receptors was based on the finding that the peptide amino-terminal sequence can be modified without affecting the bioactivity (Garbers et al., 1982a). Therefore, analogs could be radioiodinated at the amino-terminus without destroying specific binding, and respiratory and motility stimulating potential could be maintained. Surprisingly, chemical crosslinking of these radiolabeled analogs to spermatozoa identified different crosslinked proteins for speract and resact in *S. purpuratus* and *A. punctulata*, respectively (Dangott and Garbers, 1984; Shimomura et al., 1986).

The protein crosslinked to speract was identified as a glycoprotein with M_r 77,000 (SDS–PAGE, reducing conditions) whose specificity was addressed by competition with unlabeled speract, and the absence of labeling of *A. punctulata* spermatozoa. Other proteins were not identified, but low-affinity receptors may not have been detected since high concentrations of radiolabeled speract were not used.

A clone encoding this protein was subsequently obtained from an *S. purpuratus* testis cDNA library using oligonucleotides based on peptide sequence (Dangott et al., 1989). The cDNA predicted a type I transmembrane protein possessing a cysteine-rich extracellular domain, a transmembrane region, and a short, 12-amino-acid, intracellular domain. The extracellular portion contained four tandem homologous internal repeats related to the SRCR (Scavenger Receptor Cysteine-Rich) domain class of proteins (Resnick et al., 1994). This domain also has been found in several mammalian proteins associated with cells of the immune system. Recently, the membrane-proximal SRCR domain of mammalian T-cell CD6 was shown to bind the N-terminal immunoglobulin domain of ALCAM (Activated Leukocyte Cell Adhesion Molecule) providing the first direct evidence for a function of the SRCR domains (Bowen et al., 1996). Direct evidence for the binding of speract to the SRCR domains of the M_r 77,000 protein has not been reported, and therefore it remains unclear whether speract binds directly to this protein or whether the M_r 77,000 protein is associated with the true receptor.

In *A. punctulata*, similar crosslinking experiments with a radiolabeled resact analog labeled guanylyl cyclase (Shimomura et al., 1986). Thus, ligand binding and the production of a low-molecular-weight second messenger were demonstrated with the same polypeptide for the first time. A clone for guanylyl cyclase was then isolated from an *A. punctulata* testis cDNA library (Singh et al., 1988). The clone encoded a type I transmembrane protein containing a 478-amino-acid extracellular domain and a 459-amino-acid intracellular domain. Subsequently, an *S. purpuratus* testis guanylyl cyclase clone was identified with low-stringency cloning (Thorpe and Garbers, 1989). The predicted protein domain organization of this clone was similar to that of the *A. punctulata* guanylyl cyclase with a 485-amino-acid extracellular domain, and a 594-amino-acid intracellular domain. High identity between these two clones was evident from the amino-terminus through an intracellular protein kinase-like domain proximal to the membrane. However, the carboxy-termini were distinctly different in that the *S. purpuratus* sequence was highly identical to the soluble guanylyl cyclase from bovine lung whereas the *A. punctulata* sequence was not (Koesling et al., 1988).

In addition, the carboxy-terminal region was homologous to the two intracellular regions of adenylyl cyclase, suggesting that it represented the catalytic domain. Subsequent cloning of the atrial natriuretic peptide receptor/guanylyl cyclase combined with deletion mutagenesis confirmed that catalytic activity is contained in the carboxy-terminal 200–250 amino acids of guanylyl cyclases (Chinkers and Garbers, 1989; Chinkers *et al.*, 1989).

Recently, we produced an antibody to the carboxy-terminal peptide of *S. purpuratus* guanylyl cyclase: KPPPQKLTQEAIEIAANRVIPDDV (Quill and Garbers, unpublished data). This antiserum recognized sperm guanylyl cyclase from *S. purpuratus*, *L. pictus*, and surprisingly *A. punctulata*. The immunolocalization of guanylyl cyclase along the entire length of the flagellum in both *S. purpuratus* and *A. punctulata* corresponded to the region of speract binding seen using a fluorescent egg peptide analog in *L. pictus* (Cardullo *et al.*, 1994). The observation that this antibody recognized the *A. punctulata* guanylyl cyclase despite the lack of sequence homology prompted us to clone the *A. punctulata* guanylyl cyclase catalytic domain using 3' RACE. The novel sequence obtained represents the carboxy-terminal 202 amino acids of the enzyme and shows 88% identity with the *S. purpuratus* guanylyl cyclase (Fig. 2). Thus, all of the currently reported guanylyl cyclase sequences contain a conserved, consensus catalytic domain at the carboxy-terminus (Fülle *et al.*, 1995; Yang *et al.*, 1995; Yuen and Garbers, 1992).

Interestingly, in addition to the presence of guanylyl cyclase in each sea urchin species, Dangott *et al.* (1989) identified a possible homolog of the M_r 77,000 protein in *A. punctulata* by Northern blotting. This observation provides a potential explanation for the identification of two distinct egg peptide receptor candidates in that the two proteins may be in close proximity in the sperm membrane, where only one serves as the receptor. Alternatively, the two proteins could be subunits of a functional receptor complex. This second hypothesis is supported by two experimental results. First, purification of guanylyl cyclase from both *L. pictus* and *H. pulcherri-*

A.p.	TALSAASTPIQVVNMLNDLYILFDAIIANYDVYKVETIGDAYMLVSGLPIRNGDR	975
S.p.	TALSAASTPIQVVNLLNDLYTLFDAIISNYDVYKVETIGDAYMLVSGLPLRNGDR	978
A.p.	HAGQIASTAYHLLESVKNFIVPHRPDVFLKLRIGIHSGSCVAGVVGLTMPRYCLF	1030
S.p.	HAGQIASTAHHLLESVKGFIVPHKPEVFLKLRIGIHSGSCVAGVVGLTMPRYCLF	1033
A.p.	GDTVNTSSRMESNGLALKIHISPWCKEVLDRLGGYELEERGLVAMKGKGEIHTYW	1085
S.p.	GDTVNTASRMESNGLALRIHVSPWCKQVLDKLGGYELEDRGLVPMNGKGEIHTFW	1088
A.p.	LVGQDPSYKITKVKPPPQKLSQDVLDAAAARVIPDDL	1122
S.p.	LLGQDPSYKITKVKPPPQKLTQEAIEIAANRVIPDDV	1125

FIGURE 2 Comparison of the carboxy-terminal sequences of the *A. punctulata* and *S. purpuratus* guanylyl cyclases. Amino acid numbering based on the full-length sequences. Shaded residues are identical to the *S. purpuratus* sequence.

mus in the absence of denaturing agents results in the copurification of a M_r 75,000 protein (Garbers, 1976; Harumi *et al.*, 1992a). Second, fractionation of a nonionic detergent extract from *L. pictus* sperm membranes results in the loss of speract stimulatable guanylyl cyclase activity (Bentley *et al.*, 1988). However, expression of mammalian membrane guanylyl cyclases has clearly demonstrated that these enzymes serve as receptors. In addition, comparison of the guanylyl cyclase sequence identity between three sea urchin species, one which responds to resact (*A. punctulata*) and two which respond to speract (*S. purpuratus* and *H. pulcherrimus*), indicates highly conserved intracellular domains, but divergence of the extracellular domains (Fig. 3). These data support a hypothesis that guanylyl cyclase is the egg peptide receptor with divergent ligand binding domains and conserved signaling domains.

Direct identification of the egg peptide receptor has been attempted by expression of the candidate receptors alone or together in cultured mammalian or Sf9 cells. While immunoreactive protein was synthesized, no peptide binding or guanylyl cyclase activity was detected (L. Dangott, personal communication). Similar results have been obtained for several nonmammalian guanylyl cyclases. The reasons for this are unknown, but one trivial explanation is inappropriate posttranslational modification. Alternatively, expression of functional guanylyl cyclase activity may require another protein which is sufficiently different between mammalian and nonmammalian

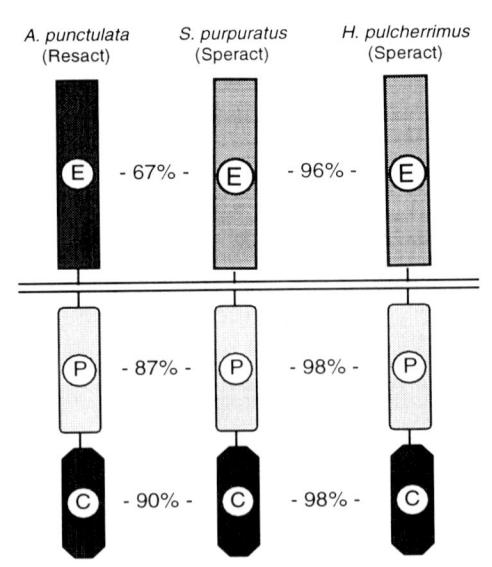

FIGURE 3 Correlation between egg peptide selectivity and guanylyl cyclase divergence across sea urchin species. Percent identity between flanking domains is shown. Extracellular domain, E; protein kinase-like domain, P; catalytic domain, C.

cells that the mammalian form is unable to regulate the sea urchin guanylyl cyclase.

Two experimental observations suggested that the sea urchin sperm guanylyl cyclase was similar to mammalian guanylyl cyclase: the kinetic characteristics and that antibody to the sea urchin sperm enzyme recognized the rat guanylyl cyclase (Lowe *et al.*, 1989). Therefore, based on the predicted conservation of the intracellular signal transducing domain, DNA probes corresponding to the intracellular region of the sea urchin sperm guanylyl cyclase were used to clone a mammalian membrane guanylyl cyclase. Subsequently, six different forms of membrane guanylyl cyclases have been reported in mammals (Fülle *et al.*, 1995; Yang *et al.*, 1995, Yuen and Garbers, 1992). Two enzymes, GC-A and GC-B, belong to the natriuretic peptide receptor family. GC-A is important for body fluid homeostasis and blood pressure regulation (Lopez *et al*, 1995). GC-C is the receptor for the bacterial heat-stable enterotoxin which causes diarrhea, and may regulate physiological fluid secretion of the gastrointestinal tract (Schulz *et al.*, 1990). The remaining three cloned mammalian guanylyl cyclases (D–F) are orphan receptors which appear to be expressed exclusively in sensory tissues (Fülle *et al.*, 1995; Yang *et al.*, 1995). The actual number of mammalian cyclase receptors remains unclear since recently Yu *et al.* (1997) have shown that at least 29 individual guanylyl cyclases exist in *Caenorhabitis elegans*. The *C. elegans* genome is about 1/30th the size of the mammalian genome.

A model of guanylyl cyclase regulation has been formulated based on the analysis of the sea urchin enzyme and GC-A. Guanylyl cyclases possess an amphipathic region between the protein kinase-like and catalytic domains (residues L830–M869 in *S. purpuratus*) which appears to mediate a ligand-independent oligomerization of the receptor (Chinkers and Wilson, 1992; Garbers, 1992). Oligomerization appears obligatory for enzymatic activity in all cyclases (Garbers *et al.*, 1994). Upon ligand binding, in the presence of ATP, a conformational change is postulated to release a protein kinase-like domain-mediated inhibition, thus producing an activation of guanylyl cyclase. That mutations of the GC-A protein kinase-like domain uncouple atrial natriuretic peptide binding from enzyme activation, and that deletion of the entire protein kinase-like domain results in elevated basal activity that is no longer regulatable, support the model (Chinkers and Garbers, 1989). Guanylyl cyclase activation is transient in that ligand binding also induces a rapid dephosphorylation, from approximately 15 to 2 mol Ser-PO_4/mol enzyme in *S. purpuratus* guanylyl cyclase, and this is coincident with a reduction in enzymatic activity (Ramarao and Garbers, 1988). The purified, dephosphorylated form of sea urchin sperm guanylyl cyclase has a fivefold lower specific activity, and loses its positive cooperative kinetic properties with respect to GTP (Bentley *et al.*, 1986b; Garbers, 1976). These changes likely contribute to the transient nature of guanylyl cyclase activation. In sea urchin, the dephosphorylation of guanylyl cyclase correlates with an increase in pH$_i$ induced by treatment with egg peptide, monensin, or NH$_4$Cl (Ramarao and Garbers, 1985; Suzuki and Garbers, 1984). In addition, inhibition of sperm cell alkalinization slows the rate of dephosphorylation of guanylyl cyclase (Harumi *et al.*, 1992b). In the presence of speract, the rate of dephosphorylation induced by elevated pH$_i$ is significantly greater than in its absence, suggesting the possibility that a conformational change provides greater access of guanylyl cyclase to a phosphatase, or that activation of a phosphatase results in the ligand-dependent dephosphorylation.

A number of questions remain concerning the regulation of guanylyl cyclase activity. For example, while oligomerization is necessary for enzymatic activity, is it required for ligand binding? With the atrial natriuretic peptide (ANP) clearance receptor, monomers appear to bind ANP with an affinity similar to that of the dimer, whereas others have suggested that only 1 ANP binds per homodimer of GC-A (Porter *et al.*, 1989; Rondeau *et al.*, 1995). A number of studies suggest that ligands do not induce dimerization of the guanylyl cyclase receptors (Chinkers and Wilson, 1992; Lowe, 1992). What is the role of ATP during guanylyl cyclase signal transduction? With GC-A, ATP is absolutely required for ligand-dependent activation, whereas the nucleotide may not be required for activation of GC-C (Chinkers *et al.*, 1991; Vaandrager *et al.*, 1993). The site of ATP binding is not known, but is presumably the protein kinase-like domain. Why is dimerization required for enzyme activity? Work on GC-A and the soluble form of guanylyl cyclase suggests that the reason is a shared GTP binding site, in that a single-point mutation in one subunit of the dimer is able to completely abolish enzyme activity (Thompson and Garbers, 1995; Yuen *et al.*, 1994). Does dephosphorylation of the receptor directly lead to desensitization? Although the decrease in phosphorylation state correlates with desensitization, there are no data to demonstrate that the cyclases can be rephosphorylated by a protein kinase, thus restoring sensitivity to ligand. Both the protein kinase(s) and protein phosphatase(s) that regulate the phosphorylation state of guanylyl cyclases remain critical enzymes to identify. Finally, careful analysis of associated proteins remains an important issue. Based on precedence with many other receptors, associated regulatory proteins are likely, raising the possibility of receptor signaling independent of cGMP.

C. Chemotaxis

Sperm chemotaxis has been described in many species distributed across several phyla (Miller, 1985). The effective range of chemoattractants appears to be limited to about a few hundred micrometers, yet would significantly increase the effective size of the egg. In the presence of limited numbers of spermatozoa at the site of fertilization, as found in many external and internal fertilizing species, the apparent increase in target size would provide a selective advantage for productive fertilization.

Only a few sperm chemoattractants have been isolated. The peptide resact, discussed above, is a potent chemoattractant for *A. punctulata* spermatozoa (Ward et al., 1985). Sea urchins that produce speract do not respond to resact, demonstrating the species-specificity of the activity. The other sea urchin egg peptides have not yet been shown to act as chemoattractants, but recent experiments demonstrate that speract in the presence of IBMX modifies the swimming pattern of *S. purpuratus* spermatozoa, similar to the effect of resact on *A. punctulata* spermatozoa (Cook et al., 1994). Thus, many or all of the other egg peptides may also possess chemotactic activity. In starfish, a synthetic peptide representing approximately the amino-terminal one-third of a M_r 13,000 protein isolated from ovaries was shown to attract spermatozoa (Miller and Vogt, 1996). The protein, named startrak, also acted in a species-specific manner. While many sperm chemoattractants appear to be peptides, other chemical classes may also act as attractant factors [e.g., sperm chemoattractants in corals are lipids (Coll et al., 1995)].

The existence of sperm chemotaxis in mammals has not been clearly demonstrated. Recently, using several different assay designs, human spermatozoa have been shown to accumulate in diluted follicular fluid, but not diluted serum (Ralt et al., 1991, 1994). The potency of the follicular fluid samples varied considerably, and a positive correlation existed between the ability of follicular fluids to cause accumulation of spermatozoa and successful *in vitro* fertilization of the corresponding egg.

The process responsible for accumulation of spermatozoa in follicular fluid could be due to chemotaxis, reduced vectorial motility, trapping (i.e., nonspecific adhesion or decreased flagellar activity), or any combination of these effects. That a selected "highly motile" population of spermatozoa migrated to a lesser extent from a well containing follicular fluid into a capillary tube containing cell medium than from the well into the capillary tube when both compartments contained either medium or follicular fluid, suggests that chemoattraction contributes to the accumulation of spermatozoa (Ralt et al., 1994). In all of the assays, only a small fraction of the sperm population (approximately 2–12%) is found to respond to follicular fluid (Cohen-Dayag et al., 1994). This subpopulation appeared to be constantly changing, and initially responsive spermatozoa gradually lose the ability to respond, while other cells gain follicular fluid responsiveness. Recently, Eisenbach and colleagues have reported a numerical and temporal correlation between the follicular fluid responsiveness of spermatozoa and the capacitation state (Cohen-Dayag et al., 1995) (spermatozoa were judged capacitated if an acrosome reaction was rapidly induced by phorbol-myristate-acetate). If a constantly changing population of capacitated spermatozoa is selectively attracted by follicular fluid, a higher probability of productive fertilization would be expected due to the appearance of fertilization-competent spermatozoa over an extended period of time.

The component(s) of the follicular fluid responsible for the above effects have not been isolated. However, two follicular fluid components have been suggested as chemotactic factors for human spermatozoa: atrial natriuretic peptide and progesterone (Anderson *et al.*, 1995; Zamir *et al.*, 1993). Binding sites for these compounds have been localized to the midpiece and sperm head, respectively (Tesarik and Mendoza, 1993). The atrial natriuretic peptide receptor was postulated as guanylyl cyclase based on an estimated M_r of 140,000 (nonreducing SDS–PAGE). However, nonexistent or extremely low levels of guanylyl cyclase activity have been found in mammalian spermatozoa (Garbers and Kopf, 1980). Therefore, atrial natriuretic peptide binding could be mediated through an alternative receptor such as the atrial natriuretic peptide clearance receptor (M_r approximately 120,000 on nonreducing SDS–PAGE), or a novel receptor (Drewett and Garbers, 1994). In the case of progesterone, chemotactic effects were detected at concentrations 10-fold higher than those that induced an acrosome reaction. The significance of progesterone-stimulated chemotaxis is unclear since induction of the acrosome reaction at some distance from the egg would likely shorten the effective fertilization life span, and thus inhibit subsequent adhesion to the egg (Brown *et al.*, 1989). In each case, no correlation was found between follicular fluid concentrations of either atrial natriuretic peptide or progesterone and the apparent chemoattractive activity (Anderson *et al.*, 1995; Ralt *et al.*, 1991). In addition, in contrast to the chemoattractants of other species, these compounds would not be predicted as species-specific, unless the ANP or progesterone receptor are exclusively expressed on the sperm cells of only a few species. Thus, other unknown follicular fluid factors may mediate chemotaxis. It is intriguing to note that a group of serpentine receptors related to the olfactory subfamily of G protein-coupled receptors have been reported in mammalian spermatogenic cells and mature spermatozoa (Parmentier *et al.*, 1992; Vanderhaeghen *et al.*, 1993). Other components of this signaling pathway, including βARK 2 and β-arrestin, also have been identified in spermatogenic cells (Dawson *et al.*, 1993; Walensky *et al.*, 1995). These receptors have been localized to the midpiece of mature spermatozoa, and therefore a potential role in motility regulation can be postulated, although no evidence for the function of these receptors exists.

Currently, the molecular mechanism of sperm chemotaxis is best understood in sea urchins. As discussed above (Section III.A) and illustrated in Fig. 1, resact binding to *A. punctulata* spermatozoa initiates a signal transduction pathway that transiently elevates Ca^{2+}_i. In the absence of Ca^{2+}_e, resact stimulates sperm respiration and motility, but increases in Ca^{2+}_i and chemotaxis are inhibited (Cook *et al.*, 1994; Ward *et al.*, 1985). Therefore, Ca^{2+} appears to be a critical signal for regulation of sperm cell orientation. In fact, *in vitro* experiments have shown that Ca^{2+} can induce an asymmetric flagellar waveform resembling the flagellar beat observed during changes in sperm cell direction (Brokaw and Nagayama, 1985). The mechanism through which Ca^{2+} alters the axoneme stroke remains to be determined, although high levels of calmodulin and calmodulin-dependent phosphatase are present.

Reorientation of spermatozoa during chemotaxis occurs when the cells no longer detect an increasing chemoattractant gradient (Miller and Brokaw, 1970). In sea urchins, it has been hypothesized that a continuously increasing gradient of chemoattractant maintains membrane hyperpolarization, which prevents Ca^{2+} channel opening, Ca^{2+} influx, and therefore turning (Cook et al., 1994). In the absence of a continuously increasing gradient, the peptide-induced increase of pH_i down-regulates the K^+ channel-mediated hyperpolarization and stimulates an increase of Ca^{2+}_i. The transient nature of the Ca^{2+}_i increase would allow the sperm cell to regain its symmetric flagellar beat and proceed in the new direction. In this model, movement in any direction other than toward the chemoattractant source would result in turning, thus increasing the probability of gamete interaction.

IV. Gamete Adhesion

Physical contact of sea urchin spermatozoa with the extracellular matrix, or jelly coat, surrounding the egg leads to gamete adhesion. The sea urchin egg jelly coat is predominantly composed of a sialoglycoprotein (~20% of mass) and a high-molecular-weight, fucose and sulfate-rich polymer, FSG (fucose sulfate glycoconjugate, ~80% of mass) (SeGall and Lennarz, 1979). Cellular adhesion appears primarily mediated by FSG which causes agglutination of live or dead sperm cells. However, this interaction seems to possess a relatively low degree of binding specificity, as many proteins and cells can adhere to the highly charged egg jelly coat. The identity of the structural component of FSG responsible for sperm binding is unknown.

A M_r 210,000 S. purpuratus sperm protein, named REJ (Receptor for Egg Jelly), binds egg jelly (Moy et al., 1996). This interaction displays at least partial species-specificity since isolated S. purpuratus REJ as well as sperm membrane vesicles bind S. purpuratus but not A. punctulata egg jelly, while A. punctulata sperm membrane vesicles bind egg jelly from both species (Podell and Vacquier, 1985). A clone encoding REJ has been obtained recently from an S. purpuratus testis cDNA library (Moy et al., 1996). The cDNA predicts a 1450-amino-acid protein containing an EGF module, two carbohydrate recognition domains (CRD), a novel "REJ" domain, and a putative transmembrane segment close to the carboxy-terminus. The presence of two CRD domains suggests that REJ binds egg jelly via oligosaccharides and that Ca^{2+} is important for the binding (Day, 1994). The REJ domain is also found in polycystin, a defective protein in autosomal dominant polycystic kidney disease I, a systemic condition affecting cellular growth, extracellular matrix composition, and fluid secretion (International Polycystic Kidney Disease Consortium, 1995). Despite this phenotype, the molecular function of the shared REJ domain is unknown. Interestingly, REJ is present over the sperm acrosome and the flagellum, suggesting it is a multifunctional protein (Trimmer et al., 1985).

The mammalian egg extracellular matrix, or zona pellucida, also functions in initial gamete adhesion (Wassarman, 1988). The murine zona pellucida consists of three microheterogeneous, sulfated glycoproteins. The molecular cloning of the cDNA encoding each of these glycoproteins from several species suggests that all three evolved from a common ancestor (Harris et al., 1994; Hedrick, 1996). As the individual zona pellucida glycoprotein sequences from the South African clawed toad, *Xenopus laevis*, are more closely related to their counterparts in mammals than to each other, the evolutionary divergence of these glycoproteins is predicted to precede the evolution of amphibians (~350 Ma). The zona pellucida glycoproteins have three common characteristics: a signal peptide, a 260-amino-acid domain called the ZP module, and a transmembrane segment near the carboxy-terminus (Dunbar et al., 1994). In each case, the transmembrane segment is removed probably as a consequence of a furin-like proteolytic processing site, just to the amino-terminus of the transmembrane segment (Yurewicz et al., 1993). The function of the ZP module is unknown, although it is found in other proteins including uromodulin and the TGF-βIII receptor (Bork and Sander, 1992). One possibility is that this domain is involved in protein–protein interactions that produce the structural organization of the zona pellucida. The zona pellucida matrix has been described in the mouse as ZP1 homodimers that crosslink extended polymers of ZP2/ZP3 heterodimers (Greve and Wassarman, 1985). Since the zona pellucida glycoproteins are conserved, the native matrix structure is likely common to most or all species.

Sperm cell binding to the zona pellucida appears mediated in part by oligosaccharides on the zona pellucida glycoproteins (Gahmberg et al., 1992). Using isolated zona pellucida glycoproteins or oligosaccharides in competition for sperm cell binding to the zona pellucida of intact eggs and two-cell embryos, Wassarman's group identified ZP3 O-linked oligosaccharides as critical for sperm cell binding in the mouse (Florman and Wassarman, 1985). Isolated ZP1 and ZP2 did not appear to compete for sperm cell binding. Subsequent expression of ZP3 cDNA in a variety of cell lines (CV-1, L fibroblasts, embryonic carcinoma) confirmed that acrosome-intact mouse spermatozoa are able to bind ZP3 (Beebe et al., 1992; Kinloch et al., 1991). Furthermore, site-directed mutagenesis of a cluster of serine residues near the carboxy-terminus of mature ZP3 resulted in a substantial loss of binding activity (Kinloch *et al.*, 1995). This region of ZP3 shows significant sequence variability between homologs found in mouse, hamster, and human, possibly reflecting a contribution to species-specific binding (Wassarman and Litscher, 1995). In the pig, the homolog of mouse ZP1, or perhaps all of the zona pellucida glycoproteins, participates in sperm binding through N-linked oligosaccharides (Nakano *et al.*, 1996). Thus, species-specificity may reflect several discrete differences in sperm–zona pellucida interaction across species as a result of posttranslational modifications. Attempts to assess the importance of ZP3 through the use of antisense RNA or gene disruption resulted in loss of the entire zona pellucida matrix and thus its specific function could not be addressed (Liu *et al.*, 1996; Tong *et al.*, 1995).

Studies in a variety of species have identified several different sperm proteins which bind to the zona pellucida (Table II). In many cases, carbohydrate appears to be the primary epitope for binding. In the mouse, three of the zona pellucida binding proteins, β-1,4-galactosyltransferase, sp56, and p95, have been reported to specifically bind ZP3 (Bleil and Wassarman, 1990; Leyton and Saling, 1989a, Miller *et al.*, 1992). Of these, sp56 has been proposed as a species-specific adhesion protein (Bookbinder *et al.*, 1995). However, only porcine zonadhesin, a minor sperm membrane glycoprotein, isolated using the native zona pellucida matrix, has been directly shown to bind the zona pellucida in a species-specific manner (Hardy and Garbers, 1994). In this study, at least one other boar sperm protein of approximately M_r 60,000 was also bound in a species-specific manner by the intact zona pellucida. The identity of this protein(s) is unknown, although it could represent a homolog of mouse β-1,4-galactosyltransferase or sp56. The identification of several zona pellucida binding proteins on spermatozoa suggests that the adhesion interaction is complex. Therefore, the relative species-specificity of gamete adhesion likely reflects the sum of several individual interactions between the zona pellucida and several sperm binding proteins.

V. Acrosome Reaction

In those species whose spermatozoa possess an acrosome overlying the apical sperm head, the acrosome reaction (an exocytotic event) is considered essential for successful fertilization. The physiologically relevant acrosome reaction generally appears to be induced by the egg extracellular matrix in a largely species-specific manner (Ward and Kopf, 1993). As a result, various hydrolytic enzymes are released (*e.g.*, acrosin, hyaluronidase) which enable

TABLE II Zona Pellucida Binding Proteins

Designation	M_r (species)	Cellular localization[a]	Comments	References
Acrosin/proacrosin	53,000/55,000 (boar)	AV	Trypsin-like protease; secondary adhesion	Hardy and Garbers (1994), Jones et al. (1988)
β-1,4-Galactosyl-transferase	60,000 (mouse)	AH	Specifically transfers galactose to ZP3 O-linked oligosaccharides; aggregation initiates acrosome reaction; cytoplasmic domain peptide associates with G_i; genetic knock-out mice are fertile	Gong et al. (1995), Macek et al. (1991), Miller et al. (1992), Lu et al. (1997)
FA-1	51,000 (human)	AH/PH, M, F	Antibody inhibits zona pellucida binding	Kadam et al. (1995), Naz et al. (1992)
Hyaluronidase	62,000–64,000 (guinea pig)	AV, PH	Two structurally related forms: GPI anchored and soluble; cumulus penetration; secondary adhesion	Hunnicutt et al. (1996a), Primakoff et al. (1985)
α-Mannosidase	115,000 (rat)	AH	Membrane associated; correlation between inhibition of enzyme activity and zona pellucida binding	Cornwall et al. (1991), Tulsiani et al. (1995)

p95	95,000, nonreduced (mouse)	AH	Bound ZP3 but not ZP2 on nitrocellulose blots; contains phosphotyrosine; may be identical to hexokinase or LL95 antigen	Kalab et al. (1994), Leyton and Saling (1989a), Leyton et al. (1995)
Rabbit sperm autoantigen (RSA)	14,000, 17,000 (rabbit)	AH	Binds zona pellucida; peripheral membrane protein; recombinant RSA 17 binds zona pellucida	Richardson et al. (1994)
sp56	56,000, reduced (mouse)	AH	Crosslinked to ZP3; peripheral membrane protein containing "sushi" domains; recognizes terminal α-galactose	Bleil and Wassarman (1990), Bookbinder et al. (1995), Cheng et al. (1994)
Spermadhesins	12,000–16,000 (boar)	AH	Binds carbohydrate; peripheral membrane protein secreted by seminal vesicles and rete testis (AWN only)	Dostalova et al. (1995), Topfer-Petersen et al. (1995)
Zona receptor kinase (ZRK)	95,000 (human)	AH	Homology to axl subfamily of receptor tyrosine kinases; controversial (see comments in *Science* **271**, 1431–1435)	Burks et al. (1995)
Zonadhesin	150,000, nonreduced (boar)	AH	Binds native zona pellucida species-specifically; homology to vWF "D" domain	Hardy and Garbers (1994, 1995)

[a] AH, anterior head; AV, acrosomal vesicle; F, flagella; M, midpiece; PH, posterior head.

penetration of the egg extracellular matrix while maintaining gamete adhesion (Hunnicutt et al., 1996b; Topfer-Petersen et al., 1990).

The sea urchin sperm cell acrosome reaction is induced by the FSG component of egg jelly (SeGall and Lennarz, 1979). This activity is potentiated by the sperm activating peptides (Section III.A), which alone cannot induce the acrosome reaction (Yamaguchi et al., 1988). The functionally active FSG structure was initially proposed as solely carbohydrate, but several treatments designed to remove the associated protein significantly reduced the biological potency (Garbers et al., 1983). Furthermore, the protein content of FSG samples correlated with increases of cAMP, a messenger whose concentrations are associated with the acrosome reaction. Thus, the protein component may have importance for biological activity. Recently, several proteins tightly associated with purified FSG have been identified in *H. pulcherrimus* and *S. purpuratus* (Keller and Vacquier, 1994; Shimizu et al., 1990). Following separation of these components from *S. purpuratus* egg jelly under denaturing conditions, acrosome reaction-inducing activity was detected in a fraction containing two proteins of M_rs 138,000 and 82,000, but not in a M_r 360,000 fucose-rich fraction. However, only a small fraction (2.4%) of the original activity was recovered after purification. Thus, as noted by the authors, additional active components may have been lost, denatured, or not detected in these experiments.

One candidate sea urchin sperm cell receptor for the egg jelly acrosome reaction-inducing factor has been identified as REJ (Section IV) (Moy et al., 1996). Preincubation of purified REJ with egg jelly inhibits the acrosome reaction-inducing activity, and a monoclonal antibody to REJ induces the acrosome reaction. The mechanism by which REJ would transduce the signal for the acrosome reaction is unclear given the predicted 15-amino-acid intracellular region. One possibility is an egg jelly-induced REJ aggregation which then results in signaling. In support of this hypothesis, divalent REJ monoclonal antibodies are approximately 200-fold more potent than monovalent Fab fragments (on a molar basis) at inducing the acrosome reaction. However, as noted above, the localization of REJ to both the sperm head and the flagellum suggests alternative or additional functions.

In mammals, by definition, capacitated spermatozoa (Section II) undergo the acrosome reaction in response to the zona pellucida (Yanagimachi, 1994). The cellular changes which occur during capacitation that result in competence for an acrosome reaction are not established, but potential mechanisms include a modification of the receptor that enables interaction with the zona pellucida, and/or intracellular events that establish a link between the receptor and the signal transduction pathway. For example, the ability to "decapacitate" spermatozoa with seminal fluid suggests that seminal plasma components can bind to the sperm cells and perhaps block subsequent zona pellucida interaction (Oliphant et al., 1985). Such factors appear to be released during the capacitation period (Fraser et al., 1990). Interestingly, recent data indicate that the major bovine seminal plasma proteins, BSPs, bind to sperm plasma membrane phospholipids over the entire cell surface (Desnoyers and Manjunath, 1992). During capacitation in the oviduct, these proteins could bind high-density lipoproteins (HDL) and be removed from the spermatozoa along with associated plasma membrane cholesterol and phospholipids. In addition, a capacitation-associated K^+-dependent hyperpolarization of mammalian spermatozoa has been described that appears to be required for subsequent zona pellucida stimulation of the acrosome reaction (Zeng et al., 1995).

The zona pellucida-stimulated acrosome reaction has been studied predominantly in the mouse. Of the three zona pellucida glycoproteins, only ZP3, either purified or expressed from cDNA, appears to induce the acrosome reaction (Bleil and Wassarman, 1983; Kinloch et al., 1991). Therefore, mouse ZP3 appears to function in both gamete adhesion and the induction of the acrosome reaction. However, while small glycopeptides and oligosaccharides generated from ZP3 can bind to spermatozoa, stimulation of acrosomal exocytosis requires larger fragments of ZP3, suggesting that the polypeptide is important (Wassarman et al., 1986). It was subsequently shown that small ZP3 glycopeptides can stimulate the acrosome reaction if the bound glycopeptides and their associated receptors are aggregated with a ZP3-specific divalent antibody (Leyton and Saling, 1989b). In fact, antibody-induced aggregation of a variety of mammalian sperm cell plasma membrane proteins results in acrosome exocytosis, possibly as a consequence of receptor coaggregation (Aarons et al., 1991; Macek et al., 1991; Tesarik et al., 1992). Thus, induction of the acrosome reaction appears to require ligand-mediated receptor aggregation/recruitment in animals as evolutionarily diverse as mammals and echinoderms.

The mammalian sperm cell receptor(s) which initiates the acrosome reaction has not been definitively identified. In the mouse, three sperm proteins that bind ZP3 have been proposed as candidate receptors for transduction of the acrosome reaction signal: β-1,4-galactosyltransferase (GalTase), sp56, and p95 (Table II) (Bleil and Wassarman, 1990; Leyton and Saling, 1989a; Miller et al., 1992). Both GalTase and p95 have been suggested to couple to intracellular signal transducing pathways, GalTase through the activation of Gi and p95 via an intrinsic protein tyrosine kinase activity (Gong et al., 1995; Leyton et al., 1992). Each of these intracellular signaling pathways appears to be involved in the acrosome reaction based on inhibitor studies (Bailey and Storey, 1994; Endo et al., 1987; Leyton et al., 1992). However, criticism of some of the studies has been reported (Bork, 1996; Kalab et al., 1994; Tsai and Silver, 1996). In addition to the ability to initiate intracellular signaling, the receptor could account for the apparent species-specificity of the acrosome reaction. Alternatively, however, other gamete recognition molecules could serve in the species-specificity role by being required for presentation of a common ligand structure to the receptor.

The mammalian sperm cell acrosome reaction can also be induced by progesterone, although poorly at concentrations of less than approximately 2 μM (Melendrez and Meizel, 1995; Meyers *et al.*, 1995; Osman *et al.*, 1989; Shi and Roldan, 1995). While the concentration of progesterone at the site of fertilization is unknown, it is unlikely that progesterone would account for the observed species-specificity of the acrosome reaction, unless the species which respond to progesterone specifically express the progesterone receptor (Meizel, 1997). The stimulatory effect of the zona pellucida may be enhanced by an initial encounter with low levels of progesterone within the cumulus oophorus surrounding the egg, and thus the steroid could also function as a cofactor in the mammalian acrosome reaction, analogous to the sperm activating peptides in sea urchins, and Co-ARIS (a steroidal saponin) in the starfish (Hoshi *et al.*, 1990; Roldan *et al.*, 1994; Yamaguchi *et al.*, 1988).

The cascade of molecular events that results in acrosomal vesicle exocytosis upon stimulation of the sperm cell receptor(s) by the egg extracellular matrix appears to be largely conserved among species (Garbers, 1989). Elevations of intracellular Ca^{2+} and pH are required to induce the acrosome reaction in all species examined. A model depicting the mechanism that produces these changes in sea urchin spermatozoa is shown in Fig. 4A. In this model, receptor activation opens a K^+ channel (different from the K^+ channel associated with the egg peptide receptor; see Section III.A), hyperpolarizing the sperm cell. Then increased pH_i occurs, possibly through activation of a Na^+/H^+ exchanger, similar to that proposed for sperm motility activation (Garbers, 1989; Gonzalez-Martinez and Darszon, 1987; Gonzalez-Martinez *et al.*, 1992). Independently, the receptor(s) appears to transiently activate a dihydropyridine-sensitive Ca^{2+} channel and this depolarizes the sperm cell (Guerrero and Darszon, 1989). As a result of the elevated pH_i, possibly in combination with the membrane depolarization, a dihydropyridine-insensitive Ca^{2+} channel is opened that further increases Ca^{2+}_i. The mechanism to elevate intracellular Ca^{2+} and pH in mammalian spermatozoa may be similar (Fig. 4B). In this model, receptor(s) stimulation activates Gi which then increases pH_i via an undefined ion channel (Arnoult *et al.*, 1996b; Ward *et al.*, 1994). Coincidentally, the receptor(s) activates a nonselective cation channel that transiently elevates Ca^{2+}_i and depolarizes the sperm cell (Florman, 1994). The combination of elevated pH_i and membrane depolarization are required to activate a T-type Ca^{2+} channel mediat-

Fertilization: Common Molecular Signaling Pathways 191

A

Egg jelly
↓
Receptor

TEA ⊣ K⁺ channel (K⁺ efflux) ↓ Na⁺/H⁺ exchanger (Na⁺ influx, H⁺ efflux) ↓ pH$_i$ increase

Ca²⁺ channel (Ca²⁺ influx, transient) ⊢ DHP ↓ Depolarization

↘ ↙ Ca²⁺ channel (Ca²⁺ influx) ↓ Exocytosis

B

Zona pellucida
↓
Receptor

PTx ⊣ Gi ↓ pH$_i$ increase

Cation channel (Ca²⁺ influx, transient) ⊢ QNB ↓ Depolarization

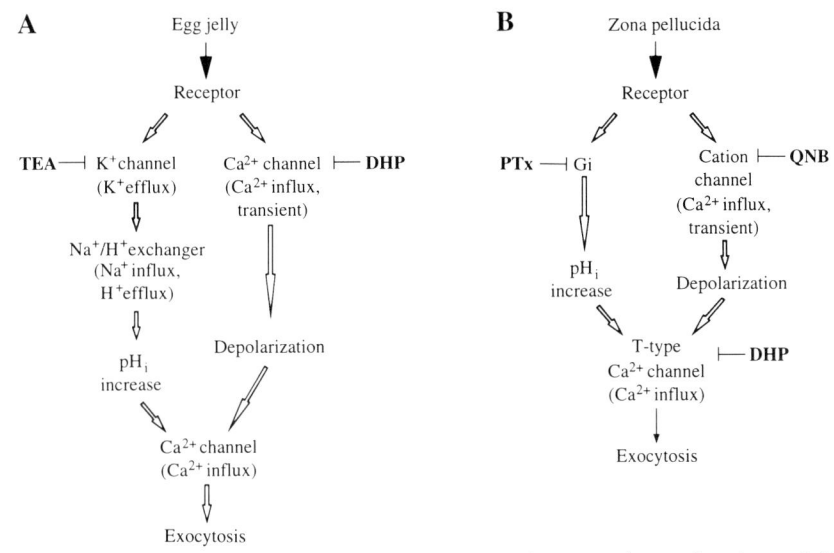

↘ ↙ T-type Ca²⁺ channel (Ca²⁺ influx) ⊢ DHP ↓ Exocytosis

FIGURE 4 Schematic representation of the signal transduction pathways that elevate Ca²⁺$_i$ and pH$_i$ leading to acrosomal exocytosis in sea urchins (A) and mammals (B). Solid arrows indicate direct activation, hollow arrows indicate activation through undefined mechanisms. Inhibitors are shown in bold. TEA, tetraethylammonium ion; DHP, dihydropyridines; Ptx, pertussis toxin; QNB, 3-quinuclidinyl benzylate.

ing additional Ca²⁺ entry (Arnoult *et al.*, 1996a; Florman *et al.*, 1992). For each of these models, no experiment showing direct protein interactions to initiate elevations of Ca²⁺$_i$ and pH$_i$ has been presented. Thus, G proteins found in sea urchin spermatozoa may participate in the acrosome reaction, as has been suggested for mammalian sperm cells (Bentley *et al.*, 1986a). Similarly, a K⁺ channel may ultimately regulate pH$_i$ in mammalian spermatozoa, as has been shown in the sea urchin. Furthermore, other G proteins (Gq/11 and Gz) which are not inhibited by pertussis toxin have been detected in mammalian spermatozoa (Glassner *et al.*, 1991; Walensky and Snyder, 1995). Whether or not these G proteins are activated upon zona pellucida binding has not been reported, although since pertussis toxin has been reported to completely inhibit zona pellucida-induced high-affinity GTPγS binding in both intact sperm cells and sperm membranes, if activated, Gq/11 and Gz may function downstream of Gi in the signal transduction pathway (Ward *et al.*, 1992; Wilde *et al.*, 1992).

In addition to the increases in Ca²⁺$_i$ and pH$_i$, the egg extracellular matrix also stimulates adenylyl cyclase and phospholipase C in both sea urchin and mammalian sperm cells, as well as sea urchin sperm phospholipase D (Domino *et al.*, 1989; Domino and Garbers, 1988; Leclerc and Kopf, 1995; Roldan *et al.*, 1994; Watkins *et al.*, 1978). The activation of each of these enzymes (with the possible exception of the mammalian adenylyl cyclase) appears to depend on an influx of Ca^{2+}, providing at least a partial explanation for the role of Ca^{2+}_e in the acrosome reaction. The activation of these enzymes precedes the exocytosis of the acrosome in both sea urchins and mammals, and thus their regulation may be directly involved in induction of an acrosome reaction. However, the consequences of the production of cAMP, inositol trisphosphate (IP_3), diacylglycerol, and phosphatidic acid in spermatozoa remain unclear. In the case of cAMP, sea urchin sperm protein kinase A is activated as a result of binding to the egg extracellular matrix; however, the substrates for this kinase have yet to be identified (Garbers et al., 1980). In mammalian spermatozoa, the IP_3 receptor, localized on the outer acrosomal membrane, and protein kinase C are likely targets of IP_3 and diacylglycerol, respectively (Breitbart et al, 1992; Rotem et al., 1992; Walensky and Snyder, 1995). The IP_3 receptor has been hypothesized to function in the release of sequestered Ca^{2+} from the acrosome, contributing to the Ca^{2+}_i elevation (Walensky and Snyder, 1995). In addition to these second messenger events, other sperm cell components that function in acrosomal exocytosis, and perhaps are related to somatic cell exocytotic factors (e.g., small GTPases and synaptotagmin), are likely to be discovered (Augustine et al., 1996; Garde and Roldan, 1996; Sudhof, 1995).

VI. Egg Plasma Membrane Interactions and Egg Activation

Plasma membrane binding and fusion represent the final intercellular interactions between the spermatozoon and the egg. In sea urchins, intercellular plasma membrane adhesion occurs at the tip of the acrosomal process in a relatively species-specific manner (Foltz and Lennarz, 1993). This adhesion appears mediated by bindin, a M_r 24,000 sperm acrosomal protein exposed following the acrosome reaction, since isolated bindin agglutinates dejellied eggs in a species-specific manner (Gao et al, 1986; Glabe and Vacquier, 1977). Clones encoding bindin have been obtained from several different sea urchin species (Gao et al., 1986; Glabe and Clark, 1991; Minor et al., 1991). Each of the cDNAs predict an approximately 51,000-Da precursor protein with a putative signal peptide, but no transmembrane segment. Prior to or during the acrosome reaction, probindin is proteolytically processed, generating a mature protein of approximately 230–240 amino acids. Comparison of bindin from four sea urchin species shows a highly conserved 70-amino-acid central region flanked by more divergent amino- and carboxy-terminal sequences (Lopez et al., 1993). Based on deletion mutations, it appears that the conserved region of bindin may function in gamete adhesion while the remainder of the protein may provide species-specificity to the interaction (Lopez et al., 1993).

The *S. purpuratus* egg receptor for bindin appears to be a disulfide-bonded homomultimer containing approximately four glycosylated subunits of M_r 350,000 each (Ohlendieck et al., 1993, 1994). A cDNA for this apparent receptor has been obtained by expression cloning and predicts an integral membrane protein with a 908-amino-acid extracellular domain, and at least one (but perhaps three) transmembrane segment(s) near the carboxy-terminus (Foltz et al., 1993). This cDNA (4.3 kb) predicts a substantially smaller protein than the isolated receptor subunit size, and in addition, a 7-kb mRNA has been identified on Northern blots. Thus, it is unclear whether the reported clone is complete. Nevertheless, expression of a portion of the clone, corresponding to the putative extracellular domain containing a region homologous to HSP70, produced a protein that specifically associated with the sea urchin sperm cell acrosomal process, as well as isolated bindin, suggesting that the clone represents the bindin receptor. Interestingly, using the receptor cDNA sequence to analyze cross-species expression suggested that the extracellular domain was divergent while the intracellular domain was conserved, consistent with species-specific adhesion. The intracellular domain of the bindin receptor shows no homology with other known proteins, yet it has been hypothesized that it also signals activation of the egg. Isolated bindin does not activate the egg, however, and therefore signaling may not be a property of the bindin receptor (Glabe et al., 1981).

In mammals, acrosome-reacted spermatozoa appear to bind and then fuse with the egg via the plasma membrane over the equatorial segment and/or posterior head (Yanagimachi, 1994). In the absence of the zona pellucida, the incidence of fertilization between species increases suggesting lower specificity at this level (Garbers, 1989). However, conspecific sperm–egg plasma membrane binding and fusion remain more efficient in most cases. A potential explanation for this is that several plasma membrane proteins may participate in these events, such that the presence of a greater number of complementary sperm and egg binding and/or fusogenic proteins within a single species yields a more efficient binding and fusion than seen when gametes from different species are mixed. Consistent with this hypothesis, several mammalian sperm cell proteins have been proposed to function during egg plasma membrane binding and fusion based on the inhibition of these events with specific antibodies (Myles, 1993). Among the sperm cell candidate plasma membrane binding and/or fusion proteins, only fertilin, a heterodimer localized on the posterior head of guinea pig spermatozoa, has been characterized in detail (Blobel et al., 1990, 1992). Recently, cDNAs representing each subunit have been obtained from a variety of species (Evans et al., 1995; Hardy and Holland, 1996; Wolfsberg et al., 1995).

Each of the subunits is synthesized as a precursor of approximately 750–800 amino acids containing a signal peptide, prodomain, metalloprotease-like domain, disintegrin-like domain, cysteine-rich domain, transmembrane segment, and an intracellular domain of approximately 30 amino acids. Thus, fertilin has been suggested as a multifunctional protein. During epididymal transit in guinea pigs, fertilin is processed, proteolytically, exposing the disintegrin domain at the amino-terminus of mature fertilin β, while most of the disintegrin domain is removed from the mature α subunit (Blobel et al., 1990). The acquisition of fertilization competence by sperm cells correlates with this maturation process. Regarding gamete plasma membrane interaction, two structural features of fertilin are intriguing. First, the α subunit of fertilin contains a potential fusogenic region in the cysteine-rich domain, which upon fertilin binding to the egg, and a postulated conformational change, could then become exposed to participate in gamete fusion (Blobel et al., 1992). Second, the identification of the disintegrin domain in mature fertilin suggests that fertilin may bind to integrins on the egg plasma membrane. In fact, synthetic peptides corresponding to the region of fertilin predicted to bind integrins do specifically block sperm–egg adhesion and fusion (Myles et al., 1994). However, genetically modified mice lacking the gene for 14 of the 23 known integrin subunits, including $\alpha 6$ and $\beta 1$, which were proposed to be important in mouse gamete adhesion, have not resulted in infertile animals, suggesting alternative or additional egg plasma membrane adhesion and fusion proteins (Almeida et al., 1995; Fassler et al., 1996). The apparent inability of spermatozoa to fuse with other cells despite the pervasive presence of integrins also indicates that gamete interaction resulting in fusion is a complex process involving other components besides fertilin (Hynes, 1992).

As a consequence of gamete plasma membrane interaction, spermatozoa trigger egg activation. Precisely how this fundamental event occurs remains unclear, although two prominent hypotheses have been proposed. One hypothesis suggests that sperm cells activate eggs through a receptor-mediated pathway. This hypothesis is supported by observations of egg activation by an acrosomal protein from *Urechis* as well as by the appropriate ligands of exogenously expressed G protein and tyrosine kinase-coupled receptors (Gould and Stephano, 1987; Moore et al., 1994; Shilling et al., 1994). The other hypothesis suggests that a soluble factor in spermatozoa diffuses into the egg upon fusion and initiates activation. One candidate diffusable factor is oscillin, an equatorial segment-associated M_r 33,000 sperm protein related to a hexose phosphate isomerase which appears to induce Ca^{2+} oscillations associated with egg activation (Parrington et al., 1996). Additional candidates include a number of soluble second messengers, such as IP_3 (Whitaker and Swann, 1993). Whether or not a soluble factor could diffuse into the egg at sufficient levels and rates to account for the speed of egg activation (ranging from approximately 10 sec to several minutes in various species) has not been demonstrated. Finally, it is not mutually exclusive that diffusable factors and receptor-mediated pathways both participate in egg activation by the spermatozoon. Thus receptor activation may prime the egg for subsequent stimulation by a diffusing component from the fertilizing sperm cell.

Acknowledgment

This work was supported in part by NIH grant HD10254.

References

Aarons, D., Boettger-Tong, H., Holt, G., and Poirier, G. R. (1991). Acrosome reaction induced by immunoaggregation of a proteinase inhibitor bound to the murine sperm head. *Mol. Reprod. Dev.* **30**, 258–264.

Almeida, E. A. C., Huovila, A.-P. J., Sutherland, A. E., Stephens, L. E., Calarco, P. G., Shaw, L. M., Mercurio, A. M., Sonnenberg, A., Primakoff, P., Myles, D. G., and White, J. M. (1995). Mouse egg integrin alpha 6 beta 1 functions as a sperm receptor. *Cell* **81**, 1095–1104.

Anderson, R. A., Jr., Feathergill, K. A., Rawlins, R. G., Mack, S. R., and Zaneveld, L. J. D. (1995). Atrial natriuretic peptide: A chemoattractant of human spermatozoa by a guanylate cyclase-dependent pathway. *Mol. Reprod. Dev.* **40**, 371–378.

Arnoult, C., Cardullo, R., A., Lemos, J. R., and Florman, H. M. (1996a). Activation of mouse sperm T-type Ca^{2+} channels by adhesion to the egg zona pellucida. *Proc. Natl. Acad. Sci. U.S.A.* **93**, 13004–13009.

Arnoult, C., Zeng, Y., and Florman, H. M. (1996b). ZP3-dependent activation of sperm cation channels regulates acrosomal secretion during mammalian fertilization. *J. Cell Biol.* **134**, 637–645.

Augustine, G. J., Burns, M. E., DeBello, W. M., Pettit, D. L., and Schweizer, F. E. (1996). Exocytosis: Proteins and perturbations. *Annu. Rev. Pharmacol. Toxicol.* **36**, 659–701.

Babcock, D. F., Bosma, M. M., Battaglia, D. E., and Darszon, A. (1992). Early persistent activation of sperm K^+ channels by the egg peptide speract. *Proc. Natl. Acad. Sci. U.S.A.* **89**, 6001–6005.

Bailey, J. L., and Storey, B. T. (1994). Calcium influx into mouse spermatozoa activated by solubilized mouse zona pellucida, monitored with the calcium fluorescent indicator, fluo-3. Inhibition of the influx by three inhibitors of the zona pellucida induced acrosome reaction: Tyrphostin A48, pertussis toxin, and 3-quinuclidinyl benzilate. *Mol. Reprod. Dev.* **39**, 297–308.

Beebe, S. J., Leyton, L., Burks, D., Ishikawa, M., Fuerst, T., Dean, J., and Saling, P. (1992). Recombinant mouse ZP3 inhibits sperm binding and induces the acrosome reaction. *Dev. Biol.* **151**, 48–54.

Beltran, C., Zapata, O., and Darszon, A. (1996). Membrane potential regulates sea urchin sperm adenylylcyclase. *Biochemistry* **35**, 7591–7598.

Bentley, J. K., Garbers, D. L., Domino, S. E., Noland, T. D., and Van Dop, C. (1986a). Spermatozoa contain a guanine nucleotide-binding protein ADP-ribosylated by pertussis toxin. *Biochem. Biophys. Res. Commun.* **138**, 728–734.

Bentley, J. K., Tubb, D. J., and Garbers, D. L. (1986b). Receptor-mediated activation of spermatozoan guanylate cyclase. *J. Biol. Chem.* **261,** 14859–14862.

Bentley, J. K., Khatra, A. S., and Garbers, D. L. (1988). Receptor-mediated activation of detergent-solubilized guanylate cyclase. *Biol. Reprod.* **39,** 639–647.

Bleil, J. D., and Wassarman, P. M. (1983). Sperm-egg interactions in the mouse: Sequence of events and induction of the acrosome reaction by a zona pellucida glycoprotein. *Dev. Biol.* **95,** 317–324.

Bleil, J. D., and Wassarman, P. M. (1990). Identification of a ZP3-binding protein on acrosome-intact mouse sperm by photoaffinity crosslinking. *Proc. Natl. Acad. Sci. U.S.A.* **87,** 5563–5567.

Blobel, C. P., Myles, D. G., Primakoff, P., and White, J. M. (1990). Proteolytic processing of a protein involved in sperm-egg fusion correlates with acquisition of fertilization competence. *J. Cell Biol.* **111,** 69–78.

Blobel, C. P., Wolfsberg, T. G., Turck, C. W., Myles, D. G., Primakoff, P., and White, J. M. (1992). A potential fusion peptide and an integrin ligand domain in a protein active in sperm-egg fusion. *Nature (London)* **356,** 248–252.

Bookbinder, L. H., Moy, G. W., and Vacquier, V. D. (1990). Identification of sea urchin sperm adenylate cyclase. *J. Cell Biol.* **111,** 1859–1866.

Bookbinder, L. H., Cheng, A., and Bleil, J. D. (1995). Tissue- and species-specific expression of sp56, a mouse sperm fertilization protein. *Science* **269,** 86–89, published erratum appears in *Science* **269,** 1120 (1995).

Bork, P. (1996). Sperm-egg binding protein or proto-oncogene? *Science* **271,** 1431-1432; discussion 1434–1435.

Bork, P., and Sander, C. (1992). A large domain common to sperm receptors (Zp2 and Zp3) and TGF-beta type III receptor. *FEBS Lett.* **300,** 237–240.

Bowen, M. A., Bajorath, J., Siadak, A. W., Modrell, B., Malacko, A. R., Marquardt, H., Nadler, S. G., and Aruffo, A. (1996). The amino-terminal immunoglobulin-like domain of activated leukocyte cell adhesion molecule binds specifically to the membrane-proximal scavenger receptor cysteine-rich domain of CD6 with a 1:1 stoichiometry. *J. Biol. Chem.* **271,** 17390–17396.

Bradley, M. P., and Garbers, D. L. (1983). The stimulation of bovine caudal epididymal sperm forward motility by bovine cumulus-egg complexes in vitro. *Biochem. Biophys. Res. Commun.* **115,** 777–787.

Breitbart, H., Lax, J., Rotem, R., and Naor, Z. (1992). Role of protein kinase C in the acrosome reaction of mammalian spermatozoa. *Biochem. J.* **281,** 473–476.

Brokaw, C. J., and Nagayama, S. M. (1985). Modulation of the asymmetry of sea urchin sperm flagellar bending by calmodulin. *J. Cell Biol.* **100,** 1875–1883.

Brown, J., Cebra-Thomas, J. A., Bleil, J. D., Wassarman, P. M., and Silver, L. M. (1989). A premature acrosome reaction is programmed by mouse t haplotypes during sperm differentiation and could play a role in transmission ratio distortion. *Development (Cambridge, UK)* **106,** 769–773.

Buck, L., and Axel, R. (1991). A novel multigene family may encode odorant receptors: A molecular basis for odor recognition. *Cell* **65,** 175–187.

Burks, D. J. (1990). "The Structures and Sites of Synthesis of the Precursors for Peptides that Stimulate Spermatozoa." Ph.D. Thesis. Vanderbilt University, Nashville, TN.

Burks, D. J., Carballada, R., Moore, H. D., and Saling, P. M. (1995). Interaction of a tyrosine kinase from human sperm with the zona pellucida at fertilization. *Science* **269,** 83–86.

Cardullo, R. A., Herrick, S. B., Peterson, M. J., and Dangott, L. J. (1994). Speract receptors are localized on sea urchin sperm flagella using a fluorescent peptide analog. *Dev. Biol.* **162,** 600–607.

Carrera, A., Gerton, G. L., and Moss, S. B. (1994). The major fibrous sheath polypeptide of mouse sperm: Structural and functional similarities to the A-kinase anchoring proteins. *Dev. Biol.* **165,** 272–284.

Cheng, A., Le, T., Palacios, M., Bookbinder, L. H., Wassarman, P. M., Suzuki, F., and Bleil, J. D. (1994). Sperm-egg recognition in the mouse: Characterization of sp56, a sperm protein having specific affinity for ZP3. *J. Cell Biol.* **125**, 867–878.

Chinkers, M., and Garbers, D. L. (1989). The protein kinase domain of the ANP receptor is required for signaling. *Science* **245**, 1392–1394.

Chinkers, M., and Wilson, E. M. (1992). Ligand-independent oligomerization of natriuretic peptide receptors. Identification of heteromeric receptors and a dominant negative mutant. *J. Biol. Chem.* **267**, 18589–18597.

Chinkers, M., Garbers, D. L., Chang, M. S., Lowe, D. G., Chin, H. M., Goeddel, D. V., and Schulz, S. (1989). A membrane form of guanylate cyclase is an atrial natriuretic peptide receptor. *Nature (London)* **338**, 78–83.

Chinkers, M., Singh, S., and Garbers, D. L. (1991). Adenine nucleotides are required for activation of rat atrial natriuretic peptide receptor/guanylyl cyclase expressed in a baculovirus system. *J. Biol. Chem.* **266**, 4088–4093.

Christen, R., Schackmann, R. W., and Shapiro, B. M. (1983). Metabolism of sea urchin sperm. Interrelationships between intracellular pH, ATPase activity, and mitochondrial respiration. *J. Biol. Chem.* **258**, 5392–5399.

Cohen-Dayag, A., Ralt, D., Tur-Kaspa, I., Manor, M., Makler, A., Dor, J., Mashiach, S., and Eisenbach, M. (1994). Sequential acquisition of chemotactic responsiveness by human spermatozoa. *Biol. Reprod.* **50**, 786–790.

Cohen-Dayag, A., Tur-Kaspa, I., Dor, J., Mashiach, S., and Eisenbach, M. (1995). Sperm capacitation in humans is transient and correlates with chemotactic responsiveness to follicular factors. *Proc. Natl. Acad. Sci. U.S.A.* **92**, 11039–11043.

Coll, J. C., Leone, P. A., Bowden, B. F., Carroll, A. R., Konig, G. M., Heaton, A., DeNys, R., Maida, M., Alino, P. M., Willis, R. H., Babcock, R. C., Florian, Z., Clayton, M. N., Miller, R. L., and Alderslade, P. N. (1995). Chemical aspects of mass spawning in corals .2. (-)-Epi-thunbergol, the sperm attractant in the eggs of the soft coral *Lobophytum crassum* (Cnidaria, Octocorallia). *Mar. Biol.* **123**, 137–143.

Cook, S. P., and Babcock, D. F. (1993a). Selective modulation by cGMP of the K^+ channel activated by speract. *J. Biol. Chem.* **268**, 22402–22407.

Cook, S. P., and Babcock, D. F. (1993b). Activation of Ca^{2+} permeability by cAMP is coordinated through the pHi increase induced by speract. *J. Biol. Chem.* **268**, 22408–22413.

Cook, S. P., Brokaw, C. J., Muller, C. H., and Babcock, D. F. (1994). Sperm chemotaxis: Egg peptides control cytosolic calcium to regulate flagellar responses. *Dev. Biol.* **165**, 10–19.

Cornwall, G. A., Tulsiani, D. R., and Orgebin-Crist, M. C. (1991). Inhibition of the mouse sperm surface alpha-D-mannosidase inhibits sperm-egg binding in vitro. *Biol. Reprod.* **44**, 913–921.

Dangott, L. J., and Garbers, D. L. (1984). Identification and partial characterization of the receptor for speract. *J. Biol. Chem.* **259**, 13712–13716.

Dangott, L. J., Jordan, J. E., Bellet, R. A., and Garbers, D. L. (1989). Cloning of the mRNA for the protein that crosslinks to the egg peptide speract. *Proc. Natl. Acad. Sci. U.S.A.* **86**, 2128–2132.

Dawson, T. M., Arriza, J. L., Jaworsky, D. E., Borisy, F. F., Attramadal, H., Lefkowitz, R. J., and Ronnett, G. V. (1993). Beta-adrenergic receptor kinase-2 and beta-arrestin-2 as mediators of odorant-induced desensitization. *Science* **259**, 825–829.

Day, A. J. (1994). The C-type carbohydrate recognition domain (CRD) superfamily. *Biochem. Soc. Trans.* **22**, 83–88.

Desnoyers, L., and Manjunath, P. (1992). Major proteins of bovine seminal plasma exhibit novel interactions with phospholipid. *J. Biol. Chem.* **267**, 10149–10155.

Dhallan, R. S., Yau, K. W., Schrader, K. A., and Reed, R. R. (1990). Primary structure and functional expression of a cyclic nucleotide-activated channel from olfactory neurons. *Nature (London)* **347**, 184–187.

Domino, S. E., and Garbers, D. L. (1988). The fucose-sulfate glycoconjugate that induces an acrosome reaction in spermatozoa stimulates inositol 1,4,5-trisphosphate accumulation. *J. Biol. Chem.* **263**, 690–695.

Domino, S. E., Bocckino, S. B., and Garbers, D. L. (1989). Activation of phospholipase D by the fucose-sulfate glycoconjugate that induces an acrosome reaction in spermatozoa. *J. Biol. Chem.* **264**, 9412–9419.

Dostalova, Z., Calvete, J. J., Sanz, L., and Topfer-Peterson, E. (1995). Boar spermadhesin AWN-1. Oligosaccaharide and zona pellucida binding characteristics. *Eur. J. Biochem.* **230**, 329–336.

Drewett, J. G., and Garbers, D. L. (1994). The family of guanylyl cyclase receptors and their ligands. *Endocr. Rev.* **15**, 135–162.

Dunbar, B. S., Avery, S., Lee, V., Prasad, S., Schwahn, D., Schwoebel, E., Skinner, S., and Wilkins, B. (1994). The mammalian zona pellucida: Its biochemistry, immunochemistry, molecular biology, and developmental expression. *Reprod. Fertil. Dev.* **6**, 331–347.

Eisenbach, M., and Ralt, D. (1992). Precontact mammalian sperm-egg communication and role in fertilization. *Am. J. Physiol.* **262**, C1095–C1101.

Emiliozzi, C., and Fenichel, P. (1997). Protein tyrosine phosphorylation is associated with capacitation of human sperm *in vitro* but is not sufficient for its completion. *Biol. Reprod.* **56**, 674–679.

Endo, Y., Lee, M. A., and Kopf, G. S. (1987). Evidence for the role of a guanine nucleotide-binding regulatory protein in the zona pellucida-induced mouse sperm acrosome reaction. *Dev. Biol.* **119**, 210–216.

Evans, J. P., Schultz, R. M., and Kopf, G. S. (1995). Mouse sperm-egg plasma membrane interactions: Analysis of roles of egg integrins and the mouse sperm homologue of PH-30 (fertilin) beta. *J. Cell Sci.* **108**, 3267–3278.

Fassler, R., Georgeslabouesse, E., and Hirsch, E. (1996). Genetic analyses of integrin function in mice. *Curr. Opin. Cell Biol.* **8**, 641–646.

Florman, H. M. (1994). Sequential focal and global elevations of sperm intracellular Ca^{2+} are initiated by the zona pellucida during acrosomal exocytosis. *Dev. Biol.* **165**, 152–164.

Florman, H. M., and Wassarman, P. M. (1985). O-linked oligosaccharides of mouse egg ZP3 account for its sperm receptor activity. *Cell* **41**, 313–324.

Florman, H. M., Corron, M. E., Kim, T. D., and Babcock, D. F. (1992). Activation of voltage-dependent calcium channels of mammalian sperm is required for zona pellucida-induced acrosomal exocytosis. *Dev. Biol.* **152**, 304–314.

Foltz, K. R., and Lennarz, W. J. (1993). The molecular basis of sea urchin gamete interactions at the egg plasma membrane. *Dev. Biol.* **158**, 46–61.

Foltz, K. R., Partin, J. S., and Lennarz, W. J. (1993). Sea urchin egg receptor for sperm: Sequence similarity of binding domain and hsp70. *Science* **259**, 1421–1425.

Fraser, L. R., Harrison, R. A., and Herod, J. E. (1990). Characterization of a decapacitation factor associated with epididymal mouse spermatozoa. *J. Reprod. Fertil.* **89**, 135–148.

Fülle, H. J., Vassar, R., Foster, D. C., Yang, R. B., Axel, R., and Garbers, D. L. (1995). A receptor guanylyl cyclase expressed specifically in olfactory sensory neurons. *Proc. Natl. Acad. Sci. U.S.A.* **92**, 3571–3575.

Gahmberg, C. G., Kotovuori, P., and Tontti, E. (1992). Cell surface carbohydrate in cell adhesion. Sperm cells and leukocytes bind to their target cells through specific oligosaccharide ligands. AMPIS *Suppl.* **27**, 39–52.

Galantino-Homer, H. L., Visconti, P. E., and Kopf, G. S. (1997). Regulation of protein tyrosine phosphorylation during bovine sperm capacitation by a cyclic adenosine 3′,5′-monophosphate-dependent pathway. *Biol. Reprod.* **56**, 707–719.

Gao, B., Klein, L. E., Britten, R. J., and Davidson, E. H. (1986). Sequence of mRNA coding for bindin, a species-specific sea urchin sperm protein required for fertilization. *Proc. Natl. Acad. Sci. U.S.A.* **83**, 8634–8638.

Garbers, D. L. (1976). Sea urchin sperm guanylate cyclase. Purification and loss of cooperativity. *J. Biol. Chem.* **251**, 4071–4077.

Garbers, D. L. (1981). The elevation of cyclic AMP concentrations in flagella-less sea urchin sperm heads. *J. Biol. Chem.* **256**, 620–624.

Garbers, D. L. (1989). Molecular basis of fertilization. *Annu. Rev. Biochem.* **58**, 719–742.

Garbers, D. L. (1992). Guanylyl cyclase receptors and their endocrine, paracrine, and autocrine ligands. *Cell* **71**, 1–4.

Garbers, D. L., and Kopf, G. S. (1980). The regulation of spermatozoa by calcium and cyclic nucleotides. *Adv. Cyclic Nucleotide Res.* **13**, 251–306.

Garbers, D. L., Tubb, D. J., and Kopf, G. S. (1980). Regulation of sea urchin sperm cyclic AMP-dependent protein kinases by an egg associated factor. *Biol. Reprod.* **22**, 526–532.

Garbers, D. L., Watkins, H. D., Hansbrough, J. R., Smith, A., and Misono, K. S. (1982a). The amino acid sequence and chemical synthesis of speract and of speract analogues. *J. Biol. Chem.* **257**, 2734–2737.

Garbers, D. L., Tubb, D. J., and Hyne, R. V. (1982b). A requirement of bicarbonate for Ca^{2+}-induced elevations of cyclic AMP in guinea pig spermatozoa. *J. Biol. Chem.* **257**, 8980–8984.

Garbers, D. L., Kopf, G. S., Tubb, D. J., and Olson, G. (1983). Elevation of sperm adenosine 3′:5′-monophosphate concentrations by a fucose-sulfate-rich complex associated with eggs: I. Structural characterization. *Biol. Reprod.* **29**, 1211–1220.

Garbers, D. L., Koesling, D., and Schultz, G. (1994). Guanylyl cyclase receptors. *Mol. Biol. Cell* **5**, 1–5.

Garde, J., and Roldan, E. R. (1996). rab 3-peptide stimulates exocytosis of the ram sperm acrosome via interaction with cyclic Amp and phospholipase A2 metabolites. *FEBS Lett.* **391**, 263–268.

Glabe, C. G., and Clark, D. (1991). The sequence of the *Arbacia punctulata* bindin cDNA and implications for the structural basis of species-specific sperm adhesion and fertilization. *Dev. Biol.* **143**, 282–288.

Glabe, C. G., and Vacquier, V. D. (1977). Species specific agglutination of eggs by bindin isolated from sea urchin sperm. *Nature (London)* **267**, 836–838.

Glabe, C. G., Buchalter, M., and Lennarz, W. J. (1981). Studies on the interaction of sperm with the surface of the sea urchin egg. *Dev. Biol.* **84**, 397–406.

Glassner, M., Jones, J., Kligman, I., Woolkalis, M. J., Gerton, G. L., and Kopf, G. S. (1991). Immunocytochemical and biochemical characterization of guanine nucleotide-binding regulatory proteins in mammalian spermatozoa. *Dev. Biol.* **146**, 438–450.

Gong, X., Dubois, D. H., Miller, D. J., and Shur, B. D. (1995). Activation of a G protein complex by aggregation of beta-1,4-galactosyltransferase on the surface of sperm. *Science* **269**, 1718–1721.

Gonzalez-Martinez, M., and Darszon, A. (1987). A fast transient hyperpolarization occurs during the sea urchin sperm acrosome reaction induced by egg jelly. *FEBS Lett.* **218**, 247–250.

Gonzalez-Martinez, M. T., Guerrero, A., Morales, E., de la Torre, L., and Darszon, A. (1992). A depolarization can trigger Ca^{2+} uptake and the acrosome reaction when preceded by a hyperpolarization in *L. pictus* sea urchin sperm. *Dev. Biol.* **150**, 193–202.

Gould, M., and Stephano, J. L. (1987). Electrical responses of eggs to acrosomal protein similar to those induced by sperm. *Science* **235**, 1654–1656.

Greve, J. M., and Wassarman, P. M. (1985). Mouse egg extracellular coat is a matrix of interconnected filaments possessing a structural repeat. *J. Mol. Biol.* **181**, 253–264.

Gross, M. K., Toscano, D. G., and Toscano, W. A., Jr. (1987). Calmodulin-mediated adenylate cyclase from mammalian sperm. *J. Biol. Chem.* **262**, 8672–8676.

Guerrero, A., and Darszon, A. (1989). Evidence for the activation of two different Ca^{2+} channels during the egg jelly-induced acrosome reaction of sea urchin sperm. *J. Biol. Chem.* **264**, 19593–19599.

Hansbrough, J. R., and Garbers, D. L. (1981a). Speract. Purification and characterization of a peptide associated with eggs that activates spermatozoa. *J. Biol. Chem.* **256,** 1447–1452.

Hansbrough, J. R., and Garbers, D. L. (1981b). Sodium-dependent activation of sea urchin spermatozoa by speract and monensin. *J. Biol. Chem.* **256,** 2235–2241.

Hardy, D. M., and Garbers, D. L. (1993). Molecular basis of signaling in spermatozoa. *In* "Molecular Biology of the Male Reproductive System" (D. M. de Kretser, ed.), pp. 233–270. Academic Press, San Diego, CA.

Hardy, D. M., and Garbers, D. L. (1994). Species-specific binding of sperm proteins to the extracellular matrix (zona pellucida) of the egg. *J. Biol. Chem.* **269,** 19000–19004.

Hardy, D. M., and Garbers, D. L. (1995). A sperm membrane protein that binds in a species-specific manner to the egg extracellular matrix is homologous to von Willebrand factor. *J. Biol. Chem.* **270,** 26025–26028.

Hardy, C. M., and Holland, M. K. (1996). Cloning and expression of recombinant rabbit fertilin. *Mol. Reprod. Dev.* **45,** 107–116.

Harris, J. D., Hibler, D. W., Fontenot, G. K., Hsu, K. T., Yurewicz, E. C., and Sacco, A. G. (1994). Cloning and characterization of zona pellucida genes and cDNAs from a variety of mammalian species: The ZPA, ZPB and ZPC gene families. *DNA Seq.* **4,** 361–393.

Harrison, R. A. P. (1996). Capacitation mechanisms, and the role of capacitation as seen in eutherian mammals. *Reprod., Fertil., Dev.* **8,** 581–594.

Harumi, T., Kurita, M., and Suzuki, N. (1992a). Purification and characterization of sperm creatine kinase and guanylate cyclase of the sea urchin *Hemicentrotus pulcherrimus*. *Dev., Growth Differ.* **34,** 151–162.

Harumi, T., Hoshino, K., and Suzuki, N. (1992b). Effects of sperm-activating peptide I on *Hemicentrotus pulcherrimus* spermatozoa in high potassium sea water. *Dev. Growth Differ.* **34,** 163–172.

Hedrick, J. L. (1996). Comparative structural and antigenic properties of zona pellucida glycoproteins. *J. Reprod. Fertil. Suppl.* **50,** 9–17.

Hildebrandt, J. D., Codina, J., Tash, J. S., Kirchick, H. J., Lipschultz, L., Sekura, R. D., and Birnbaumer, L. (1985). The membrane-bound spermatozoal adenylyl cyclase system does not share coupling characteristics with somatic cell adenylyl cyclases. *Endocrinology (Baltimore)* **116,** 1357–1366.

Hoshi, M., Amano, T., Okita, Y., Okinaga, T., and Matsui, T. (1990). Egg signals for triggering the acrosome reaction in starfish spermatozoa. *J. Reprod. Fertil. Suppl.* **42,** 23–31.

Hunnicutt, G. R., Mahan, K., Lathrop, W. F., Ramarao, C. S., Myles, D. G., and Primakoff, P. (1996a). Structural relationship of sperm soluble hyaluronidase to the sperm membrane protein PH-20. *Biol. Reprod.* **54,** 1343–1349.

Hunnicutt, G. R., Primakoff, P., and Myles, D. G. (1996b). Sperm surface protein PH-20 is bifunctional—one activity is a hyaluronidase and a second, distinct activity is required in secondary sperm-zona binding. *Biol. Reprod.* **55,** 80–86.

Hynes, R. O. (1992). Integrins: Versatility, modulation, and signaling in cell adhesion. *Cell* **69,** 11–25.

International Polycystic Kidney Disease Consortium (1995). Polycystic kidney disease: The complete structure of the PKD1 gene and its protein. *Cell* **81,** 289–298.

Jones, D. T., and Reed, R. R. (1989). G$_{olf}$: an olfactory neuron specific-G protein involved in odorant signal transduction. *Science* **244,** 790–795.

Jones, R., Brown, C. R., and Lancaster, R. T. (1988). Carbohydrate-binding properties of boar sperm proacrosin and assessment of its role in sperm-egg recognition and adhesion during fertilization. *Development (Cambridge, UK)* **102,** 781–792.

Kadam, A. L., Fateh, M., and Naz, R. K. (1995). Fertilization antigen (FA-1) completely blocks human sperm binding to human zona pellucida: FA-1 antigen may be a sperm receptor for zona pellucida in humans. *J. Reprod. Immunol.* **29,** 19–30.

Kalab, P., Visconti, P., Leclerc, P., and Kopf, G. S. (1994). p95, the major phosphotyrosine-containing protein in mouse spermatozoa, is a hexokinase with unique properties. *J. Biol. Chem.* **269,** 3810–3817.

Keller, S. H., and Vacquier, V. D. (1994). The isolation of acrosome-reaction-inducing glycoproteins from sea urchin egg jelly. *Dev. Biol.* **162**, 304–312.

Kinloch, R. A., Mortillo, S., Stewart, C. L., and Wassarman, P. M. (1991). Embryonal carcinoma cells transfected with ZP3 genes differentially glycosylate similar polypeptides and secrete active mouse sperm receptor. *J. Cell Biol.* **115**, 655–664.

Kinloch, R. A., Sakai, Y., and Wassarman, P. M. (1995). Mapping the mouse ZP3 combining site for sperm by exon swapping and site-directed mutagenesis. *Proc. Natl. Acad. Sci. U.S.A.* **92**, 263–267.

Koesling, D., Herz, J., Gausepohl, H., Niroomand, F., Hinsch, K. D., Mulsch, A., Bohme, E., Schultz, G., and Frank, R. (1988). The primary structure of the 70 kDa subunit of bovine soluble guanylate cyclase. *FEBS Lett.* **239**, 29–34.

Kopf, G. S., and Vacquier, V. D. (1984). Characterization of a calmodulin-stimulated adenylate cyclase from abalone spermatozoa. *Adv. Exp. Med. Biol.* **259**, 7590–7596.

Kopf, G. S., Tubb, D. J., and Garbers, D. L. (1979). Activation of sperm respiration by a low molecular weight egg factor and by 8-bromoguanosine 3′,5′-monophosphate. *J. Biol. Chem.* **254**, 8554–8560.

Kopf, G. S., Lewis, C. A., and Vacquier, V. D. (1983). Regulation of abalone sperm cyclic AMP concentrations and the acrosome reaction by calcium and methylxanthines. *Dev. Biol.* **98**, 28–36.

Krupinski, J., Coussen, F., Bakalyar, H. A., Tang, W.-J., Feinstein, P. G., Orth, K., Slaughter, C., Reed, R. R., and Gilman, A. G. (1989). Adenylyl cyclase amino acid sequence: Possible channel- or transporter-like structure. *Science* **244**, 1558–1564.

Leclerc, P., and Kopf, G. S. (1995). Mouse sperm adenylyl cyclase: General properties and regulation by the zona pellucida. *Biol. Reprod.* **52**, 1227–1233.

Leclerc, P., Delamirande, E., and Gagnon, C. (1996). Cyclic adenosine -3′,5′-monophosphate-dependent regulation of protein tyrosine phosphorylation in relation to human sperm capacitation and motility. *Biol. Reprod.* **55**, 684–692.

Lee, H. C. (1988). Internal GTP stimulates the speract receptor mediated voltage changes in sea urchin spermatozoa membrane vesicles. *Dev. Biol.* **126**, 91–97.

Lee, H. C., and Garbers, D. L. (1986). Modulation of the voltage-sensitive Na^+/H^+ exchange in sea urchin spermatozoa through membrane potential changes induced by the egg peptide speract. *J. Biol. Chem.* **261**, 16026–16032.

Lefebvre, R., and Suarez, S. S. (1996). Effect of capacitation on bull sperm binding to homologous oviductal epithelium. *Biol. Reprod.* **54**, 575–582.

Lefebvre, R., Chenoweth, P. J., Drost, M., LeClear, C. T., MacCubbin, M., Dutton, J. T., and Suarez, S. S. (1995). Characterization of the oviductal sperm reservoir in cattle. *Biol. Reprod.* **53**, 1066–1074.

Leyton, L., and Saling, P. (1989a). 95 kd sperm proteins bind ZP3 and serve as tyrosine kinase substrates in response to zona binding. *Cell* **57**, 1123–1130.

Leyton, L., and Saling, P. (1989b). Evidence that aggregation of mouse sperm receptors by ZP3 triggers the acrosome reaction. *J. Cell Biol.* **108**, 2163–2168.

Leyton, L., LeGuen, P., Bunch, D., and Saling, P. M. (1992). Regulation of mouse gamete interaction by a sperm tyrosine kinase. *Proc. Natl. Acad. Sci. U.S.A.* **89**, 11692–11695.

Leyton, L., Tomes, C., and Saling, P. (1995). LL95 monoclonal antibody mimics functional effects of ZP3 on mouse sperm: Evidence that the antigen recognized is not hexokinase. *Mol. Reprod. Dev.* **42**, 347–358.

Liu, C., Litscher, E. S., Mortillo, S., Sakai, Y., Kinloch, R. A., Stewart, C. L., and Wassarman, P. M. (1996). Targeted disruption of the mZP3 gene results in production of eggs lacking a zona pellucida and infertility in female mice. *Proc. Natl. Acad. Sci. U.S.A.* **93**, 5431–5436.

Lopez, A., Miraglia, S. J., and Glabe, C. G. (1993). Structure/function analysis of the sea urchin sperm adhesive protein bindin. *Dev. Biol.* **156**, 24–33.

Lopez, M. J., Wong, S. K., Kishimoto, I., Dubois, S., Mach, V., Friesen, J., Garbers, D. L., and Beuve, A. (1995). Salt-resistant hypertension in mice lacking the guanylyl cyclase-A receptor for atrial natriuretic peptide. *Nature (London)* **378**, 65–68.

Lowe, D. G. (1992). Human natriuretic peptide receptor-A guanylyl cyclase is self-associated prior to hormone binding. *Biochemistry* **31,** 10421–10425.
Lowe, D. G., Chang, M. S., Hellmiss, R., Chen, E., Singh, S., Garbers, D. L., and Goeddel, D. V. (1989). Human atrial natriuretic peptide receptor defines a new paradigm for second messenger signal transduction. *EMBO J.* **8,** 1377–1384.
Lu, Q., Hasty, P., and Shur, B. D. (1997). Targeted mutation in β-1,4-galactosyltransferase leads to pituitary insufficiency and neonatal lethality. *Dev. Biol.* **181,** 257–267.
Macek, M. B., Lopez, L. C., and Shur, B. D. (1991). Aggregation of beta-1,4-galactosyltransferase on mouse sperm induces the acrosome reaction. *Dev. Biol.* **147,** 440–444.
Meizel, S. (1997). Amino acid neurotransmitter receptor/chloride channels of mammalian sperm and the acrosome reaction. *Biol. Reprod.* **56,** 569–574.
Melendrez, C. S., and Meizel, S. (1995). Studies of porcine and human sperm suggesting a role for a sperm glycine receptor/Cl^- channel in the zona pellucida-initiated acrosome reaction. *Biol. Reprod.* **53,** 676–683.
Meyers, S. A., Overstreet, J. W., Liu, I. K., and Drobnis, E. Z. (1995). Capacitation *in vitro* of stallion spermatozoa: Comparison of progesterone-induced acrosome reactions in fertile and subfertile males. *J. Androl.* **16,** 47–54.
Miller, D. J., Macek, M. B., and Shur, B. D. (1992). Complementarity between sperm surface beta-1,4-galactosyltransferase and egg-coat ZP3 mediates sperm-egg binding. *Nature (London)* **357,** 589–593.
Miller, R. L. (1985). Sperm chemo-orientation in the metazoa. *Biol. Fertil.* **2,** 276–340.
Miller, R. L., and Brokaw, C. J. (1970). Chemotactic turning behavior of *Tubularia* spermatozoa. *J. Exp. Biol.* **52,** 699–706.
Miller, R. L., and Vogt, R. (1996). An N-terminal partial sequence of the 13 kDa *Pycnopodia helianthoides* sperm chemoattractant startrak possesses sperm-attracting activity. *J. Exp. Biol.* **199,** 311–318.
Minor, J. E., Fromson, D. R., Britten, R. J., and Davidson, E. H. (1991). Comparison of the bindin proteins of *Strongylocentrotus franciscanus, S. purpuratus,* and *Lytechinus variegatus:* Sequences involved in the species specificity of fertilization. *Mol. Biol. Evol.* **8,** 781–795.
Moore, G. D., Ayabe, T., Visconti, P. E., Schultz, R. M., and Kopf, G. S. (1994). Roles of heterotrimeric and monomeric G proteins in sperm-induced activation of mouse eggs. *Development (Cambridge, UK)* **120,** 3313–3323.
Moy, G. W., Mendoza, L. M., Schulz, J. R., Swanson, W. J., Glabe, C. G., and Vacquier, V. D. (1996). The sea urchin sperm receptor for egg jelly is a modular protein with extensive homology to the human polycystic kidney disease protein, PKD1. *J. Cell Biol.* **133,** 809–817.
Myles, D. G. (1993). Molecular mechanisms of sperm-egg membrane binding and fusion in mammals. *Dev. Biol.* **158,** 35–45.
Myles, D. G., Kimmel, L. H., Blobel, C. P., White, J. M., and Primakoff, P. (1994). Identification of a binding site in the disintegrin domain of fertilin required for sperm-egg fusion. *Proc. Natl. Acad. Sci. U.S.A.* **91,** 4195–4198.
Nakano, M., Yonezawa, N., Hatanaka, Y., and Noguchi, S. (1996). Structure and function of the N-linked carbohydrate chains of pig zona pellucida glycoproteins. *J. Reprod. Fertil. Suppl.* **50,** 25–34.
Naz, R. K., Brazil, C., and Overstreet, J. W. (1992). Effects of antibodies to sperm surface fertilization antigen-1 on human sperm-zona pellucida interaction. *Fertil. Steril.* **57,** 1304–1310.
Ohlendieck, K., Dhume, S. T., Partin, J. S., and Lennarz, W. J. (1993). The sea urchin egg receptor for sperm: Isolation and characterization of the intact, biologically active receptor. *J. Cell Biol.* **122,** 887–895.
Ohlendieck, K., Partin, J. S., and Lennarz, W. J. (1994). The biologically active form of the sea urchin egg receptor for sperm is a disulfide-bonded homo-multimer. *J. Cell Biol.* **125,** 817–824.

Okamura, N., Tajima, Y., Soejima, A., Masuda, H., and Sugita, Y. (1985). Sodium bicarbonate in seminal plasma stimulates the motility of mammalian spermatozoa through direct activation of adenylate cyclase. *J. Biol. Chem.* **260**, 9699–9705.

Okamura, N., Tajima, Y., Onoe, S., and Sugita, Y. (1991). Purification of bicarbonate-sensitive sperm adenylylcyclase by 4-acetamido-4′-isothiocyanostilbene-2,2′-disulfonic acid-affinity chromatography. *J. Biol. Chem.* **266**, 17754–17759.

Oliphant, G., Reynolds, A. B., and Thomas, T. S. (1985). Sperm surface components involved in the control of the acrosome reaction. *Am. J. Anat.* **174**, 269–283.

Osman, R. A., Andria, M. L., Jones, A. D., and Meizel, S. (1989). Steroid induced exocytosis: The human sperm acrosome reaction. *Biochem. Biophys. Res. Commun.* **160**, 828–833.

Parks, J. E., and Ehrenwald, E. (1990). Cholesterol efflux from mammalian sperm and its potential role in capacitation. *In* "Fertilization in Mammals" (B. Bavister, J. Cummins, and E. R. S. Roldan, eds.), pp. 155–167. Serono Symposia USA, Norwell, MA.

Parmentier, M., Libert, F., Schurmans, S., Schiffmann, S., Lefort, A., Eggerickx, D., Ledent, C., Mollereau, C., Gerard, C., Perret, J., Grootegoed, A., and Vassart, G. (1992). Expression of members of the putative olfactory receptor gene family in mammalian germ cells. *Nature (London)* **355**, 453–455.

Parrington, J., Swann, K., Shevchenko, V. I., Sesay, A. K., and Lai, F. A. (1996). Calcium oscillations in mammalian eggs triggered by a soluble sperm protein. *Nature (London)* **379**, 364–368.

Pitt, G. S., Milona, N., Borleis, J., Lin, K. C., Reed, R. R., and Devreotes, P. N. (1992). Structurally distinct and stage-specific adenylyl cyclcase genes play different roles in Dictyostelium development. *Cell* **69**, 305–315.

Podell, S. B., and Vacquier, V. D. (1985). Purification of the M_r 80,000 and M_r 210,000 proteins of the sea urchin sperm plasma membrane. Evidence that the M_r 210,000 protein interacts with egg jelly. *J. Biol. Chem.* **260**, 2715–2718.

Porter, J. G., Scarborough, R. M., Wang, Y., Schenk, D., McEnroe, G. A., Kang, L. L., and Lewicki, J. A. (1989). Recombinant expression of a secreted form of the atrial natriuretic peptide clearance receptor. *J. Biol. Chem.* **264**, 14179–14184.

Primakoff, P., Hyatt, H., and Myles, D. G. (1985). A role for the migrating sperm surface antigen PH-20 in guinea pig sperm binding to the egg zona pellucida. *J. Cell Biol.* **101**, 2239–2244.

Ralt, D., Goldenberg, M., Fetterolf, P., Thompson, D., Dor, J., Mashiach, S., Garbers, D. L., and Eisenbach, M. (1991). Sperm attraction to a follicular factor(s) correlates with human egg fertilizability. *Proc. Natl. Acad. Sci. U.S.A.* **88**, 2840–2844.

Ralt, D., Manor, M., Cohen-Dayag, A., Tur-Kaspa, I., Ben-Shlomo, I., Makler, A., Yuli, I., Dor, J., Blumberg, S., Mashiach, S., and Eisenbach, M. (1994). Chemotaxis and chemokinesis of human spermatozoa to follicular factors. *Biol. Reprod.* **50**, 774–785.

Ramarao, C. S., and Garbers, D. L. (1985). Receptor-mediated regulation of guanylate cyclase activity in spermatozoa. *J. Biol. Chem.* **260**, 8390–8396.

Ramarao, C. S., and Garbers, D. L. (1988). Purification and properties of the phosphorylated form of guanylate cyclase. *J. Biol. Chem.* **263**, 1524–1529.

Ramarao, C. S., Burks, D. J., and Garbers, D. L. (1990). A single mRNA encodes multiple copies of the egg peptide speract. *Biochemistry* **29**, 3383–3388.

Repaske, D. R., and Garbers, D. L. (1983). A hydrogen ion flux mediates stimulation of respiratory activity by speract in sea urchin spermatozoa. *J. Biol. Chem.* **258**, 6025–6029.

Resnick, D., Pearson, A., and Krieger, M. (1994). The SRCR superfamily: A family reminiscent of the Ig superfamily. *Trends Biochem. Sci.* **19**, 5–8.

Richardson, R. T., Yamasaki, N., and O'Rand, M. G. (1994). Sequence of a rabbit sperm zona pellucida binding protein and localization during the acrosome reaction. *Dev. Biol.* **165**, 688–701.

Roldan, E. R., Murase, T., and Shi, Q. X. (1994). Exocytosis in spermatozoa in response to progesterone and zona pellucida. *Science* **266**, 1578–1581.

Rondeau, J. J., McNicoll, N., Gagnon, J., Bouchard, N., Ong, H., and De Lean, A. (1995). Stoichiometry of the atrial natriuretic factor-R1 receptor complex in the bovine zona glomerulosa. *Biochemistry* **34,** 2130–2136.

Rotem, R., Paz, G. F., Homonnai, Z. T., Kalina, M., Lax, J., Breitbart, H., and Naor, Z. (1992). Ca(2+)-independent induction of acrosome reaction by protein kinase C in human sperm. *Endocrinology (Baltimore)* **131,** 2235–2243.

Schackmann, R. W., and Chock, P. B. (1986). Alteration of intracellular [Ca2+] in sea urchin sperm by the egg peptide speract. Evidence that increased intracellular Ca^{2+} is coupled to Na^+ entry and increased intracellular pH. *J. Biol. Chem.* **261,** 8719–8728.

Schultz, J. E., Klumpp, S., Benz, R., Schürhoff-Goeters, W. J. C., and Schmid, A. (1992). Regulation of adenylyl cyclase from Paramecium by an intrinsic potassium conductance. *Science* **255,** 600–603.

Schulz, S., Green, C. K., Yuen, P. S., and Garbers, D. L. (1990). Guanylyl cyclase is a heat-stable enterotoxin receptor. *Cell* **63,** 941–948.

SeGall, G. K., and Lennarz, W. J. (1979). Chemical characterization of the component of the jelly coat from sea urchin eggs responsible for induction of the acrosome reaction. *Dev. Biol.* **71,** 33–48.

Shi, Q. X., and Roldan, E. R. (1995). Evidence that a GABAA-like receptor is involved in progesterone-induced acrosomal exocytosis in mouse spermatozoa. *Biol. Reprod.* **52,** 373–381.

Shilling, F. M., Carroll, D. J., Muslin, A. J., Escobedo, J. A., Williams, L. T., and Jaffe, L. (1994). Evidence for both tyrosine kinase and G-protein-coupled pathways leading to starfish egg activation. *Dev. Biol.* **162,** 590–599.

Shimizu, T., Kinoh, H., Yamaguchi, M., and Suzuki, N. (1990). Purification and characterization of the egg jelly macromolecules, sialoglycoprotein and fucose sulfate glycoconjugate, of the sea urchin *Hemicentrotus pulcherrimus*. *Dev., Growth Differ.* **32,** 473–487.

Shimomura, H., and Garbers, D. L. (1986). Differential effects of resact analogues on sperm respiration rates and cyclic nucleotide concentrations. *Biochemistry* **25,** 3405–3410.

Shimomura, H., Dangott, L. J., and Garbers, D. L. (1986). Covalent coupling of a resact analogue to guanylate cyclase. *J. Biol. Chem.* **261,** 15778–15782.

Singh, S., Lowe, D. G., Thorpe, D. S., Rodriguez, H., Kuang, W. J., Dangott, L. J., Chinkers, M., Goeddel, D. V., and Garbers, D. L. (1988). Membrane guanylate cyclase is a cell-surface receptor with homology to protein kinases. *Nature (London)* **334,** 708–712.

Smith, T. T., and Yanagimachi, R. (1989). Capacitation status of hamster spermatozoa in the oviduct at various times after mating. *J. Reprod. Fertil.* **86,** 255–261.

Snell, W. J., and White, J. M. (1996). The molecules of mammalian fertilization. *Cell* **85,** 629–637.

Sudhof, T. C. (1995). The synaptic vesicle cycle: A cascade of protein-protein interactions. *Nature (London)* **375,** 645–653.

Sunahara, R. K., Dessauer, C. W., and Gilman, A. G. (1996). Complexity and diversity of mammalian adenylyl cyclases. *Annu. Rev. Pharmacol. Toxicol.* **36,** 461–480.

Suzuki, N. (1995). Structure, function and biosynthesis of sperm-activating peptides and fucose sulfate glycoconjugate in the extracellular coat of sea urchin eggs. *Zool. Sci.* **12,** 13–27.

Suzuki, N., and Garbers, D. L. (1984). Stimulation of sperm respiration rates by speract and resact at alkaline extracellular pH. *Biol. Reprod.* **30,** 1167–1174.

Suzuki, N., and Yoshino, K. (1992). The relationship between amino acid sequences of sperm-activating peptides and the taxonomy of echinoids. *Comp. Biochem. Physiol.* **102,** 679–690.

Suzuki, N., Nomura, K., Ohtake, H., and Isaka, S. (1981). Purification and the primary structure of sperm-activity peptides from the jelly coat of sea urchin eggs. *Biochem. Biophys. Res. Commun.* **99,** 1238–1244.

Tash, J. S. (1989). Protein phosphorylation: The second messenger signal transducer of flagellar motility. *Cell Motil. Cytoskel.* **14,** 332–339.

Tash, J. S., and Bracho, G. E. (1994). Regulation of sperm motility: Emerging evidence for a major role for protein phosphatases. *J. Androl.* **15**, 505-509.

Tash, J. S., and Means, A. R. (1988). cAMP-dependent regulatory processes in the acquisition and control of sperm flagellar movement. *Prog. Clin. Biol. Res.* **267**, 335-355.

Tash, J. S., Krinks, M., Patel, J., Means, R. L., Klee, C. B., and Means, A. R. (1988). Identification, characterization, and functional correlation of calmodulin-dependent protein phosphatase in sperm. *J. Cell Biol.* **106**, 1625-1633.

Tesarik, J., and Mendoza, C. (1993). Insights into the function of a sperm-surface progesterone receptor: Evidence of ligand-induced receptor aggregation and the implication of proteolysis. *Exp. Cell Res.* **205**, 111-117.

Tesarik, J., Mendoza Oltras, C., and Testart, J. (1990). Effect of the human cumulus oophorus on movement characteristics of human capacitated spermatozoa. *J. Reprod. Fertil.* **88**, 665-675.

Tesarik, J., Mendoza, C., Moos, J., Fenichel, P., and Fehlmann, M. (1992). Progesterone action through aggregation of a receptor on the sperm plasma membrane. *FEBS Lett.* **308**, 116-120.

Thompson, D. K., and Garbers, D. L. (1995). Dominant negative mutations of the guanylyl cyclase-A receptor. Extracellular domain deletion and catalytic domain point mutations. *J. Biol. Chem.* **270**, 425-430.

Thorpe, D. S., and Garbers, D. L. (1989). The membrane form of guanylate cyclase. Homology with a subunit of the cytoplasmic form of the enzyme. *J. Biol. Chem.* **264**, 6545-6549.

Tong, Z. B., Nelson, L. M., and Dean, J. (1995). Inhibition of zona pellucida gene expression by antisense oligonucleotides injected into mouse oocytes. *J. Biol. Chem.* **270**, 849-853.

Topfer-Petersen, E., Cechova, D., Henschen, A., Steinberger, M., Friess, A. E., and Zucker, A. (1990). Cell biology of acrosomal proteins. *Andrologia* **1**, 110-121.

Topfer-Petersen, E., Calvete, J. J., Sanz, L., and Sinowatz, F. (1995). Carbohydrate-and heparin-binding proteins in mammalian fertilization. *Andrologia* **27**, 303-324.

Trimmer, J. S., Trowbridge, I. S., and Vacquier, V. D. (1985). Monoclonal antibody to a membrane glycoprotein inhibits the acrosome reaction and associated Ca^{2+} and H^+ fluxes of sea urchin sperm. *Cell* **40**, 697-703.

Tsai, J. Y., and Silver, L. M. (1996). Sperm-egg binding protein or proto-oncogene? *Science* **271**, 1432-1434; discussion: 1434-1435.

Tulsiani, D. R., NagDas, S. K., Skudlarek, M. D., and Orgebin-Crist, M. C. (1995). Rat sperm plasma membrane mannosidase: Localization and evidence for proteolytic processing during epididymal maturation. *Dev. Biol.* **167**, 584-595.

Uguz, C., Vredenburgh, W. L., and Parrish, J. J. (1994). Heparin-induced capacitation but not intracellular alkalinization of bovine sperm is inhibited by Rp-adenosine-3',5'-cyclic monophosphorothioate. *Biol. Reprod.* **51**, 1031-1039.

Vaandrager, A. B., van der Wiel, E., and de Jonge, H. R. (1993). Heat-stable enterotoxin activation of immunopurified guanylyl cyclase C. Modulation by adenine nucleotides. *J. Biol. Chem.* **268**, 19598-19603.

Vanderhaeghen, P., Schurmans, S., Vassart, G., and Parmentier, M. (1993). Olfactory receptors are displayed on dog mature sperm cells. *J. Cell Biol.* **123**, 1441-1452.

Visconti, P. E., Bailey, J. L., Moore, G. D., Pan, D., Olds-Clarke, P., and Kopf, G. S. (1995a). Capacitation of mouse spermatozoa. I. Correlation between the capacitation state and protein tyrosine phosphorylation. *Development (Cambridge, UK)* **121**, 1129-1137.

Visconti, P. E., Moore, G. D., Bailey, J. L., Leclerc, P., Connors, S. A., Pan, D., Olds-Clarke, P., and Kopf, G. S. (1995b). Capacitation of mouse spermatozoa. II. Protein tyrosine phosphorylation and capacitation are regulated by a cAMP-dependent pathway. *Development (Cambridge, UK)* **121**, 1139-1150.

Walensky, L. D., and Snyder, S. H. (1995). Inositol 1,4,5-trisphosphate receptors selectively localized to the acrosomes of mammalian sperm. *J. Cell Biol.* **130**, 857-869.

Walensky, L. D., Roskams, A. J., Lefkowitz, R. J., Snyder, S. H., and Ronnett, G. V. (1995). Odorant receptors and desensitization proteins colocalize in mammalian sperm. *Mol. Med.* **1**, 130–141.
Ward, C. R., and Kopf, G. S. (1993). Molecular events mediating sperm activation. *Dev. Biol.* **158**, 9–34.
Ward, C. R., Storey, B. T., and Kopf, G. S. (1992). Activation of a Gi protein in mouse sperm membranes by solubilized proteins of the zona pellucida, the egg's extracellular matrix. *J. Biol. Chem.* **267**, 14061–14067.
Ward, C. R., Storey, B. T., and Kopf, G. S. (1994). Selective activation of Gi1 and Gi2 in mouse sperm by the zona pellucida, the egg's extracellular matrix. *J. Biol. Chem.* **269**, 13254–13258.
Ward, G. E., Brokaw, C. J., Garbers, D. L., and Vacquier, V. D. (1985). Chemotaxis of *Arbacia punctulata* spermatozoa to resact, a peptide from the egg jelly layer. *J. Cell Biol.* **101**, 2324–2329.
Wassarman, P. M. (1988). Zona pellucida glycoproteins. *Annu. Rev. Biochem.* **57**, 415–442.
Wassarman, P. M. (1995). Towards molecular mechanisms for gamete adhesion and fusion during mammalian fertilization. *Curr. Opin. Cell Biol.* **7**, 658–664.
Wassarman, P. M., and Litscher, E. S. (1995). Sperm–egg recognition mechanisms in mammals. *Curr. Top. Dev. Biol.* **30**, 1–19.
Wassarman, P. M., Bleil, J. D., Florman, H. M., Greve, J. M., Roller, R. J., and Salzmann, G. S. (1986). Nature of the mouse egg's receptor for sperm. *Adv. Exp. Med. Biol.* **207**, 55–77.
Watkins, H. D., Kopf, G. S., and Garbers, D. L. (1978). Activation of sperm adenylate cyclase by factors associated with eggs. *Biol. Reprod.* **19**, 890–894.
Whitaker, M., and Swann, K. (1993). Lighting the fuse at fertilization. *Development (Cambridge, UK)* **117**, 1–12.
Wilde, M. W., Ward, C. R., and Kopf, G. S. (1992). Activation of a G protein in mouse sperm by the zona pellucida, an egg-associated extracellular matrix. *Mol. Reprod. Dev.* **31**, 297–306.
Wolfsberg, T. G., Straight, P. D., Gerena, R. L., Huovila, A. P., Primakoff, P., Myles, D. G., and White, J. M. (1995). ADAM, a widely distributed and developmentally regulated gene family encoding membrane proteins with a disintegrin and metalloprotease domain. *Dev. Biol.* **169**, 378–383.
Yamaguchi, M., Niwa, T., Kurita, M., and Suzuki, N. (1988). The participation of speract in the acrosome reaction of *Hemicentrotus pulcherrimus*. *Dev., Growth Differ.* **30**, 159–167.
Yanagimachi, R. (1994). Mammalian fertilization. *In* "The Physiology of Reproduction" (E. Knobil and J. D. Neill, eds.), Vol. 2, pp. 189–317. Raven Press, New York.
Yang, R. B., Foster, D. C., Garbers, D. L., and Fülle, H. J. (1995). Two membrane forms of guanylyl cyclase found in the eye. *Proc. Natl. Acad. Sci. U.S.A.* **92**, 602–606.
Yao, X., Segal, A. S., Welling, P., Zhang, X., McNicholas, C. M., Engel, D., Boulpaep, E. L., and Desir, G. V. (1995). Primary structure and functional expression of a cGMP-gated potassium channel. *Proc. Natl. Acad. Sci. U.S.A.* **92**, 11711–11715.
Yu, S., Avery, L., Baude, E., and Garbers, D. L. (1997). Guanylyl cyclase expression in specific sensory neurons: A new family of chemosensory receptors. *Proc. Natl. Acad. Sci. U.S.A.* **94**, 3384–3387.
Yuen, P. S., and Garbers, D. L. (1992). Guanylyl cyclase-linked receptors. *Annu. Rev. Neurol.* **15**, 193–225.
Yuen, P. S., Doolittle, L. K., and Garbers, D. L. (1994). Dominant negative mutants of nitric oxide-sensitive guanylyl cyclase. *J. Biol. Chem.* **269**, 791–793.
Yurewicz, E. C., Hibler, D., Fontenot, G. K., Sacco, A. G., and Harris, J. (1993). Nucleotide sequence of cDNA encoding ZP3 alpha, a sperm-binding glycoprotein from zona pellucida of pig oocyte. *Biochim. Biophys. Acta* **1174**, 211–214.

Zamir, N., Riven-Kreitman, R., Manor, M., Makler, A., Blumberg, S., Ralt, D., and Eisenbach, M. (1993). Atrial natriuretic peptide attracts human spermatozoa in vitro. *Biochem. Biophys. Res. Commun.* **197**, 116–122.

Zeng, Y., Clark, E. N., and Florman, H. M. (1995). Sperm membrane potential: Hyperpolarization during capacitation regulates zona pellucida-dependent acrosomal secretion. *Dev. Biol.* **171**, 554–563.

Zeng, Y., Oberdorf, J. A., and Florman, H. M. (1996). pH regulation in mouse sperm: Identification of Na(+)-, Cl(-)-, and HCO3(-)-dependent and arylaminobenzoate-dependent regulatory mechanisms and characterization of their roles in sperm capacitation. *Dev. Biol.* **173**, 510–520.

Audrey Minden*
Michael Karin†

*Columbia University
Biological Sciences Department, 2460
New York, New York 10027

†Department of Pharmacology
Program in Biomedical Sciences
UCSD School of Medicine
La Jolla, California 92093

The JNK Family of MAP Kinases: Regulation and Function

Mammalian cells respond to extracellular stimuli by activating signaling cascades that lead to long-term changes in gene expression. In many cases, these signal transduction cascades are mediated by serine/threonine kinases of the mitogen-activated protein (MAP) kinase family. MAP kinases mediate responses to extracellular stimuli by phosphorylating various substrates, including transcription factors. This usually leads to activation of transcription factors, which in turn regulate expression of specific sets of genes. The products of these genes presumably mediate a specific response to the stimulus. The basic organization of signal transduction pathways leading to MAP kinase activation is highly conserved in different organisms, ranging from yeast to mammalian cells. Several subgroups of MAP kinases were identified in mammalian cells, which differ in their substrate specificities and regulation. Members of the JNK subgroup of MAP kinases phosphorylate specific sites on the amino-terminal transactivation domain of transcription factor c-Jun. The JNKs mediate responses to diverse extracellular stimuli, including UV irradiation, proinflammatory cytokines, and certain mitogens. Phosphorylation of the amino-terminal sites of c-Jun stimulates its activity as a transcription factor. This review provides an overview of MAP kinase activation, with particular emphasis on the signal transduction pathways leading to activation of the JNK subgroup of MAP kinases. We also discuss the possible functions of the JNK activation pathway.

Introduction

Normal eukaryotic cells must be able to respond to a diverse array of extracellular stimuli. In some cases, they also need to adapt to adverse and stressful conditions. Exposure to certain extracellular stimuli can trigger cell growth and division, while exposure to other stimuli can induce cell differentiation or even programmed cell death. Exposure to environmental stresses results in activation of various stress responses which help the cells withstand adverse environmental conditions. An intriguing question is how external stimuli that signal through cell surface receptors elicit changes in cell phenotype and morphology. One common way for executing such responses is by turning on cascades of biochemical events in response to receptor activation which transmit information from the cell surface to the transcriptional machinery in the nucleus. By modulating the activities of sequence-specific transcription factors and combinations thereof, extracellular stimuli activate or repress the expression of specific sets of genes. In this way, exposure to extracellular stimuli or cellular stresses can lead to long-term changes in gene expression patterns. The products of genes that are induced by extracellular stimuli are the ultimate mediators of the phenotypic responses. Understanding the mechanisms by which signal transduction pathways transmit information to the nucleus is important in order to develop a better understanding for how normal cells respond to extracellular stimuli. Such information will help in deciphering what goes wrong in a variety of clinical disorders. In diseases such as cancer or diabetes, for example, cells fail to respond properly to extracellular stimuli, resulting in loss of normal growth control, or improper regulation of energy metabolism.

Among the major types of signal transduction pathways in eukaryotic cells are protein kinase cascades which lead to the activation of mitogen-activated protein (MAP) kinases (1–4). MAP kinases are found in nearly all eukaryotic cells that have been studied. Although different MAP kinases vary in their substrate specificities and responses to extracellular stimuli, they are all highly conserved in their primary structure and mode of activation (see Fig. 1). All MAP kinases are proline-directed serine/threonine kinases that phosphorylate their substrates on serine residues followed by a proline. The MAP kinases themselves are activated by phosphorylation, in response to a signal transduction cascade that is triggered by an extracellular stimulus. This signal transduction cascade results in the phosphorylation of MAP kinases on threonine and tyrosine residues, separated by a single amino acid. The kinase that phosphorylates the MAP kinases is a dual-specificity kinase called a MAP kinase kinase (MAPKK), MAPKKs also form a highly

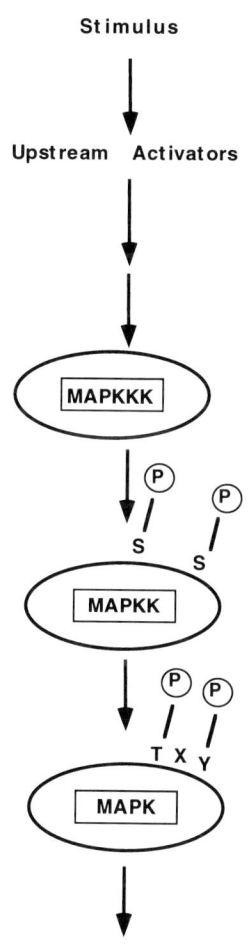

```
              Stimulus
                 |
                 v
        Upstream Activators
                 |
                 v
              ( MAPKKK )
                 |
                 v   (P)  (P)
                  \ /    /
                   S    S
               ( MAPKK )
                 |
                 v   (P) (P)
                  \ /   /
                   T X Y
                ( MAPK )
                 |
                 v
```

FIGURE 1 Signaling pathway leading to activation of MAP Kinases. MAP Kinase Kinase Kinases are regulated by protein kinases or GTP binding proteins in response to a stimulus. Once activated, MAP Kinase Kinase Kinases phosphorylate MAP Kinase Kinases, which in turn phosphorylate and activate MAP Kinases.

conserved group that is activated through phosphorylation of conserved serine and/or threonine residues by a somewhat more diverse group of MAP kinase kinase kinases (MAPKKKs). The latter group receives signals from cell surface receptors through a variety of intermediates, including other protein kinases and small GTP binding proteins. In the different MAP kinase pathways the MAPK, MAPKK, and MAPKKKs are usually highly conserved in their catalytic domains, but can differ considerably in their regulatory domains and their substrate specificities.

The first mammalian MAP kinase cascade to be characterized in detail was the pathway leading to activation of the Extracellular signal-Regulated

Kinase (ERK) subgroup of MAP kinases (5). Recently, a great deal of effort has gone into characterization of another mammalian MAP kinase cascade, leading to activation of the Jun kinases (JNKs). The JNKs, also known as stress-activated protein kinases (SAPKs) (6–8), were first identified by their ability to phosphorylate specific sites on the amino-terminal transactivation domain of the transcription factor c-Jun following exposure to UV irradiation, growth factors, or expression of transforming oncogenes (6,7). By phosphorylating these sites, the JNKs stimulate the ability of c-Jun to activate transcription of target genes. More recently, a third subgroup of mammalian MAP kinases, collectively known as p38 or Mpk2, was identified (9,10). p38 was originally identified as a kinase that is activated in response to exposure to lipopolysaccharide (LPS), and it is activated by many of the same stimuli that activate JNK (10). Unlike JNK, however, the p38 subgroup does not phosphorylate the activation domain of c-Jun. p38 thus differs from JNK in its substrate specificity, and possibly function.

Regulation of c-Jun Expression and Activity

c-Jun, encoded by the *c-jun* proto-oncogene, is a sequence-specific transcription factor whose function has been implicated in various cellular events ranging from cell proliferation and differentiation to neoplastic transformation (11). c-Jun is a component of the sequence-specific transcriptional activator AP-1 (11). AP-1 is a collection of dimers composed of members of the Jun, Fos, or ATF families of bZIP (basic region-leucine zipper) DNA binding proteins. These dimers bind to a common cis-acting element known as the TRE (TPA response element), also known as the AP-1 site, found in the promoters of target genes (11). While many of the AP-1 factors are transcriptional activators, certain AP-1 complexes can function as transcriptional repressors (11,12). The exact function and potency of AP-1 complexes is determined by their composition. The composition of the AP-1 complex may also determine the specificity of target gene activation, as suggested by the distinct phenotypes observed in response to inactivation of genes encoding specific AP-1 components (13–15). Part of the specificity of AP-1 complexes may be determined by interaction with other transcription factors belonging to the Ets (16) or Rel (17) families. In addition, the activity of AP-1 components is modulated by interaction with coactivator proteins, such as the Jun activation domain binding protein 1 (JAB1; 18).

The abundance and activity of c-Jun are modulated when cells are exposed to extracellular stimuli. In addition to c-Jun, the Jun family of proto-oncogenic transcription factors contains JunB (19) and JunD (20). It seems likely that JunB and JunD are also regulated, although we know less about how they respond to extracellular stimuli. The amount of c-Jun in the cell is regulated primarily at the transcriptional level. The *c-jun* gene is an immediate early gene (21), and its expression is quite low in nonstimulated logarithmic growing cells, and even lower in serum-deprived quiescent cells. Like other immediate early genes, *c-jun* transcription is rapidly stimulated, independently of *de novo* protein synthesis, after cells are exposed to a variety of extracellular stimuli. Such stimuli include growth factors, proinflammatory cytokines, UV irradiation, and many other stimuli (21–24). This results in production of more *c-jun* mRNA and ultimately an increase in the amount of c-Jun protein in the nucleus.

A second level of regulation occurs posttranscriptionally through c-Jun phosphorylation. Phosphorylation of sites within the N-terminal activation domain of c-Jun, serines 63 and 73, has a most pronounced effect on the activity of the protein (25–27). In addition, phosphorylation of the same sites was suggested to increase the stability of c-Jun, resulting in a modest increase in its steady state level (28). As mentioned above, nonstimulated cells express low levels of c-Jun protein and the N-terminal sites of c-Jun in these cells are mostly not phosphorylated. In this case, c-Jun's ability to activate transcription of target genes is quite low. Exposure of cells to growth factors, proinflammatory cytokines, UV irradiation (22,29), or other stimuli such as the alkylating agent methylmethane sulfonate (MMS; 30) results in a rather rapid increase in phosphorylation of serine 73 and to a lesser extent serine 63 (29). Interestingly, the same stimuli which enhance the phosphorylation of these N-terminal sites also cause induction of *c-jun* transcription (22,23,29). This correlation supports the autoregulatory model for regulation of *c-jun* transcription (21). Once c-Jun is activated by phosphorylation, it can activate transcription of the c-jun gene through the modified TRE element in the c-jun promoter.

As discussed above, phosphorylation of serines 63 and 73 enhances the ability of c-Jun to activate transcription (25–27). Both of these sites are specifically phosphorylated by the JNK subgroup of MAP kinases. However, it is possible to construct altered specificity mutants of c-Jun in which these sites are phosphorylated by protein kinase A (31). Phosphorylation of the two sites by protein kinase A still stimulates the transcriptional activity of c-Jun (31). Phosphorylation of these two sites, therefore, is critical for c-Jun activation. A likely mechanism underlying the enhancement of c-Jun activity in response to phosphorylation is the ability of N-terminal phosphorylated c-Jun to interact with coactivators such as CBP (32).

Phosphorylation of the c-Jun Amino-Terminal Sites by JNK

The JNKs were identified by virtue of their ability of phosphorylate the N-terminal sites of c-Jun. The use of affinity columns containing the N-terminal activation domain of c-Jun led to identification of at least two kinase activities, 46 and 55 kDa in size, which bind to this region of c-Jun and phosphorylate it on serines 63 and 73 (6). Most importantly, the regulation of these activities parallels the regulation of c-Jun phosphorylation in intact cells (29). The JNKs were also independently identified as cyclohexamide-activated protein kinases and were named SAPKs (32). Only after isolation of the corresponding DNA clones was it realized that the JNKs and SAPKs are the same (7,8). The JNK subgroup includes the products of three related genes: *JNK1* (7), *JNK2* (33), and *JNK3* (34). Alternatively, they are referred to as SAPKα, SAPKβ and SAPKγ (8). Each of these genes directs the production of several JNK polypeptides due to alternative splicing (K. Yoshioka and M. Karin, unpublished results). While JNK1 and JNK2 are expressed in most if not all cell types (33), expression of JNK3 appears to be limited to neuronal cells (34). All the JNK isoforms are regulated identically in response to extracellular stimuli (33,34).

Although the major products of the *JNK1* and *JNK2* genes respond to the exact same extracellular stimuli, they differ in their ability to phosphorylate c-Jun. Using c-Jun activation domain affinity beads it was shown that the 55-kDa JNK2 polypeptide has a much higher affinity toward c-Jun than the 46-kDa JNK1 polypeptide (33). As a result, JNK2 is a more efficient c-Jun kinase than JNK1, having a lower K_m and higher V_{max} (33). The two kinases are, however, identical in their ability of phosphorylate murine p53, and it is assumed that some substrates may be more efficiently phosphorylated by JNK1. Using a series of chimeras, the differences in the ability of JNK1 and JNK2 to bind and phosphorylate c-Jun were mapped to a 23-amino-acid segment in the carboxyl-terminal region of JNK, located next to the catalytic pocket of the enzyme (33). Interestingly, this region is encoded by a separate exon which, in the case of the JNK2 gene, has been duplicated and is subject to alternative splicing (K. Yoshioka, T. Kallunki, and M. Karin, unpublished results). Thus, the major polypeptide encoded by the JNK2 gene is an efficient c-Jun kinase, while the minor product is a low-affinity c-Jun kinase similar to JNK1 (K. Yoshioka, T. Kallunki, and M. Karin, unpublished results). A mutant of JNK1 in which the 23-amino-acid segment is derived from JNK2 is as efficient in c-Jun phosphorylation as wild-type JNK2 (33). This region of the enzyme is likely to be an important determinant of substrate specificity. This short segment of JNK serves to anchor the enzyme onto a specific docking site on its substrate c-Jun. The docking site on c-Jun is located between amino acids 30 and 60 in the N-terminal activation domain (35).

Although JNK2 and c-Jun interact very tightly *in vitro* (33), all attempts to coimmunoprecipitate the two proteins from cells have failed (M. Hibi and M. Karin, unpublished results). There are at least two reasons for this failure. First, like other MAP kinases, the JNKs are excluded from the nucleus prior to activation (36), and therefore cannot bind c-Jun in quiescent cells. Second, once the cells are stimulated and c-Jun is phosphorylated by the activated form of JNK in the nucleus (36), the c-Jun:JNK complex is destabilized (6). Despite the failure to isolate c-Jun:JNK complexes from cells, the importance of JNK docking to c-Jun is strongly supported by mutational analysis. Mutations that abolish or decrease JNK binding to c-Jun *in vitro* result in a large decrease or complete inhibition of JNK-mediated c-Jun phosphorylation in intact cells (35).

In addition to a docking site located between amino acids 30 and 60 of c-Jun, efficient phosphorylation of c-Jun by JNK requires specific residues located around the phosphoacceptor sites (35). Most important is a proline at the P+1 position, directly following the phosphoacceptor site. This requirement is common to JNK and all other MAP kinases (35, and references therein). In addition, efficient c-Jun phosphorylation by JNK requires a positively charged residue located several amino acids C-terminal to the phosphoacceptor site (35). While these sequences are absolutely essential for JNK-mediated c-Jun phosphorylation, they do not affect the binding of JNK to c-Jun (35). Finally, the c-Jun docking site, in addition to ensuring high efficiency of phosphorylation, is responsible for selection of the phosphoacceptor site. Docking-deficient mutants of c-Jun are phosphorylated on multiple sites in addition to serines 63 and 73. The studies demonstrate the importance of the docking site in directing JNK phosphorylation of c-Jun to the correct phosphoacceptor sites (35).

Collectively the studies described above suggest the following mechanism for phosphorylation of c-Jun by JNK. JNK, which is found in the cytoplasm of quiescent cells, translocates to the nucleus once it is activated in response to extracellular stimuli. There it forms a transient complex with the N-terminal activation domain of c-Jun. The docking of JNK to c-Jun serves to increase the local concentration of the enzyme next to its substrate. Once the initial complex dissociates, JNK interacts with the phosphoacceptor site of c-Jun via its catalytic pocket. This classical form of enzyme–substrate interaction results in c-Jun phosphorylation and dissociation of the enzyme–substrate complex.

The strict sequence requirements explain the specificity of c-Jun phosphorylation by JNK, and explains why other subgroups of mammalian MAP kinases do not efficiently phosphorylate serines 63 and 73 of c-Jun. This remarkable substrate specificity stands in contrast to the results of peptide selection experiments. Using an ingenious method based on phosphorylation of a random collection of serine–proline-containing peptides (37), Cantely and co-workers found that the ERK, JNK, and p38 MAP kinases phosphorylate the same peptide sequence (L. Cantely, personal communication). Thus, the interaction of MAP kinases with the phosphoacceptor peptide, although critical for substrate phosphorylation, does not account for the differences in substrate specificities. The actual specificity of these enzymes is most likely determined by the same mechanism described above for JNK-mediated c-Jun phosphorylation.

Of the different Juns, only c-Jun is an efficient JNK substrate. Although JunB contains an effective JNK-docking site, similar in its efficiency to the JNK docking site of c-Jun, JunB is not phosphorylated by JNK at all. This can be explained by the fact that JunB does not contain prolines at the P+1 position following its equivalents of serines 63 and 73 (35). Although JunB, like c-Jun, contains cryptic JNK phosphorylation sites, the presence of the efficient docking site localizes the kinase to the N-terminal activation domain of JunB and suppresses its ability to phosphorylate other sites.

Unlike c-Jun and JunB, JunD does not contain an effective JNK docking site. The phosphoacceptor region of JunD, however, is almost identical in its sequence to the phosphoacceptor region of c-Jun, including the prolines at the P+1 positions. The phosphorylation of JunD by JNK, measured *in vitro* or in intact cells, is considerably less efficient than phosphorylation of c-Jun by JNK. JunD is phosphorylated more efficiently than docking-defective mutants of c-Jun, however (35). Phosphorylation of JunD by JNK was found to be facilitated by the heterodimerization of JunD with either c-Jun or JunB, both of which have efficient docking sites. In this case, JNK is first recruited to JunD containing dimers by docking to the partner protein. As in the case of c-Jun, this results in a higher local concentration of JNK, thereby facilitating the phosphorylation of the docking-defective protein in JunD (35). In addition to explaining how the JNKs can discriminate between the different Jun proteins, these results illustrate that heterodimerization between transcription factors can affect their targeting by MAP kinases and possibly other signal-regulated protein kinases.

Signal Transduction Pathways Leading to Activation of JNK and Other MAP Kinases

Like other MAP kinases, JNK binding is stimulated in response to a signal transduction cascade that is triggered when cells are exposed to extracellular stimuli. Many aspects of the signaling cascade leading to JNK activation are conserved in other MAP kinase cascades, ranging from yeast to mammalian cells. In fact, MAP kinase cascades in yeast have provided ideal model systems for studying MAP kinase cascades in higher eukaryotes such as mammals. In the budding yeast *Saccharomyces cerevisiae*, several signaling cascades lead to activation of different MAP kinases which mediate distinct biological responses (38, and Fig. 2). The first yeast MAP kinase pathway to be identified is triggered when haploid yeast respond to mating pheromones (39). This pathway is necessary for activation of specific genes required for yeast mating. Other yeast MAP kinase pathways include a MAP kinase cascade that is necessary for maintenance of cell wall integrity (40), and a MAP kinase cascade that is triggered by osmotic stress (41). All of these cascades are distinct and are mediated by different MAP kinases.

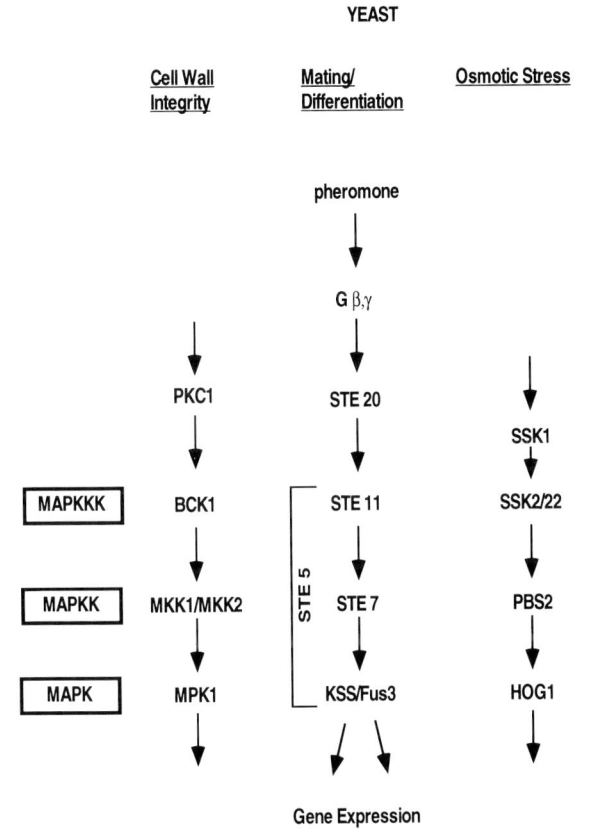

FIGURE 2 At least five different MAP Kinase pathways have been studied in yeast. Three of them are outlined here. While they all mediate different types of responses, the different MAP Kinases are regulated by a similar mechanism.

The different MAP kinases in these pathways all phosphorylate different substrates, which include transcription factors that control distinct sets of target genes. Despite these differences, all of these MAP kinase cascades are organized in a similar fashion, consisting of MAPKKK→MAPKK→MAPK, where MAPKKK is activated by other kinases and GTP binding proteins. Recently, the pheromone–responsive MAP kinase cascade was also found to be organized by a scaffolding protein, STE5, which is necessary for its proper function (42). STE5 has separate binding sites for the MAPK, MAPKK, and MAPKKK. In this way STE5 organizes the components of the pathway into a functional module. This allows for their efficient activation (42), and insulates them from components of other modules (42). STE5 was also shown to physically interact with the different GTP binding proteins that affect this cascade (43), as well as an upstream protein kinase STE20

(44,45). Although the organization of the different components into a tightly linked module greatly enhances the specificity of the cascade, such an organization is a hindrance to signal amplification.

Studies of mammalian MAP kinase cascades have revealed a striking evolutionary conservation from yeast to mammals. The signaling pathways leading to activation of the ERK, JNK, and p38 MAP kinases are illustrated in Fig. 3, and described below, with emphasis placed on the JNK activation pathway. The signaling pathway leading to ERK activation was the first MAP kinase pathway to be studied in mammalian cells. The ERK subgroup of MAP kinases is activated most potently by mitogenic stimuli such as growth factors. The pathway leading to ERK activation usually begins when growth factors bind to cell surface receptors that contain intrinsic tyrosine kinase activity (46). Ligand-induced receptor clustering results in autophosphorylation of the receptor cytoplasmic domain on tyrosine residues. This is followed by the recruitment of SH2 domain containing adapter

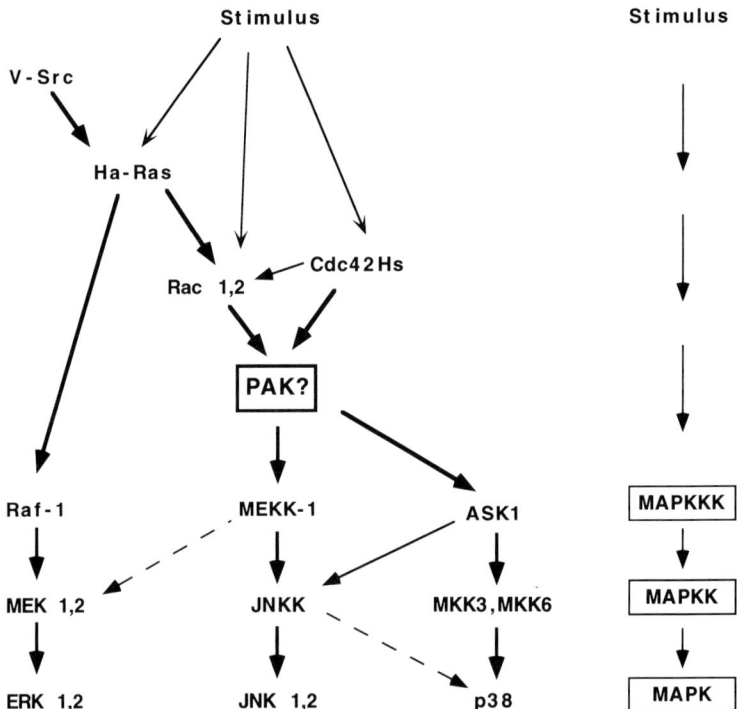

FIGURE 3 The signaling pathways leading to activation of three mammalian MAP Kinases, ERK, JNK, and p38, are outlined.

proteins such as GRB2 or SHC, to the activated receptor. This leads to the recruitment of additional proteins, which interact with the adapter (47). The recruitment of the exchange factor, Sos, places it in the membrane next to its substrate, the small GTP binding protein Ha-Ras (48). Sos activates Ras by catalyzing the exchange of GDP for GTP (49,50). Once Ha-Ras is activated it binds and recruits the serine/threonine kinase Raf-1 to the membrane (51). While membrane translocation of Raf-1 is essential for its activation, a second step, leading to Raf-1 phosphorylation, is also required (52). The mechanism of Raf-1 phosphorylation is still unknown. Raf-1 functions as the MAPKKK in the ERK cascade, although by sequence it is not a classical MAPKKK. Once Raf-1 is activated it phosphorylates and activates the MAPKKs, MEK1 and MEK2 (53,54). Either MEK1 or MEK2 can directly activate the ERKs by phosphorylating their conserved threonine and tyrosine activation sites (55). Once the ERKs are activated a fraction of the enzyme translocates to the nucleus where they phosphorylate and thereby regulate transcription factors which mediate changes in gene expression (56). The mechanism responsible for nuclear translocation of activated ERK or other activated MAP kinases is still unknown. It is possible, however, that nuclear entry occurs by passive diffusion. Physical association with upstream components of their activation cascades, such as MAPKKs may exclude the inactive forms of MAP kinases from the nucleus. Transcription factors that are phosphorylated and activated by the ERKs include various members of the TCF family, such as Elk1 (57) and SAP1 (58), as well as Ets1 and Ets2 (59,60).

The signaling pathway leading to JNK activation was the first mammalian MAP kinase pathway to be investigated as thoroughly as the ERK pathway. In fact, it was the first demonstration that different types of MAP kinase cascades operate in mammalian cells. Although it contains the conserved features of classical MAP kinase cascades, the JNK pathway is distinct from the pathway that activates the ERKs. The JNKs are activated by a variety of different types of extracellular stimuli including growth factors, cytokines, and cellular stresses such as heat shock, hyperosmolarity, and UV irradiation (29,33,61). Like the ERKs, activation of the JNKs by growth factors such as Epidermal Growth Factor (EGF) or Nerve Growth Factor (NGF), that bind to receptor tyrosine kinases, is dependent on the GTP binding protein Ha-Ras (62). However, the Ha-Ras dependent pathway leading to JNK activation differs substantially from the Ha-Ras dependent pathway leading to ERK activation. In the case of ERK, Raf-1 is a critical intermediate between Ras and ERK (53,54). Raf-1, however, does not directly affect the JNK pathway. Instead, another MAPKKK, MEKK1, leads to JNK activation (62,63). Although MEKK1 was first identified as a protein kinase that phosphorylates MEK (64), MEKK1 is an inefficient activator of the ERK pathway (62). The mechanism that controls MEKK1 activity in response to Ras activation or other inputs is not known. Furthermore, it is not even clear whether such inputs actually stimulate MEKK1 catalytic activity. Unlike Raf, MEKK1 isolated from nonstimulated cells is constitutively active (Y. Xia and M. Karin, unpublished results). Furthermore, expression of wild-type MEKK1 is sufficient for JNK activation (B. Su, Y. Xia, and M. Karin, unpublished), but does not lead to activation of the ERK cascade (62). This suggests that MEKK by itself may be constitutively active. One possibility is that upstream stimuli block an MEKK inhibitor in the cell, rather than stimulate MEKK activity. MEKK1 and the related MEKK2 and MEKK3 (65) are similar in sequence to the yeast MAPKKK STE11 (64) (see Fig. 2). Like STE11, MEKK1 functions as a MAPKKK in the JNK pathway (62). MEKK2 can also activate the JNK cascade (65), while MEKK3 preferentially activates the ERK cascade (65). Activation of JNK by MEKK1 is mediated by the direct JNK activating kinase JNKK1, also known as SEK1 or MKK4 (66–68). JNKK1 directly phosphorylates JNK on threonine and tyrosine, the conserved MAP kinase activation sites (63,66). JNKK1 itself is activated by MEKK1 through phosphorylation on serine and threonine residues (63).

These findings indicate that Ha-Ras can trigger at least two diverging MAP kinase cascades. One is mediated by Raf-1 activation and transmits information through MEK1 and -2, resulting in ERK activation. The other cascade operates through MEKK1 and JNKK to activate the JNKs. The mechanism of Ha-Ras-dependent activation of the JNK cascade is not clear. Although Ras-GTP was reported to bind MEKK1 (69), a similar interaction was not found in the case of the yeast MEKK1 homologue STE11 (70,71, and M. Wigler, personal communication). Recently, two other members of the Ras superfamily, namely Rac (both Rac-1 and Rac-2) and Cdc42Hs, were found to activate the JNK cascade in response to signals transmitted through receptor and nonreceptor tyrosine kinases (72–76). These small GTP binding proteins until recently were mostly known for controlling organization of the actin cytoskeleton (77–79). Rac mediates actin polymerization into lamellipodia or membrane ruffles in response to growth factors and Ha-Ras activation (77), while overexpression of activated Cdc42Hs induces filopodia formation (78,79). The finding that these GTP binding proteins also activate the JNK pathway indicated that they are also involved in transmitting signals from membrane receptors to the nucleus and regulating gene expression. While Rac and Cdc42Hs both activate the JNK MAP kinase cascade, so far they have not been shown to be involved in activation of the ERK pathway (72–76). Experiments using dominant inhibitory mutants place Rac downstream to receptor and nonreceptor tyrosine kinases and Ha-Ras and upstream to MEKK1 in the pathway leading to JNK activation (see Fig. 3) (72). Such a placement is consistent with previous findings that Rac mediates cytoskeletal changes caused by growth factors and activated Ha-Ras (77) although Rac may also mediate JNK activation by a Ha-Ras-independent pathway. It is not yet clear how Cdc42Hs fits into the pathway. Cdc42Hs may activate Rac, leading to Rac-dependent signaling events (78,79). Some workers suggested that Cdc42Hs may respond to Ras-independent stimuli (78). One possibility is that Cdc42Hs responds to stimuli, such as TNF-α, that lead to JNK activation. However, TNF-α does not induce fillopodia and lamelipodia formation, the hallmarks of Cdc42Hs and Rac activation (Z. G. Liu, J. Feramisco, and M. Karin, unpublished results).

Although Rac and Cdc42Hs activate the JNK pathway, they have not been shown to activate MEKK1 directly. It was suggested that the Rac-responsive serine/threonine kinase PAK, or one of its close relatives, acts between Rac and MEKK (74,80–83). These protein kinases bind GTP-loaded Rac and Cdc42Hs through a specific interaction domain, and are strongly activated by them in a GTP-dependent manner (80,81). Human PAK1,2 and PAK65, which are nearly identical, are the mammalian homologues of the yeast protein kinase STE20. In *S. cerevisiae*, STE20 functions upstream to STE11 in the pheromone-responsive MAP kinase pathway (38). Furthermore, the founding member of the Rho family, Cdc42Sc, originally identified by its role in bud site selection [i.e., a cytoskeletal reorganization event (84)] was recently found to be involved in activation of the pheromone-responsive MAP kinase cascade (38,85; Fig. 2). However, it is not clear how STE20 affects STE11 activity and, in fact, it has not even been shown to have a direct effect on STE11 activity. Considering the conservation in organization of yeast and mammalian MAP kinase cascades, and the fact that PAK1,2 and PAK65 are activated directly by Rac and Cdc42Hs, it is tempting to place the PAK family of proteins downstream of the small GTP binding proteins in the pathway leading to JNK activation. Although it was reported that some of the PAKs may activate JNK or p38 (74–76), the effects were rather small and were found to be cell-type dependent (A. Minden and M. Karin, unpublished results). Thus, the steps between Rac and Cdc42Hs activation and JNK activation remain to be identified.

Likewise, it is not yet clear how Ha-Ras leads to Rac activation. One possibility is that the lipid kinase phosphoinositide-3 kinase (PI3 kinase) has a role. Recent work has shown that a constitutive active form of the catalytic subunit of PI3 kinase, p110, activates the JNK pathway but not the ERK pathway. Activation of JNK by p110 is blocked by a dominant negative Rac mutant, suggesting that PI3 kinase may function upstream to Rac (86). Consistent with this, PI3 kinase is necessary for PDGF-stimulated membrane ruffling, a function known to be mediated by Rac (87). Interestingly, the p110 subunit of PI3 kinase also binds to, and is activated by, Ha-Ras, in a GTP-dependent manner (88–90). Taken together, these data suggest that Rac activation by Ha-Ras could be mediated at least in part by PI3 kinase.

The signaling pathway described above mediates JNK activation by extracellular stimuli such as growth factors which bind to receptor tyrosine kinases. Likewise, nonreceptor tyrosine kinases of the Src family also appear to activate JNK through the same pathway (72). JNK is also activated by other types of stimuli, however. For example, thrombin and charbacol, which activate heterotrimeric G protein coupled receptors, were reported to activate JNK (91–95). Consistent with this, constitutively active mutants of the α subunits of heterotrimeric G proteins Gα12 and Gα13 were shown to activate JNK by a pathway that involves Rac, MEKK1, and JNKK1 (95–97). However, it is not clear whether and how Gα12 and Gα13 lead to Rac activation. Other stimuli that activate the JNKs include the proinflammatory cytokines TNF-α and IL-1, as well as UV irradiation, heat shock, and hyperosmolarity (29,34,61,98,99). Interestingly, TNFα and IL-1 activate JNK by a Ha-Ras-independent pathway (62–100). In the case of TNFα, this appears to depend on recruitment of two signaling molecules RIP and TRAF2, which bind to the cytoplasmic domain of the TNF receptor (TNFR1) (101).

The third mammalian MAP kinase pathway is the one leading to p38 activation. The p38 MAP kinase is activated by many of the same extracellular stimuli that activate the JNKs (102). Like the JNKs, p38 is activated by Rac and Cdc42Hs (72,76), but it is not activated by MEKK1, -2, or -3 (65,66). Recently, the serine threonine kinase ASK1 was shown to be an effective MAPKKK in the p38 as well as the JNK pathways (103). p38 can be directly phosphorylated and activated by JNKK (66–68). In addition, p38 is phosphorylated by two other MAPKKs, MKK3 and MKK6 (68,104,105). MKK3 and MKK6 do not phosphorylate or activate JNK, and therefore seem to be specific activators of the p38 pathway.

Interestingly, many of the signaling enzymes involved in mammalian MAP kinase cascades are related to yeast signaling enzymes. As described above, MEKK is a homologue of the mammalian signaling enzyme STE11, a MAPKKK in the yeast pheromone response pathway. Likewise, JNKK1 is structurally and functionally homologous to the yeast MAPKK PBS2. In yeast, PBS2 phosphorylates and activates the hyperosmolarity-responsive HOG1 MAP kinase (41,106). Consistent with this, JNKK1 can partially substitue for PBS2 in PBS2-deficient yeast strains (66). Furthermore, both the JNKs and p38 can also partially substitute for HOG1, and they are activated by PBS2 after yeast cells are exposed to hyperosmolarity (99; A. Lin and M. Karin unpublished). These results are somewhat surprising, because only p38 is a structural homologue of the HOG1 MAP kinase.

Other Substrates for the JNKs

Like other MAP kinases, the JNKs phosphorylate and regulate transcription factors and thereby lead to long-term changes in gene expression. Different MAP kinases have different, although sometimes overlapping, substrate specificities. Activation of different signaling pathways can thus lead to regulation of distinct sets of target genes. JNK was originally identified as a kinase that phosphorylates and activates transcription factor c-Jun. So far, JNK is the only MAP kinase known to phosphorylate the amino-terminal sites of c-Jun. Consistent with the role for JNK in phosphorylating and activating c-Jun, signals that activate the JNK pathway also stimulate c-Jun transcriptional activity (62–107) and presumably induce expression of c-Jun target genes. In addition to c-Jun, however, JNK can phosphorylate and thereby stimulate the activities of other transcription factors. One of these is the transcription factor ATF2 (108–110). Interestingly, c-Jun and ATF2 form heterodimers which bind to the nonconsensus TRE in the *c-jun* promoter, thereby simulating expression of the *c-jun* gene (111). By activating both c-Jun and ATF2, therefore, JNK has the potential to regulate both the abundance and the activity of c-Jun.

Another direct target for JNK, is TCF/Elk-1 (36,112,113). ELK-1 is a transcription factor that is involved in the induction of the c-fos gene (114). Interestingly, c-Fos is also a component of AP-1. Like *c-jun, c-fos* is an immediate early gene regulated by several cis-acting elements in its promoter/enhancer region (115). Induction of the c-fos gene by many stimuli, such as growth factors, requires the presence of the serum response element (SRE), found in the c-fos promoter/enhancer region (115). At least two types of transcription factors bind simultaneously to the SRE. These include the serum response factor (SRF), which binds as a homodimer, and the ternary complex factors (TCF) (115,116). While several different TCFs have been identified, the best characterized TCF is Elk-1 (117). Like c-Jun, Elk-1 activity is regulated by phosphorylation (56,57,118,119). Phosphorylation of Elk-1 at specific sites in its transactivation domain dramatically stimulates its ability to activate transcription of genes such as the c-fos gene. Unlike c-Jun, Elk-1 is phosphorylated and activated by more than one MAP kinase. Originally, growth factor activation of ELK-1 was shown to be a result of its phosphorylation by the ERKs (56,57,119,120). However, c-fos transcription is enhanced by many stimuli, some of which do not activate the ERKs efficiently. These include UV irradiation, IL-1, and TNF-α. c-fos induction by these stimuli is also dependent on the SRE (36,121). This can be explained by recent findings that, like the ERKs, JNK can also phosphorylate Elk-1 on a subset of its positive regulatory sites, thereby enhancing the ability of ELK-1 to activate transcription (36,57,80,113,121).

p38 can phosphorylate some of the same substrates as JNK, although it does not phosphorylate the amino-terminal sites of c-Jun. Like JNK, p38 can phosphorylate and activate AFT2, and ELK-1 (105,108,122). Interestingly, p38 also appears to be required for induction of the c-fos and c-jun genes in response to certain stimuli. This is demonstrated by the finding that the specific p38 inhibitor SB203580 (123,124) blocks *c-fos* and *c-jun* expression in response to UV irradiation and the protein synthesis inhibitor anisomycin (125). These experiments should be interpreted carefully, however, because although in vitro SB203580 exhibits specificity to p38, at higher concentrations it also inhibits JNK activity. In future work it will be interesting to see what other transcription factors are targets of the JNKs and of other MAP kinases. Importantly, determining what genes are regulated by these transcription factors will provide important new information about how MAP kinase pathways mediate specific biological responses.

Biological Functions of JNK and Other MAP Kinase Pathways

Although much has been learned about the components of MAP kinase pathways and their organization, much more remains to be learned about the biological functions of MAP kinases, especially in higher eukaryotes. The JNK pathway was proposed to serve as a major "on" switch for programmed cell death (126–129). However, it is not clear whether the JNKs themselves are directly involved in this process (130). In the case of TNF-α-induced apoptosis, it was shown that activation of the JNK pathway can be prevented without interfering with induction of apoptosis (101). Furthermore, lymphoid cells deficient in JNKK1 were found to be more sensitive to induction of apoptosis rather than more resistant. This would tend to suggest that the JNK pathway may actually play a protective role against apoptosis (131). JNK activation leading to c-Jun induction also most likely mediates a protective function in response to UV irradiation (F. Piu and M. Karin, unpublished results). It will be interesting to see whether JNK has a protective role against other types of environmental stresses.

Other work has suggested that the JNK pathway could have a role in cell proliferation and oncogenic transformation. JNK is strongly activated by onco-proteins such as v-Src (72,132), and previous work has suggested that the JNK pathway may be involved in oncogenic transformation of immortalized fibroblasts by v-Src (132). More recently JNK activation was suggested to be important for transformation by the Met tyrosine kinase (133). In many cells JNK is also activated by growth factors (2,72), and in the liver epithelial cell line GN4, the mitogen angiotensin 2 activates JNK very efficiently without exerting much of an effect on ERK activity (94). Likewise, JNK is activated by Rac and Cdc42Hs, GTP binding proteins which appear to play key roles in cell proliferation, progression through the cell cycle, and oncogenic transformation (134–139). Overexpression of constitutively activated Rac1V12 or Cdc42HsV12 mutants causes quiescent fibroblasts to enter the cell cycle and proliferate (134). Rac1V12 induces a transformed phenotype in Rat1 cells (135), and guanine nucleotide exchange factors which directly activate these proteins are themselves proto-oncoproteins (136,140–142). It is therefore essential to determine more directly whether JNK activation plays a critical role in these mitogenic and oncogenic pathways. Also of interest is whether improper regulation of the JNK pathway can contribute to uncontrolled proliferation and neoplastic transformation.

Interestingly, JNK and p38 are strongly activated by inflammatory cytokines. They are also activated by infectious agents such as the endotoxin lipopolysaccharide (LPS) which leads to a potent inflammatory response. It seems likely, therefore, that the JNK and p38 signaling pathways may be involved in mediating inflammatory responses. Indeed, pharmacological inhibitors of p38 block some primary inflammatory responses, such as the production of more cytokines (123,143,144). Future work will indicate whether the JNK and/or p38 pathways are involved in mediating other cellular activities necessary for mediating an inflammatory response.

Studying the functions of signal transduction pathways is complicated by the fact that a number of signal transduction pathways are probably activated at any given time in a cell in response to stimulation of a single cell surface receptor. The combinatorial effects of two or more signaling pathways are probably necessary for mediating certain responses. A good example comes from studying the signaling cascades mediated by Ha-Ras. Ha-Ras is constitutively active in many tumors, and transforms rodent fibroblasts *in vitro* (145). By using dominant negative mutants, the ERK pathway was shown to be necessary for oncogenic transformation by Ha-RasV12, a constitutive active mutant of Ha-Ras. Likewise, activation of the ERK pathway by overexpressing constitutively active MEK mutants can also transform rodent fibroblasts (146,147). Recent work has shown, however, that the ERK pathway may not be sufficient for the full transforming effect of Ha-Ras V12 (135,148). Several studies suggest that signaling pathways mediated by Rho family members, including Rac, are also necessary for Ha-Ras-induced transformation (135–138). Consistent with this, the Rac-mediated pathway may synergize with the ERK pathway to produce oncogenic foci in fibroblasts (135). Since Rac activates the JNK and p38 pathways, it will be interesting to determine whether these signaling pathways play a role in the proliferative effects of the GTP binding proteins. Likewise, it will be interesting to determine whether the JNK or p38 pathways act in concert with the ERK pathway or other signaling pathways to produce a proliferation response. Studies using various effector mutants of Rac, however, indicate that Rac-mediated JNK activation can be separated from Rac-mediated changes in cell proliferation and transformation, and from changes in the organization of the actin cytoskeleton (149,150). These studies suggest that the JNK pathway may not be directly involved in mediating the mitogenic effects of Rac, and that other signaling pathways may mediate proliferation and oncogenic transformation in response to the Rho family of GTP binding proteins.

In some cases the same signaling pathway can have different functions depending on the cell context. One of the best examples comes from studying the ERK pathway in the rat pheochromocytoma cell line, PC12. PC12 cells proliferate when exposed to EGF. However, when they are exposed to NGF, they stop growing and differentiate into neuron-like cells (151). Interestingly, the ERK activation pathway has been shown to be directly involved in neuronal differentiation in PC12 cells. This is in sharp contrast to fibroblasts, where the ERK pathway is involved in proliferation (144, 152). In addition to cell type, other factors may also determine the functions of MAP kinases. For example, the magnitude and duration of the response can play a significant role in the outcome. In PC12 cells, for example, a prolonged activation of the ERK pathway is associated with differentiation, while a transient activation is associated with proliferation (153).

Studies such as those described above implicate MAP kinase pathways in a variety of different types of cellular responses. The most informative and rigorous analysis of MAP kinase function, however, comes from genetic studies of lower eukaryotes. In the nematode *Caenorhabitis elegans,* the Sur-1/MPK MAP kinases, which are similar to mammalian ERKs (154–156), were shown to mediate the differentiation of vulval precursor cells. The *C. elegans* MAP kinase pathway is triggered by the Let23 receptor tyrosine kinase, which is activated by the EGF-related ligand Lin3 (157,158). Activation of Sur-1/MPK in response to receptor activation is mediated by the Ras homologue Let60. Mutations in the genes that encode components of the Sur-1/MPK MAP kinase pathway result in either a vulvaless or a multivulval phenotype (158,159).

In drosophila, several MAP kinase pathways have been described. The most extensively characterized drosophila MAP kinase pathway is involved in eye development. Undifferentiated R7 precursor cells in the drosophila compound eye receive signals from neighboring cells which cause them to differentiate into photoreceptor cells. Neuronal differentiation of the R7 precursor cell is initiated when the bride of sevenless (BOSS) ligand interacts with the sevenless (SEV) tyrosine kinase receptor on the R7 cells. Activation of the receptor by the ligand leads to activation of the rolled MAP kinase by a pathway that is dependent on drosophila Ras1. Rolled, which is similar to the mammalian ERKs, phosphorylates various transcription factors necessary for normal eye development. Mutations in genes that encode components of this pathway disrupt normal eye development (160–166).

Recently, a role for drosophila JNK (DJNK) has also been described. DJNK was shown to be essential for embryonic development. Loss of DJNK inhibits the movement of lateral epithelial cells during mid-embryogenesis, and blocks dorsal closure, processes that involve changes in cell shape and migration (167,168). In addition to its role in drosophila development and morphogenesis, DJNK is activated by LPS, which initiates an insect immune response (167). Thus, in insects as well as mammals, JNK may have a role in mediating immune and inflammatory responses.

In summary, the JNK pathway is activated by a wide range of different types of stimuli. These include not only growth factors, but also proinflammatory cytokines, heat shock, short-wavelength UV radiation, and numerous other stimuli. These stimuli elicit very different types of cellular responses, ranging from cell growth to cell death (30,169). It seems likely, therefore, that the JNK pathway, as well as other MAP kinase pathways, may have entirely different functions under different conditions. The exact function of JNK is likely to be revealed by genetic analysis akin to the studies conducted in yeast, *C. elegans*, and *Drosophila melanogaster*.

Conclusions and Future Considerations

A great deal of progress has been made in recent years toward understanding how MAP kinases such as JNK mediate responses to extracellular stimuli. This work has opened up intriguing new questions that need to be addressed in the future. For example, how do cells coordinate responses from different stimuli, and how do they make the decision to activate one signaling pathway over another? Another interesting question is why so many signaling enzymes are members of families of closely related proteins. For example, recent work demonstrated that MEKK and PAK are parts of large families of related proteins (74,80–83). Do these enzymes simply have redundant functions, or do they have distinct functions, such as mediating and integrating responses from different types of extracellular stimuli? Other questions relate to the specific functions of signaling pathways. As discussed above, different signaling pathways can probably act together to some extent. In some cases one pathway can even activate another by an autocrine loop mechanism (62), but in other cases one pathway may negatively regulate the other. For example, activation of cyclic AMP (cAMP)-dependent protein kinase A (PKA) can down-regulate the ERK pathway by inhibiting Raf activity. This inhibition weakens Raf-1 interaction with Ha-Ras and may also directly inhibit Raf-1 activity by phosphorylation of its kinase domain (170–176). It will be important to determine how different signaling pathways interact with each other in order to mediate and coordinate specific biological responses. Another important area of investigation is the identification of target genes whose expression is regulated by MAP kinases. In addition, it will be of extreme importance to determine whether and how MAP kinases regulate steps other than transcription initiation in the overall process of gene expression. These include regulation of mRNA stability and translational efficiency (177). Finally, the biological functions of specific signaling pathways need to be investigated further, both in mammalian cells and in lower eukaryotes. When studying the functions of various signal transduction pathways including the JNK pathway, it will be important to consider the possible contributions of many factors. These include cell type, the type of stimulus, the duration and magnitude of the response, and the activation of other signaling pathways in the cell.

References

1. Marshall, C. J. (1994). *Curr. Opin. Genet. Dev.* **4**, 82–89.
2. Cano, E., and Mahadevan, L. C. (1995). *Trends Biochem. Sci.* **20**, 117–122.
3. Waskiewicz, A. J., and Cooper, J. (1995). *Curr. Opin. Cell Biol.* **7**, 798–805.
4. Karin, M. (1996). *Philos. Trans. R. Soc. London, Ser. B* **351**, 127–134.
5. Cobb, M. H., Boulton, T. G., and Robbins, D. T. (1991). *Cell Regul.* **2**, 965–978.
6. Hibi, M., Lin, A., Smeal, T., Minden, A., and Karin, M. (1993). *Genes Dev.* **7**, 2135–2148.
7. Derijard, B., Hibi, M., Wu, I.-H., Barrett, T., Su, B., Deng, T., Karin, M., and Davis, R. (1994). *Cell (Cambridge, Mass.)* **76**, 1025–1037.
8. Kyriakis, J. M., Banerjee, P., Nikolakaki, E., Dai, T., Rubie, E. A., Ahmad, M. F., Avruch, J., and Woodgett, J. R. (1994). *Nature (London)* **369**, 156–160.
9. Han, J., Lee, J. D., Bibbs, L., and Ulevitch, R. J. (1994). *Science* **265**, 808–811.
10. Rouse, J., Cohen, P., Trigon, S., Morange, M., Alonso-Llamazares, A., Zamanillo, D., Hunt, T., and Nebreda, A. R. (1994). *Cell (Cambridge, Mass.)* **78**, 1027–1037.
11. Angel, P., and Karin, M. (1991). *Biochim. Biophys. Acta* **1072**, 129–157.
12. Chiu, R., Angel, P., and Karin, M. (1989). *Cell (Cambridge, Mass.)* **59**, 979–986.
13. Johnson, R. S., van Lingen, B., Papaioannou, V. E., and Spiegelman, B. M. (1993). *Genes Dev.* **7**, 1309–1317.
14. Hilberg, F., Aguzzi, A., Howells, N., and Wagner, E. F. (1993). *Nature (London)* **365**, 179–181.
15. Johnson, R. S., Spiegelman, B. M., and Papaioannou, U. P. (1992). *Cell (Cambridge, Mass.)* **71**, 577–586.
16. Wasylyk, B., Wasylyk, C., Flores, P., Begue, A., Leprince, D., and Stehelin, D. (1990). *Nature (London)* **346**, 191–193.
17. Stein, B., Baldwin, A. S., Jr., Ballard, D. W., Greene, W. C., Angel, P., and Herrlich, P. (1993). *EMBO J.* **12**, 3879–3891.
18. Claret, F. X., Hibi, M., Dhut, S., Toda, T., and Karin, M. (1996). *Nature (London)* **383**, 453–457.
19. Ryder, K., Lau, L. F., and Nathans, D. (1988). *Proc. Natl. Acad. Sci. U.S.A.* **85**, 1487–1491.
20. Ryder, K., Lanahan, A., Perez-Albuerne, E., and Nathans, D. (1989). *Proc. Natl. Acad. Sci U.S.A.* **86**, 1500–1503.
21. Angel, P., Hattori, K., Smeal, T., and Karin, M. (1988). *Cell (Cambridge, Mass.)* **55**, 875–885.
22. Devary, Y., Gottlieb, R. A., Lau, L. F., and Karin, M. (1991). *Mol. Cell. Biol.* **11**, 2804–2811.
23. Lamph, W. W., Wamsley, P., Sassone-Corsi, and Verma, I. M. (1988). *Nature (London)* **344**, 629–631.
24. Brenner, D. A., O'Hara, M., Angel, P., Chojkier, M., and Karin, M. (1989). *Nature (London)* **337**, 661–663.
25. Smeal, T., Binetruy, B. Mercola, D., Birrer, M., and Karin, M. (1991). *Nature (London)* **354**, 494–496.
26. Pulverer, B. J., Kyriakis, J. M., Avruch, J., Nikolakaki, E., and Woodgett, J. R. (1991). *Nature (London)* **353**, 670–674.

27. Smeal, T., Binetruy, B., Mercola, D., Grover-Bardwick, A., Heidecker, G., Rapp, U. R., and Karin, M. (1992). *Mol. Cell. Biol.* **12**, 3507–3513.
28. Musti, A. M., Treier, M., and Bohmann, D. (1997). *Science* **275**, 400–402.
29. Minden, A., Lin, A., Smeal, T., Derijard, B., Cobb, M., Davis, R., and Karin, M. (1994). *Mol. Cell. Biol.* **14**, 6683–6688.
30. Liu, Z.-G., Baskaran, R., Lea-Chou, E. T., Wood, L. D., Chen, Y., Karin, M., and Wang, J. Y. (1996). *Nature (London)* **384**, 273–276.
31. Smeal, T., Hibi, M., and Karin, M. (1994). *EMBO J.* **13**, 6006–6010.
32. Bannister, A. J., Oehler, T., Wilhelm, D., Angel, P., and Kouzarides, T. (1995). *Oncogene* **11**, 2509–2514.
33. Kallunki, T., Su, B., Tsigelny, I., Sluss, H. K., Derijard, B., Moore, G., Davis, R., and Karin, M. (1994). *Genes Dev.* **8**, 2996–3007.
34. Gupta, S., Barrett, T., Whitmarsh, A. J., Cavanagh, J., Sluss, H. K., Derijard, B., and Davis, R. J. (1996). *EMBO J.* **15**, 2760–2770.
35. Kallunki, T., Deng, T., Hibi, M., and Karin, M. (1996). *Cell (Cambridge, Mass.)* **87**, 929–939.
36. Cavigelli, M., Dolfi, F., Clater, F.-X., and Karin, M. (1995). *EMBO J.* **14**, 5957–5964.
37. Songyang, Z., Blechner, S., Hoagland, N., Hoekstra, M. F., Piwnica-Worms, H., and Cantley, L. C. (1994). *Curr. Biol.* **4**, 973–982.
38. Levin, D. E., and Errede, B. (1995). *Curr. Opin. Cell Biol.* **7**, 197–202.
39. Herskowitz, I. (1995). *Cell (Cambridge, Mass.)* **80**, 187–197.
40. Lee, K. S., Irie, K., Gotoh, Y., Watanabe, Y., Araki, H., Nishida, E., Matsumoto, K., and Levin, D. E. (1993). *Mol. Cell. Biol.* **13**, 3067–3075.
41. Brewster, J. L., De Valoir, T., Dwyer, N. D., Winter, E., and Gustin, M. C. (1993). *Science* **259**, 1760–1763.
42. Choi, K. Y., Satterberg, B., Lyons, D. M., and Elion, E. A. (1994). *Cell (Cambridge, Mass.)* **78**, 499–512.
43. Whiteway, M. S., Wu, C., Leeuw, T., Clark, K., Fourest-Lieuvin, A., Thomas, D. Y., and Leberer, E. (1995). *Science* **269**, 1572–1575.
44. Leberer, E., Dignard, D., Harcus, D., Hougan, L., Whiteway, M., and Thomas, D. Y. (1993). *Mol. Gen. Genet.* **241**, 241–254.
45. Leberer, E., Dignard, D., Harcus, D., Thomas, D. Y., and Whiteway, M. (1992). *EMBO J.* **11**, 4815–4824.
46. Cadena, D., and Gill, G. N. (1992). *FASEB J.* **6**, 2332–2337.
47. Schlessinger, J. (1994). *Curr. Opin. Genet. Dev.* **4**, 25–30.
48. Aronheim, A., Engelberg, D., Li, N., al-Alawi, N., Schlessinger, J., and Karin, M. (1994). *Cell (Cambridge, Mass.)* **78**, 949–961.
49. Chardin, P., Camonis, J. H., Gale, N. W., van Aelst, L., Schlessinger, J., Wigler, M. H., and Bar-Sagi, D. (1993). *Science* **260**, 1338–1343.
50. Gale, N. W., Kaplan, S., Lowenstein, E. J., Schlessinger, J., and Bar-Sagi, D. (1993). *Nature (London)* **363**, 88–92.
51. Marais, R., Light, Y., Paterson, H. F., and Marshall, C. J. (1995). *EMBO J.* **14**, 3136–3145.
52. Vojtek, A. B., Hollenberg, S. M., and Cooper, J. A. (1993). *Cell (Cambridge, Mass.)* **74**, 205–214.
53. Dent, P., Haser, W., Haystead, T. A., Vincent, L. A., Roberts T. M., and Sturgill, T. W. (1992). *Science* **257**, 1404–1406.
54. Kyriakis, J. M., App, H., Zhang, X. F., Banerjee, P., Brautigan, D. L., Rapp, U. R., and Avruch, J. (1992). *Nature (London)* **358**, 417–421.
55. Ahn, N. G., Seger, R., and Krebs, E. G. (1992). *Curr. Opin. Cell Biol.* **4**, 992–999.
56. Marais, R., Wynne, J., and Treisma, R. (1993). *Cell (Cambridge, Mass.)* **73**, 381–393.
57. Gille, H., Kortenjann, M., Thomae, O., Moomaw, C., Slaughter, C., Cobb, M. H., and Shaw, P. E. (1995). *EMBO J.* **14**, 951–962.

58. Janknecht, R., Ernst, W. H., and Nordheim, A. (1995). *Oncogene* 10, 1209–1216.
59. Coffer, P., de Jonge, M., Mettouchi, A., Binetruy, B., Ghysdael, J., and Kruijer, W. (1994). *Oncogene* 9, 911–921.
60. Yang, B.-S., Hauser, C. A., Henkel, G., Colman, M. S., van Beveren, C., Stacey, K. J., Hume, D. A., Maki, R. A., and Ostrowski, M. C. (1996). *Mol. Cell. Biol.* 16, 538–547.
61. Matsuda, S., Kawasaki, H., Moriguchi, T., Gotoh, Y., and Nishida, E. (1995). *J. Biol. Chem.* 270, 12781–12786.
62. Minden, A., Lin, A., McMahon, M., Lange-Carter, C., Derijard, B., Davis, R. J., Johnson, G. L., and Karin, M. (1994). *Science* 266, 1719–1723.
63. Yan, M., Dai, T., Deak, J. C., Kyriakis, J. M., Zon, L. I., and Woodgett, J. R. (1994). *Nature (London)* 372, 798–800.
64. Blank, J. L., Gerwins, P., Elliott, E. M., Sather, S., and Johnson, G. L. (1996). *J. Biol. Chem.* 271, 5361–5368.
65. Lange-Carter, C. A., Pleiman, C. M., Gardner, A. M., Blumer, K. J., and Johnson, G. L. (1993). *Science* 260, 315–319.
66. Lin, A., Minden, A., Martinetto, H., Claret, F. X., Lange-Carter, C., Mercurio, F., Johnson, G. L., and Karin, M. (1995). *Science* 268, 286–290.
67. Sanchez, I., Hughes, R. T., Mayer, B. J., Yee, K., Woodgett, J. R., Avruch, J., Kyriakis, J. M., and Zon, L. I. (1994). *Nature (London)* 372, 794–798.
68. Derijard, B., Raingeaud, J., Barrett, T., Wu, I. H., Han, J., Ulevitch, R. J., and Davis, R. J. (1995). *Science* 267, 682–685.
69. Russell, M., Lange-Carter, C. A., and Johnson, G. L. (1995). *J. Biol. Chem.* 270, 11757–11760.
70. Van Aelst, L., White, M. A., and Wigler, M. H. (1994). *Cold Spring Harbor Symp. Quant. Biol.* 59, 181–186.
71. Van Aelst, L., Barr, M., Marcus, S., Polverino, A., and Wigler, M. (1993). *Proc. Natl. Acad. Sci. U.S.A.* 90, 6213–6217.
72. Minden, A., Lin, A., Claret, F. X., Abo, A., and Karin, M. (1995). *Cell (Cambridge, Mass.)* 81, 1147–1157.
73. Coso, O. A., Chiariello, J.-C., Yu, J. C., Crespo, P., Xu, N., Miki, T., and Gutkind, J. S. (1995). *Cell (Cambridge, Mass.)* 81, 1137–1146.
74. Brown, J. L., Stowers, L., Baer, M., Trejo, J., Coughlin, S., and Chant, J. (1996). *Curr. Biol.* 6, 598–605.
75. Bagrodia, S., Derijard, B., Davis, R. J., and Cerione, R. A. (1995). *J. Biol. Chem.* 270, 27995–27998.
76. Zhang, S., Han, J., Sells, M. A., Chernoff, J., Knaus, U. G., Ulevitch, R. J., and Bokoch, G. M. (1995). *J. Biol. Chem.* 270, 23934–23936.
77. Ridley, A. J., Paterson, H. F., Johnson, C. L., Diekman, D., and Hall, A. (1992). *Cell (Cambridge, Mass.)* 70, 401–410.
78. Kozma, R., Ahmed, S., Best, A., and Lim, L. (1995). *Mol. Cell. Biol.* 15, 1942–1952.
79. Nobes, C. D., and Hall, A. (1995). *Cell (Cambridge, Mass.)* 81, 53–62.
80. Martin, G. A., Bollag, G., McCormick, F., and Abo, A. (1995). *EMBO J.* 14, 1970–1978.
81. Manser, E., Leung, T., Salihuddin, H., Zhao, Z. S., and Lim, L. (1994). *Nature (London)* 367, 40–46.
82. Manser, E., Chong, L., Zhuo-Shen, Z., Leung, T., Michael, G., Hall, C., and Lim, L. (1995). *J. Biol. Chem.* 270, 25070–25078.
83. Teo, M., Manser, E., and Lim, L. (1995). *J. Biol. Chem.* 270, 26690–26697.
84. Adams, A. E., Johnson, D. I., Longnecker, R. M., Sloat, B. F., and Pringle, J. R. (1990). *J. Cell Biol.* 111, 131–142.
85. Zhao, Z. S., Leung, T., Manser, E., and Lim, L. (1995). *Mol. Cell. Biol.* 15, 5246–5257.
86. Klippel, A., Reinhard, C., Kavanaugh, W. M., Apell, G., Escobedo, M.-A., and Williams, L. T. (1996). *Mol. Cell. Biol.* 16, 4117–4127.

87. Wennstrom, S., Hawkins, P., Cooke, F., Hara, K., Yonezawa, K., Kasuga, M., Jackson, T., Claesson-Welsh, L., and Stephens, L. (1994). *Curr. Biol.* **4**, 385–393.
88. Kodaki, T., Wosholski, R., Hallberg, B., Rodriguez-Viciana, P., Downward, J., and Parker, P. J. (1994). *Curr. Biol.* **4**, 798–806.
89. Rodriguez-Viciana, P., Warne, P. H., Dhand, R., Vanhaesebroeck, B., Gout, I., Fry, M. J., Waterfield, M. D., and Downward, J. (1994). *Nature (London)* **370**, 527–532.
90. Rodriguez-Viciana, P., Warne, P. H., Vanhaesebroeck, B., Waterfield, M. D., and Downward, J. (1996). *EMBO J.* **15**, 2442–2451.
91. Coso, O. A., Chiariello, M., Kalinec, G., Kyriakis, J. M., Woodgett, J., and Gutkind, J. S. (1995). *J. Biol. Chem.* **270**, 5620–5624.
92. Coso, O. A., Teramoto, H., Simonds, W. F., and Gutkind, J. S. (1996). *J. Biol. Chem.* **271**, 3963–3966.
93. Mitchell, F. M., Russell, M., and Johnson, G. L. (1995). *Biochem. J.* **309**, 381–384.
94. Zohn, I. E., Yu, H., Li, X., Cox, A. D., and Earp, H. S. (1995). *Mol. Cell. Biol.* **15**, 6160–6168.
95. Collins, L., Minden, A., Karin, M., and Brown, J. H. (1996). *J. Biol. Chem.* **271**, 17349–17353.
96. Heasley, L. E., Storey, B., Fanger, G. R., Butterfield, L., Zamarripa, J., Blumberg, D., and Maue R. A. (1996). *Mol. Cell. Biol.* **16**, 648–656.
97. Prasad, M. V., Dermott, J. M., Heasley, L. E., Johnson, G. L., and Dhanasekaran, N. (1995). *J. Biol. Chem.* **270**, 18655–18659.
98. Adler, V., Schaffer, A., Kim, J., Dolan, L., and Ronai, Z. (1995). *J. Biol. Chem.* **270**, 26071–26077.
99. Galcheva-Gargova, Z., Derijard, B., Wu, I. H., and Davis, R. J. (1994). *Science* **265**, 806–808.
100. Bird, T. A., Kyriakis, J. M., Tyshler, L., Gayle, M., Milne, A., and Virca, G. D. (1994). *J. Biol. Chem.* **269**, 31836–31844.
101. Liu, Z.-G., Hsu, H., Goeddel, D. V., and Karin, M. (1996). *Cell (Cambridge, Mass.)* **87**, 565–576.
102. Kyriakis, J. M., and Avruch, J. (1996). **18**, 567–577.
103. Ichijo, H., Nishida, E., Irie, K., ten Dijke, P., Saitoh, M., Moriguchi, T., Takagi, M., Matsumoto, K., Miyazono, K., and Gotoh, Y. (1997). *Science* **275**, 90–94.
104. Han, J., Lee, J. D., Jiang, Y., Li, Z., Feng, L., and Ulevitch, R. J. (1996). *J. Biol. Chem.* **271**, 2886–2891.
105. Raingeaud, J., Whitmarsh, A. J., Barrett, T., Derijard, B., and Davis, R. J. (1996). *Mol. Cell. Biol.* **16**, 1247–1255.
106. Boguslawski, G., and Polazzi, J. O. (1987). *Proc. Natl. Acad. Sci. U.S.A.* **84**, 5848–5852.
107. Lin, A., Frost, J., Deng, T., Smeal, T., al-Alawi, N., Kikkawa, V., Hunter, T., Brenner, D., and Karin, M. (1992). *Cell (Cambridge, Mass.)* **70**, 777–789.
108. Gupta, S., Campbell, D., Derijard, B., and Davis, R. J. (1995). *Science* **267**, 389–393.
109. Livingstone, C., Patel, G., and Jones, N. (1995). *EMBO J.* **14**, 1785–1797.
110. van Dam, H., Wilhelm, D., Herr, I., Steffen, A., Herrlich, P., and Angel, P. (1995). *EMBO J.* **14**, 1798–1811.
111. van Dam, H., Duyndam, M., Rottier, R., Bosch, A., de Vries-Smits, L., Herrlich, P., Zantema, A., Angel, P., and van der Eb, A. J. (1993). *EMBO J.* **12**, 479–487.
112. Gille, H., Strahl, T., and Shaw, P. E. (1995). *Curr. Biol.* **5**, 1191–1200.
113. Zinck, R., Cahill, M. A., Kracht, M., Sachsenmaier, C., Hipskind, R. A., and Nordheim, A. (1995). *Mol. Cell Biol.* **15**, 4930–4938.
114. Treisman, R. (1995). *EMBO J.* **14**, 4905–4913.
115. Treisman, R. H. (1992). *Trends Biochem. Sci.* **17**, 423–426.
116. Treisman, R. H. (1994). *Curr. Opin. Genet. Dev.* **4**, 96–107.
117. Hipskind, R. A., Rao, V. N., Mueller, C. G., Reddy, E. S., and Nordheim, A. (1991). *Nature (London)* **354**, 531–534.

118. Hill, C. S., Marais, R., John, S., Wynne, J., Dalton, S., and Treisman, R. (1993). *Cell (Cambridge, Mass.)* **73**, 395–406.
119. Gille, H., Sharrocks, A., and Shaw, P. (1992). *Nature (London)* **358**, 414–417.
120. Zinck, R., Hipskind, R. A., Pingoud, V., and Nordheim, A. (1993). *EMBO J.* **12**, 2377–2387.
121. Whitmarsh, A. J., Shore, P., Sharrocks, A. D., and Davis, R. J. (1995). *Science* **269**, 403–407.
122. Price, M. A., Cruzalegui, F. H., and Treisman, R. (1996). *EMBO J.* **15**, 6552–6563.
123. Lee, J. C., Laydon, J. T., McDonnel, P. C., Gallagher, T. F., Kumar, S., Green, D., McNulty, D., Blumenthal, M. J., Heys, J. R., Landvatter, S. W., Strickler, J. E., McLaughlin, M. M., Siemens, I. R., Fisher, S., Livi, G., White, J., Adams, J. L., and Young, P. R. (1994). *Nature (London)* **372**, 739–746.
124. Cuenda, A., Rouse, J., Doza, Y. N., Meier, R., Cohen, P., Gallagher, T. F., Young, P. R., and Lee, J. C. (1995). *FEBS Lett.* **364**, 229–233.
125. Hazzalin, C. A., Cano, E., Cuenda, A., Barratt, M. J., Cohen, P., and Mahadevan, L. C. (1996). *Curr. Biol.* **6**, 1028–1031.
126. Chen, Y. R., Meyer, C. F., and Tan, T. H. (1996). *J. Biol. Chem.* **271**, 631–634.
127. Verheij, M., Bose, R., Lin, X. H., Yao, B., Jarvis, W. D., Grant, S., Birrer, M. J., Szabo, E., Zon, L. I., Kyriakis, J. M., Haimovitz-Friedman, A., Fuks, Z., and Kolesnick, R. N. (1996). *Nature (London)* **380**, 75–79.
128. Wilson, D. J., Fortner, K. A., Lynch, D. H., Mattingly, R. R., Macara, I. G., Posada, J. A. and Budd, R. C. (1996). *Eur. J. Immunol.* **26**, 989–994.
129. Xia, Z., Dickens, M., Raingeaud, J., Davis, R. J., and Greenberg, M. E. (1995). *Science* **270**, 1326–1331.
130. Lassignal Johnson, N., Gardner, A., Diener, K., Lange-Carter, C., Gleavy, J., Jarpe, M., Minden, A., Karin, M., Zon, L., and Johnson, G. (1996). *J. Biol. Chem.* **271**, 3229–3237.
131. Nishina, H., Fischer, K., Radvanyi, L., Shahinian, A., Hakem, R., Rubie, E., Bernstein, A., Mak, T. W., Woodgett, J. R., and Penninger, J. M. (1997). *Nature (London)* **385**, 350–353.
132. Xie, W., and Herschman, H. R. (1985). *J. Biol. Chem.* **270**, 27622–27628.
133. Rodrigues, G. A., Park, M., Schlessinger, J. (1997). *EMBO J.* **16**, 2643–2645.
134. Olson, M. F., Ashworth, A., and Hall, A. (1995). *Science* **269**, 1270–1272.
135. Qiu, R. G., Chen, J., Kirn, D., McCormick, F., and Symons, M. (1995). *Nature (London)* **374**, 457–459.
136. van Leeuwen, F. N., van der Kammen, R. A., Habets, G. G., and Collard, J. G. (1995). *Oncogene* **11**, 2215–2221.
137. Qiu, R. G., Chen, J., McCormick, F., and Symons, M. (1995). *Proc. Natl. Acad. Sci. U.S.A.* **92**, 11781–11785.
138. Prendergast, G. C., Khosravi-Far, R., Solski, P. A., Kurzawa, H., Lebowitz, P. F., and Der, C. J. (1995). *Oncogene* **10**, 2289–2296.
139. Michiels, F., Habets, G. G., Stam, J. C., van der Kammen, R. A., and Collard, J. G. (1995). *Nature (London)* **375**, 338–340.
140. Quilliam, L. A., Khosravi-Far, R., Huff, S. Y., and Der, C. J. (1995). *BioEssays* **17**, 395–404.
141. Khosravi-Far, R., Chrzanowska-Wodnicka, M., Solski, P. A., Eva, A., Burridge, K., and Der, C. J. (1994). *Mol. Cell. Biol.* **14**, 6848–6857.
142. Horii, Y., Beeler, J. F., Sakaguchi, K., Tachibana, M., and Miki, T. (1994). *EMBO J* **13**, 4776–4786.
143. Lee, J. C., and Young, P. R. (1996). *J. Leukocyte Biol.* **59**, 152–157.
144. Beyeart, R., Cuenda, A., Vanden Berghe, W., Plaisance, S., Lee, J. C., Haegerman, G., Cohen, P., and Fiers, W. (1996). *EMBO J.* **15**, 1914–1923.
145. Bourne, H. R., Sanders, D. A., and McCormick, F. (1990). *Nature (London)* **348**, 125–132.
146. Cowley, S., Paterson, H., Kemp, P., and Marshall, C. J. (1994). *Cell (Cambridge, Mass.)* **77**, 841–852.

147. Mansour, S. J., Matten, W. T., Hermann, A. S., Candia, J. M., Rong, S., Fukasawa, K., Vande Woude, G. F., and Ahn, N. G. (1994). *Science* **265**, 966–970.
148. White, M. A., Nicolette, C., Minden, A., Polverino, A., Van Aelst, L., Karin, M., and Wigler, M. H. (1995). *Cell (Cambridge, Mass.)* **80**, 533–541.
149. Joneson, T., McDonough, M., Bar-Sagi, D., and Van Aelst, L. (1996). *Science* **274**, 1374–1476.
150. Lamarche, N., Tapon, N., Stowers, L., Burbelo, P. D., Aspenström, P., Bridges, T., Chant, J., and Hall, A. (1996). *Cell (Cambridge, Mass.)* **87**, 519–529.
151. Greene, L. A., and Tischler, A. S. (1976). *Proc. Natl. Acad. Sci. U.S.A.* **73**, 2424–2428.
152. Wood, K. W., Qi, H., D'Arcangelo, G., Armstrong, R. C., Roberts, T. M., and Halegoua, S. (1993). *Proc. Natl. Acad. Sci. U.S.A.* **90**(11), 5016–5020.
153. Traverse, S., Gomez, N., Paterson, H., Marshall, C., and Cohen, P. (1992). *Biochem. J.* **288**, 351–355.
154. Lackner, M. R., Kornfeld, K., Miller, L. M., Horvitz, H. R., and Kim, S. K. (1994). *Genes Dev.* **8**, 160–173.
155. Wu, Y., and Han, M. (1994). *Genes. Dev.* **8**, 147–159.
156. Kayne, P. S., and Sternberg, P. W. (1995). *Curr. Opin. Genet. Dev.* **5**, 38–43.
157. Aroian, R. V., Koga, M., Mendel, J. E., Ohshima, Y., and Sternberg, P. W. (1990). *Nature (London)* **348**, 693–699.
158. Hill, R. J., and Sternberg, P. W. (1992). *Nature (London)* **358**, 470–476.
159. Horvitz, H. R., and Sternberg, P. W. (1991). *Nature (London)* **351**, 535–541.
160. Simon, M. A., Bowtell, D. D., Dodson, G. S., Laverty, T. R., and Rubin, G. M. (1991). *Cell (Cambridge, Mass.)* **67**, 701–716.
161. Zipurski, S. L., and Rubin, G. M. (1994). *Annu. Rev. Neurosci.* **17**, 373–397.
162. Rogge, R. D., Karlovich, C. A., and Banerjee, U. (1991). *Cell (Cambridge, Mass.)* **64**, 39–48.
163. Fortini, M. E., Simon, M. A., and Rubin, G. M. (1992). *Nature (London)* **355**, 559–561.
164. Biggs, W. H., Zavitz, K. H., Dickson, B., van der Straten, A., Brunner, D., Hafen, E., and Zipurski, S. L. (1994). *EMBO J.* **13**, 1628–1635.
165. Brunner, D., Oellers, N., Szabad, J., Biggs, W. H., Zipursky, S. L., and Hafen, E. (1994). *Cell (Cambridge, Mass.)* **76**, 875–888.
166. Wasserman, D. A., Therrien, M., and Rubin, G. M. (1995). *Curr. Opin. Genet. Dev.* **5**, 44–50.
167. Sluss, H. K., Han, Z., Barrett, T., Davis, R. J., and Ip, Y. T. (1996). *Genes Dev.* **10**, 2745–2758.
168. Riesgo-Escovar, J. R., Jenni, M., Fritz, A., and Hafen, R. (1996). *Genes Dev.* **10**, 2759–2768.
169. Canman, C. E., and Kastan, M. B. (1996). *Nature (London)* **384**, 213–214.
170. Hafner, S., Adler, H., Mischak, H., Janosch, P., Heidecker, G., Wolfman, A., Pippig, S., Lohse, M., Ueffing, M., and Kolch, W. (1994). *Mol. Cell. Biol.* **14**, 6696–6703.
171. Burgering, B. M., Pronk, G. J., van Weeren, P. C., Chardin, P., and Bos, J. L. (1993). *EMBO J.* **12**, 4211–4220.
172. Graves, L. M., Bornfeldt, K. E., Raines, E. W., Potts, B. C., MacDonald. S. G., Ross, R., and Krebs, E. G. (1993). *Proc. Natl. Acad. Sci. U.S.A.* **90**, 10300–10304.
173. Hordijk, P. L., Verlaan, I., Jalink, K., van Corven, E. J., and Moolenaar, W. H. (1994). *J. Biol. Chem.* **269**, 3534–3538.
174. Russell, M. S., Winitz, S., and Johnson, G. L. (1994). *Mol. Cell. Biol.* **14**, 2343–2351.
175. Sevetson, B. R., Kong, X., and Lawrence, J. C., Jr. (1993). *Proc. Natl. Acad. Sci. U.S.A.* **990**, 10305–10309.
176. Wu, J., Dent, P., Jelinek, T., Wolfman, A., Weber, M. J., and Sturgill, T. W. (1993). *Science* **262**, 1065–1069.
177. Lin, T. A., Kong, X., Haystead, T. A., Pause, A., Belsham, G., Sonenberg, N., and Lawrence, J. C., Jr. (1994). *Science* **266**, 542–543.

Pallavi R. Devchand
Walter Wahli

Institut de Biologie Animale, Bâtiment de Biologie
Université de Lausanne
CH-1015 Lausanne, Switzerland

PPARα: Tempting Fate with Fat

An imbalance in energy and lipid homeostasis is symptomatic of prevalent diseases such as diabetes, obesity, cardiovascular disease, hyperlipidemia, and some inflammatory disorders. Peroxisome proliferator-activated receptors (PPARs) are transcription factors that regulate the metabolism of lipids, eicosanoids, and many xenobiotics. We have recently reported the PPARα subtype as a nuclear receptor for eicosanoids and hypolipidemic drugs. This review reevaluates what we know about PPARα in the context of three of its functions: control of inflammation, detoxification, and lipid lowering. We propose a simple model in which diverse compounds interact directly with PPARα to determine their own fate and to modulate overall lipid metabolism. Although we are only beginning to understand the mechanisms and functions of this transcription factor, it is already quite clear that PPARα is a good candidate for drug and diet intervention in many metabolic disease states.

A major preoccupation of our society today is the link between nutrition and health. With the advent of the fast-food lifestyle, industrialized countries have to contend with many consequences of high-fat diets. Obesity, hyperlipidemia, and the resulting cardiovascular disease are among the prominent disorders that are associated with an imbalance between lipid storage, production, and degradation. While fats have gained notoriety for their adverse effects, some fatty acids, such as polyunsaturated fatty acids (PUFAs), are beneficial for the health. High-PUFA diets can act as anti-inflammatory agents and also effectively slow down nerve degeneration associated with some peroxisomal disorders. In rodents, high-PUFA diets result in proliferation of peroxisomes in the liver and kidney. This phenomenon is also observed as a response to exposure to various xenobiotics and lipid-lowering drugs; hence, the association of peroxisomal proliferation with a detoxification function. Research on ligand-activated transcription factors called Peroxisome Proliferator-Activated Receptors (PPARs) has led to the unveiling of key mechanisms that control many adaptive responses.

We refer readers to extensive reviews devoted to PPARs as members of the nuclear hormone receptor superfamily (Lemberger *et al.*, 1996a; Desvergne and Wahli, 1995) and to the role of PPARγ in adipogenesis (Tontonoz *et al.*, 1995), obesity, and associated diabetes (Spiegelman and Flier, 1996).

We have recently reported the identification of some ligands of PPARα and the potential role of this transcription factor in inflammation, detoxification, and lipid lowering (Devchand *et al.*, 1996). In light of these new findings, this review will focus primarily on PPARα. We will attempt to place PPARα into the context of a network of activities where cross-talk exists at each level: from the multiple functions of peroxisomal enzymes to the promiscuity of transcription factors. We concentrate on the ability of PPARα to trigger induction of the ω- and β-oxidation pathways in response to a broad array of signals, and the implications of this function in different cell types and under varying physiological conditions.

I. PPARα as a Ligand-Activated Transcription Factor

PPARα was first identified from a screen for a receptor that could mediate the pleiotropic response to a group of chemicals called peroxisome proliferators (Issemann and Green, 1990). With the isolation and characterization of the three PPAR subtypes from *Xenopus* came the recognition that PPARs form a distinct subfamily of the nuclear hormone receptor superfamily (Dreyer *et al.*, 1992). To date, the three PPAR subtypes (α, β/δ or FAAR, and γ) have also been isolated from many rodents (Issemann and Green, 1990; Kliewer *et al.*, 1992; Amri *et al.*, 1995; Chen *et al.*, 1993; Zhu *et al.*, 1993; Aperlo *et al.*, 1995) and man (Greene *et al.*, 1995; Schmidt *et al.*, 1992; Sher *et al.*, 1993). The structure of PPARs can be envisioned as two functional regions, the DNA-binding domain (DBD) and the ligand

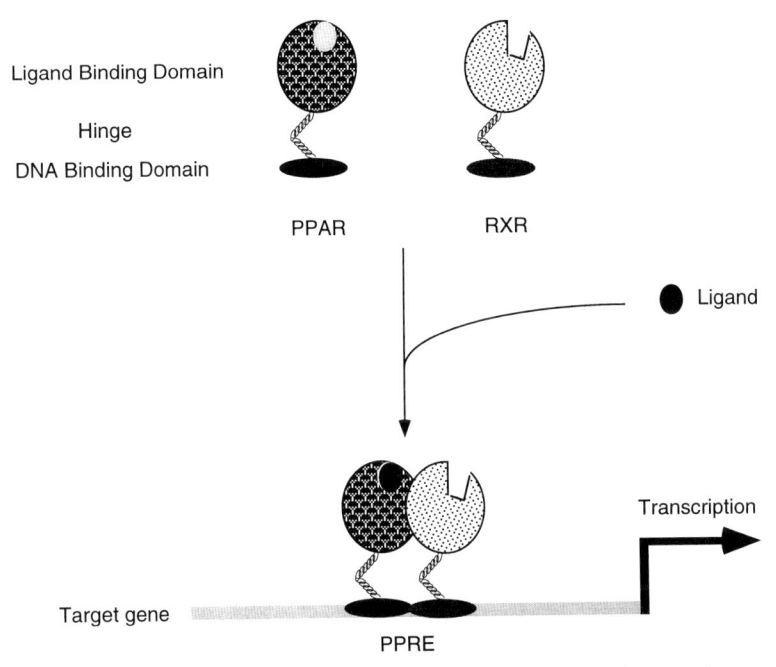

FIGURE 1 Schematic of PPAR action. PPAR forms heterodimers with RXR, the receptor for 9-cis retinoic acid. In the presence of a PPAR ligand, the PPAR–RXR complex activates transcription from a DNA response element (PPRE) in the promoter of target genes.

binding domain (LBD), connected by a hinge region (see Fig. 1). Sequence homology to the thyroid (TR) and retinoic acid (RAR) receptors suggested that PPARs could act as ligand-activated transcription factors. In the presence of ligand, the PPARs exert their effects as heterodimers with RXR, the 9-cis retinoic acid receptor (Kliewer et al., 1992; Keller et al., 1993; Gearing et al., 1993). The PPAR–RXR heterodimers bind to DNA elements termed Peroxisome Proliferator Response Elements (PPREs) and induce transcription from the promoters of target genes (Dreyer et al., 1992; Tugwood et al., 1992). The target genes of PPARs are reflective of, but not limited to, the broad range of homeostatic functions associated with microsomes and peroxisomes (for an expansive list of target genes, see Wahli et al., 1995).

II. Lipid Homeostasis: How Complex Can It Be?

We start with a simple equation to illustrate lipid homeostasis (Fig. 2). The net lipid level in a given cell is equal to the difference between lipid accumulation (uptake and synthesis) and lipid disposal (catabolism and secretion). What are the variables?

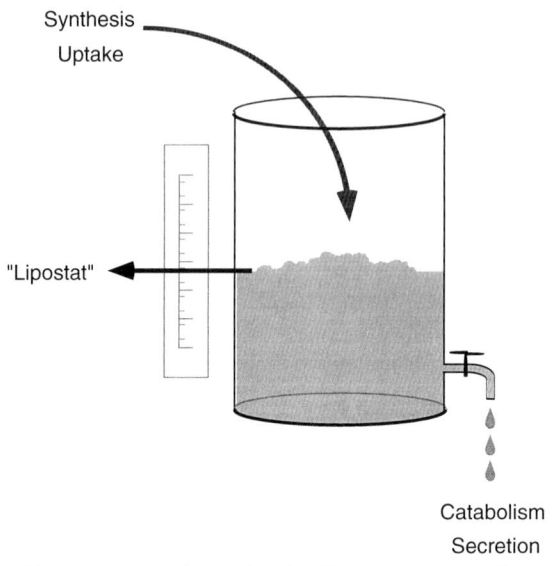

FIGURE 2 Lipid homeostasis made simple. The "lipostat" indicates the capacity for lipids. The amount of lipid in a cell is the difference between accumulation and disposal.

A. Cell and Lipid Type

Obviously, the capacity for total lipid levels differs between cell types. By virtue of its function, an adipose cell has a higher capacity to store fatty acids than a liver or muscle cell. The type of lipid content is important. While adipose cells are high in triglycerides, neural cells contain high levels of very-long-chain fatty acids (VLCFA). Furthermore, since many specialized signaling molecules, such as eicosanoids, are fatty acid derivatives, the ability to produce, secrete, take up, and/or catabolize these molecules varies with cell type. For example, neutrophils both produce and catabolize large amounts of the chemotactic agent leukotriene B_4 (LTB_4), whereas hepatocytes, as part of their detoxification function, will primarily clear and catabolize LTB_4 from the bloodstream (see Ford-Hutchinson, 1990, for review). Thus, the "lipostat" for a given cell is dependent on the cell type, function, and environment. Since PPARs regulate the fate of lipids in the cell, the expression patterns of these transcription factors give an important clue to their tissue-specific functions. Many tissues that express PPARα have a high capacity for fatty acid catabolism (Braissant *et al.*, 1996).

B. Peroxisomes Are Multifunctional Organelles

Peroxisomes represent a hub of metabolic activity ranging from the detoxification of xenobiotics to the biosynthesis of cholesterol (see Masters, 1996 for comprehensive review). We focus only on a few aspects of peroxisomal activities to illustrate some functions of PPARα and to highlight the different levels of cross-talk that are involved in adaptation and homeostasis.

1. Energy Homeostasis

Peroxisomes play a prominent role in both anabolic and catabolic pathways of lipid metabolism. In terms of energy homeostasis, one might reduce the PPAR system to a simple balance between anabolic events potentiated by PPARγ and catabolic events that are induced by PPARα. The expression patterns, target genes, activators, and ligands of these transcription factors are consistent with this view: PPARγ is expressed primarily in adipose tissue where it potentiates storage and also acts as a master gene in adipogenesis (Tontonoz et al., 1994), and PPARα is highly expressed in many tissues that characteristically catabolize fatty acids and their derivatives (Braissant et al., 1996). The contribution of the third isotype, PPARβ, to energy homeostasis is not yet reported.

2. Many Functions of the Peroxisomal β-Oxidation Pathway

The peroxisomal β-oxidation pathway (Fig. 3) is used in the catabolism of a variety of molecules (Fig. 4) ranging from many types of fatty acids (e.g., branched, carboxylic, dicarboxylic, saturated, and polyunsaturated fatty acids) to important fatty acid derivatives such as the eicosanoid signaling molecules (e.g., prostanoids and leukotrienes) to a variety of xenobiotic compounds including structurally unrelated lipid-lowering drugs. It follows, then, that this catabolic pathway presents a basis for cross-talk between dietary fatty acids, processes that are regulated by eicosanoids (like inflammation), and also detoxification of xenobiotic compounds.

The reaction catalyzed by acyl coA-oxidase (ACO) is the rate-limiting step of the β-oxidation pathway. Expression of the ACO protein seems to be regulated primarily at the transcriptional level as the levels of mRNA are indicative of protein levels in the cell (Lemberger et al., 1996b). Thus monitoring of the ACO mRNA levels is reflective of the levels of the β-oxidation pathway. A primary function of PPARα is to induce transcription of the peroxisomal β-oxidation pathway. Given that this pathway represents the junction for cross-talk between catabolism of fatty acids, their eicosanoid derivatives, and also many chemical toxins, the array of signals that trigger PPARα activity is indeed remarkable.

3. Redundancy of Function

The peroxisome is far from an isolated organelle. There are many levels of interaction with other cellular compartments, both in terms of overlap of function and shuttling of metabolites. As an example we follow the catabolism of very long chain fatty acids (VLCFAs). The VLCFAs undergo ω-oxidation in the microsomes and are then imported into the peroxisomes. After many rounds of the peroxisomal β-oxidation pathway, the substrates

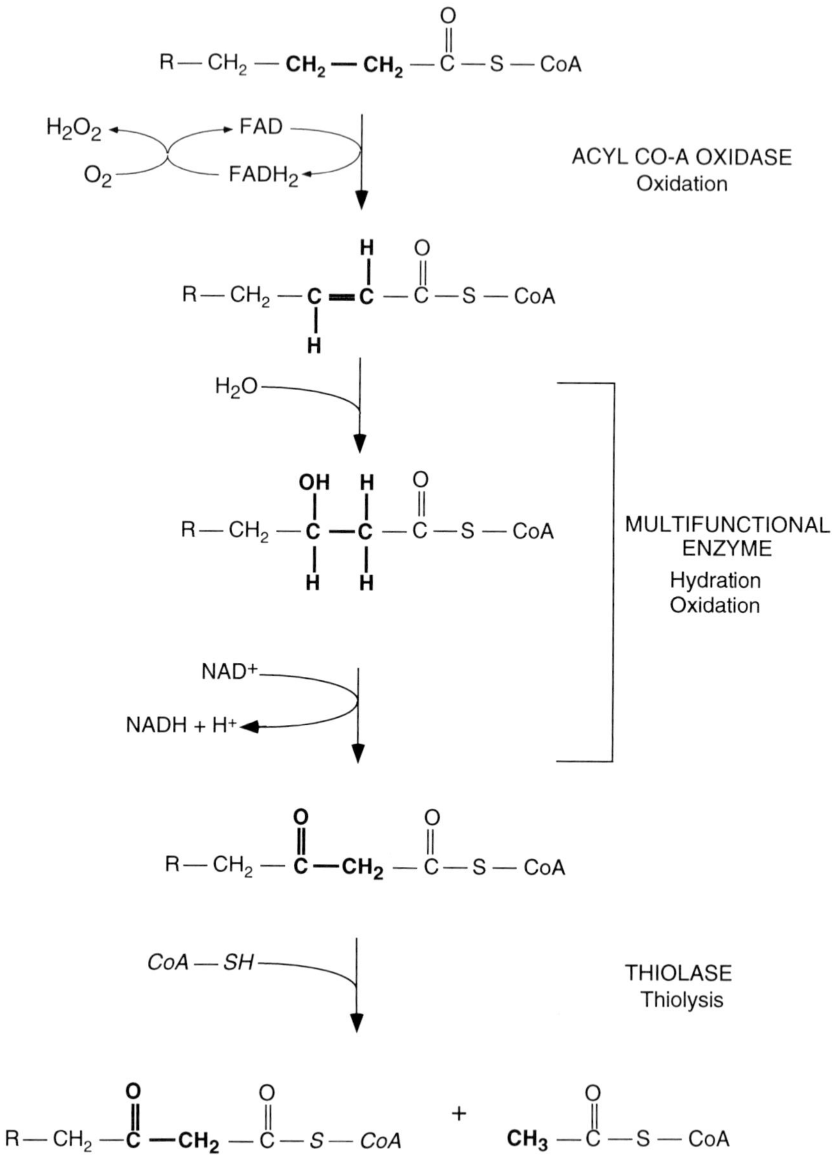

FIGURE 3 The peroxisomal β-oxidation pathway.

are catabolized to medium-chain fatty acids (MCFAs; $C \geq 12$). Depending on cell context, these MCFAs may have many fates: they may be shuttled to the mitochondria where they undergo further β-oxidation by a similar enzymatic pathway, or they may be used for the production of ketone bodies or, alternatively, for the synthesis of complex lipids via the formation of

FATTY ACIDS

Saturated fatty acids

Palmitic acid

Monounsaturated fatty acids

Oleic acid

Polyunsaturated Fatty acids (PUFAs)

Arachidonic Acid

Branched fatty acids

Phytanic acid

Dicarboxylic fatty acids

Hexadecanedioic acid

XENOBIOTICS

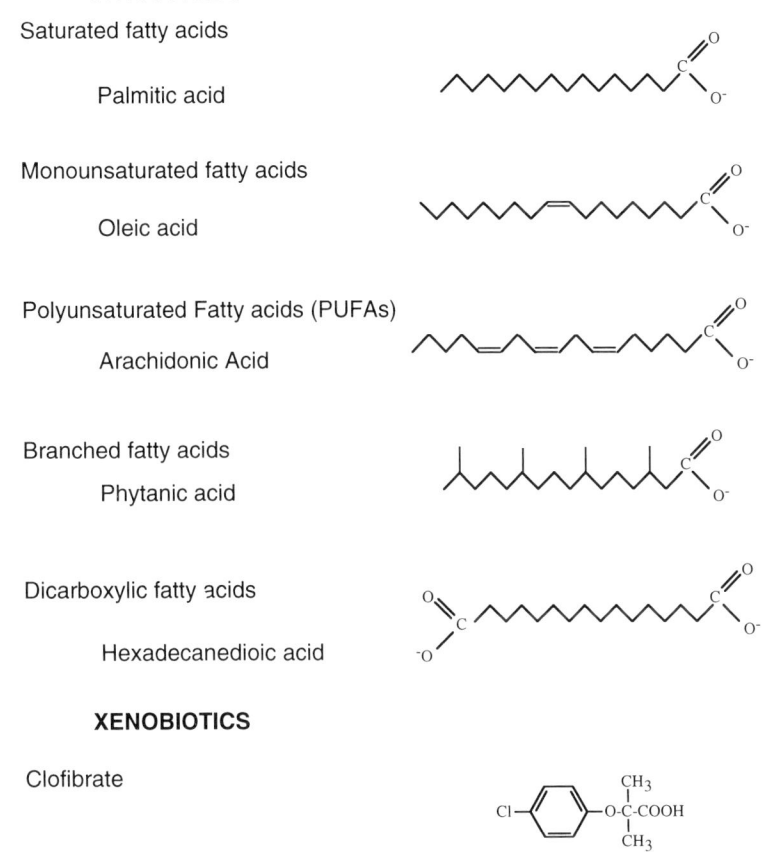

Clofibrate

FIGURE 4. Some substrates for the ω- and β-oxidation pathways.

acetyl coA. The complexity of this cross-talk at the level of enzymatic function is also apparent at the transcriptional level, as indicated by the target genes of PPARα.

C. Cross-Talk at the Transcription Factor Level

The involvement of PPARs in energy homeostasis extends beyond the simple monitoring and modulation of fatty acids. For the system to be functional, there must be considerable cross-talk from other pathways involving energy mobilization or storage. The links that have been identified to date reveal an intricate and elegant intertwining of signaling pathways at various levels. Below, we highlight only a few examples of the types of mechanisms involved.

The activity of PPARα can be modulated directly either at the transcription level or at the protein level. For example, glucocorticoids, hormones that are indicative of stress conditions, up-regulate the expression of PPARα mRNA (Lemberger et al., 1996b). The net result would be the desired increase in mobilization of fat via enhanced PPARα activity. In the presence of insulin, this effect of glucocorticoids on PPARα is reduced (Steineger et al., 1994). Recent studies have also shown that insulin modulates the activity of the PPARγ subtype via phosphorylation of the nuclear hormone receptor (Hu et al., 1996). The most apparent display of cross-talk between the glucose and lipid signaling pathways is in the finding that the antidiabetic drugs, thiazolidinediones, are activators and ligands for PPARγ (Lehmann et al., 1995; Forman et al., 1995).

A significant level of cross-talk occurs at the level of the active transcription complex. PPARα activates transcription as a heterodimer with RXR (see Section I). Inherent in this system is cross-talk between the retinoid and lipid signaling pathways. The three PPAR subtypes exhibit overlapping expression patterns and can influence each other's activities. For example, overexpression of PPARβ can effectively down-regulate PPARα-mediated transcription (Jow and Mukerjee, 1995). Furthermore, any other nuclear hormone receptor that interacts with RXR, for example the thyroid hormone receptor (TR; Chu et al., 1995; Juge-Aubry et al., 1995), will indirectly modulate the availability and formation of active PPARα–RXR complexes.

Recent studies also indicate synergism between the CAAT-Enhancer Binding Proteins (C/EBP) and PPAR transcription factors during adipogenesis (Wu et al., 1996). Many other tissues, like the liver, express different combinations of C/EBP and PPAR subtypes, suggesting that the communication between the two families of transcription factors is probably not restricted to adipose cells.

We are only beginning to understand just how much communication takes place between the different signaling pathways at the level of transcription factors. This review is biased toward PPARα and its role in energy homeostasis and adaptation in the adult. With time, we will no doubt explore the mechanisms and dynamics required for channeling of lipids during development.

III. PPARα Expression during Development and Adulthood

Northern blot analyses indicate that in *Xenopus*, PPARα is expressed throughout oogenesis (Dreyer et al., 1992). The maternal transcripts persist up to the gastrula stages and zygotic transcripts start to accumulate at tailbud stages. Developmental expression patterns of PPARα in mammals have yet to be documented. In the adult of *Xenopus* and rodents, PPARα is expressed primarily in the liver, kidney, brown adipose tissue, eye, digestive tract, and immune and genital systems. However, it should be noted that no tissue strictly expresses only one single subtype of PPAR (Braissant et *al.*, 1996). Since PPARs are involved in the dynamics of homeostasis, one might expect that the expression of the different PPARs varies with physiological and dietary conditions. Indeed PPARα expression modulates with adaptation events ranging from diurnal rhythm to more extreme conditions like stress and fasting (Lemberger *et al.*, 1996b).

IV. PPARα and Its Activators: Open Relationships?

Progress in the identification of activators has been rapid. The first rounds of activator screening were based on the association of PPARs with peroxisomal function and also on the ability of PPARs to mediate the response to xenobiotics such as fibrates and other structurally unrelated peroxisome proliferators. These methodical searches have for the most part been very fruitful. Unfortunately, they have also produced confusing and often misleading data that primarily reflect the nature of the PPAR activators themselves, the assay systems used, or a combination of both.

A. Assays for PPAR Activators

Transient transfection assays are the standard method for the evaluation of PPAR responsiveness to various compounds. The theory behind this assay is simple (Fig. 5). Cells are transfected with three plasmids: an expression plasmid that produces a sensor for the ligand, a reporter plasmid that allows measure of the response from the sensor, and an internal standard to normalize between experiments. In this case, the sensor PPAR is a ligand-dependent transcription factor, and therefore the amount of reporter produced is indicative of the potency of the PPAR activator present. The sensitivity of such a screening assay is inherently dependent on cell type and reporter system. For stable compounds such as some synthetic peroxisome proliferators, sensitivity of the assay system is usually not an issue. However, for chemically and metabolically unstable compounds such as prostaglandins and leukotrienes, the sensitivity of the assay determines the success of the screen. Thus the assay systems developed in different research groups vary in almost all aspects: cell type, internal standard, detection, and reporter systems (Fig. 5b). Despite the use of different systems, in the final analysis there is a general consensus on most of the PPAR activators identified to date.

These transfection assays identify compounds that can activate PPARs through one or many mechanisms: (i) directly, by binding to the nuclear receptor as a ligand; (ii) indirectly, through the production of a ligand or secondary modification of the receptor; or (iii) by a combination of both direct and indirect events. Distinguishing between the direct and indirect mechanisms of PPAR activation requires some kind of ligand identification assay.

a

SIGNAL

SENSOR

INTERNAL STANDARD

transcription

sensor binding site

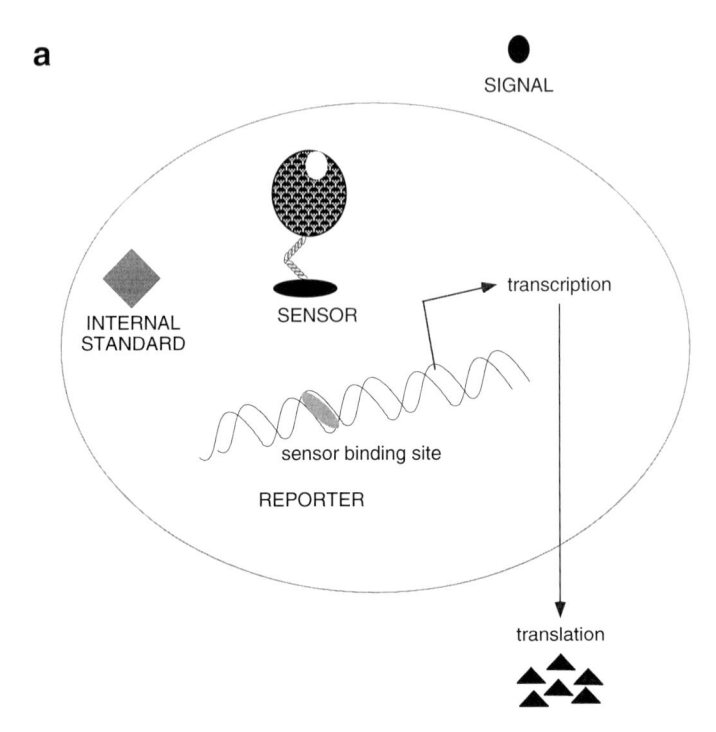

REPORTER

translation

b

SENSOR	REPORTER		CELL TYPE	INTERNAL STANDARD	REFERENCE
	Sensor Binding Site	Gene			
FULL LENGTH PPAR	PPRE containing:				
	-promoter fragment eg.ACO	CAT	HeLa	β-Gal	Dreyer et al, 1992
	aP2	CAT	CV-1	luc	Lehmann et al, 1995
	-oligomer eg.CYP4A6Z-Pal	CAT	HeLa	luc	Devchand et al, 1996
FUSION PROTEINS:					
Gal4 $_{DBD}$- PPAR$_{LBD}$	5xGal4-	CAT	CV-1	luc	Lehmann et al, 1995
TetR-PPAR$_{LBD}$	TetO	luc	U205		Yu et el, 1995
ER $_{DBD}$- PPAR$_{LBD}$	ERE	CAT	HeLa	β-Gal	Dreyer et al, 1992
GR $_{DBD}$- PPAR$_{LBD}$	MMTV	PAP	CHO	CAT	Gottlicher et al, 1992

FIGURE 5 (a) Transfection assays for PPAR activity. Cells are transfected with three constructs encoding a sensor, reporter, and internal standard. In the presence of a signal, the sensor is activated, binds to the sensor binding site, and induces transcription of the reporter gene. The effect of the signal on sensor activity is quantitated by evaluation of the reporter product. An internal standard is used to normalize transfection efficiency between different experiments. (b) Different transfection systems developed for PPAR. Examples of PPAR transfection systems used by different research groups. The systems vary in every aspect: cell type, sensor–reporter pairs, and internal standards.

B. Fatty Acids

Many groups have reported the activation of PPARs by a broad spectrum of fatty acids (Göttlicher *et al.*, 1992, 1993; Keller *et al.*, 1993; Kliewer *et al.*, 1994; Yu *et al.*, 1995). Activation usually requires high micromolar concentrations of fatty acids. Physiologically, these concentrations are relevant. However, technically this can pose problems since very high concentrations of some fatty acids are toxic or can lead to detachment of cells from the culture dishes.

All three PPAR subtypes can be activated by fatty acids. PPARα is the least discriminating and the most responsive: it can be activated by the two essential fatty acids (linoleic and linolenic acid), ω-3 and ω-6 polyunsaturated fatty acids (including arachidonic acid), and some monounsaturated fatty acids. In terms of total activation, PPARβ and PPARγ do respond, but to a lesser extent. Saturated and dicarboxylic fatty acids are not effective activators of PPARs. The activation of PPARs by fatty acids supports the role of PPARs as key regulators in lipid homeostasis: they sense fatty acid levels and activate the appropriate downstream processes. Whether fatty acids are the ligands for PPARs, and thus act as hormones that direct their own fate, is an interesting question that remains to be answered (see *Note added in proof*).

C. Channeling Arachidonic Acid

The metabolic fate of arachidonic acid is determined by three major pathways: cyclo-oxygenase, lipoxygenase, or epoxygenase. The net result is the production of important metabolites like prostanoids, lipoxins, and leukotrienes, which signal a spectrum of functions ranging from apoptosis to recruitment of the immune system to cellular differentiation. It is not surprising, then, that production, uptake, response, and degradation of these signals are cell- and tissue-specific functions. The first problem with evaluating these compounds as PPAR activators in transfection assays is evident: the uptake, metabolic fate, and stability of each compound will depend upon the cell-type being used. In other words, two different cell types using similar reporter systems will potentially produce conflicting results for the same compound: the compound might score as an activator in one cell line where its uptake is efficient or the environment renders it metabolically stable, whereas no activation will be detected in a different cell line where the compound is degraded rapidly or its uptake is inefficient. To further complicate the situation many of these compounds are chemically unstable. Thus, if a reporter system is not sensitive enough, an activation response might not be apparent.

1. Metabolic Inhibitors

Many anti-inflammatory medications and other synthetic drugs have been developed to target key enzymes in the arachidonic acid metabolic pathway. Theoretically, these drugs can be used to inhibit the cyclo-oxygenase, lipoxygenase, epoxygenase pathways or a combination thereof, and effectively favor the formation of selected arachidonic acid metabolites (Fig. 6). These types of compounds have been used in attempts to identify which arm(s) of arachidonic acid catabolism results in the activation of PPARs (Keller et al., 1993; Yu et al., 1995). At face value, the logic is simple. In transfection assays, cells are exposed to arachidonic acid (or a precursor), with or without a given inhibitor. If the response obtained is lower in the presence than in the absence of the inhibitor, then the pathway inhibited is involved in the activation of PPAR. The results from these types of experiments have been baffling. For example, exposure of cells to ETYA, an inhibitor of both cyclo-oxygenase and lipoxygenase pathways, results in appreciable activation of PPARs, even without the external addition of arachidonic acid (Keller et al., 1993). More surprisingly, despite the fact that both PPARα and -γ can be activated by prostaglandins (Yu et al., 1995; Kliewer et al., 1995; Forman et al., 1995), no reduction in activation of PPARα or -γ is observed when indomethacin, a cyclo-oxygenase inhibitor, is used to channel arachidonic acid catabolism away from prostaglandin synthesis (Yu et al., 1995). The whole scenario seems incredibly messy and easy to dismiss. However, the information obtained is far from trivial. A recent report partially clarifies the unforeseen complication in these experi-

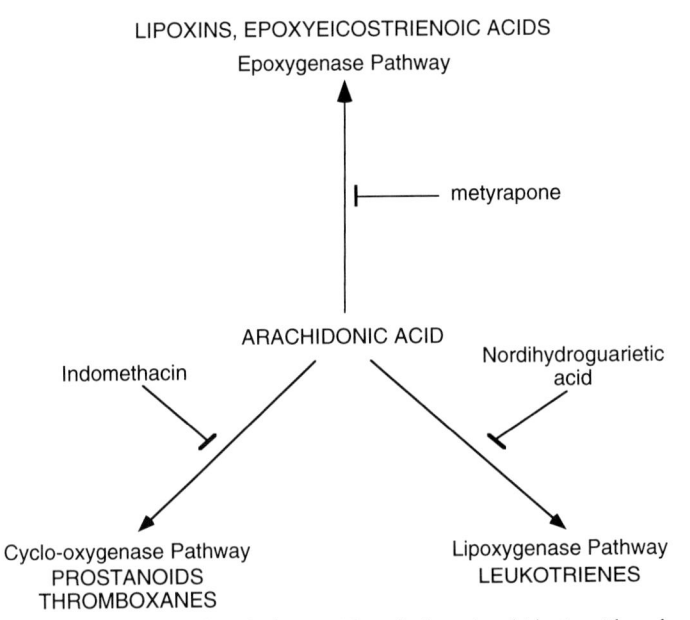

FIGURE 6 Main pathways of arachidonic acid catabolism. Arachidonic acid can be catabolized to form important signaling molecules. Examples of synthetic inhibitors of the three pathways are indicated.

ments: many cyclo-oxygenase inhibitors such as some NSAIDs (nonsteroid anti-inflammatory drugs), are themselves PPARγ ligands and activators (Lehmann *et al.*, 1997). Thus, even though a given NSAID channels arachidonic acid away from the production of activators, the drug itself activates PPARγ. Hence the observed result: the NSAID does not reduce the total activation of PPARγ by arachidonic acid. In light of the recently identified role of PPARα in the inflammatory response, we can certainly expect that many anti-inflammatory drugs and perhaps other metabolic inhibitors are also ligands and activators of PPARα.

2. Arachidonic Acid Metabolites

The sensitivity of an activation assay system and subtle variations between different systems, such as cell types and reporters, are two important factors in the evaluation of arachidonic acid metabolites. Many groups have reported the differential activation of the three PPARs by prostaglandins (PGs; see Fig. 7). These results were observed from experiments in two cell lines: monkey kidney cells (CV-1; Forman *et al.*, 1995; Kliewer *et al.*, 1995) and human osteosarcoma cells (U2OS; Yu *et al.*, 1995). While all three PPAR subtypes (α, β, and γ) can be activated by PGA_1, PGA_2, PGD_1, and PGD_2, the selectivity becomes more apparent with metabolites further downstream: PGJ_2 preferentially activates PPARα and -γ; and the derivative 15d-$\Delta^{12,14}$-PDJ_2 activates PPARγ more effectively than PPARα. These prosta-

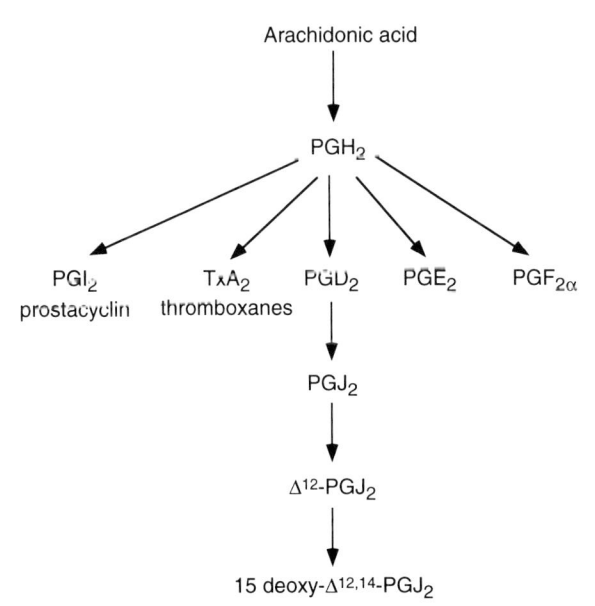

FIGURE 7 Cyclo-oxygenase pathways. The PGD_2 pathway which has been shown to be involved in PPAR activation is delineated (Kliewer *et al.*, 1995; and Forman *et al.*, 1995).

glandins are only effective activators in micromolar concentrations. Carbaprostacyclin (cPGI) has been reported to activate PPARα, -β, and also -γ to a lesser extent (Hertz et al., 1996; Brun et al., 1996). In HeLa cells, the stable cPGI analogue and anti-inflammatory agent iloprost activates PPARα preferentially (Devchand et al., 1996). Activation by iloprost or cPGI can be detected in the nanomolar range.

Evaluation of products of the lipoxygenase pathway (Fig. 8) as PPAR activators has highlighted the differences in assay systems between different research groups. The most important difference is probably the cell types used, since the sensitivity of these systems has proved sufficient in identifying activation by various prostaglandins. In U2OS cells, the inflammatory mediator 8-HETE preferentially activates PPARα in a stereospecific manner; activation is observed with the S but not the R enantiomer (Yu et al., 1995). In the same assay system, activation is not observed with the 5, 11, or 15 HETEs. Activation of PPARs by leukotrienes has been observed in HeLa but not CV-1 cells (Devchand et al., 1996; Forman et al., 1995), although

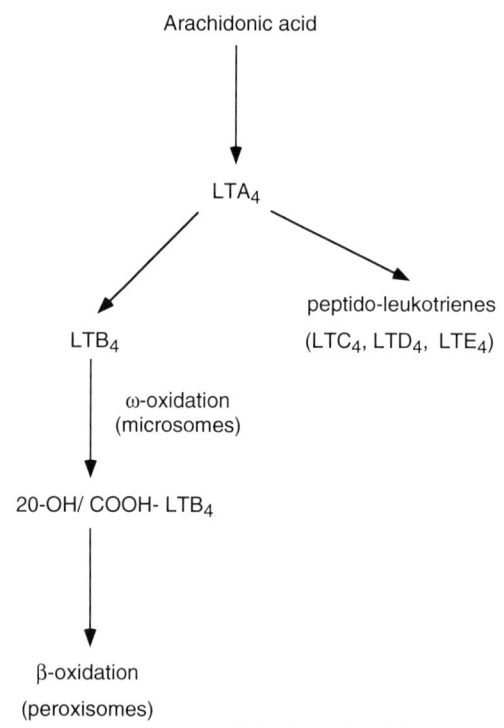

Arachidonic acid
↓
LTA$_4$
↙   ↘
LTB$_4$ peptido-leukotrienes
              (LTC$_4$, LTD$_4$, LTE$_4$)
↓ ω-oxidation
(microsomes)
20-OH/ COOH- LTB$_4$
↓
β-oxidation
(peroxisomes)

FIGURE 8 The 5-lipoxygenase pathway. LTA$_4$ is produced from arachidonic acid and can be further catabolized to either the peptido-leukotrienes or the potent chemotactic agent LTB$_4$. Induced catabolism of LTB$_4$ via the ω- and β-oxidation pathways can be mediated by PPARα (Devchand et al., 1996).

it is noteworthy that in CV-1 cells, the leukotriene D4 receptor antagonist LY 171883 does activate PPARs (Kliewer et al., 1994). So far, one leukotriene, LTB$_4$, has been reported as a preferential activator of PPARα (Devchand et al., 1996). Analogous to the prostaglandins, activation by LTB$_4$ occurs in the micromolar range. In the same assay system, no significant activation was observed with either of the other PPAR subtypes, β or γ. Furthermore, the LTB$_4$ ω-oxidation metabolites, 20-OH and 20-COOH LTB$_4$, do not result in significant activation of PPARα.

Despite the hurdles of working with arachidonic acid metabolites, including technical difficulties and cost, significant progress has been made over a very short period of time. We can confidently expect that more PPAR effectors will be added to the above list, and that these compounds will lead us to some unexpected functions of PPARs.

D. Peroxisome Proliferators

A compound is classified as a peroxisome proliferator (PP) if, when fed to an organism, it induces a massive increase in the number of peroxisomes in hepatic and/or renal tissue (for review, see Desvergne and Wahli, 1995). In some cases prolonged exposure to the compound leads to the formation of liver tumors. Many natural and synthetic compounds are PPs. This list can be roughly classified into four groups: (i) toxic substances such as herbicides, plasticizers, solvents, and industrial compounds; (ii) dietary polyunsaturated fatty acids (PUFAs); (iii) hypolipidemic drugs such as fibrates and pirinixic acid; and (iv) anti-inflammatory agents such as aspirin. There is considerable overlap between these functional groups; for example, some dietary PUFAs are PPs that are also associated with anti-inflammatory and hypolipidemic functions. This strongly suggests that a common mechanistic link(s) exists between peroxisome proliferation, dietary fats, hypolipidemic agents, and anti-inflammatory drugs.

Many fibrates and xenobiotic hypolipidemic drugs preferentially activate PPARα in transient transfection assays (Dreyer et al., 1992). However, PPARβ and -γ also respond to these PPs, albeit to a lower extent. It should be noted, however, that these PPs are not very potent activators of PPARα: activation often requires millimolar quantities of the compound. The PPAR activation profiles of dietary fatty acids and anti-inflammatory drugs have been discussed above.

E. Thiazolidinediones

Thiazolidinediones (TZDs) are a class of anti-diabetic drugs that preferentially activate the PPARγ subtype (Lehmann et al., 1995). Of the TZD compounds reported, the most potent PPARγ activator is BRL49653, a compound which, at higher doses, can induce some activation of PPARα.

F. Species Differences

In general, the activation of the three PPAR subtypes by natural compounds is conserved between the different species tested: *Xenopus,* rodents, and man. This is consistent with conservation of the biological functions of PPAR from lower vertebrates to man. In contrast, activation by synthetic compounds is not necessarily as consistent: although the subtype specificity is usually maintained, the effectiveness of a compound often differs by as much as 100-fold between species. One manifestation of this difference is that many xenobiotic compounds induce peroxisome proliferation in a species-dependent manner. Thus, when targeting PPARs, species-dependent differences are an important factor to be considered during both drug development and drug testing. One might take advantage of these species differences to elucidate structure–function relations in PPAR receptors.

G. Summary of PPARα Activation Profile

The transfection assays have revealed many important points. The relationships between PPARα and its activators are for the most part open on both sides. PPARα is a promiscuous nuclear hormone receptor; it responds to many structurally different compounds such as dietary fatty acids, antiinflammatory agents, some arachidonic acid metabolites, many hypolipidemic drugs, peroxisome proliferators, and other xenobiotics. The compounds themselves can also potentially activate more than one PPAR subtype. The activation profiles indicate that the nuclear receptors are phamacologically distinct, but there is still tremendous overlap. So, even though PPARα responds better to some compounds, this activation is exclusive to the PPARα subtype only within a given concentration range. Consistent with this, PPARα also responds to a lesser extent to TZDs, the antidiabetic drugs that show a preference for PPARγ. However, certain activators do show a more restricted activation pattern (for example, the chemotactic inflammation mediators LTB_4).

Admittedly, the degree of promiscuity between the PPAR subtypes and their activators was unexpected. However, the two main messages from these activation studies are certainly not a surprise. First, at the simplest level, the transfection analyses emphasize that the PPARα activation response in a given cell is a result of both the relative amounts of each PPAR subtype and the combination of PPAR activators present. In other words, activation of PPARα is dependent on cell context. Second, the activators themselves provide an avenue of cross-talk between PPARα and the other PPAR subtypes.

V. PPARα Ligands

We have recently reported two ligands for PPARα: the inflammation mediator, leukotriene B$_4$, and a hypolipidemic drug, Wy 14,643. (Fig. 9a; Devchand *et al.*, 1996). These are structurally unrelated ligands suggesting a high degree of flexibility in the PPARα receptor or the ligand, or a combination of both. As indicated above, many additional compounds activate PPARα. Is each activator a PPARα ligand, or is the activation a result of a secondary effect such as the accumulation of a PPARα ligand via metabolism of the test compound? Finding answers to these questions has, for the most part, been difficult.

The most obvious approach would be direct binding assays of radiolabeled ligand to purified fusion protein (Fig. 9b). Inevitably, there is the standard technical setback associated with the overexpression of proteins in bacteria: the mammalian PPARα fusion proteins have a tendency to be insoluble. Second, and probably the more intimidating, is the nature of the PPARα activators themselves. Based on the activation profiles, the potential synthetic ligands identified to date are not readily available in the radiolabeled form. Many would presumably have a low binding affinity (e.g., clofibrate, EC$_{50}$ in transfection assays in the millimolar range). Furthermore, many arachidonic acid metabolites are chemically unstable and expensive.

Undoubtedly, we will find a way around the technical and costly problems. A high-affinity stable synthetic PPARα ligand would be one route to the analysis of PPARα activators by classical binding assays. An alternate route would be the development of some kind of ligand-dependent semifunctional assay. Obviously, the search for further natural and synthetic ligands of PPARα is intense and industrious—we need not look too far into the future for some more interesting answers (see *Note added in proof*).

VI. PPARα Functions

The activation profiles, ligands, and target genes of PPARα predict an exciting future. Progress in identifying functions of PPARα has been rapid and greatly facilitated by the PPARα knock-out mouse (Lee *et al.*, 1995). Below, we summarize the published reports to date on some surprising functions of PPARα.

A. Evaluation of PPARα Function *in Vitro*

Two systems have been established as *in vitro* assays for PPARα function.

1. Inducible β-Oxidation

PPARα is expressed in the liver, where it plays an active role in many homeostatic processes. One global function of PPARα is to mediate inducible

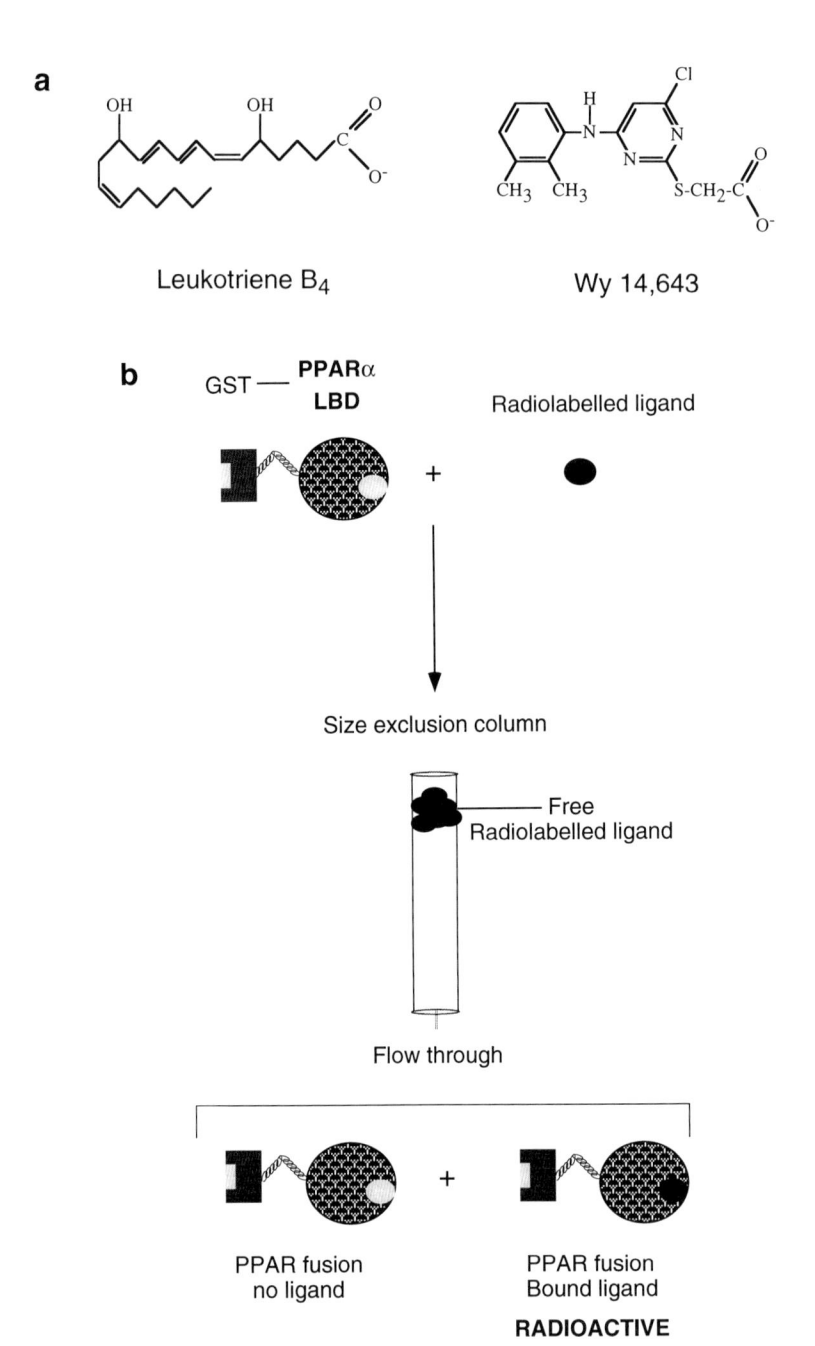

FIGURE 9 (a) Chemical structures of PPARα ligands. The two PPARα ligands identified to date, LTB₄ and Wy 14,643 (also called pirixinic acid), have very different structures. (b) Classical binding assays. Bacterially expressed fusion proteins (for example, GST-PPAR$_{LBD}$) are incubated with radiolabeled ligand. Bound ligand is separated from free ligand on a size exclusion column: the free ligand stays on the column, whereas all the protein is eluted in the void volume. Radiolabeled ligand bound to the protein is quantitated by scintillation counting.

β-oxidation at the transcription level. A reliable and standard evaluation of PPARα activity is the measurement of ACO mRNA (see Section II.B.2) by RNase protection assays in rat primary hepatocytes (Lemberger *et al.*, 1996b).

2. Adipogenesis

In cell culture, PPARα can induce the adipogenesis program, albeit less efficiently than PPARγ (Brun *et al.*, 1996). Although PPARα is not expressed at appreciable levels in white adipose tissue, the ability of PPARα to induce the adipogenesis program in other tissues might be reflective of some pathophysiological conditions.

B. PPARα Knock-Out Mice

PPARα knock-out mice were generated by targeted disruption of the ligand binding domain (Lee *et al.*, 1995). These mice develop normally, suggesting either that PPARα does not play a prominent role during development or that an alternate mechanism compensates for the lack of PPARα activity. Under laboratory conditions, the mice are apparently healthy and reproduce normally. It is always difficult to analyze the "defect" in knock-out mice when there is no overt phenotype. In many cases, "no overt phenotype" is synonymous with "functional redundancy." The lack of apparent abnormalities might at first be disheartening. Fortunately, the disappointment is short-lived, and we begin to discover the many hidden treasures of the PPARα knock-out mouse.

1. Peroxisome Proliferation

Analyses of gene expression in the liver indicate that PPARα knock-out mice do not display the normal pleiotropic response to PPs such as clofibrate and Wy14,643 (Lee *et al.*, 1995). Although they exhibit constitutive basal levels of enzymes involved in the ω- and β-oxidation pathways for fatty acid degradation, these mice are unable to further increase expression of these enzymes when exposed to PPs. As a result, when fed on a diet containing PPs, these mice accumulate lipid droplets in the liver (Lee *et al.*, 1995). This strongly supports a role for PPARα in the maintenance of hepatic lipid homeostasis.

2. The PPARα Avenue for Lipid Lowering

Many peroxisome proliferators can act as hypolipidemic (lipid-lowering) drugs. Wy 14,643 is the first such drug reported as a ligand for PPARα (Devchand *et al.*, 1996). Based on this finding, it is conceivable that the lipid-lowering effects of drugs like Wy 14,643 are a final result of the activation of PPARα, which in turn up-regulates transcription of a network of genes involved in many aspects of lipid metabolism. Together with the expression of PPARα in many cell and tissue types (Braissant *et al.,* 1996), this suggests that the role of PPARα in lipid homeostasis extends beyond the hepatic system. Biochemical analyses will reveal if the PPARα knock-out mice have abnormal lipid levels in the blood, and whether these mice do indeed have difficulty in maintaining homeostasis under various physiological and dietary conditions.

3. Nuclear Eicosanoid Receptor That Controls Inflammation

PPARα binds to the natural eicosanoid LTB_4 and induces transcription of genes involved in the degradation of this chemotactic inflammatory agent, the ω- and β-oxidation pathways (Fig. 10; Devchand *et al.,* 1996). Consistent with this finding, PPARα knock-out mice show a prolonged inflammatory response when challenged by LTB_4 or its precursor, arachidonic acid. Thus the lack of ability to induce the ω- and β-oxidation pathways is not restricted to the liver and PPs, but is potentially a global defect associated with the response to many PPARα-specific activators and ligands (see Section V).

The model for feedback mechanism predicts that some ligands of the LTB membrane receptor should also be ligands of PPARα, and vice versa. A reevalutaion of the growing banks of effectors for PPARα (lipid-lowering drugs such as fibrates and xenobiotics) and for the LTB_4 membrane receptor (synthetic antagonists and agonists) will no doubt reveal compounds that have interesting properties. This cross-talk at the level of ligands signals caution in drug therapy: a compound directed at the nuclear receptor PPARα might inadvertently also affect an eicosanoid membrane receptor.

Based on the LTB_4 model it is tempting to predict that many other PPARα activators control their own metabolic fate via a similar feedback mechanism (Devchand *et al.,* 1996; Serhan, 1996). Since the list of PPARα

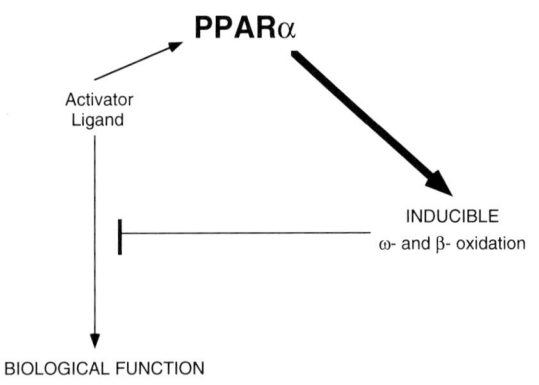

FIGURE 10 Feedback mechanism of PPARα. Ligands and activators of PPARα, such as fatty acids or xenobiotics, control their own biological fate via a feedback mechanism. For example, at the site of inflammation, the fatty acid derivative LTB_4 triggers its own degradation by ω- and β-oxidation by inducing PPARα activity. Thus the role of PPARα in this scenario is to control inflammation. In the liver, the same feedback mechanism facilitates the detoxification function of PPARα.

activators that are catabolized by the β-oxidation includes key signaling molecules, we can only anticipate more exciting times ahead.

VII. Conclusion

PPARα is a transcription factor that directs traffic of lipid metabolic pathways to maintain homeostasis. The few PPARα functions identified to date are reflective of the ability of PPARα to induce the ω- and β-oxidation pathways, probably via a feedback mechanism. Depending on cell type and context, this seemingly simple task has implications in many processes including energy homeostasis, detoxification of harmful chemicals (natural and foreign), monitoring and maintaining lipid levels in the blood, and control of inflammation. One might view PPARα as a promising target for intervention in prevalent disorders such as obesity, cardiovascular disease, diabetes, and many inflammatory diseases. In understanding PPARα we will realize more fully the mechanisms that link diet and health. The challenge then is to unmask the complex and elegant mechanisms hidden behind the deceptively simple adage: "You are what you eat!"

Acknowledgments

We thank Dr. L. Michalik for critical reading of the manuscript. Work performed in the authors' laboratory has been supported by the Etat de Vaud and Swiss National Science Foundation.

Note added in proof. For recent reports evaluating fatty acids, fibrates, and eicosanoids as PPAR ligands see: Dowell, P., *et al* (1997) *J. Biol. Chem.* **272**, 2013–2020; Forman, B. M., *et al.* (1997) *Proc. Natl. Acad. Sci. USA* **94**, 4312–4317; Kliewer, S. A., *et al.* (1997) *Proc. Natl. Acad. Sci. USA* **94**, 4312–4317; and Krey, G., *et al.* (1997) *Mol. Endocrinol.* **11**, 779–791.

References

Amri, E. Z., Bonino, F., Ailhaud, G., Abumrad, N. A., and Grimaldi, P. A. (1995). *J. Biol. Chem.* **270**, 2367–2371.
Aperlo, C., Pognonec, P., Saladin, R., Auwerx, J., and Boulukos, K. E. (1995). *Gene* **162**, 297–302.
Braissant, O., Foufelle, F., Scotto, C., Dauça, M., and Wahli, W. (1996). *Endocrinology (Baltimore)* **137**, 354–366.
Brun, R. P., Tontonoz, P., Forman, B. M., Ellis, R., Chen, J., Evans, R. M., and Spiegelman, B. M. (1996). *Genes Dev.* **10**, 974–984.
Chen, F., Law, S. W., and O'Malley, B. W. (1993). *Biochem. Biophys. Res. Commun.* **196**, 671–677.
Chu, R. Y., Madison, L. D., Lin, Y. L., Kopp, P., Rao, M. S., Jameson, J. L., and Reddy, J. K. (1995). *Proc. Natl. Acad. Sci. U.S.A.* **92**, 11593–11597.
Desvergne, B., and Wahli, W. (1995). *In* "Inducible Transcription" (P. Bæuerle, ed.), Vol. 1; pp. 142–176. Birkhäuser, Boston.

Devchand, P. R., Keller, H., Peters, J. M., Vasquez, M., Gonzalez, F., and Wahli, W. (1996). *Nature* (*London*) **384**, 39–43.
Dreyer, C., Krey, G., Keller, H., Givel, F., Helftenbein, G., and Wahli, W. (1992). *Cell* (*Cambridge, Mass.*) **68**, 879–887.
Ford-Hutchinson, A. W. (1990). *Crit. Rev. Immunol.* **10**, 1–11.
Forman, B. M., Tontonoz, P,, Chen, J., Brun, R. P., Spiegelman, B. M., and Evans, R. M. (1995). *Cell* (*Cambridge, Mass.*) **83**, 803–812.
Gearing, K. L., Göttlicher, M., Teboul, M., Widmark, E., and Gustafsson, J. Å. (1993). *Proc. Natl. Acad. Sci. U.S.A.* **90**, 1440–1444.
Göttlicher, M., Widmark, E., Li, Q., and Gustafsson, J. Å. (1992). *Proc. Natl. Acad. Sci. U.S.A.* **89**, 4653–4657.
Göttlicher, M., et al. (1993). *Biochem. Pharmacol.* **46**, 2177–2184.
Greene, M. E., et al. (1995). *Gene Expression* **4**, 281–299.
Hertz, R., Berman, I., Keppler, D., and Bar-Tana, J. (1996). *Eur. J. Biochem.* **235**, 242–247.
Hu, E. D., Kim, J. B., Sarraf, P., and Spiegelman, B. M. (1996). *Science* **274**, 2100–2103.
Issemann, I., and Green, S. (1990). *Nature* (*London*) **347**, 645–650.
Jow, L., and Mukherjee, R. (1995). *J. Biol. Chem.* **270**, 3836–3840.
Juge-Aubry, C. E., et al. (1995). *J. Biol. Chem.* **270**, 18117–18122.
Keller, H., Dreyer, C., Medin, J., Mahfoudi, A., Ozato, K., and Wahli, W. (1993). *Proc. Natl. Acad. Sci. U.S.A.* **90**, 2160–2164.
Kliewer, S. A., Umesono, K., Noonan, D. J., Heyman, R. A., and Evans, R. M. (1992). *Nature* (*London*) **358**, 771–774.
Kliewer, S. A., et al. (1994). *Proc. Natl. Acad. Sci. U.S.A.* **91**, 7355–7359.
Kliewer, S. A., Lenhard, J. M., Wilson, T. M., Patel, I., Morris, D. C., and Lehmann, J. M. (1995). *Cell* (*Cambridge, Mass.*) **83**, 813–819.
Lee, S. S.-T., Pineau, T., Drago, J., Lee, E. J., Owens, J. W., Kroetz, D. L., Fernandez-Salguero, P. M., Westphal, H., and Gonzalez, F. J. (1995). *Mol. Cell. Biol.* **15**, 3012–3022.
Lehmann, J. M., Moore, L. B., Smith-Oliver, T. A., Wilkison, W. O., Willson, T. M., and Kliewer, S. A. (1995). *J. Biol. Chem.* **270**, 12953–12956.
Lehmann, J. M., Lenhard, J. M., Ringold, G. M., and Kliewer, S. A. (1997). *J Biol. Chem.* **272**, 3406–3410.
Lemberger, T., Staels, B., Saladin, R., Desvergne, B., Auwerx, J., and Wahli, W. (1994). *J. Biol. Chem.* **269**, 24527–24530.
Lemberger, T., et al. (1996a). *J. Biol. Chem.* **271**, 1764–1769.
Lemberger, T., Desvergne, B., and Wahli, W. (1996b). *Annu. Rev. Cell Dev. Biol.* **12**, 335–363.
Masters, C. J. (1996). *Cell. Signal.* 8(3), 197–208.
Schmidt, A., Endo, N., Rutledge, S. J., Vogel, R., Shinar, D., and Rodan, G. A. (1992). *Mol. Endocrinol.* **6**, 1634–1641.
Serhan, C. (1996). *Nature* (*London*) **384**, 23–24.
Sher, T., Yi, H. F., McBride, O. W., and Gonzalez, F. J. (1993). *Biochemistry* **32**, 5598–5604.
Spiegelman, B. M., and Flier, J. S. (1996). *Cell* (*Cambridge, Mass.*) **87**, 377–389.
Steineger, H. H., et al. (1994). *Eur. J. Biochem.* **225**, 967–974.
Tontonoz, P., Hu, E., and Spiegelman, B. M. (1994). *Cell* (*Cambridge, Mass.*) **79**, 1147–1156.
Tontonoz, P., Hu, E., and Spiegelman, B. M. (1995). *Curr. Opin. Genet. Dev.* **5**, 571–576.
Tugwood, J. D., Issemann, I., Anderson, R. G., Bundell, K. R., McPheat, W. L., and Green, S. (1992). *EMBO J.* **11**, 433–439.
Wahli, W., Braissant, O., and Desvergne, B. (1995). *Chem. Biol.* **2**, 261–266.
Wu, Z., Bucher, N. L. R., and Farmer, S. R (1996). *Mol. Cell. Biol.* **16**, 4128–4136.
Yu, K., Bayona, W., Kallen, C. B., Harding, H. P., Ravera, C. P., McMahon, G., Brown, M., and Lazar, M. A. (1995). *J. Biol. Chem.* **270**, 23975–23983.
Zhu, Y., Alvares, K., Huang, Q., Rao, M. S., and Reddy, J. K. (1993). *J. Biol. Chem.* **268**, 26817–26820.

Sandeep Robert Datta
Michael E. Greenberg

Division of Neuroscience
Department of Neurology
Children's Hospital
Harvard Medical School
Boston, Massachusetts 02115

Molecular Mechanisms of Neuronal Survival and Apoptosis

The death of neurons during the development of the nervous system was first appreciated almost 100 years ago. Since then a great deal has been learned about the mechanisms that underlie the differential survival or death of various populations of neurons *in vivo* and *in vitro*. In the developing and adult nervous systems, neuronal life and death decisions are regulated by both extracellular stimuli—in the form of trophic factors, neurotransmitters and depolarizing agents—and by an intrinsic, genetically determined cell death programs. Extracellular stimuli capable of promoting survival recruit distinct but overlapping signal transduction pathways to mediate the transmission of survival or death signals. This review focuses on our current understanding of the involvement of particular signal transduction molecules or pathways in mediating neuronal survival and death. In addition, the potential mechanisms by which signal transduction molecules may effect survival or death is discussed.

The development of complex multicellular organisms requires mechanisms by which an individual cell can communicate with its surrounding environment. In molecular terms, such mechanisms involve changes at the plasma membrane—such as alterations in electrical potential or the binding of a ligand to a receptor—that initiate the activation of cascades of intracellular second messengers that are capable of inducing changes in cellular phenotypes. These signal transduction pathways are critical to the genesis of architecturally complex anatomic structures during development, and dynamic adaptive responses during adulthood.

Two fundamental developmental and adaptive cell phenotypes are survival and death. Perhaps nowhere is the delicate compromise between these polar opposites more critical than in the nervous system. During ontogeny, the nervous system must both generate a wide number of cell types and connect them in ways that allow the useful integration of information whose sources are frequently spread over meters within response times measured in milliseconds. During adulthood, evaluation of diverse stimuli demands continual adaptation by the central and peripheral nervous systems. To meet these rigorous demands, the nervous system has evolved mechanisms that promote the survival of functionally appropriate neurons and the elimination of neurons that are incapable of developmental maturation, have migrated incorrectly, have formed inefficient or improper synaptic connections, are produced in functional excess, are injured, or are defective in damage repair.

The survival or death of particular neurons in higher vertebrates is in part genetically determined, as revealed by studies of nervous system development and of numerous neuropathologies. Specific genetic mutations in several model organisms result in defects in cell-autonomous neuron survival during development, and many human diseases have been identified whose etiologies may involve genetic perturbation of the regulatory mechanisms governing physiologic cell death (Appel, 1981). These diseases include Alzheimer's disease, Huntington's disease, Parkinson's disease, amytrophic lateral sclerosis, Down's syndrome, familial dysautonomia, spinal muscular atrophy, ataxia-telangiectasia, and retinitis pigmentosa (see Table I) (Breakfield, 1982; Busiglio and Yankner, 1995; Roy et al., 1995; Thompson, 1995). A great deal of evidence exists that neuron survival or death also occurs as part of the normal physiology of developing and adult organisms

TABLE I Neuronal Populations That Undergo Pathologic Apoptosis

Alzheimer's disease	Basal nuclei of the forebrain, cortical neurons
Huntington's disease	Striatial neurons
Parkinson's disease	Substantia nigra
Amytrophic lateral sclerosis	Spinal and brainstem motoneurons
Down's syndrome	Cortical neurons
Familial dysautonomia	Small dorsal root ganglia neurons and sympathetic neurons
Spinal muscular atrophies	Spinal motoneurons
Retinitis pigmentosa	Retinal photoreceptors

in response to extracellular stimuli, and is therefore also influenced by non-cell-autonomous factors. For example, early studies demonstrated wide variations in numbers and densities of neurons in particular ganglia taken from individual animals of the same species (Williams and Herrup, 1988), suggesting that the survival and death of neurons is part of a physiologic program of nervous system adaptation to unique stimuli.

It is likely that both genetically determined and adaptive neuronal survival or death are mediated by regulation of a cell-intrinsic, biochemical mechanism whose activation causes cell death. Genetic studies of cell survival and death have identified a number of proteins critical to the cell-intrinsic machinery mediating this suicide program. In contrast, such studies have largely failed to identify the molecules that transmit cell-extrinsic signals to neurons to mediate adaptive survival or death. Given that a great deal of progress has been made in understanding the second messenger systems activated by the key factors that influence survival and death in the nervous system, namely trophic factors and trans-synaptic activity, a major goal is to understand the links between signal transduction molecules and neuronal survival or death.

This review will summarize our understanding of the mechanistic links between extracellular stimuli and neuronal survival or death. In light of the many fine reviews on the general subjects of cell survival and death, this work will focus on the regulation of these phenotypes in the nervous system. An overview of the basic neurobiology that underlies many studies in this field, including the role for trophic factors and electrical activity in neuronal survival, will be provided. The signal transduction pathways that have been implicated in neuronal survival and death will then be discussed, as will mechanisms by which they possibly interact. Finally, the cell-intrinsic mechanisms of neuron survival and death and their role as possible final common pathways regulated by survival and death stimuli will be summarized.

The Neurotrophic Theory

It has been long established that neurons die during normal development, and for nearly half a century that cell death plays a major role in organismal patterning in general and in the nervous system in particular (Collin, 1906; Hamburger and Levi-Montalcini, 1949; Glucksmann, 1951). Neuronal cell death has been observed in nearly every vertebrate species examined, ranging from the tree shrew to the electric fish (reviewed in Oppenheim, 1991). In many species more than 50% of generated neurons die at some point during development (Oppenheim, 1991). In fact, almost no neuronal structure has been described in which developmental cell death, a type of programmed cell death, is not observed (Lockshin and Williams, 1964).

These observations have been largely explained under the rubric of the neurotrophic theory (Korsching, 1993). This model, based on pioneering studies by Levi-Montalcini, Hamburger, and Cohen, postulates that during development neurons are produced in excess, and then whittled away by cell death due to competition for specific soluble trophic factors synthesized by neuronal targets (Hamburger and Levi-Montalcini, 1949; Cohen et al., 1954; Levi-Montalcini and Angeletti, 1968; Levi-Montalcini, 1987; Hamburger, 1992). An attractive aspect of the neurotrophic theory is that the survival of neurons is correlated with their functional fitness; neurons that lack appropriate and useful connections—for example, neurons that do not reach any target, or innervate targets already adequately innervated—fail to obtain trophic support, and subsequently die.

One primary prediction of the neurotrophic theory is that developmental neuron death can be controlled in a non-cell-autonomous manner. This prediction has been borne out by a variety of studies demonstrating that expansion or reduction of the number of target cells causes proportional increases or decreases in the number of innervating neurons (Oppenheim, 1981). The first target-derived factor found to mediate cell survival in such a manner was nerve growth factor (NGF), which in early studies was found to support the survival of sympathetic neurons and a subset of DRG neurons *in vivo* (Bueker, 1948; Hamburger and Levi-Montalcini, 1949; Levi-Montalcini and Hamburger, 1951, 1953; Levi-Montalcini and Angeletti, 1968; Levi-Montalcini, 1987). For example, depletion of NGF by introduction of anti-NGF antiserum into rat fetuses was found to lead to degeneration of sympathetic ganglia, whereas these ganglia were found to be hypertrophic after injection of NGF into rats (Levi-Montalcini and Booker, 1960; Levi-Montalcini and Angeletti, 1966; Gorin and Johnson, 1979). It is now known that NGF is the prototype of a family of related dimeric peptide trophic factors, whose members are called neurotrophins. These include brain-derived neurotrophic factor (BDNF), neurotrophin-3 (NT-3), neurotrophin-4/5 (NT-4/5), and neurotrophin-6 (NT-6) (Snider and Johnson, 1989; Eide et al., 1993; Korsching, 1993; Gotz et al., 1994; Snider, 1994).

Generation of mouse strains deleted for various neurotrophin genes by homologous recombination has revealed that members of this family play critical and unique roles in promoting neuronal survival throughout the nervous system. Commensurate with early studies, mice in which the NGF gene is disrupted develop dramatic peripheral nervous system (PNS) defects, with severe losses in neuron number in their dorsal root and sympathetic ganglia (Crowley et al., 1994). Similarly, mice lacking NT-3 display reductions in neuron number in nearly all of their sympathetic and sensory ganglia (Farinas et al., 1994).

These analyses have also revealed functional roles for neurotrophins in the central nervous system (CNS). For example, mice deficient in BDNF exhibit defects in both the CNS and the PNS; histological analysis demonstrates reductions in the number of neurons in their mesencephalic trigeminal nucleus and their trigeminal, geniculate, vestibular, petrosal-nodose, and dorsal root ganglia (Ernfors et al., 1994; Jones et al., 1994). Interestingly, mice that are genetically deficient in combinations of neurotrophins reveal that while certain neuronal populations are strictly dependent on a single neurotrophin, other populations require unique combinations of neurotrophins for survival (Conover et al., 1995).

The *in vivo* analysis of the role of neurotrophins in neuronal survival has been complemented by a variety of *in vitro* experiments. Since the turn of the century it has been possible to isolate individual neuronal populations from embryonic or postnatal animals and to support their survival in defined culture media (Harrison, 1907). A wide variety of neuronal populations, ranging from midbrain dopaminergic neurons to retinal ganglion cells, have been successfully cultured in this manner, although a few in particular—dorsal root ganglion sensory neurons, sympathetic ganglion neurons (Martin et al., 1992), and cerebellar granule cells (D'Mello et al., 1993)—have provided particularly useful systems for examining aspects of survival and death. Studies of survival have also been aided by immortal cell lines, especially the PC12 rat pheochromocytoma cell line that, when exposed to NGF, acquires many properties of sympathetic neurons (Greene and Tischler, 1976). One particular advantage of these survival paradigms is the tractability of genetic analysis; in each case, a technology exists for the introduction of exogenous DNA into the cells, thereby allowing the genetic dissection of signal transduction pathways underlying survival and death.

Using these *in vitro* culture systems it has been possible to define a number of peptide factors that promote neuronal survival. These include protein growth factors such as the neurotrophins NGF, BDNF, and NT-3, glial-derived neurotrophic factor (GDNF), platelet-derived growth factor (PDGF), insulin-like growth factor I (IGF-1), basic and acidic fibroblast growth factor (a and bFGF), cytokines such as ciliary neurotrophic factor (CNTF), leukemia inhibitory factor (LIF) and transforming growth factor-beta (TGF-β), and peptide factors like vasoactive intestinal peptide (VIP) (Rydel and Greene, 1987; Barde, 1989; Pincus et al., 1990; Smits et al., 1991; Thoenen, 1991; Torres-Aleman et al., 1992; DiCicco-Bloom et al., 1993; Poulsen et al., 1994; Segal and Greenberg, 1996).

Apoptosis—A Means of Neuronal Death

The withdrawal of trophic agents from defined *in vitro* culture systems triggers stereotyped neuronal death. This type of neuronal death largely occurs via a process termed apoptosis, from the Greek apo (away from) and ptosis (fall) (Kerr et al., 1972). Originally used to describe toxicity-induced death in hepatocytes, apoptosis is a noninflammatory form of cell death distinct from the classically described necrosis (Cotran et al., 1994). The morphologic signs of apoptosis, observed under light and electron microscopy, are cell shrinkage, the condensation of the nucleus and cytoplasm, DNA margination to nuclear membranes, and cytoplasmic blebbing, followed by rapid fragmentation of the cell into bodies endocytosed by adjacent engulfing cells (Wyllie et al., 1984; Kerr et al., 1987; Steller, 1995).

A number of characteristic biochemical markers of neuronal apoptosis have also been defined, the principal of which is the fragmentation of genomic DNA into oligonucleosomal-sized fragments by regulated endonucleases (Batistatou and Greene, 1991; Pittman, et al., 1993; Dudek et al., 1997). Unique biochemical processes required for apoptosis in specific neuronal populations have also been described. For example, apoptosis due to the withdrawal of trophic support is blocked in postmitotic neuronal populations *in vivo* and *in vitro* by pharmacologic inhibition of transcription or translation, whereas more developmentally immature neurons or neuronal cell types—such as undifferentiated PC12 cells—continue to die in the presence of such inhibitors. In contrast, survival promoted by trophic factors or trans-synaptic activity, regardless of the mitogenic state of the cell population, is not blocked by inhibition of transcription or translation (Oppenheim, 1991; Rukenstein et al., 1991; Koike, 1992; Martin et al., 1992; Mesner et al., 1992; D'Mello et al., 1993; Pittman et al., 1993; Busiglio and Yankner, 1995).

It is thought that developmental and adaptive death *in vivo* largely take place by the process of apoptosis. Although phagocytosis of apoptotic bodies is rapid (with clearance half-times measured in hours), careful observation has made possible the direct visualization of the pyknotic nuclei characteristic of developmental apoptosis *in vivo*. In addition to morphologic criteria, biochemical markers of apoptosis, including DNA fragmentation in the developing nervous system, have been observed, thereby establishing definitive evidence for *in vivo* neuronal apoptosis (Wood et al., 1993).

Activity, Intracellular Calcium, and Survival

Soluble peptides are not unique in their ability to promote neuronal survival and prevent apoptosis. A number of studies suggest that afferent input supports neuron survival both *in vivo* and *in vitro*. Blockade of afferent input increases cell death in a number of neuronal populations, including the cochlear nuclei, motoneurons, ciliary ganglia, lateral geniculate nuclei, sympathetic neurons, parasympathetic neurons, and nucleus magnocellularis neurons (Wright, 1981; Born and Rubel, 1988; Maderdrut et al., 1988; reviewed in Oppenheim, 1991). These results have been recapitulated *in vitro* in a number of experimental systems, including cultures of hypothalamic neurons, retinal ganglion cells, and spinal cord neurons (Brenneman and Eiden, 1986; Lipton, 1986; Ling *et al.,* 1991), in which pharmacological blockade of activity via agents such as tetrodotoxin abrogates *in vitro* survival.

These results have been complemented by the finding that chronic depolarization of cultured neurons by elevating extracellular potassium levels promotes survival (Scott and Fisher, 1970; Scott, 1971; Franklin and Johnson, 1992). Depolarizing levels of potassium are thought to mediate survival at least in part by opening voltage-sensitive calcium channels, as pharmacologic inhibitors of VSCCs inhibit potassium-promoted neuronal survival (Nishi and Berg, 1981; Gallo *et al.,* 1987; Collins and Lile, 1989; Koike *et al.,* 1989; Collins *et al.,* 1991; Franklin and Johnson, 1992; Franklin *et al.,* 1995; Galli *et al.,* 1995). Survival of neurons is also potentiated by culturing neurons in the presence of pharmacologic agents that potentiate calcium flux through VSCCs (Gallo *et al.,* 1987; Koike *et al.,* 1989; Galli *et al.,* 1995).

It is also clear that the increase in intracellular calcium itself is responsible for increased survival, as depolarizing agents like choline and carbamoylcholine, or excitatory amino acids which cause increases in intracellular calcium in a manner independent of VSCCs, promote survival of rat sympathetic neurons and cerebellar neurons (Koike *et al.,* 1989; Hack *et al.,* 1993). Intracellular calcium promotes survival in a dose-dependent manner: levels of intracellular calcium and levels of neuron survival are positively related in a biphasic manner (Collins *et al.,* 1991; Franklin *et al.,* 1995). Depolarization-induced survival is also probably the result of sustained increases in intracellular calcium, as demonstrated by detailed measurement of cytoplasmic calcium levels following application of potassium chloride to sympathetic neurons (Collins *et al.,* 1991). *In vitro* depolarization is therefore thought to mimic the trophic action of afferent action *in vivo* by promoting sustained increases in levels of intracellular calcuim by opening VSCCs.

The Calcium Set-Point Hypothesis

Although the sufficiency of individual trophic factors to maintain neuronal survival has been established, it is likely that *in vivo* survival requirements for neurons are more complex. Even *in vitro,* optimal survival of neurons is promoted by combinations of trophic factors, and the types and combinations of factors required for neuronal survival continually change as developmental maturation progresses (Lazarus *et al.,* 1976; Meyer-Franke *et al.,* 1995). Sympathetic neurons, for instance, are dependent on neurotrophins early in development, but relatively insensitive to trophic-factor deprivation later in development. The related observation that during the maturation of sympathetic neurons basal levels of intracellular calcium steadily increase—in inverse proportion to their degree of neurotrophin dependence—led to the proposal of the "calcium set-point hypothesis" (Collins et al., 1991; Franklin and Johnson, 1992). According to this model, a low level of intracellular calcium is absolutely required for survival, but is not sufficient; neuron survival in these circumstances requires the additional presence of neurotrophic factors. At a higher level of intracellular calcium, however, neurotrophic factors are dispensable and calcium itself is sufficient to promote neuronal survival (Franklin and Johnson, 1992). It should be noted that the relative contributions to survival of growth factors and calcium in particular neuronal populations at particular stages of development remains unknown (Koike et al., 1989; Koike and Tanaka, 1991).

In sum these studies demonstrate that both soluble trophic factors and afferent activity mediate neuronal survival *in vivo*. The actions of these factors have been recapitulated and mimicked *in vitro*, allowing experimental dissection of the mechanisms of apoptosis and its prevention in neurons.

Mechanisms of Survival: Neurotrophin Receptors and Second Messengers

Many of the actions promoted by neurotrophins can be attributed to specific interactions between members of this ligand family and specific transmembrane receptors whose cytoplasmic domains contain tyrosine kinase activity (van der Geer et al., 1994). In general, binding of growth factors to their cognate receptors causes dimerization of receptor subunits; this dimerization, in turn, causes phosphorylation of receptor subunits. Although most neurotrophins can weakly bind most members of the Trk family. TrkA specifically recognizes NGF, TrkB specifically recognizes BDNF and NT-4/5, and TrkC specifically binds NT-3 (Cordon-Cardo et al., 1991; Glass et al., 1991; Hempstead et al., 1991; Kaplan et al., 1991a,b; Klein et al., 1991a,b; Lamballe et al., 1991; Soppet et al., 1991; Squinto et al., 1991). In addition, all of the neurotrophins bind a second receptor, the p75 low-affinity neurotrophin receptor. p75 is structurally unrelated to the Trks, but is related to the tumor necrosis factor alpha (TNF-α) and Fas family of receptors. p75 lacks tyrosine kinase activity but may modulate signaling from the Trk receptors (Berg et al., 1991).

A number of lines of evidence suggest that receptor activation mediates the trophic effects of neurotrophins and growth factors. First, genomic deletion of various Trk receptors via homologous recombination in mouse models reveals that the survival of a number of populations of neurons relies on the presence of intact receptors (Klein et al., 1993, 1994; Smeyne et al., 1994; Minichiello and Kline, 1996). For example, mice homozygous for deletion of the murine TrkB receptor demonstrate nervous system deficits in both the CNS and the PNS, including smaller brains and deficiencies in the trigeminal and dorsal root ganglia, motor neurons, and facial nuclei

(Klein et al., 1993). Like mice deleted for combinations of neurotrophins, mice homozygous for combinations of Trk receptors reveal distinct but overlapping requirements for these receptors in promoting neuronal survival of various populations (Minichiello and Klein, 1996; Pinon et al., 1996).

Reconstitution of Trk activation and phosphorylation in neuronal and nonneuronal cell types that do not express neurotrophin receptors has also demonstrated that survival is dependent on Trk function. NGF does not promote survival of mutant PC12 cell lines that only express the p75 low-affinity receptor. Exogenous expression of TrkA is sufficient to rescue the survival defect in these cells (Loeb et al., 1991; Rukenstein et al., 1991). Introduction of TrkA cDNAs via microinjection and treatment with NGF is also sufficient to mediate survival in some types of NGF-independent primary neuronal cells (Allsopp et al., 1993a). Pharmacologic inhibition of Trk receptor phosphorylation, using drugs such as the Trk kinase inhibitor K-252a and the general kinase inhibitor staurosporine, is sufficient to block Trk-mediated survival in neurotrophin-responsive populations (Rukenstein et al., 1991; Borasio et al., 1993; Nobes et al., 1996). Interestingly, ectopic expression of the PDGF-β receptor in PC12 cells and cerebellar granule cells, which are normally not responsive to PDGF, is sufficient to promote their survival in the presence of PDGF (Yao and Cooper, 1995; H. Dudek, S. R. Datta, and M. E. Greenberg, personal communication, 1997). Thus, a number of distinct receptor tyrosine kinases have the capacity to promote survival in the presence of their cognate ligands.

Ligation of neurotrophin receptors results in the activation of second messenger systems. These signal transduction cascades are thought to mediate the biologic actions of the neurotrophins. The activation of second messenger systems by the neurotrophins or various other growth factors is, in general, a result of ligand-induced association of various receptor subunits. In the case of the Trks and related receptors such as the insulin, IGF-1, and PDGF receptors, the ligand induces homodimerization of receptor subunits. Such dimerization is thought to activate the intrinsic tyrosine kinase activity of these receptors and result in phosphorylation of the intracellular domains of the receptors on multiple tyrosines. These phosphorylated tyrosines, in turn, serve as docking sites for other molecules involved in transducing growth factor signals via a modular protein domain known as the SH2 (src homology 2) domain (Cohen et al., 1995). This domain, identified first in the proto-oncogene pp60^{c-src}, binds to specific peptide motifs that include phosphotyrosine (p-Tyr) residues, and directly recruits signaling molecules to activated receptor multimers. Binding of these proteins to an individual receptor allows them to be phosphorylated by the receptor's kinase domain; in addition, the interaction between signaling proteins and the activated receptor serves to bring these proteins in the proximity of other enzymes and substrates and/or the plasma membrane.

Most receptors are phosphorylated at multiple tyrosines and recruit specific signaling molecules to specific tyrosines; the specificity of phosphotyrosine–SH2 interaction, which determines which tyrosines recruit which molecules, derives largely from sequences C-terminal to the p-Tyr. A novel phosphotyrosine interaction motif, the protein tyrosine binding domain (PTB domain) is also found on a number of signaling molecules; unlike SH2 domains, the binding determinant of PTB domains is specified by sequences N-terminal to the p-Tyr. Recruitment of SH2 or PTB domain-containing signaling intermediates are thought to be the initial step in the activation of cascades of kinases and signaling molecules, which lead through transcriptionally dependent and independent mechanisms to changes in long-term phenotypes, including survival.

A number of molecules typically associate with activated neurotrophin and growth factor receptors, including adapter molecules like the Shc protooncogene and enzymes like phospholipase-C gamma (PLC-γ), which hydrolyzes plasma membrane phosphotidylinositol bisphosphate. Initial evidence suggested that each of these proteins activated distinct, discrete downstream signaling pathways, and that specific neurotrophin and growth factor-dependent phenotypes result from the activation of specific pathways or subsets of pathways. As will be discussed below, it is now known that there is a great deal of crosstalk between signal transduction cascades; it may be more appropriate to describe the events that follow receptor ligation as neurotrophin-induced second messenger programs instead of pathways.

These signaling cascades ultimately effect long-term changes in a number of cell phenotypes by modulating programs of immediate-early and late gene expression. Although immediate-early genes regulated by these second messengers play uncertain roles in mediating survival or apoptosis, they have served as reliable molecular readouts of intact neurotrophin signaling mechanisms. Still, induction of immediate-early genes such as the c-fos proto-oncogene in response to specific neurotrophins is developmentally coincident with the ability of neurotrophins to promote survival, suggesting that common second messengers are responsible for cellular responses such as survival (Segal *et al.*, 1992; Nobes *et al.*, 1996).

The Ras–MAPK Pathway

The Ras–MAPK pathway includes a series of second messengers—including adapter molecules, exchange factors, and kinases—whose sequential activation terminates in the phosphorylation of transcription factors and changes in gene expression, including the activation of c-fos transcription (see Fig. 1). Although a number of mechanisms exist to modulate Ras–MAPK pathway activity, the best studied mechanism is initiated by receptor activation. Shc recruitment to an activated receptor results in the

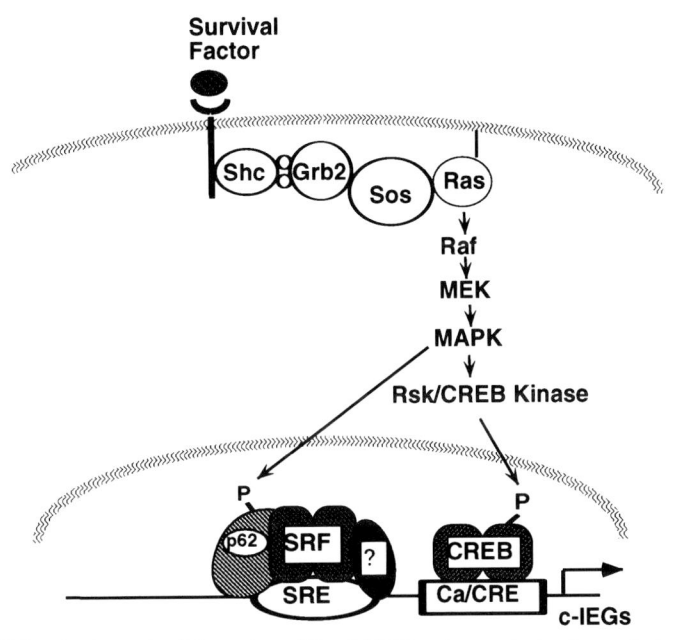

FIGURE 1 A schematic of the activated Ras-MAPK pathway. Activation and phosphorylation of a receptor leads to recruitment via protein–protein interactions of the Sos guanidine nucleotide exchange factor to the membrane-bound Ras proto-oncogene. GTP-bound ras then initiates a phosphorylation cascade that results in activation of a number of cytoplasmic signal transduction molecules. Terminal members of this cascade (which include MAPK and Rsk/CREB kinase) translocate to the nucleus and regulate transcription via the phosphorylation of transcription factors.

phosphorylation of Shc on tyrosine. This event causes Shc association with the adapter protein Grb2 via an SH2 domain present in Grb2 (Rozakis-Adcock *et al.*, 1992). Grb 2 contains an additional domain, a src homology 3 domain (SH3), which is capable of binding polyproline-rich peptides (Cohen *et al.*, 1995). This domain mediates a constitutive association between Grb2 and a proline-rich region within the nucleotide exchange factor Sos (son of sevenless).

Recruitment of the Grb2–Sos complex to the plasma membrane brings Sos in proximity to its target, the small G protein Ras, which is membrane-bound via a farnesyl posttranslational modification. Sos then activates Ras by promoting exchange of GDP for GTP; GTP-bound Ras in turn interacts with and activates the serine–threonine kinase Raf (Wood *et al.*, 1992; Moodie *et al.*, 1993; Vojtek *et al.*, 1993; McCormick, 1994). Raf phosphorylates and activates the dual-specificity kinase MEK1, which phosphorylates and stimulates the mitogen-activated protein kinases ERK1 and ERK2 (Payne *et al.*, 1991; Jaiswal *et al.*, 1994; Lange-Carter and Johnson, 1994).

ERK activation enables the ERKs to phosphorylate and activate members of the serine–threonine kinase family p90rsk, which were initially identified as ribosomal S6 subunit kinases. Upon activation both ERKs and p90rsk family members, which when inactive largely reside in the cytoplasm, translocate to the nucleus where they influence gene expression by directly phosphorylating a number of transcription factors (Hill and Treisman, 1995). For example, Rsk2 is responsible for the growth-factor-mediated phosphorylation of the transcription factor CREB, which is thought to play a critical role in neuronal adaptive responses (Xing et al., 1996).

Many of the components of the Ras–MAPK pathway were originally characterized as proto-oncogenes. Because one model postulates that oncogenesis is a consequence of disregulated apoptosis (Williams, 1991), and because the Ras–MAPK pathway remains the most clearly delineated serine–threonine kinase cascade, a great deal of effort has been focused on elucidating the possible role of the Ras–MAPK pathway in cell survival.

A number of studies suggest that activated Ras is sufficient to promote survival in the absence of trophic support. Evidence of Ras involvement in survival was obtained in many early studies by use of an experimental protocol involving cell trituration to introduce various proteins into neurons. Using this method, the role of wild-type and mutant Ras proteins in neuronal survival has been analyzed. In the pioneering study in this series, introduction of Ras protein or a constitutively active viral form of Ras protein into chick dorsal root ganglion, nodose ganglion, or ciliary ganglion cells (which are dependent for survival on NGF, BDNF, and CNTF, respectively) was found to be largely sufficient to promote the survival of these populations in the absence of relevant trophic support (Borasio et al., 1989). This finding contrasts with the failure of inactive mutant Ras proteins, which lack the protein sequence required for palmitoylation, to sustain neuronal survival. Such mutants have been demonstrated to fail to activate downstream signaling components, suggesting that wild-type Ras promotes survival in these contexts via engagement of endogenous signaling pathways (Willumsen et al., 1984; Casey, 1995).

These findings have been extended to show that Ras is sufficient to support NGF-dependent human and rat sympathetic neurons, and human dorsal root ganglion neurons (Borasio et al., 1996; Nobes et al., 1996). In addition, introduction of v-Ras protein into rat sympathetic neurons is sufficient to promote survival in neurons treated with NGF but whose Trk phosphorylation is blocked by staurosporine (Nobes et al., 1996). Definitive genetic evidence implicating Ras in survival has been obtained from mice homologous for a deletion of the gene encoding neurofibromin 1(NF1), a GTPase activating protein that down-regulates the activity of Ras. In such mice, which presumably have hyperactivated endogenous Ras, neurotrophin-dependent populations of DRG, nodose ganglion, trigeminal ganglion, and sympathetic neurons survive in the absence of the normally required trophic support (Vogel *et al.*, 1995).

The finding that Ras is sufficient promote neuronal survival has been complemented by other experiments exploring the requirement for endogenous Ras activity in the promotion of survival by physiologic trophic factors. Introduction of neutralizing anti-Ras antibody fragments (Fabs) into E9 chick dorsal root ganglion neurons blocks survival promoted by NGF, suggesting that Ras activity is required for NGF-mediated survival (Borasio *et al.*, 1993). In fact, anti-Ras Fabs have been found to block the survival of rat sympathetic neurons normally promoted by a wide variety of neurotrophic factors, including NGF, CNTF, and LIF (Nobes and Tolkovsky, 1995).

While the experiments with sympathetic neurons described above demonstrate the sufficiency and the necessity of Ras activity for neuronal survival in particular experimental paradigms, additional experiments suggest that Ras may not play a universal role in promoting survival. The survival of growth factor-deprived chick sympathetic neurons is not promoted by introduction of wild-type of viral Ras (Borasio *et al.*, 1993). Growth factor-dependent survival of chick sympathetic and ciliary ganglion neurons is also Ras-independent, as introduction of anti-Ras Fabs does not block their survival (Borasio *et al.*, 1993).

Although these findings may also be explained by species specificity, with chick neurons behaving differently than rat or human, the mixed evidence regarding the Ras–MAPK pathway and survival are interesting in light of studies demonstrating that in some contexts Ras activity actually potentiates apoptotic death (discussed further below). In support of these findings, knock-out mice deleted for both copies of rasGAP, a GTPase activating protein that, like NF1, down-regulates Ras activity, show dramatic increases in apoptosis in the developing nervous system (Scheid *et al.*, 1995). Intriguingly, rasGAP/NF1 double mutants demonstrate a massive developmental failure of the midbrain and forebrain, but reveal focal areas of neural overgrowth within the neural tube (Scheid *et al.*, 1995). It is possible that various populations of neurons are differentially responsive to Ras activity such that Ras is capable of promoting both survival and death.

To date only one downstream effector of Ras has been positively identified as mediating neurotrophin-dependent survival. Transfection of PC12 cells with a vector overexpressing a constitutively activated form of MEK1, the kinase which regulates ERK1 and ERK2, is sufficient to rescue the cells from NGF withdrawal-mediated apoptosis (Xia *et al.*, 1995). This finding supports the idea that Ras–MAPK pathway activation is sufficient for neuron survival. However, activation of known proteins downstream of MEK1 may or may not be required for cell survival. Although NGF withdrawal clearly reduces ERK phosphorylation in PC12 cells, activation of the ERKs does not directly correlate with survival of rat sympathetic neurons, as stimuli which activate the ERKs do not always promote survival (Virdee and Tolkovsky, 1995).

Further experiments have established that sustained application of a drug that inhibits MEK1 activity, which results in a loss of most of the NGF-induced MAPK activity, has no effect on NGF-mediated survival in PC12 cells or in rat sympathetics (Park et al., 1996a; Virdee and Tolkovsky, 1996). The specificity of this drug, however, is not entirely clear; in addition, the importance of residual activation of MAPK is unknown; it may be that a low basal level of MAPK activity is sufficient for survival. Definitive genetic experiments establishing the role of ERK1, ERK2, and their substrates remain to be undertaken.

Taken as a whole, these studies provide strong evidence that Ras plays a role in neuronal survival in certain circumstances. Ras may promote survival via a number of pathways, including the canonical Raf–MEK–ERK pathway, via a pathway involving a novel MEK substrate, or via other pathways that do not involve Raf and MEK. One such possible pathway, the phosphatidylinositol-3 kinase pathway/Akt pathway, is discussed below. Still, additional work is required to assess the *in vivo* relevance of particular kinases downstream of Ras in Ras-mediated survival.

The Phosphatidylinositide-3′-OH Kinase/Akt Pathway

Another kinase cascade that has been implicated in neuron survival is named after one of its members, the phosphatidylinositide-3′-OH kinase (PI3′K) (see Fig. 2). PI3′K is both a protein and a lipid kinase, capable of phosphorylating the D-3 position of phosphoinositides to generate phosphatidylinositol-3-phosphate (Ptd 3-P), PtdIns $3,4-P_2$, or PtdIns $3,4,5-P_3$ (Nishizuka, 1992). There are several isoforms of mammalian PI3′K, but the best-characterized form is a heterodimer, composed of an SH2-domain containing a 85-kDa regulatory domain, and a 110 kDa catalytic domain (Pons et al., 1995; Moriya et al., 1996). The SH2 domain of p85 couples PI3′K to a wide variety of tyrosine-phosphorylated molecules, including activated Trk receptors and adapter proteins such as IRS-1 (Insulin Receptor Substrate-1). In addition, PI3′K has been found to bind to other effector molecules, like the small G protein Ras. The lipid kinase activity of PI3′K has been shown to influence a wide variety of cell biological functions, including retrograde transport, mitogenesis, and actin dynamics, and it may play an important role in a number of neuronal functions (Plyte et al., 1992). The protein kinase activity of PI3′K is not thought to play a major role in the function of PI3′K; the PI3′K regulatory subunit p85 is the major protein substrate for the p110 catalytic subunit. A number of downstream targets of PI3′K have been discovered, including the small G protein rac, the p70 ribosomal S6 kinase, various protein kinase C (PKC) isoforms, and the

Survival Factor

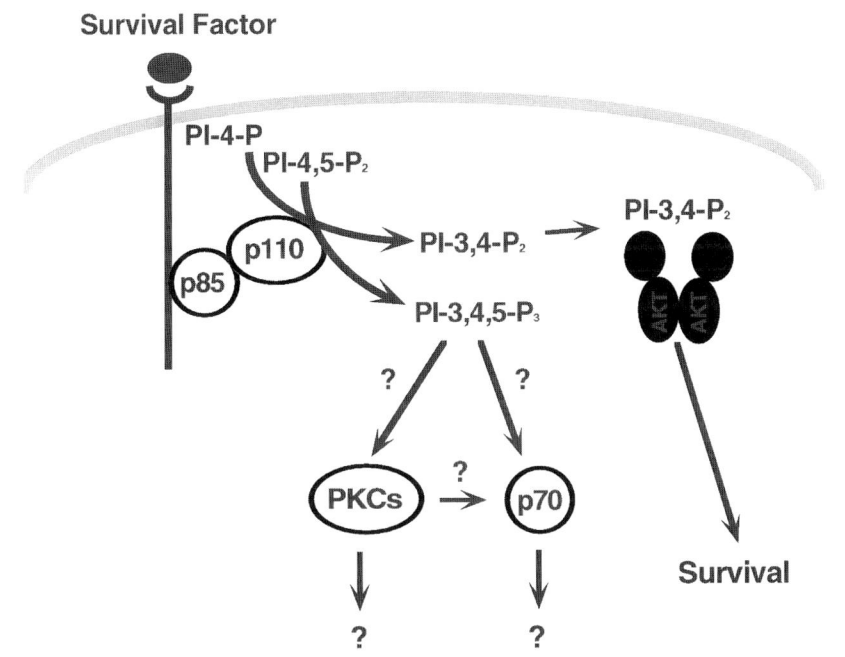

FIGURE 2 Regulation of protein kinases by the products of phosphatidylinositol 3-OH kinase. The activated, membrane-localized p110 subunit phosphorylates phosphoinositides, generating three products: PtdIns3-P, PtdIns3,4-P, and PtsIns3,4,5-P. All three of these products are capable of regulating signal transduction molecules. Both PKCs and p70 may be directly or indirectly regulated by these products. The Akt proto-oncogene may be regulated in part by direct binding to PtdIns3,4-P.

serine/threonine kinase Akt (Chung *et al.*, 1994; Burgering and Coffer, 1995; Franke *et al.*, 1995; Akimoto *et al.*, 1996; Moriya *et al.*, 1996). The exact mechanism of activation of these targets by PI3'K is not clear, as none of the aforementioned proteins are directly phosphorylated by PI3'K. Recently, evidence has been obtained suggesting that D-3 phospholipids directly bind to and induce multimerization of proteins such as Akt that contain a pleckstrin homology domain, a lipid-binding domain first identified in the PKC substrate pleckstrin (Franke *et al.*, 1997).

Recently experiments have demonstrated that PI3'K can mediate survival in PC12 cells (Yao and Cooper, 1995). Pharmacologic inhibition of PI3'K activity—which in PC12 cells has little influence on MAPK activity (S. R. Datta, H. Dudek, and M. E. Greenberg, unpublished results, 1997)— promotes apoptosis in PC12 cells even in the presence of NGF. Genetic evidence establishing the involvement of this pathway was obtained by transfecting PC12 cells, which are normally not PDGF responsive, with wild-type and mutant PDGF receptor derivatives. NGF-deprived cells transfected with wild-type PDGF receptors survive in the presence of PDGF, while cells transfected with mutant PDGF receptors deleted for the tyrosines that mediate receptor-p85 interaction undergo apoptosis. Conversely, NGF-deprived PC12 cells transfected with mutant PDGF receptors deleted for the tyrosines known to interact with SH2 domains, but with the p85-interacting tyrosine "added back," survive in the presence of PDGF. Subsequently these PC12 cell findings have been verified in a wide variety of cell types, including fibroblasts, hematopoetic cell lines, oligodendrocytes, and hippocampal cells (D'Mello et al., 1997; Chao, 1995; Scheid et al., 1995; Vemuri and McMorris, 1996; Yao and Cooper, 1996).

The finding that PI3'K promotes survival has also been confirmed in primary neurons, and a kinase that may mediate the PI3'K pathway survival has been identified. In cerebellar granule cells, IGF-1 acts as a potent survival factor; however, by a number of criteria, IGF-1 fails to activate downstream components of the Ras–MAPK pathway. IGF-1 does, however, cause dramatic and prolonged increases in IRS-1/p85 association and PI3'K activity, suggesting that PI3'K may mediate IGF-1 survival in these cells. Cerebellar neurons supported by IGF-1 undergo synchronous apoptosis in the presence of pharmacologic PI3'K inhibitors. In addition, IGF-1 potently induces activation of the PI3'K targets p70 and Akt. In these and previous studies, pharmacologic block of p70 activity with the macrolide immunosuppressant rapamycin had no effect on neuronal survival (Yao and Cooper, 1996; D'Mello et al., 1997; Dudek et al., 1997). In contrast, transfection of cerebellar neurons with wild-type and dominant-negative Akt constructs demonstrated that Akt activity is both sufficient and necessary for trophic factor-mediated survival of granule neurons (Dudek et al., 1997). Because PI'3K has been implicated in the survival of a number of cell types, Akt is likely an important mediator of survival in neurons and other cells (Dudek et al., 1997; Kauffmann-Zeh et al., 1997; Kulik et al., 1997).

The role for Akt in survival may not be surprising given Akt's previously identified role as a proto-oncogene (Bellacosa et al., 1991). However, very little is known either about the regulation of Akt activity, or of the substrates of activated Akt. One possible effector of Akt-mediated survival is the glycogene synthase kinase-3, the only defined *in vivo* target of Akt (Cross et al., 1995). GSK-3 is also downstream of a critical developmental signal transduction pathway, named for its most upstream component in *Drosophila melanogaster*, wingless (wg) (Plyte et al., 1992); knock-outs of a mammalian isoform of wg reveal dramatic defects in central nervous system development, allowing the conjecture that GSK-3 is a general mediator of survival (Thomas and Capecchi, 1990). Preliminary studies have suggested, however, that Akt is likely to have multiple *in vivo* targets, and that these proteins may play a significant role in mediating PI3'K-promoted survival in neurons (S. R. Datta, H. Dudek, and G. E. Greenberg, unpublished results, 1997). Other targets for PI3'K that are sufficient to mediate survival may also exist; several of the atypical PKC isoforms, which are directly activated by PI3'K generated phospholipids, have been shown to promote survival in fibroblasts (Diaz-Meco et al., 1996).

The definition of a pathway from PI3'K to Akt that mediates neuronal survival represents an advance in our understanding of signaling mechanisms regulating cell survival. Because components of this signaling pathway are still being identified, a great deal remains to be learned about the role that particular molecules of the PI3'K signaling pathway play in neuronal survival.

Protein Kinase C

Protein kinase Cs are members of a large family of proteins with highly conserved catalytic domains that are responsive to a variety of lipid metabolites generated as signaling intermediates. Several of these metabolites are lipids generated by PI3'K or by the enzymatic activity of phospholipase C gamma, which hydrolyzes phosphoinositol-4,5-bisphosphate to generate diacylglycerol (DAG) and inositol triphosphate (IP_3) (Nishizuka, 1992; Akimoto et al., 1996). IP_3 binds to a receptor located on intracytoplasmic membranes, and causes increases in the concentration of intracellular calcium (further discussed below). The best known members of the PKC family, the classical PKCs, are activated in response to increases in intracellular calcium and diacylglycerol. Nonclassical PKCs are also DAG responsive, and atypical PKC family members are thought to be regulated primarily by PI3'K-generated lipids (Moriya et al., 1996).

Because PKCs are generally activated by lipid metabolites, most studies exploring the role of these kinases in neuron survival have relied on pharmacologic activators or inhibitors to manipulate PKC activity (Castagna et al., 1982). Studies in chick sympathetic neurons have shown that phorbol esters like 12-O-tetradecanoylphorbol 13-acetate (TPA), a tumor promoter which mimics DAG, is capable of promoting survival in the absence of trophic support (Wakade et al., 1988). Such studies are usually complicated by the fact that long-term treatment with phorbol esters has been shown to downregulate activity of PKC isoforms. However, in chick sympathetic neurons, after 2 days of TPA treatment PKC activity does not appear to be reduced (Wakade et al., 1988).

The promotion of survival by TPA has also been observed in rat cerebellar granule neurons. Although TPA is toxic to granule cells at high doses, within a narrow dosage range TPA promotes survival of cerebellar neurons in the absence of trophic support. In these experiments, the specificity of the trophic effect of TPA was verified by blocking the survival-promoting activity of TPA with the pharmacologic PKC inhibitor calphostin C; in addition, calphostin C blocked BDNF-mediated survival, suggesting that endogenous PKC activity is involved in survival of cerebellar granule cells.

However, there is a wide variety of evidence suggesting that PKCs are not generally important in neuronal survival. At least in certain circumstances long-term treatment with PMA, another phorbol ester, does not promote survival nor does it block (presumably via PKC down-regulation) NGF-promoted survival in NGF-dependent PC12 cells (Rukenstein et al., 1991). In addition, PKC independence of trophic support or cell death has also been reported for rat sympathetic neurons and for cerebellar granule cells (Martin et al., 1992; Nobes et al., 1996). Importantly, although PMA does not influence the survival of rat sympathetic neurons, treatment with PMA is sufficient to induce the expression of the immediate early gene c-fos, indicating that PMA-regulated signal transduction pathways are intact in these cells (Nobes et al., 1996). Because experiments to date studying PKC largely involve application of pharmacologic inhibitors and activators, the manipulation of PKC activity by genetic means is likely to be highly informative in establishing the role of this kinase family in neuron survival.

Neurotrophin-Dependent Pathway Crosstalk

In the above discussion of survival-promoting second messengers, signal transduction pathways were described as they were initially characterized: linear and unrelated pathways whose activation elicits differing cellular responses. It is clear, however, that there is a tremendous amount of interaction between different classes of second messengers. Examples demonstrating the complexity of interaction between pathways previously thought to be linear are plentiful. One of the first such interactions that was discovered was between protein kinase C family members and the Ras–MAPK pathway; PKCs have been shown to directly phosphorylate and activate Raf-1 (Kolch et al., 1993). PI3'K provides a subtle example of signal transduction crosstalk, having been placed both upstream and downstream of the Ras–MAPK pathway in various cell contexts (Sjölander et al., 1991; DePaolo et al., 1996; Klinghoffer et al., 1996). The plurality of genetic evidence suggests that activated Ras directly binds and activates PI3'K, and that Ras activity can play an important role in regulating PI3'K pathway activity. Because crosstalk between signal transduction pathways is almost certainly cell- and species-specific, detailed characterization of neuronal signaling molecules in general is crucial to a comprehensive understanding of neurotrophin-induced survival.

Calcium-Dependent Kinase Pathways

As mentioned previously, depolarization—which potently promotes survival—increases intracellular calcium via voltage-sensitive calcium channels. Such increases have been shown to activate a number of signaling molecules, including the classical PKC isoforms, the phosphatase calcineurin, the calcium-dependent adenylate cyclases, and members of the calcium-calmodulin dependent (CaM) kinase family (Ghosh and Greenberg, 1995). These proteins contain conserved domains responsible for the binding of calcium or complexes of calcium and the ubiquitous calcium-binding protein calmodulin (Hanson and Schulman, 1992; Nishizuka, 1992).

Molecular analysis of the involvement of these calcium-dependent proteins in neuronal survival has been limited. Proteins dependent on calcium/calmodulin complexes are likely involved in depolarization-mediated survival, as calmodulin inhibitors such as trifluoperazine, calmidazolium, and W7 block the KCl and excitatory amino acid-induced survival of cerebellar granule neurons and rat sympathetic neurons (Gallo et al., 1987; Hack et al., 1993; Franklin et al., 1995). However, these drugs are somewhat nonspecific, and at high concentrations can decrease calcium influx through VSCCs (Franklin et al., 1995). Pharmacologic inhibition of calcineurin activity with the immunosuppressants FK506 or cyclosporin A has been shown to promote survival of neurons subject to ischemic insult, but the ability of these compounds to promote survival *in vitro*, specifically in trophic-factor withdrawal models, has not been extensively characterized (Sharkey and Butcher, 1994). Changes in intracellular levels of cAMP after depolarization have been reported in some neuronal populations (although not in sympathetic neurons), and therefore calcium-calmodulin-dependent adenylate cyclases may play a role in neuronal survival (Franklin and Johnson, 1992; Meyer-Franke et al., 1995). PKC regulation by calcium is not likely to play a role in survival because, as mentioned previously, in general phorbol esters do not promote survival in the absence of trophic support, and long-term treatment with phorbol esters and more nonspecific PKC inhibitors like H-7, polymixin B, and gangliosides has not been found to decrease the survival of depolarized neurons (Hack et al., 1993).

The most likely candidates to date to mediate calcium-promoted survival are therefore the CaM kinases, which have already been implicated in a wide variety of adaptive phenotypes including hippocampal long-term potentiation (Bliss and Collingridge, 1993). These kinases, which have overlapping substrate specificity with PKA and PKC family members, phosphorylate a wide number of substrates including transcription factors that influence immediate-early gene expression (Hanson and Schulman, 1992). There is currently no genetic evidence implicating these protein kinases in survival. However, the availability of pharmacologic inhibitors of CaM kinases, including KN-62 and KN-93, have made possible molecular analysis of the role of CaM kinases in *in vitro* survival models. At doses that show no blockade of VSCCs, KN-62 has been found to block KCl and excitatory amino acid-induced survival in cerebellar granule cells (Hack et al., 1993; S. R. Datta, H. Dudek, and M. E. Greenberg, personal communication,

1997). However, the specificity of KN-62, which inhibits CaM kinases by blocking the calcium/calmodulin binding site, is not sufficient to discriminate between CaM kinase family members. Because depolarization is frequently the most powerful single trophic factor in a number of *in vitro* survival systems, and neuronal activity may play an important role in neuronal survival *in vivo*, further dissection of functional mediators of calcium-dependent survival is of tremendous interest.

Interaction of Calcium and Neurotrophin Signaling in Survival

Traditionally, regulation and activity of calcium-activated proteins (with the noted exception of calcium-responsive protein kinase C isoforms) were thought to be independent of growth-factor-induced signaling pathways, and vice versa. However, in neurons there is accumulating evidence that interactions between these pathways can occur at the level of ligand—with neurotrophins inducing influxes of extracellular calcium, and increased levels of intracellular calcium inducing increased expression of paracrine-acting neurotrophins. In addition, convincing evidence has been obtained suggesting that calcium and neurotrophin second messenger systems themselves interact, resulting in complex interrelationships between various classes of signal transduction molecules.

Conditioned media from retinal neurons whose survival is dependent on electrical activity are sufficient to promote survival of sister cultures whose electrical activity is pharmacologically blocked, suggesting that activity boosts levels of paracrine survival factors in neuronal culture media, and that these paracrine factors may functionally mediate activity-induced survival (Lipton, 1986). One possible mechanism to explain such a finding comes from the discovery that in cultures of cortical neurons, calcium influx through VSCCs causes pronounced increases in BDNF message. This increase has also been found to play a functional role in promoting neuronal survival: addition of neutralizing anti-BNDF antisera blocks the survival of the cultured neurons (Ghosh *et al.*, 1994; Ghosh and Greenberg, 1995). In addition to BDNF, NGF has also been found to be regulated by neuronal activity in a calcium-dependent manner (Gall and Isackson, 1989; Zafra *et al.*, 1990, 1991, 1992; da Penha Berzaghi *et al.*, 1993).

The presence of paracrine factors has also been suggested by the observation that survival of cerebellar granule cell cultures in density-dependent (H. Dudek, S. R. Datta, and M. E. Greenberg, personal communication, 1997; Ohga *et al.*, 1996). However, data suggesting that such paracrine mechanisms do not exclusively explain the effects of calcium on survival come from the *in vitro* observation that potassium-mediated depolarization remains a potent survival factor for sympathetic neurons regardless of culture cell densities (Franklin and Johnson, 1992).

Increases in intracellular levels of calcium, from both intracellular and extracellular sources, have been demonstrated after growth factor treatment in a wide range of cell types, including fibroblasts, cerebellar granule cells, hippocampal neurons, cortical neurons, and PC12 cells (Nikodijevic and Guroff, 1991; Berninger et al., 1993; Zirrgiebel et al., 1995; Peppelenbosch et al., 1996; S. Finkbeiner and M. E. Greenberg, unpublished results, 1997). Although calcium may mediate important neurotrophin effects in certain circumstances, evidence that increases in calcium mediate most neurotrophin effects is limited. Electrical blockade of cerebellar neuron cultures does not affect BDNF induction of c-fos message, which correlates with (although probably does not cause) survival of these cells (Segal et al., 1992). NGF has a minimal effect on cytoplasmic calcium levels, and in sympathetic neurons NGF fails to cause detectable increases in intracellular calcium (Tolkovsky et al., 1990). Moreover, the functional relevance of small calcium influxes is questionable, as pharmacologic calcium channel blockers fail to interfere with the survival of chick ciliary, sympathetic, or DRG neurons by CNTF, NGF, or bFGF (Collins and Lile, 1989).

Although the interaction between calcium and neurotrophin survival pathways is currently under investigation, it is already known that increases in intracellular calcium can activate a number of the signaling molecules traditionally defined as mediators of the responses to trophic factors. Therefore, it is highly likely that survival conferred by increases in intracellular calcium and neurotrophins involve distinct but overlapping groups of second messengers. Calcium influx through VSCCs has been found to potently activate components of the Ras–MAPK pathway, including Ras, MEK, and MAPK (Rosen et al., 1994; Finkbeiner and Greenberg, 1996). One aspect of this activation may involve the calcium-dependent exchange factor Ras–GRF, which has been shown to potentiate GTP loading and activation of Ras (Shou et al., 1992). In addition, influx of extracellular calcium has been found, in PC12 cells, to activate the Ras–MAPK pathway indirectly by inducing phosphorylation of the EGF receptor and formation of the Shc/Grb2/Sos complex (Rosen and Greenberg, 1996). It is thought that phosphorylation of the EGF receptor is mediated by a pp60^{c-src} family member, as the EGFR is a src substrate and calcium has been shown to lead to src activation (Rusanescu et al., 1995). The mechanism of calcium-dependent src activation remains undefined, but may involve the recently identified calcium-regulated kinase PYK2 (Lev et al., 1995; Finkbeiner and Greenberg, 1996). Calcium activation of Src has also been demonstrated to lead to Ras–MAPK pathway activation (Rusanescu et al., 1995). Through multiple mechanisms, then, calcium appropriates classic growth factor receptor pathways to mediate activation of signal transduction pathways. Therefore it is likely that at least some aspect of calcium-induced neuronal survival is the result of recruitment of trophic-factor-regulated second messenger systems.

Mechanisms of Death

Because neuronal survival is controlled in a manner that largely reflects non-cell-autonomous influences, such as the presence of neurotrophic factors, it is appropriate to consider the induction of its opposite—neuronal death—a consequence of change in extracellular stimuli. One well-supported model postulates death as a "default" phenotype in the absence of trophic support. At some level, death induced by neurotrophin or activity withdrawal probably involves the failure to activate signaling molecules normally turned on by survival stimuli; for example, P13'K activity falls in IGF-starved cerebellar granule cells, in which P13'K activation is sufficient to promote survival (S. R. Datta, H. Dudek, and M. E. Greenberg, unpublished results, 1997; Dudek et al., 1997).

It is evident that neuron death also involves the activation of certain second messenger pathways, largely distinct from those mediating neuronal survival that are regulated by death-inducing stimuli. Very little is known about these death pathways, but because their activation is the result of a broad range of insults, including trophic factor withdrawal and ligation of "death receptors," it is likely that these second messenger pathways play key roles in modulating neuronal death.

Upstream Events in Neuronal Death and p75

The relevant upstream events responsible for the induction of apoptosis after physiologic withdrawal of neurotrophins or electrical activity are not yet known. One possibility is "default" inactivation of required factor-sensitive second messenger pathways. Another possibility is that withdrawal of trophic support is a "stressful" stimulus, much like exposure to other pro-apoptotic stimuli such as ultraviolet light and osmotic stress. It has been shown that such stress stimuli induce apoptosis by causing the clustering and subsequent trans-activation of the TNF, interleukin-1 (IL-1), and epidermal growth factor (EGF) receptors in HeLa cells (Rosette and Karin, 1996). Activation of the TNF-α and IL-1 receptors by addition of their cognate ligands has also been shown to induce apoptosis in a number of cell types (Hannun, 1996). Stress stimuli—and possibly trophic factor withdrawal—can therefore induce apoptosis via activation of death-inducing receptors in *trans*, much as increases in intracellular calcium results in the activation of MAPK via transphosphorylation of the EGF receptor.

Activation of specific death-regulating receptors may play a role in neuronal cell death *in vivo*. This paradigm is well established in the immune system, whose cells express receptors that cause cell death upon ligation. Although no novel ligands have been identified which specifically promote physiologic death in the mammalian nervous system, data suggest that factors capable of promoting survival in certain circumstances can promote death in other contexts. For example, TGF-β, which is neurotropic for sympathetic, dopaminergic, and lesioned neurons, promotes the cell death of cultured cerebellar granule cells (de Luca *et al.*, 1996). CNTF and LIF, both cytokines that are trophic for motor neurons, promote apoptosis of cultured sympathetic neurons at specific developmental stages (Kessler *et al.*, 1993; Burnham *et al.*, 1994).

Neurotrophins are also apparently capable of promoting both neuronal survival and, in particular contexts, neuronal death. Treatment with NGF has been shown to promote the death of axotomized motoneurons, of retinal neurons, and of neurons of the isthmo-optic nucleus (Sendtner *et al.*, 1992; von Bartheid *et al.*, 1994; Frade *et al.*, 1996). These effects have been attributed largely to NGF binding to the p75 neurotrophin receptor. As mentioned previously, p75 binds to all of the classical neurotrophins with equal efficiency, and may play a functional role in recruiting these ligands to receptors of the Trk family (Hempstead *et al.*, 1991; Kaplan *et al.*, 1991a; Barker and Shooter, 1994). It is not clear precisely what this role may be; while intact p75 cytoplasmic domains may modulate NGF-mediated signal transduction (Berg *et al.*, 1991), p75 is neither necessary nor sufficient for TrkA signaling (Loeb *et al.*, 1991; Weskamp and Reichardt, 1991; Ibanez *et al.*, 1992; Marsh *et al.*, 1993; Barker and Shooter, 1994). However, it appears that p75 may itself have distinct signaling capabilities. The p75 gene encodes a single-pass transmembrane protein lacking intrinsic tyrosine kinase activity but containing an extracellular domain homologous to the TNF family of receptors, which includes two forms of the TNF receptor (TNFR) and the Fas reeptor, all of which have been implicated in apoptosis of various cell types (Smith *et al.*, 1994). Members of this receptor family contain the 70-amino-acid "death domain," a protein–protein interaction domain involved in recruiting a number of other death-domain-containing proteins, including TRADD, TRAF1, TRAF2, and MORT/FADD, all of which are thought to be functionally involved in apoptosis (Cleveland and Ihle, 1995).

Despite the homology between p75 and TNFR family members, none of these death domain proteins have been demonstrated to interact with the cytoplasmic domain of p75. In fact, it is likely that there are substantial cell-type and receptor-specific differences in signaling initiated by TNFR family members. Nevertheless, p75 has been implicated in the control of apoptosis both *in vitro* and *in vivo*. Overexpression of p75 accentuates the cell death that results from serum deprivation of a conditionally immortalized neuronal cell line. In additional, treatment of PC12 cells with a monoclonal antibody that binds the extracellular domain of p75 is sufficient to block NGF deprivation-induced apoptosis (Rabizadeh et al., 1993). Introduction of antisense oligonucleotides to p75 into E19 and P2 DRG neuron induces substantial protection from NGF deprivation-induced apoptosis (Barrett and Bartlett, 1994), implying that p75 promotes death at postinnervation stages of development.

Such findings have been complemented by *in vivo* studies of the E4 chick retina, which expresses p75 and NGF but not TrkA. In this population, widespread cell death is evident—presumably from NGF stimulation of a p75-mediated death signal—and is blocked by treatment with neutralizing antibodies to either NGF or to p75. However, all retinal neurons that express p75 in this context do not die, and injection of NGF does not cause increased levels of cell death, suggesting that downstream components of the p75 death signaling machinery may be rate-limiting (Frade et al., 1996).

Interestingly, p75 is most highly expressed in the CNS in the nucleus basalis of Meynert, which degenerates in Alzheimer's dementia. p75 levels are also induced in these cells in the context of Alzheimer's disease, and in certain neuronal populations after axotomy (Taniuchi et al., 1988; Dobrowsky et al., 1994). Thus p75 may also regulate apoptosis during pathology and injury, as well as during development. Interestingly, a novel p75 ligand has been identified in mollusks but its biological relevance in vertebrates is as yet undefined (Fainzilber et al., 1996).

Despite these findings, many aspects of p75 biology support its role as a receptor that supports neuronal survival. Trk signaling apparently does not require a direct interaction between NGF and p75, as an NGF mutant protein that cannot interact with p75, and binds to TrkA with low affinity, is sufficient to promote survival of cultured sympathetic neurons (Ibanez et al., 1992; Barker and Shooter, 1994). Although the role for p75 in Trk-mediated signal transduction is still under intense debate, it is likely that p75 at least modulates neurotrophin–Trk interactions. Consistent with a role for p75 in neurotrophin function is the finding that mice lacking the p75 receptor have reduced innervation of sympathetic neuron targets like pineal and sweat glands and a loss of dermal innervation from sensory neurons (Lee et al., 1992, 1994; Davies et al., 1993). p75 has also been implicated in promoting the survival of neurons *in vitro*. Cultured sensory neurons taken from p75 knock-out mice require higher concentrations of NGF to promote their survival compared to sister cultures of wild-type mice (Davies et al., 1993). NGF-mediated survival is also blocked by introduction of anti-p75 oligonucleotides into trophic factor-dependent E12 and E15 DRG neurons (Barrett and Bartlett, 1994).

Given its positive role in Trk signal transduction, it is paradoxical that *in vivo* and *in vitro* experiments have also defined a role for p75 in the promotion of cell death. It may be that complex interactions between signaling pathways activated by the Trks and p75 are capable of inducing both cell survival and death.

Death and Second Messengers: Ceramide and JNK/p38

The second messengers activated downstream of p75, and other death-inducing stimuli, are not clearly defined, but a picture is emerging of death pathways composed of both lipid and phosphorylation-regulated kinase cascades similar in general organization to the survival pathways. p75 directly interacts with proteins that are associated with kinase activity, although these kinases have not yet been definitively identified (Canossa et al., 1996). Downstream signaling events subsequent to the activation of p75 may be largely due to the generation of ceramide, a lipid second messenger (Chao, 1995; Greene and Kaplan, 1995; Bothwell, 1996). Intracellular ceramide levels increase upon hydrolysis of the ubiquitous structural sphingolipid sphingomyelin by acidic and neutral sphingomyelinases. Such enzymes are thought to be regulated by members of the TNF superfamily including p75 (Hannun, 1996). The mechanisms linking TNF family receptors and sphingomyelinase activation are not yet clearly defined.

Neurotrophin binding to p75 in the absence of Trk receptors may induce p75's ability to generate ceramide. NGF treatment of T9 glioma cells, which express p75 but not TrkA, or 3T3 fibroblasts overexpressing p75 is sufficient to induce the hydrolysis of sphingomyelin and the liberation of ceramide (Dobrowsky et al., 1994). Binding to p75 by other members of the neurotrophin family also result in increases in ceramide levels (Dobrowsky et al., 1995). This regulation of sphingomyelinases is dependent on the cytoplasmic domain of p75; EGF treatment of 3T3 fibroblasts overexpressing an EGF–p75 receptor chimera composed of the EGF extracellular domain and the p75 intracellular domain also results in the generation of ceramide (Dobrowsky et al., 1994).

Other stimuli which result in neuronal apoptosis also have been shown to affect ceramide levels. Immortalized hippocampal and DRG neurons undergo apoptosis upon treatment with the kinase inhibitor staurosporine; such apoptosis has been correlated with dramatic increases in ceramide levels (Dobrowsky et al., 1995). Similar findings have also been demonstrated using primary cultures of embryonic chick neurons (Wiesner and Dawson, 1996a). In addition, withdrawal of trophic support from leukemia cells has been associated with dramatic and prolonged elevations in intracellular ceramide; it is possible that neurotrophin deprivation *in vivo* may lead to increases in ceramide levels in neurons (Andrieu et al., 1994; Jayadev et al., 1995; Tepper et al., 1995).

Treatment of cultured neurons with synthetic ceramide analogs or blockade of ceramidase activity increases levels of intracellular ceramide. These treatments potently induce apoptosis in a number of neuronal and nonneuronal cell types, including immortalized hippocampal and DRG neurons, primary chick cerebral neurons, and mesencephalic neurons (Obeid et al., 1993; Brugg et al., 1996; Wiesner and Dawson, 1996 a,b). These results suggest that, as has been found for immune cells, ceramide may play a key role in regulating death pathways in neurons.

A number of enzymes and transcription factors have been identified whose activities are regulated at least in part by ceramide. These include the ceramide-activated protein kinase, the ceramide-activated protein phosphatase, and the highly related protein phosphatase 2A. However, the role of these proteins in the regulation of apoptosis is unknown (Mathias et al., 1991; Dobrowsky et al., 1993). Ceramide also plays an unclear role in MAPK regulation, although it has been shown to both positively and negatively regulate this kinase (Raines et al., 1993; Westwick et al., 1995). Some of the effects of ceramide may be mediated by the transcription factor NF-κB (Westwick et al., 1995). Although the exact epistatic relationship between these factors is unclear, the generation of ceramide and NF-κB activation have been phenomenologically linked (Hannun, 1996). NF-κB and its family members are transcription factors that are important in morphogenesis and cytokine responses. NF-κB may be important in death induction in neurons, as NGF binding to p75, in the absence of TrkA, causes activation (as demonstrated by nuclear translocation) of NF-κB (Carter et al., 1996). However, NF-κB has been shown to prevent apoptosis in some cell types, probably by modulating the expression of genes important for cell survival and death, and may be part of a negative feedback loop during ceramide-induced apoptosis (Beg and Baltimore, 1996; Van Antwerp et al., 1996; Wang et al., 1996). Because the prosurvival function of NF-κB is thought to require new gene expression, its relevance in mature neurons—whose survival in general does not depend on transcription or translation—is not certain.

Ceramide also has been shown to regulate the components of mammalian MAPK cascades (see Fig. 3). Three MAPK cascades have been identified; the first is the Ras–MAPK pathway, which includes an upstream small G protein (Ras) coupled to a MAPK kinase kinase (Raf1), a MAPK kinase (MEK1), and MAPKs (ERK1 and ERK2). The second cascade consists of a number of different small G proteins that regulate the MAPKK SEK1/MKK4, which in turn activates the Stress Activated Protein Kinase (SAPK)/Jun N-terminal Kinase (JNK) (Davis, 1994). The third cascade consists of a number of small G proteins which regulate the MAPKKSs MKK3 and MKK6, which then phosphorylate and activate the MAPK p38 (Ichiio et al., 1996). Like ERK1 and ERK2, both JNK and p38 are capable of phosphorylating transcription factors and modulating programs of gene expression (Davis, 1995).

Both the SAPK/JNK and p38 pathways are activated by stress stimuli and by cytokines known to generate ceramide and cause apoptosis (Derijard

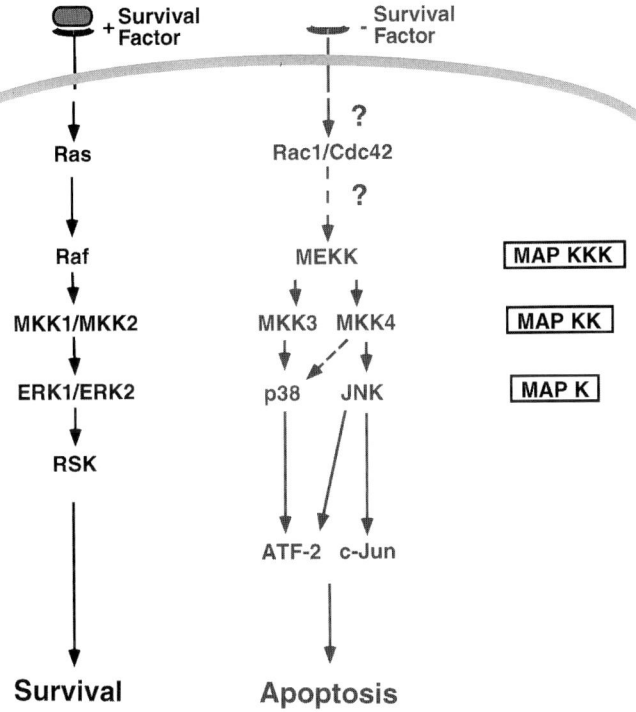

FIGURE 3 Summary of MAPK cascades. MAPK cascades typically include a MAPKKK, a MAPKK and a MAPK. The pathways have been implicated in both survival (the Ras-MAPK pathway) and in death (the JNK and p38 pathways).

et al., 1994; Galcheva-Gargova *et al.*, 1994; Raingeaud *et al.*, 1995; Hannun, 1996; Ichijo *et al.*, 1996). Although the mechanistic link between ceramide and the JNK and p38 cascades remains to be defined, these results suggest that it may involve a ceramide-activated protein kinase (Joseph *et al.*, 1994). Ceramide generation is sufficient for the activation of the JNK and p38 cascades, as exogenous addition of ceramide analogues or sphingomyelinases to promyelocytic cells is sufficient to activate SAPK/JNK (Westwick *et al.*, 1995; Verheij *et al.*, 1996).

Importantly, the SAPK/JNK and p38 pathways have themselves been implicated in the control of apoptosis. One recently cloned MAPKKK regulating SAPK/JNK and p38 pathways is the kinase ASK1. Overexpression of ASK1 is sufficient to induce apoptosis in mink lung epithelial cells; conversely, a dominant negative mutant form of ASK1 blocks apoptosis in embryonic kidney cells and Jurkat T cells (Ichijo *et al.*, 1996). Expression of a dominant negative mutant of the ASK1 target SEK1 is also sufficient to block ceramide-induced apoptosis (Verheij *et al.*, 1996). ASK1 activity has not yet been examined in neurons; however, the JNK and p38 pathways have been found to play critical roles in NGF-deprivation and drug-induced apoptosis of PC12 cells. The activation of JNK and p38 correlates with neuronal death; staurosporine treatment or trophic factor withdrawal causes induction of both JNK and p38 activity in PC12 cells that precedes the onset of morphologic apoptosis (Xia et al., 1995).

The activity of both the JNK and p38 pathways may have functional consequences. Expression of a constitutively active form of MEKK1, which induces JNK activity, was sufficient to induce apoptosis of NGF-differentiated PC12 cells; conversely, the NGF-deprivation-induced death of these cells was blocked by expression of a dominant-negative form of MKK4, p38 also may play a role in PC12 cell apoptosis. Transfection of PC12 cells with a constitutively active MKK3, which activates p38, induces their apoptosis; p38 activity is also required for apoptosis, as introduction of a dominant-negative p38 into NGF-deprived PC12 cells rescues them from death (Xia et al., 1995). Recent experiments suggest that JNK and p38 may not mediate all forms of induced apoptosis (Natoli et al., 1996). However, taken together current data suggest that p75 activation or *in vivo* trophic factor withdrawal leads to the generation of ceramide and activation of components of the JNK and p38 pathways that are critical to regulating a program of cell death.

In many populations of mature neurons, growth factor withdrawal-mediated apoptosis is blocked by inhibition of transcription or translation. This observation has led to the notion that activation of death-promoting second messenger cascades ultimately influences the activity of transcription factors that may control expression of genes required for cell death. The p38 targets ATF-2 and Elk-1, both transcription factors known to play important roles in immediate-early gene expression, have not been extensively studied in this context (Davis, 1995). However, the predominant JNK target, c-Jun, has been implicated in the control of cell death in neurons. c-Jun and members of the Jun family (including Jun B and Jun D) are themselves immediate-early gene products and homodimerize or heterodimerize with Fos family members (which include c-Fos, Fos B, Fra-1, and Fra-2) to form AP-1 transcription factor complexes. Phosphorylation of c-Jun by its *in vivo* kinase JNK promotes transcriptional activation of target genes by AP-1. While JNK phosphorylates all three Jun family members, it phosphorylates JunB and JunD much less efficiently than c-Jun. Moreover, the phosphorylated forms of JunB and JunD fail to activate transcription as robustly as phosphorylated c-Jun.

Neuronal insults that result in apoptosis trigger the induction of c-Jun. c-Jun mRNA and protein levels have been found to rise subsequent to hypoxic, excitotoxic, and axotomy-induced CNS injury (Herdegen et al., 1993; Dragunow et al., 1994; Kaminska et al., 1994). c-Jun protein levels also rise selectively in cerebellar granule cells and Purkinje neurons undergoing apoptosis (Gillardon et al., 1995; Miller and Johnson, 1996). In addition to these results demonstrating up-regulation of c-Jun upon stressful stimuli, data suggest that c-Jun phosphorylation levels rise after NGF withdrawal in sympathetic neurons (Ham et al., 1995). Thus c-Jun may be regulated transcriptionally, translationally, and posttranslationally by apoptotic stimuli.

A number of studies implicate c-Jun as functional mediator of apoptosis. Treatment of neuronal and nonneuronal cells with antisense oligonucelotides directed against c-Jun message or neutralizing antibodies to c-Jun blocks trophic factor withdrawal-mediated apoptosis, suggesting that c-Jun is necessary for apoptosis (Colotta et al., 1992; Estus et al., 1994). These findings have been extended by the demonstration that microinjection of expression vectors encoding a c-Jun dominant negative mutant into sympathetic neurons protects these neurons from NGF withdrawal-mediated death. c-Jun may also be sufficient to promote apoptosis, as injection of expression vectors encoding wild-type c-Jun promotes the death of sympathetic neurons even in the presence of NGF (Ham et al., 1995). Similar findings have been demonstrated in PC12 cells, where transfection with expression vectors encoding c-Jun dominant-negative mutants that lack the JNK binding site protects cells from apoptosis induced either by NGF withdrawal or by cotransfection with MEKK1 (Xia et al., 1995).

Regulation of the c-Jun partner c-Fos may also play a role in promoting apoptosis. c-Fos expression has been found to be increased in CNS neurons undergoing genetically programmed and excitotoxic death, and DRG and spinal cord neurons following axotomy. These findings may have functional significance, as overexpression of c-fos in serum-deprived fibroblasts is sufficient to promote apoptosis (Smeyne et al., 1993). These findings are paradoxical in that both c-Fos and c-Jun expression are also induced by a number of survival-promoting stimuli. In addition, similar increases in c-Fos levels have not been noted in a number of *in vitro* neuronal culture systems in which trophic factor withdrawal triggers apoptosis, and no direct evidence exists as yet suggesting a functional role for c-Fos expression in promoting apoptosis in neurons (Martin et al., 1992; Miller and Johnson, 1996). These conflicting results reveal the subtlety of immediate-early gene function in mediating adaptive responses that include both survival and death.

Signal Integration: The Balance between Life and Death

The existence of multiple pathways involved in survival and death have suggested a model in which the balance between pro-life and pro-death signals determines neuronal survival or apoptosis. This model is supported by data suggesting that both loss of survival pathway activation and induction of death pathway activation are required for neuronal apoptosis; for example, the withdrawal of NGF from PC12 cells, which causes apoptosis, both activates JNK and p38 and induces the loss of ERK activity (Xia et al., 1995).

The mechanistic interaction between these pathways is unclear, but available evidence suggests that life and death pathways may directly regulate each other. In the appropriate context TrkA kinase activity, for example, blocks generation of ceramide by p75 (Dobrowsky et al., 1995). On the other hand calcium influx, a powerful survival stimulus, activates the PYK2 kinase, which has been shown to be activated by stress stimuli and to lie upstream of JNK (Tokiwa et al., 1996).

Such findings are complemented by studies demonstrating interdependence of signaling in the life and death pathway; for example, activation of the death-inducing sphingomyelinases has been shown to lead to activation of the life-promoting MAPK cascade; on the other hand, activation of the life-promoting small G protein ras is required for activity of the small G proteins rac and cdc42, which regulate the death-inducing kinases JNK and p38. Crosstalk between life and death pathways may be direct: treatment of leukemia cells with ionizing radiation induces the formation of a complex between P13'K and JNK (Kharbanda et al., 1995), and activated ras has similarly been found to directly bind to JNK and c-Jun (Adler et al., 1996). Thus feedback and feedforward mechanisms may exist to balance the relative levels of endogenous life and death signals.

Another possibility is that life and death signals compete for influence over a "final common pathway" for cell death. Such a model integrates data demonstrating cell-extrinsic control of signaling pathways with evidence for a cell-intrinsic program of cell death. The cell-cycle machinery is a plausible candidate for such a point of integration. Molecules involved in regulating cell-cycle progression have already been shown to effectively integrate signals—conveyed by growth factor-regulated kinases and phosphatases—promoting cell-cycle arrest and cell-cycle progression. The cell-cycle machinery thereby effectively meets the complex demand of coordinating cell-cycle progression in response to diverse, temporally distinct signals.

That the cell-cycle machinery may integrate survival and death signals in neurons is especially tantalizing because neurons, on the whole, are postmitotic. This fact, coupled with the phenomenological observation that a number of phenotypic aspects of mitogenesis and apoptosis are similar (e.g., chromatin condensation), has suggested a model whereby apoptosis is the result of a abortive attempt of a postmitotic neuron to enter the cell cycle (Heintz, 1993; Rubin et al., 1993). Such a model has also been used to explain the ability of transcription and translation inhibitors to suppress apoptosis of NGF-differentiated, but not naive, PC12 cells. Naive, cycling PC12 cells presumably contain the complement of proteins required for cell-cycle progression, while differentiated PC12 cells lack basic cell-cycle components necessary for an abortive cell-cycle entry, and thus must synthesize them *de novo* (Hammang *et al.*, 1993; Ferrari and Greene, 1994). Consistent with this idea, mature CNS neurons and NGF-supported sympathetic neurons have been shown to lack protein kinases required for cell-cycle progression such as cdc2 and cdk2 (Hayes *et al.*, 1991; Freeman *et al.*, 1994).

Evidence that apoptosis is related to abortive cell-cycle entry derives from the observation that expression of proteins known to cause cell-cycle progression can induce apoptosis in cells that are growth-arrested. Forced expression of the c-myc proto-oncogene, which is known to drive fibroblasts into the cell cycle, is sufficient to promote apoptosis of serum-starved cells. These data suggest that discordance of inputs to the cell-cycle regulatory machinery (serum starvation promoting cell-cycle withdrawal while concomitant c-myc expression suggesting cell cycle reentry) is sufficient to promote apoptosis (Evan *et al.*, 1992). The finding that c-fos overexpression also causes apoptosis in the context of serum-starvation, and that the blockade of c-myc expression is sufficient to prevent T-cell apoptosis, provided additional evidence for this model (Smeyne *et al.*, 1993). However, like c-fos, c-myc itself has not been found to be induced upon induction of apoptosis in a number of neuronal models (Miller and Johnson, 1996).

A number of lines of experimental evidence support the concept that neuronal apoptosis during development may also involve aberrant cell cycle entry. Unregulated cell cycle entry is sufficient to cause apoptosis of certain populations of neurons *in vivo*. Mice with a targeted disruption of the Retinoblastoma (Rb) tumor suppresser, which regulates cell-cycle progression, are embyonic lethal and demonstrate broad neuron death in the CNS, suggesting that neurons that abnormally enter the cell cycle undergo apoptosis *in vivo*. Mice containing transgenes overexpressing SV40 T antigen, which promotes cell-cycle entry, exhibit neuronal apoptosis in a number of neuronal populations, including retinal photoreceptors, cerebellar Purkinje cells, and retinal horizontal cells (Al-Ubaidi *et al.*, 1992; Feddersen *et al.*, 1992).

Cell-cycle-related mechanisms may also be involved in trophic factor withdrawal-mediated death. NGF withdrawal was found to induce tritiated thymidine incorporation into PC12 cells, further suggesting that cell-cycle reentry plays a role in regulation of apoptosis (Ferrari and Greene, 1994). Pharmacologic blockade of G1/S progression also prevents apoptosis of sympathetic neurons, naive PC12 cells, and NGF-differentiated PC12 cells upon trophic factor withdrawal (Farinelli and Greene, 1996). However, blocking cell-cycle entry by pharmacologically inhibiting cyclin-dependent kinases with the drugs flavopiridol and olomoucine only protects differentiated neurons such as cultured rat sympathetics and NGF-treated PC12 cells and not naive PC12 cells (Park *et al.*, 1996a); such results suggest that cell-cycle entry may not be a universal mode of inducing apoptosis.

Although the Ras–MAPK pathway may play critical roles in mediating survival, it also may mediate the death of neurons by promoting abortive cell-cycle reentry. Such a role would be consonant with the well-defined function of the Ras–MAPK pathways in the mitogenesis of mitotic cell populations. Consistent with this possibility, disruption of Ras function has been shown to mediate survival of neurons in some circumstances. For example, inducible expression of dominant negative mutants of Ras is sufficient to prevent the NGF-deprivation-mediated death of both naive and NGF-differentiated PC12 cells (Ferrari and Greene, 1994). In addition, expression of dominant negative mutants of Ras in NGF-deprived PC12 cells largely blocks the tritiated thymidine incorporation that accompanies cell death.

One attractive candidate for an apoptosis-related component of the cell-cycle machinery is cyclin D1. Cyclin D1 is induced in some dying neuronal populations. For example, cyclin D1 expression is selectively induced in sympathetic neurons after NGF withdrawal (Freeman *et al.*, 1994). Cyclin D1 protein levels also rise in a neuroblastoma cell line undergoing apoptosis (Kranenburg *et al.*, 1996). Overexpression of cyclin D1 induces apoptosis of neurons regardless of the availability of trophic support, and block of cyclin D1 associated-kinase activity by overexpression of the cdk inhibitor p16INK4A is sufficient to rescue neuroblastoma cells from serum-deprivation-induced apoptosis (Kranenburg *et al.*, 1996). However, cyclin D1 is not induced in cerebellar granule cells undergoing apoptosis (Miller and Johnson, 1996). In addition, published analysis of cyclin D1 knock-out mice did not reveal any gross nervous system abnormality outside of the retina. Thus, it remains to be demonstrated that cyclin D1 plays a role in the apoptosis of *in vivo* neuronal populations (Siciniski *et al.*, 1995). Still, the possibility that deregulation of components of the cell-cycle machinery affects the balance between survival and death signals remains an attractive mechanism of neuronal apoptosis.

Caspases and the Bcl-2 Family

The proteins that are ultimately responsible for inducing or blocking neuronal death may be those that compose the cell-intrinsic death machinery. As such, these proteins may be a point of integration of life and death signal transduction cascades. Much of our current knowledge regarding the molecules that mediate cell-intrinsic death arises from studies in the nematode *Caenorhabditis elegans,* a genetically amenable worm in which a significant number of cells undergo cell-autonomous programmed cell death during development (Ellis and Horvitz, 1986). Two genes, ced-9 and ced-3, are thought to be major players in cell-intrinsic death programs. Genetic analysis has revealed that ced-9 acts to support survival in a cell-autonomous manner, and to antagonize the action of ced-3, which has been found to be promote cell death in a cell-autonomous manner.

Both ced-9 and ced-3 have important mammalian homologues. The ced-9 protein is homologous to proteins in the Bcl-2 family, and ced-3 is homologous to caspase-1 (formerly known as interleukin-converting enzyme-1, or ICE), the prototype of a family of cysteine proteases. Bcl-2, the first member of the eponymous family to be described, was originally isolated due to its association with a translocation in a B-cell follicular lymphoma, and was initially characterized as a proto-oncogene. It is now known that Bcl-2 family members, which encompass a growing number of proteins, including Bax, Bad, and two splice variants of Bcl-x (a short form, Bcl-X_S, and a long form, Bcl-X_L), can both support cell survival and promote cell death. Bcl-2 family members are widely expressed, and many of them have been demonstrated to be expressed in the nervous system (Boise et al., 1993; Oltvai et al., 1993; Merry et al., 1994; Gonzalez-Garcia et al., 1995). It is unclear how Bcl-2 family members influence cell survival, but recent evidence suggests that they function as homo- and heterodimers. The relative abundance of Bcl-2 family members in a cell can control the composition of homo- and heterodimers; the relative size of these homo- and heterodimeric populations may determine the survival or death of the cell.

Initial characterization of Bcl-2 family members consisted mostly of overexpression studies, in which the effects of these genes on cell survival and death were assessed in a number of cell types. Because Bcl-2 family members are proposed to be downstream effectors of survival and death signals, in theory Bcl-2 and prosurvival family members like Bcl-X_L should protect neuronal populations from physiologic insults. In fact, a great deal of data have demonstrated a broad role for Bcl-2 in protecting neurons from trophic factor withdrawal-mediated apoptosis. PC12 cells stably overexpressing Bcl-2 are protected from death caused by serum deprivation, and fail to generate the characteristic oligonucleosomal ladder associated with apoptosis upon serum withdrawal (Batistatou et al., 1993; Mah et al., 1993). Conditionally immortalized nigral and hippocampal cell lines that overexpress Bcl-2 are also resistant to apoptosis induced by growth factor and serum withdrawal (Zhong et al., 1993; Eves et al., 1996). In a similar fashion, microinjection of constructs encoding Bcl-2 and Bcl-X_L rescues neurotrophin withdrawal-induced death of primary sensory and sympathetic neurons (Garcia et al., 1992; Allsopp et al., 1993b; Gonzalez-Garcia et al., 1995; Greenlund et al., 1995).

Experiments utilizing knock-out and transgenic technology have verified that Bcl-2 promotes neuronal survival. Bcl-2-overexpressing transgenic mice have mesencephalic and DRG neuron populations 30% larger than normal. DRG neurons isolated from such mice are resistant to apoptosis upon neurotrophin deprivation. Motor neurons in these animals are also resistant to apoptosis caused by sciatic nerve transection, and facial neurons from similar mice overexpressing Bcl-2 are also resistant to axotomy-induced apoptosis (Dubois-Dauphin et al., 1994; Farlie et al., 1995). Analysis of mice that have undergone homologous disruption of Bcl-2 have confirmed that Bcl-2 plays an important role in promoting neuronal survival. Although the brains of such knock-out mice are grossly normal, cultured sympathetic neurons from these mice die more readily upon neurotrophin deprivation (Veis et al., 1993; Greenlund et al., 1995). Other Bcl-2 family members may also be required for survival, as mice in which the Bcl-X_L gene is disrupted die before birth, possibly due to massive nervous system apoptosis. Structures disrupted in these mice include the brain, brainstem, spinal cord, and DRG (Motoyama et al., 1995).

The caspases, the mammalian homologs of ced-3, have also been found to play a important role in neuronal survival and death. The caspases encompass a large family of cysteine proteases that specifically cleave substrates C-terminal to aspartate residues. Caspases are thought to promote death via a cascade of proteolytic events, which eventually either directly cause cell death or activate, through cleavage, an ultimate death effector. A number of physiological substrates for the caspases have been identified, including poly (ADP ribose) polymerase, nuclear lamins, and topoisomerase I (Martin and Green, 1995). Surprisingly mice in which the gene for the prototypical caspase ICE is disrupted do not demonstrate a dramatic perturbation of nervous system morphology. However, the requirement for caspase activity in neuronal apoptosis has been demonstrated by inhibition of caspases with naturally occurring and synthetic inhibitors. The cowpox serine protease inhibitor CrmA was initially characterized by its ability to prevent ICE-mediated apoptosis. Microinjection of constructs encoding CrmA into DRG neurons protects these cells from NGF deprivation-induced death (Gagliardini et al., 1994). Treatment of motoneurons with peptide ICE inhibitors also blocks apoptosis upon trophic factor withdrawal (Milligan et al., 1995). The effects of ICE inhibitors are also seen *in vivo*, as treatment of developing chick embryos with peptide caspase inhibitors blocks naturally occurring apoptosis in the lumbar spinal cord (Milligan et al., 1995).

Taken together, these data demonstrate *in vitro* and *in vivo* roles for caspases and Bcl-2 family members in downstream processes of neuronal cell survival and death. However, several studies in neurons have indicated that there are types of neuronal apoptosis that are not influenced by Bcl-2 or caspase family members. The first is the demonstration that chick ciliary ganglion neurons, which are CNTF-dependent, are not protected from CNTF withdrawal by microinjection of a Bcl-2 expression vector (Allsopp et al., 1993b). In a similar vein, treatment of chick embryos with peptide inhibitors of caspases does not block all developmental neuronal death, suggesting that physiologic death mechanisms may not rely exclusively on caspase activity (Milligan et al., 1995).

Additional studies also suggest that Bcl-2 may not universally regulate apoptosis. Sympathetic neurons isolated from Bcl-2 knock-out mice, like their wild-type counterparts, become independent of neurotrophins over time in culture (Greenlund et al., 1995). This suggests that neurotrophin-independent survival mechanisms (which in this case correlate with increases in intracellular calcium levels) do not require Bcl-2 activity. It is important to note that the multiplicity of caspase and Bcl-2 family members implies functional redundancy. This has been verified by a number of seemingly paradoxical observations, including the sufficiency of both Bcl-2 and general caspase inhibitors to promote neuronal survival *in vitro* but the modest neuronal phenotype of Bcl-2 and ICE knock-out animals.

Cell-Extrinsic Signaling Meets Cell-Intrinsic Survival and Death Mechanisms

Little is known about the potential coupling between the fundamental survival and death signal transduction pathways and Bcl-2/caspase death effector mechanisms. One possible mode of regulation is at the level of expression. TGF-β, a trophic factor for rat hippocampal neurons, has been shown to increase expression of Bcl-2 (Prehn et al., 1994). Conversely, NGF withdrawal from neurotrophin-dependent sympthetic neurons results in a drop in expression of both Bcl-2 and Bcl-X_L. Increases in cell density, which improve survival in rat cerebellar neurons, have also been shown to increase expression of Bcl-2 and Bcl-x (Ohga et al., 1996).

However, in most neuron populations, expression levels of Bcl-2 family members do not vary dramatically after development, suggesting that post-translational modification of Bcl-2 family members or caspases may regulate the activity of these proteins in modulating apoptosis. Both Bcl-2 and the pro-death Bcl-2 family member Bad are phosphorylated, and phosphorylation is thought to modulate the effect of these proteins on survival. The kinases that mediate these effects have not been identified, but Bcl-2 family members have been shown to interact physically with a number of signaling molecules, either directly or indirectly. Bcl-2 has been shown to form a complex with the GTPase R-ras, which has been shown to promote apoptosis upon its activation. Raf, the kinase downstream of the small G protein ras, has also been shown to directly associate with Bcl-2. Bcl-2 also interacts with Nip1 and Nip2, proteins homologous to Ca2+/calmodulin-dependent phosphodi-esterase and RhoGAP, respectively. Other Bcl-2 family members have also been demonstrated to interact with molecules involved in signal transduction. The adapter protein 14-3-3, which binds to such proteins as Raf-1 and PI3'K, can directly associate with death-promoting Bcl-2 family member Bad. In addition, Bcl-2 overexpression has been correlated with blockade of JNK activity, suggesting that Bcl-2 can modulate the activity of kinases involved in death (Park *et al.*, 1996b).

Signal transduction molecules that are involved in promoting neuronal apoptosis also may modulate Bcl-2 family member and caspase activity. For example, p75 may directly induce caspase activity. Although p75 has not been shown to directly interact with any specific death domain-containing molecules, the p75 homolog FasR recruits a specific caspase to the plasma membrane via interaction with the death domain-containing protein FADD (Boldin *et al.*, 1996; Muzio *et al.*, 1996). The activity of this caspase, MACH/FLICE, may underlie FasR-mediated death. These findings raise the exciting possibility that the two mechanisms most commonly associated with vertebrate signal transduction, protein–protein interactions and phosphorylation, may be responsible for linking cell-extrinsic survival and death signals with the cell-intrinsic death machinery.

Frontiers in Neuronal Survival and Death

There is clearly a great deal left to learn about the mechanisms by which neuronal survival and death are regulated. Current experiments will likely build on the paradigms established to date, namely that trans-synaptic stimulation and neurotrophic factors activate signaling cascades capable of influencing survival and death. This review has described several of the stimuli capable of promoting survival and death, and has considered many of the signal transduction molecules known to be activated by those stimuli. It is likely that there are many ligands yet to be discovered that play crucial *in vivo* roles in survival and death phenomena in the nervous system. Our view of the signaling pathways that underlie survival is also largely incomplete. It remains unclear how various signal transduction molecules actually influence neuronal decisions to die or to live. Future research will hopefully further define the molecular players in these critical processes, and in doing so may identify promising targets for therapies for a wide range of neuropathologies.

Acknowledgments

This work was supported by National Institute of Health Grant CA43855 (M.E.G.), American Cancer Society Faculty Research Award FRA-379 (M.E.G.) and Mental Retardation Research Center Grant NIH P30-HD18655. S.R.D. thanks Henryk Dudek and Azad Bonni for stimulating discussions regarding neuronal survival mechanisms.

References

Adler, V., Pincus, M. R., Brandt-Rauf, P. W., and Ronai, Z. (1996). Complexes of p21ras with JUN N-terminal kinase and JUN proteins. *Proc. Natl. Acad. Sci. U.S.A.* **92**, 10585–10589.

Akimoto, K., Takahashi, R., Moriya, S., Nishioka, N., Takayanagi, J., Kimura, K., Fukui, Y., Osada, S.-I., Mizuno, K., Hirai, S.-I., Kazlauskas, A., and Ohno, S. (1996). EGF or PDGF receptors activate atypical PKC-lambda through phosphatidylinositol 3-kinase. *EMBO J.* **15**, 788-798.

Allsopp, T. E., Robinson, M., Wyattt, S., and Davies, A. M. (1993a). Ectopic trkA expression mediates NGF survival response in NGF-independent sensory neurons but not in parasympathetic neurons. *J. Cell Biol.* **123**, 1555-1566.

Allsopp, T. E., Wyattt, S., Paterson, H. F., and Davies, A. M. (1993b). The proto-oncogene bcl-2 can selectively rescue neurotrophic factor-dependent neurons from apoptosis. *Cell (Cambridge, Mass.)* **73**, 295-307.

Al-Ubaidi, M. R., Hollyfield, J. G., Overbeek, P. A., and Baehr, W. (1992). Photoreceptor degenration induced by the expression of simian virus 40 large tumor antigen in the retina of transgenic mice. *Proc. Natl. Acad. Sci. U.S.A.* **89**, 1194-1198.

Andrieu, N., Salvayre, R., and Levade, T. (1994). Evidence against involvement of the acid lysosomal sphingomyelinase in the tumor-necrosis-factor and interleukin-1-induced sphingomyelin cycle and cell proliferation in human fibroblasts. *Biochem. J.* **303**, 341-305.

Appel, S. H. (1981). A unifying hypothesis for the cause of amytrophic lateral sclerosis, Parkinsonism and Alzheimer disease. *Ann. Neurol.* **10**, 499-505.

Barde, Y.-A. (1989). Trophic factor and neuronal survival. *Neuron* **2**, 1525-1534.

Barker, P. A., and Shooter, E. M. (1994). Disruption of NGF binding to the low affinity neurotrophin receptor p75LNTR reduces NGF binding to TrkA on PC12 cells. *Neuron* **13**, 203-215.

Barrett, G. L., and Bartlett, P. F. (1994). The p75 nerve growth factor receptor mediates survival or death depending on the stage of sensory neuron development. *Proc. Natl. Acad. Sci. U.S.A.* **91**, 6501-6505.

Batistatou, A., and Greene, L. A. (1991). Aurincarboxylic acid rescues PC12 cells and sympathetic neurons from cell death caused by nerve growth factor deprivation: Correlation with suppression of endonuclease activity. *J. Cell Biol.* **115**, 461-471.

Batistatou, A., Merry, D. E., Korsmeyer, S. J., and Greene, L. A. (1993). Bcl-2 affects survival but not neuronal differentiation of PC12 cells. *J. Neurosci.* **13**, 4422-4428.

Beg, A. A., and Baltimore, D. (1996). An essential role for NF-kappaB in preventing TNF-alpha induced cell death. *Science* **274**, 782-784.

Bellacosa, A., Testa, J. R., Staal, S. P., and Tsichlis, P. N. (1991). A retroviral oncogene, akt, encoding a serine threonine kinase containing an SH2-like region. *Science* **254**, 274-277.

Berg, M. M., Sternberg, D. W., Hempstead, B. L., and Chao, M. V. (1991). The low-affinity p75 nerve growth factor (NGF) receptor nediates NGF-induced tyrosine phosphorylation. *Proc. Natl. Acad. Sci. U.S.A.* **88**, 7106-7110.

Berninger, B., Garcia, D. E., Inagaki, N., Hahnel, C., and Lindholm, D. (1993). BDNF and NT-3 induce intracellular CA2+ elevation in hippocampal neurones. *NeuroReport* **4**, 1303-1306.

Bliss, T. V. P., and Collingridge, G. L. (1993). A synaptic model of memory: Long-term potentiation in the hippocampus. *Nature (London)* **361**, 31-39.

Boise, L. H., Gonzalez-Garcia, M., Postema, C. E., Ding, L., Lindsten, T., Turka, L. A., Mao, X., Nunez, G., and Thompson, C. B. (1993). bcl-x, a bcl-2-related gene that functions as a dominant regulator of apoptotic cell death. *Cell (Cambridge, Mass.)* **74**, 597-608.

Boldin, M. P., Goncharov, T. M., Goltsev, Y. V., and Wallach, D. (1996). Involvement of MACH, a novel MORT-1/FADD-interacting protease, in Fas/APO-1 and TNF receptor-induced cell death. *Cell (Cambridge, Mass.)* **85**, 803-815.

Borasio, G. D., John, J., Wittinghofer, A., Barde, Y.-A., Sendtner, M., and Heumann, R. (1989). ras p21 protein promotes survival and fiber outgrowth of cultures embryonic neurons. *Neuron* **2**, 1087-1096.

Borasio, G. D., Markus, A., Wittinghofer, A., Barde, Y.-A., and Heumann, R. (1993). Involvement of ras p21 in neurotrophin-induced response of sensory, but not sympathetic neurons. *J. Cell Biol.* **121**, 665-672.

Borasio, G. D., Merkus, A., Heumann, R., Ghezzi, C., Sampietro, A., Wittinghofer, A., and Silani, V. (1996). Ras p21 protein promotes survival and differentiation of human embryonic neuronal crest-derived cells. *Neuroscience* **73**, 1121–1127.

Born, D. E., and Rubel, E. W. (1988). Afferent influences on brain stem auditory nuceli of the chicken: Presynaptic action potentials regulate protein synthesis in nucleus magnocellularis neurons. *J. Neurosci.* **8**, 901–919.

Bothwell, M. (1996). p75NTR: A receptor after all. *Science* **272**, 506–507.

Breakfield, X. O. (1982). Altered nerve growth factor in familial dysautonomia: Disovering the molecular basis of an inherited neurologic disease. *Neurosci. Comment* **1**, 28–32.

Brenneman, D. E., and Eiden, E. L. (1986). Vasoactive intestinal peptide and electrical activity influence neuronal survival. *Proc. Natl. Acad. Sci. U.S.A.* **83**, 1159–1162.

Brugg, B., Michel, P. P., Agid, Y., and Ruberg, M. (1996). Ceramide induces apoptosis in cultured mesencephalic neurons. *J. Neurochem.* **66**, 733–739.

Bueker, E. D. (1948). Implantation of tumors in the hind limb field of the embryonic chick and the developmental response of the lubosacral nervous system. *Anat. Rec.* **102**, 369–390.

Burgering, B. M., and Coffer, P. J. (1995). Protein kinase B (c-Akt) in phosphatidylinosotol-3-OH kinase signal transduction *Nature (London)* **376**, 599–602.

Burnham, P., Louis, J.-C., Magal, E., and Varon, S. (1994). Effects of ciliary neurotrophic factor on the survival and response to nerve growth factor of cultured rat sympathetic neurons. *Dev. Biol.* **161**, 96–106.

Busiglio, J., and Yankner, B. A. (1995). Apoptosis and increased generation of reactive oxygen species in Down's syndrome neurons in vitro. *Nature (London)* **378**, 776–779.

Canossa, M., Twiss, J. L., Verity, A. N., and Shooter, E. M. (1996). p75 (NGFR) and TrkA receptors collaborate to rapidly activate a p75 (NGFR)-associated protein kinase. *EMBO J.* **15**, 3369–3376.

Carter, B. D., Kalischmidt, C., Kaltschmidt, B., Offernhauser, N., Bohm-Matthaei, R., Baeuerle, P. A., and Barde, Y.-A. (1996). Selective activation of NF-kappaB by nerve growth factor through the neurotrophin receptor p75. *Nature (London)* **272**, 542–545.

Casey, P. (1995). Protein lipidation in cell signaling. *Science* **268**, 221–224.

Castagna, M., Takai, Y., Kaibuchi, K., Sano, K., Kikkawa, U., and Nishizuka, Y. (1982). Direct activation of calcium activated, phospholipid-dependent protein kinase by tumor-promoting phorbol esters. *J. Biol. Chem.* **257**, 7847–7851.

Chao, M. V. (1995). Ceramide: A potential second messenger in the nervous system. *Mol. Cell. Neurosci.* **6**, 91–96.

Chung, J., Grammer, T. C., Lemon, K. P., Kazlauskas, A., and Blenis, J. (1994). PDGF- and insulin-dependent pp70S6k activation mediated by phosphatidylinositol-3-OH kinase. *Nature (London)* **370**, 71–75.

Cleveland, J. L., and Ihle, J. H. (1995). Contenders in FasL/TNF death signaling. *Cell (Cambridge, Mass.)* **81**, 479–482.

Cohen, G. B., Ren, R., and Baltimore, D. (1995). Modular binding domains in signal transduction proteins. *Cell (Cambridge, Mass.)* **80**, 237–248.

Cohen, S. R., Levi-Montalcini, R., and Hamburger, V. (1954). A nerve growth-stimulating factor isolated from sarcomas 37 and 180. *Proc. Natl. Acad. Sci. U.S.A.* **40**, 1014–1018.

Collin, R. (1906). Recherches cytologiques sur le développement de la cellule nerveuse. *Nevraxe* **8**, 181–307.

Collins, F., and Lile, J. D. (1989). The role of dihydropyridine-sensitive voltage-gated calcium channels in potassium-mediated neuronal survival. *Brain Res.* **502**, 99–108.

Collins, F., Schmidt, M. F., Guthrie, P. B., and Kater, S. B. (1991). Sustained increase in intracellular calcium promotes neuronal survival. *J. Neurosci.* **11**, 2582–2587.

Colotta, F., Polentaruti, N., Sironi, M., and Mantovani, A. (1992). Expression and involvement of c-fos and c-jun protooncogenes in programmed cell death induced by growth factor deprivation in lympoid cell lines. *J. Biol. Chem.* **267**, 18278–18283.

Conover, J. C., Erickson, J. T., Katz, D. M., Cianchi, L. M., Poueymirou, W. T., McClain, J., Pan, L., Helgren, M., Ip, N. Y., Boland, P., Freidman, B., Weigland, S., Vejsada, R., Kato, A. C., DeChiara, T. M., and Yancopoulos, G. D. (1995). Neuronal deficits, not involving motor neurons, in mice lacking BDNF and/or NT4. *Nature (London)* **375**, 235–238.

Cordon-Cardo, C., Tapley, P., Jing, S. Q., Nanduri, V., O'Rourke, E., Lamballe, F., Kovary, K., Klein, R., Jones, K. R., Reichardt, L. F., and Barbacid, M. (1991). The trk tyrosine protein kinase mediates the mitogenic proterties of nerve growth factor. *Cell (Cambridge, Mass.)* **66**, 173–183.

Cotran, R. S., Kumar, V., and Robbins, S. L. (1994). *In* "Pathologic Basis of Disease" (F. J. Schoen, ed.), 5th ed. Saunders, Philadelphia.

Cross, D. A., Alessi, D. R., Cohen, P., Andjelkovich, M., and Hemmings, B. (1995). Inhibition of glycogen synthase kinase-2 by insulin mediated by protein kinase B. *Nature (London)* **378**, 785–789.

Crowley, C., Spencer, S. D., Nishimura, M. C., Chen, K. S., Pitts-Meek, S., Armanini, M. P., Ling, L. H., McMahon, S. B., Shelton, D. L., Levinson, A. D., and Phillips, H. S. (1994). Mice lacking nerve growth factor display perinatal loss of sensory and sympathetic neurons yet develop basal forebrain cholinergic neurons. *Cell (Cambridge, Mass.)* **76**, 1001–1011.

da Penha Berzaghi, M., Cooper, J., Castren, E., Zafra, F., Sofroniew, M., Thoenen, H., and Lindholm, D. (1993). Cholinergic regulation of brain-derived neurotrophic factor (BDNF) and nerve growth factor (NGF) but not neurotrophin-3 (NT-3) mRNA levels in the developing rat hippocampus. *J. Neurosci.* **13**, 3818–3826.

Davies, A. M., Lee, K.-F., and Jaenisch, R. (1993). p-75 deficient trigeminal sensory neurons have an altered response to NGF but not to other neurotrophins. *Neuron* **11**, 565–574.

Davis, R. J. (1994). MAPKs: New JNK expands the group. *Trends Biochem. Sci.* **19**, 470–473.

Davis, R. J. (1995). Transcriptional regulation of MAP kinases. *Mol. Reprod. Dev.* **42**, 459–467.

DePaolo, D., Reusch, J. E.-B., Carel, K., Bhuripanyo, P., Leitner, J. W., and Draznin, B. (1996). Functional interaction of phosphatidylinositol 3-kinase with GTPas-activating protein in 3T3-L1 adipocytes. *Mol. Cell. Biol.* **16**, 1450–1457.

Derijard, B., Hibi, M., Wu, I.-H., Barrett, T., Su, B., Deng, T., Karin, M., and Davis, R. J. (1994). JNK1: A protein kinase stimulated by UV light and Ha-Ras that binds and phosphorylates the c-Jun activation domain. *Cell (Cambridge, Mass.)* **76**, 1025–1037.

Diaz-Meco, M. T., Municio, M. M., Frutos, S., Sanchez, P., Lozano, J., Sanz, L., and Moscat, J. (1996). The product of par-4, a gene induced during apoptosis, interacts selectively with the atypical isoforms of protein kinase C. *Cell (Cambridge, Mass.)* **86**, 777–786.

DiCicco-Bloom, E., Freidman, W. J., and Black, I. B. (1993). NT-3 stimulates sympathetic neuroblast proliferation by promoting precursor survival. *Neuron* **11**, 1101–1111.

D'Mello, S. R., Galli, C., Ciotti, T., and Calissano, P. (1993). Induction of apoptosis in cerebellar granule neurons by low potassium: Inhibition of death by insulin-like growth factor I and cAMP. *Proc. Natl. Acad. Sci. U.S.A.* **90**, 10989–10993.

D'Mello, S. R., Borodezt, K., and Soltoff, S. P. (1997). Insulin-like growth factor and potassium depolarization maitain neuronal survivla by distinct pathways: Possible involvement of PI3-kinase in IGF-1 signaling. *J. Neurosci.* **17**, 1548–1560.

Dobrowsky, R. T., Kamibayashi, C., Mumby, M. C., and Hannun, Y. A. (1993). Ceramide activates heterotrimeric protein phosphatase 2A. *J. Biol. Chem.* **268**, 15523–15530.

Dobrowsky, R. T., Werner, M. H., Castellino, A. M., Chao, M. V., and Hannun, Y. A. (1994). Activation of the sphingomyelin cycle through the low affinity neurotrophin receptor. *Science* **265**, 1596–1599.

Dobrowsky, R. T., Jenkins, G. M., and Hannun, Y. A. (1995). Neurotrophins induce sphingomyelin hydrolysis: Modulation by coexpression of p75 with Trk receptors. *J. Biol. Chem.* **270**, 22135–22142.

Dragunow, M., Beilharz, E., Sirimanne, E., Lawlor, P., Williams, C., Bravo, R., and Gluckman, P. (1994). Immediate-early gene protein expression in neurons undergoing delayed death, but not necrosis, following hypoxic-ischaemic injury to the young rat brain. *Mol. Brain Res.* **25**, 19–33.

Dubois-Dauphin, M., Frankowski, H., Tsujimoto, Y., Huarte, J., and Martinou, J.-C. (1994). Neonatal motoneurons overexpressing the bcl-2 protooncogene in transgenic mice are protected from axotomy-induced cell death. *Proc. Natl. Acad. Sci. U.S.A.* **91**, 3309–3313.

Dudek, H., Datta, S. R., Franke, T. F., Birnbaum, M. J., Yao, R., Cooper, G. M., Segal, R. S., Kaplan, D. R., and Greenberg, M. E. (1997). Regulation of neuronal survival by the serine-threonine protein kinase Akt. *Science* **275**, 661–665.

Eide, E. E., Lowenstein, D. H., and Reichardt, L. F. (1993). Neurotrophins and their receptors—current concepts and implications for neurologic disease. *Exp. Neurol.* **121**, 200–214.

Ellis, H., and Horvitz, H. R. (1986). Genetic control of programmed cell death in the nematode *C. elegans. Cell (Cambridge, Mass.)* **44**, 817–829.

Ernfors, P., Lee, K.-F., and Jaenisch, R. (1994). Mice lacking brain-derived neurotrophic factor develop with sensory deficits. *Nature (London)* **368**, 147–150.

Estus, S., Zaks, W. J., Freeman, R. S., Gruda, M., Bravo, R., and Johnson, E. M., Jr. (1994). Altered gene expression in neurons during programmed cell death: Identification of c-jun as necessary for neuronal apoptosis. *J. Cell Biol.* **127**, 1717–1727.

Evan, G. I., Wyllie, A. H., Gilbert, C. S., Littlewood, T. D., Land, H., Brooks, M., Waters, C. M., Penn, L. Z., and Hancock, D. C. (1992). Induction of apoptosis in fibroblasts by c-myc protein *Cell (Cambridge, Mass.)* **69**, 119–128.

Eves, E. M., Boise, L. H., Thompson, C. B., Wagner, A. J., Hay, N., and Rosner, M. R. (1996). Apoptosis induced by differentiation or serum deprivation in an immortalized central nervous system neuronal cell line. *J. Neurochem.* **67**, 1908–1920.

Fainzilber, M., Smit, A. B., Syed, N. I., Wildering, W. C., Hermann, P. M., van der Schors, R. C., Jimenez, C., Li, K. W., van Minnen, J., Bulloch, A. G. M., Ibanez, C. F., and Geraerts, W. P. M. (1996). CRNF, a molluscan neurotrophic factor I that interacts with the p75 neurotrophin receptor. *Science* **274**, 1540–1543.

Farinas, I., Jones, K. R., Backus, C., Wang, X.-Y., and Reichardt, L. F. (1994). Severe sensory and sympathetic deficits in mice lacking neurotrophin-3. *Nature (London)* **369**, 658–661.

Farinelli, S. E., and Greene, L. A. (1996). Cell cycle blockes mimosine, ciclopirox, and deferoxamine prevent the death of PC12 cells and postmitotic sympathetic neurons after removal of trophic support. *J. Neurosci.* **16**, 1150–1162.

Farlie, P. G., Dringern, R., Rees, S. M., Kannourakis, G., and Bernard, O. (1995). bcl-2 transgene expression can protect neurons against developmental and induced cell death. *Proc. Natl. Acad. Sci. U.S.A.* **92**, 4397–4401.

Feddersen, R. M., Ehlenfeldt, R., Yunis, W. S., Clark, H. B., and Orr, H. T. (1992). Disrupted cerebellar cortical development and progressive degeneration of Purkinje cells in SV40 T antigen transgenic mice. *Neuron* **9**, 955–966.

Ferrari, G., and Greene, L. A. (1994). Proliferative inhibition by dominant-negative Ras rescues naive and neuronally differentiated PC12 cells from apoptotic death. *EMBO J.* **13**, 5922–5928.

Finkbeiner, S., and Greenberg, M. E. (1996). Ca2+-dependent routes to Ras: Mechanisms for neuronal survival, differentiation, and plasticity? *Neuron* **16**, 233–236.

Frade, J. M., Rodriguez-Tebar, A., and Barde, Y.-A. (1996). Induction of cell death by endogenous nerve growth factor through its p75 receptor. *Science* **383**, 166–168.

Franke, T. F., Yang, S. I., Chan, T. O., Datta, K., Kazlauskas, A., Morrison, D. K., Kaplan, D. R., and Tsichlis, P. N. (1995). The protein kinase encoded by the Akt proto-oncogene is a target of the PDGF-activated phosphatidlyinositol 3-kinase. *Cell (Cambridge, Mass.)* **81**, 727–736.

Franke, T. F., Kaplan, D. R., Cantley, L. C., and Toker, A. (1997). Direct regulation of the Akt proto-oncogene product by phosphatidylinositol-3,4-bisphosphate. *Science* **275**, 665–668.

Franklin, J. L., and Johnson, E. M., Jr. (1992). Suppression of programmed neuronal death by sustained elevation of cytoplasmic calcium. *Trends Neurosci.* **15,** 501–508.

Franklin, J. L., Sanz-Rodriguez, C., Juhasz, A., Deckworth, T. L., and Johnson, E. M., Jr. (1995). Chronic depolarization prevents programmed death of sympathetic neurons in vitro but does not support growth: Requirement for Ca2+ influx but not Trk activation. *J. Neurosci.* **15,** 643–664.

Freeman, R. S., Estus, S., and Johnson, E. M., Jr. (1994). Analysis of cell cycle-related gene expression in postmititic neurons: Selective induction of cyclin D1 during programmed cell death. *Neuron* **12,** 343–355.

Gagliardini, V., Fernandez, P.-A., Lee, R. K. K., Drexler, H. C. A., Rotello, R. J., Fishman, M. C., and Yuan, J. (1994). Prevention of vertebrate neuronal death by the crmA gene. *Science* **263,** 826–828.

Galcheva-Gargova, Z., Derijard, B., Wu, I.-H., and Davis, R. J. (1994). An osmosensing signal transduction cascade in mammalian cells. *Science* **265,** 806–808.

Gall, C. M., and Isackson, P. J. (1989). Limbic seizures increase neuronal production of messenger RNA for nerve growth factor. *Science* **245,** 758–761.

Galli, C., Meucci, O., Scorziello, A., Werge, T. M., Calissano, P., and Schettini, G. (1995). Apoptosis in cerebellar granule cells in blocked by high KCl, Forskolin, and IGF-1 through distinct mechanisms of action: The involvement of intracellular calcium and RNA synthesis. *J. Neurosci.* **15,** 1172–1179.

Gallo, V., Kingsbury, A., Balazs, R., and Jorgensen, O. S. (1987). The role of depolarization in the survival and differentiation of cerebellar granule cells in culture. *J. Neurosci.* **7,** 2203–2213.

Garcia, I., Martinou, I., Tsujimoto, Y., and Martinou, J.-C. (1992). Prevention of programmed cell dath of sympathetic neurons by the bcl-2 proto-oncogene. *Science* **258,** 302–304.

Ghosh, A., and Greenberg, M. E. (1995). Calcium signaling in neurons: Molecular mechanisms and cellular condsequences. *Science* **265,** 239–247.

Ghosh, A., Carnahan, J., and Greenberg, M. E. (1994). Requirement for BDNF in activity-dependent survival of cortical neurons. *Science* **263,** 1618–1623.

Gillardon, F., Baurle, J., Grusser-Cornehls, U., and Zimmerman, M. (1995). DNA fragmentation and activation of c-Jun in the cerebellum of mutant mice (weaver, Purkinje cell degeneration). *NeuroReport* **6,** 1766–1768.

Glass, D. J., Nye, S. H., Hazantopolous, P., Macchi, M. J., Squinto, S. P., Goldfarb, M., and Yancopoulos, G. D. (1991). TrkB mediates BDNF/NT3-dependent survival and proliferation of fibroblasts lacking the low affinity NGF receptor. *Cell (Cambridge, Mass.)* **66,** 405–413.

Glucksmann, A. (1951). Cell deaths in normal vertebrate ontogeny. *Biol. Rev. Cambridge Philos. Soc.* **26,** 59–86.

Gonzalez Garcia, M., Garcia, I., Ding, L., O'Shea, S., Boise, L. H., Thompson, C. B., and Nunez, G. (1995). bcl-x is expressed in embryonic and postnatal neural tissues and functions to prevent neuronal death. *Proc. Natl. Acad. Sci. U.S.A.* **92,** 4304–4308.

Gorin, P. D., and Johnson, E. M., Jr. (1979). Experimental autoimmune model of nerve growth factor deprivation: Effects on developing sympathetic and sensory neurons. *Proc. Natl. Acad. Sci. U.S.A.* **76,** 5382–5386.

Gotz, R., Koster, R., Winkler, C., Raulf, F., Lattspeich, F., Schartzl, M., and Thoenen, H. (1994). Neurotrophin-6 is a new member of the nerve growth factor family. *Nature (London)* **372,** 266–269.

Greene, L. A., and Kaplan, D. R. (1995). Early events in neurotrophin signalling via Trk and p75 receptors. *Curr. Opin. Neurobiol.* **5,** 579–587.

Greene, L. A., and Tischler, A. S. (1976). Establishment of a noradrenergic clonal line of rat adrenal pheochromocytoma cells which respond to nerve growth factor. *Proc. Natl. Acad. Sci. U.S.A.* **73,** 2424–2428.

Greenlund, L. J. S., Korsmeyer, S. J., and Johnson, E. M., Jr. (1995). Role of bcl-2 in the survival and function of developing and mature neurons. *Neuron* **15,** 649–661.

Hack, N., Hidaka, H., Wakefield, M. J., and Balazs, R. (1993). Promotion of granule cell survival by high K+ or excitatory amino acid treatment and Ca+/calmodulin-dependent protein kinase activity. *Neuroscience* **57,** 9–20.

Ham, J., Babij, C., Whitfield, J., Pfarr, C. M., Lallemand, D., Yaniv, M., and Rubin, L. L. (1995). A c-Jun dominant negative mutant protects sympathetic neurons against programmed cell death. *Neuron* **15,** 927–939.

Hamburger, V. (1992). History of the discovery of neuronal death in embryos. *J. Neurobiol.* **23,** 1116–1123.

Hamburger, V., and Levi-Montalcini, R. (1949). Proliferation, differentiation and degeneration in the spinal ganglia of the chick embryo under normal and experimental conditions. *J. Exp. Zool.* **111,** 457–501.

Hammang, J. P., Behringer, R. R., Baetge, E. E., Palmiter, R. D., Brinster, R. L., and Messing, A. (1993). Oncogene expression in retinal horizontal cells of transgenic mice results in a cascade of neurodegeneration. *Neuron* **10,** 1197–1209.

Hannun, Y. A. (1996). Functions of ceramide in coordinating cellular responses to stress. *Science* **274,** 1855–1859.

Hanson, P. I., and Schulman, H. (1992). Neuronal Ca2+/Calmodulin-dependent protein kinases. *Annu. Rev. Biochem.* **61,** 559–601.

Harrison, R. G. (1907). Observations on living developing nerve fibres. *Proc. Soc. Exp. Biol. Med.* **4,** 140–143.

Hayes, T. E., Valtz, N. L., and McKay, R. D. G. (1991). Downregulation of CDC2 upon terminal differentiation of neurons. *New Biol.* **3,** 259–269.

Heintz, N. (1993). Cell death and the cell cycle: A relationship between transformation and neurodegeneration? *Trends Biochem. Sci.* **18,** 157–159.

Hempstead, B. L., Martin-Zanca, D., Kaplan, D. R., Parada, L. R., and Chao, M. V. (1991). High-affinity NGF binding requires coexpression of the trk proto-oncogene and the low-affinity NGF receptor. *Nature (London)* **350,** 678–683.

Herdegen, T., Brecht, S., Mayer, B., Leah, J., Kummer, W., Bravo, R., and Zimmermann, M. (1993). Long-lasting expression of JUN and KROX transcirption factors and nitric oxide synthase in intrinsic neurons of the rat brain following axotomy. *J. Neurosci.* **13,** 4130–4145.

Hill, C. S., and Treisman, R. (1995). Transcriptional regulation by extracellular signals: Mechanisms and specificity. *Cell (Cambridge, Mass.)* **80,** 199–211.

Ibanez, C. F., Ebendal, T., Barbany, G., Murray-Rust, J., Blundell, T. L., and Persson, H. (1992). Disruption of the low affinity receptor binding site in NGF allows neuronal survival and differentiation by binding to the trk gene product. *Cell (Cambridge, Mass.)* **69,** 329–341.

Ichijo, H., Nishida, E., Irie, K., ten Dijke, P., Saitoh, M., Moriguchi, T., Takagi, M., Matsumoto, K., Miyazono, K., and Gotoh, Y. (1996). Induction of apoptosis by ASK1, a mammalian MAPKKK that activated SAPK/JNK and p38 signalling pathways. *Science* **275,** 90–94.

Jaiswal, R. K., Moodie, S. A., Wolfman, A., and Landreth, G. E. (1994). The mitogen-activated protein kinase cascade is activated by B-Raf in response to nerve growth factor through interaction with p21ras. *Mol. Cell. Biol.* **14,** 6944–6953.

Jayadev, S., Liu, B., Bielawska, A. E., Lee, J. Y., Nazaire, F., Pushkareva, M. Y., Obeid, L. M., and Hannun, Y. A. (1995). Role for ceramide in cell cycle arrest. *J. Biol. Chem.* **270,** 2047–2052.

Jones, K. R., Farinas, I., Backus, C., and Reichardt, L. F. (1994). Targeted disruption of the BDNF gene perturbs brain and sensory neuron development but not motor neuron development. *Cell (Cambridge, Mass.)* **76,** 989–999.

Joseph, C. K., Byun, H. S., Bittman, R., and Kolesnick, R. N. (1994). Substrate recognition by ceramide -activated protein kinase: Evidence that kinase activity is proline-directed. *J. Biol. Chem.* **268,** 20002–20006.

Kaminska, B., Filipkowski, R. K., Zurkowska, G., Lason, W., Przewlokci, R., and Kaczmarek, L. (1994). Dynamic changes in the composition of the AP-1 trnascription factor DNA-binding activity in rat brain following kainate-induced seizures and cell death. *Eur. J. Neurosci.* **6**, 1558–1566.

Kaplan, D. R., Hempstead, B. L., Martin-Zanca, D., Chao, M. V., and Parada, L. F. (1991a). The trk proto-oncogene product: A signal transducing receptor for nerve growth factor. *Science* **252**, 554–558.

Kaplan, D. R., Martin-Zanca, D., and Parada, L. F. (1991b). Tyrosine phosphorylation and tyrosine kinase activity of the trk proto-oncogene product induced by NGF. *Nature (London)* **350**, 158–160.

Kauffmann-Zeh, A., Rodriguez-Vicana, P., Ulrich, E., Gilbert, C., Coffer, P., Downward, J., and Evan, G. (1997). Suppression of c-Myc-induced apoptosis by Ras signalling through PI(3)K and PKB. *Nature (London)* **385**, 544–548.

Kerr, J. F. R., Wylliw, A. H., and Currie, A. R. (1972). Apoptosis: A basic biological phenomenon with wide-ranging implication in tissue kinetics. *Br. J. Cancer* **26**, 239–257.

Kerr, J. F. R., Searle, J., Harmon, B. V., and Bishop, C. J. (1987). Apoptosis. *In* "Perspectives in Mammalian Cell Death" (C. Potten, ed.), pp. 93–128. Oxford University Press, Oxford.

Kessler, J. A., Ludlam, W. H., Freidin, M. M., Hall, D. H., Michaelson, M. D., Spray, D. C., Dougherty, M., and Batter, D. K. (1993). Cytokine-induced programmed death of cultured sympathetic neurons. *Neuron* **11**, 1123–1132.

Kharbanda, S., Saleem, A., Shafman, T., Emoto, Y., Taneja, N., Rubin, E., Weichselbaum, R., Woodgett, J., Avruch, J., Kyriakis, J., and Kufe, D. (1995). Ionizing radiation stimulated a Grb2-mediated association of the stress-activated protein kinase with phosphatidylinositol 3-kinase. *J. Biol. Chem.* **270**, 18871–18874.

Klein, R., Jing, S. Q., Nanduri, V., O'Rourke, E., and Barbacid, M. (1991a). The trk proto-oncogene encodes a receptor for nerve growth factor. *Cell (Cambridge, Mass.)* **65**, 189–197.

Klein, R., Nanduri, V., Jing, S. A., Lamballe, F., Tapley, P., Bryant, S., Cordon-Cardo, C., Jones, K. R., Reichardt, L. F., and Barbacid, M. (1991b). The TrkB tyrosine kinase is a receptor for brain-derived neurotrophic factor and neurotrophin-3. *Cell (Cambridge, Mass.)* **66**, 395–403.

Klein, R., Smeyne, R. J., Wurst, W., Long, L. K., Auerbach, B. A., Joyner, A. L., and Barbacid, M. (1993). Targeted disruption of the trkB neurotrophin receptor gene results in nervous system lesions and neonatal death. *Cell (Cambridge, Mass.)* **75**, 113–122.

Klein, R., Silos-Santiago, S., Smeyne, R. J., Lira, S. A., Brambilla, R., Bryant, S., Zhang, L., Snider, W., and Barbacid, M. (1994). Disruption of the neurotrophin-3 receptor gene trkC eliminates Ia muscle afferents and results in abnormal movements. *Nature (London)* **368**, 249–251.

Klinghoffer, R. A., Duckworth, B., Valius, M., Cantley, L., and Kazlauskas, A. (1996). Platelet-derived growth factor-dependent activation of phosphatidylinositol 3-kinase is regulated by receptor binding of SH2-domain containing proteins which influence Ras activity. *Mol. Cell. Biol.* **16**, 5905–5914.

Koike, T. (1992). Molecular and cellular mechanism of neuronal degeneration caused by nerve growth factor deprivation approached through PC12 cell culture. *Prog. Neuro-Psychopharmacol Biol. Psychiaty* **16**, 95–106.

Koike, T., and Tanaka, S. (1991). Evidence that nerve growth factor dependence of sympathetic neurons for survival in vitro may be determined by levels of cytoplasmic free Ca2+. *Proc. Natl. Acad. Sci. U.S.A.* **88**, 3892–3896.

Koike, T., Martin, D. P., and Johnson, E. M., Jr. (1989). Role of Ca2+ channels in the ability of membrane depolarization to prevent neuronal death induced by trophic-factor deprivation: Evidence that levels of internal Ca2+ determine nerve growth factor dependence of sympathetic ganglion cells. *Proc. Natl. Acad. Sci. U.S.A.* **86**, 6421–6425.

Kolch, W., Heldecker, G., Kochs, G., Hummerl, R., Vahidi, H., Mischak, H., Finkenzeller, G., Marme, D., and Rapp, U. R. (1993). Protein kinase C-alpha activated RAF-1 by direct phosphorylation. *Nature (London)* **364,** 249–252.

Korsching, S. (1993). The neurotrophic factor concept: A reexamination. *J. Neurosci.* **13,** 2739–2748.

Kranenburg, O., van der Eb, A. J., and Zantema, A. (1996). Cyclin D1 is an essential mediator of apoptotic neuronal cell death. *EMBO J.* **15,** 46–54.

Kulik, G., Klippel, A., and Weber, M. J. (1997). Antiapoptotic signaling by the insulin-like growth factor I repector, phosphatidlyinositol 3-kinase and Akt. *Mol. Cell. Biol.* **17,** 1595–1601.

Lamballe, F., Klein, R., and Barbacid, M. (1991). TrkC, a new member of the trk facimly of tyrosine protein kinases, is a receptor for neurotrophin-3. *Cell (Cambridge, Mass.)* **66,** 967–979.

Lange-Carter, C. A., and Johnson, G. L. (1994). Ras-dependent growth factor regulation of MEK kinase in PC12 cells. *Science* **265,** 1458–1461.

Lazarus, K. J., Bradshaw, R. A., West, N. R., and Bunge, R. (1976). Adaptive survival of rat sympathetic neurons cultured without supporting cells or exogenous nerve growth factor. *Brain Res.* **113,** 159–164.

Lee, K.-F., Li, E., Huber, L. J., Landis, S. C., Sharpe, A. H., Chao, M. V., and Jaenisch, R. (1992). Targeted mutation of the gene encoding the low affinity NGF receptor leads to deficits in the peripheral sensory nervous system. *Cell (Cambridge, Mass.)* **69,** 737–749.

Lee, K.-F., Bachman, K., Landis, S., and Jaenisch, R. (1994). Dependence on p75 for innervation of some sympathetic targets. *Science* **263,** 1447–1449.

Lev, S., Moreno, H., Martinez, R., Canoll, P., Peles, E., Musacchio, J. M., Plowman, G. D., Rudy, B., and Schlessinger, J. (1995). Protein tyrosine kinase PYK2 involved in Ca(2+)-induced regulation of ion channel and MAP kinase functions. *Nature (London)* **376,** 737–745.

Levi-Montalcini, R. (1987). The nerve growth factor 35 years later. *Science* **237,** 1154–1162.

Levi-Montalcini, R., and Angeletti, P. U. (1966). Immunosympathectomy. *Pharmacol. Rev.* **18,** 619–628.

Levi-Montalcini, R., and Angeletti, P. U. (1968). Nerve growth factor. *Physiol. Rev.* **48,** 534–569.

Levi-Montalcini, R., and Booker, B. (1960). Destruction of sympathetic ganglia in mammals by an antiserum to nerve growth factor protein. *Proc. Natl. Acad. Sci. U.S.A.* **46,** 384–391.

Levi-Montalcini, R., and Hamburger, V. (1951). Selective growth stimulating effects of mouse sarcoma on the sensory and sympathetic system of the chick embryo. *J. Exp. Zool.* **116,** 321–362.

Levi-Montalcini, R., and Hamburger, V. (1953). A diffusible agent of mouse sarcoma, producing hyperplasia of sympathetic ganglia and hyperneurotization of viscera in the chick embryo. *J. Exp. Zool.* **123,** 233–288.

Ling, D. S. F., Petroski, R. E., and Geller, H. M. (1991). Both survival and development of spontaneously active rat hypothalamic neurons in dissociated culture are dependent on membrane depolarization. *Dev. Brain Res.* **59,** 99–103.

Lipton, S. A. (1986). Blockade of electrical activity promotes the death of mammalian retinal ganglion cells in culture. *Proc. Natl. Acad. Sci. U.S.A.* **83,** 9774–9778.

Lockshin, R. A., and Williams, C. M. (1964). Programmed cell death. II. Endocrine potentiation of the breakdown of the intersegmental muscles of silkmoths. *J. Insect Physiol.* **10,** 643–649.

Loeb, D. M., Maragos, J., Martin-Zanec, D., Chao, M. V., Parada, L. F., and Greene, L. A. (1991). The trk proto-oncogene rescues NGF responsiveness in mutant NGF-nonresponsive PC12 cell lines. *Cell (Cambridge, Mass.)* **66,** 961–966.

Maderdrut, J. L., Oppenheim, R. W., and Prevette, D. (1988). Enhancement of naturally-occuring cell death in the sympathetic and parasympathetic ganglia of the chicken embryo following blockade of ganglionic transmission. *Brain Res.* **444,** 189–194.

Mah, S. P., Zhong, L. T., Liu, Y., Roghani, A., Edwards, R. H., and Bredesen, D. E. (1993). The protooncogene bcl-2 inhibits apoptosis in PC12 cells. *J. Neurochem.* **60,** 1183–1186.

Marsh, H. N., Scholz, W. K., Lamballe, F., Klein, R., Nanduri, V., Barbacid, M., and Palfrey, H. C. (1993). Signal transuction events mediated by the BDNF receptor gp145(trkB) in primary hippocampal pyramidal cell culture. *J. Neurosci.* **13,** 4281–4292.

Martin, D. P., Ito, A., Horigome, K., Lampe, P. A., and Johnson, Jr., E. M. (1992). Biochemical characterization of programmed cell death in NGF-deprived sympathetic neurons. *J. Neurobiol.* **23,** 1205–1220.

Martin, S. J., and Green, D. R. (1995). Protease activation during apoptosis: Death by a thousand cuts? *Cell (Cambridge, Mass.)* **82,** 349–352.

Mathias, S., Dressler, K. A., and Kolesnick, R. N. (1991). Characterization of a ceramide-activated protein kinase: stimulation by tumor necrosis factor alpha. *Proc. Natl. Acad. Sci. U.S.A.* **88,** 10009–10013.

McCormick, F. (1994). Activators and effectors of ras p21 proteins. *Curr. Opin. Genet. Dev.* **4,** 71–76.

Merry, D. E., Veis, D. J., Hickey, W. F., and Korsmeyer, S. J. (1994). bcl-2 protein expression in widespread in the developing nervous system and retained in the adult PSN. *Development (Cambridge, UK)* **120,** 301–311.

Mesner, P. W., Winters, T. R., and Green, S. H. (1992). Nerve growth factor-withdrawl induced cell death in neuronal PC12 cells resembles that in sympathetic neurons. *J. Cell Biol.* **119,** 1669–1680.

Meyer-Franke, A., Kaplan, M. R., Pfrieger, F. W., and Barres, B. A. (1995). Characterization of signaling interaction that promote the survival and growth of developing retinal ganglion cells in culture. *Neuron* **15,** 805–819.

Miller, T. M., and Johnson, E. M., Jr. (1996). Metabolic and genetic analysis of apoptosis in potassium/serum-deprived rat cerebellar granule cells. *J. Neurosci.* **16,** 7487–7495.

Milligan, C. E., Prevette, D., Yaginuma, H., Homma, S., Cardwell, C., Fritz, L. C., Tomaselli, K. J., Oppenheim, R. W., and Schawtz, L. M. (1995). Peptide inhibitors of the ICE protease family arrest programmed cell death of motoneurons in vivo and in vitro. *Neuron* **15,** 385–393.

Minichiello, L., and Klein, R. (1996). TrkB and TrkC neurotrophin receptors cooperate in promoting survival of hippocampal and cerebellar granule neurons. *Genes Dev.* **10,** 2849–2858.

Moodie, S. A., Willumsen, B. M., Weber, M. J., and Wolfman, A. (1993). Complexes of Ras-GTP with Raf-1 and mitogen-activated protein kinase. *Science* **260,** 1658–1661.

Moriya, S., Kazlauskas, A., Akimoto, K., Hirai, S., Mizuno, K., Takenawa, T., Fukui, Y., Watanabe, Y., Ozaki, S., and Ohno, S. (1996). PDGF activates protein kinase C epsilon through redundant and independent signaling pathways involving phospholipase C-gamma or phosphatidylinositol 3-Kinase. *Proc. Natl. Acad. Sci. U.S.A.* **93,** 151–155.

Motoyama, N., Wang, F., Roth, K. A., Sawa, H., Nakayama, K.-i., Nakayama, K., Negishi, I., Senju, S., Zhang, Q., Fujii, S., and Loh, D. Y. (1995). Massive cell death of immature hematapoetic cells and neurons in bcl-x deficient mice. *Science* **267,** 1506–1510.

Muzio, M., Chinnaiyan, A. M., Kischkel, F. C., O'Rourke, K., Shevchenko, A., Ni, J., Scaffidi, C., Bretz, J. D., Zhang, M., Gentz, R., Mann, M., Krammer, P. H., Peter, M. E., and Dixit, V. M. (1996). FLICE, a novel FADD-homologous ICE/CED-3-like protease, is recruited to the CD95 (Fas/APO-1) death-inducing signaling complex. *Cell (Cambridge, Mass.)* **85,** 817–827.

Natoli, G., Costanzo, A., Ianni, A., Templeton, D. J., Woodgett, J. R., Balsano, C., and Levero, M. (1996). Activation of SAPK/JNK by TNF receptor 1 through a noncytotoxic TRAF2-dependent pathway. *Science* **275,** 200–203.

Nikodijevic, B., and Guroff, G. (1991). Nerve growth factor-induced increase in calcium uptake by PC12 cells. *J. Neurosci. Res.* **28,** 192–199.

Nishi, R., and Berg, D. K. (1981). Effects of high K+ concentrations on the growth and development of ciliary ganglion neurons in cell culture. *Dev. Biol.* **87**, 301–307.

Nishizuka, Y. (1992). Intracellular signaling by hydrolysis of phospholipids and activation of protein kinase C. *Science* **258**, 607–614.

Nobes, C., and Tolkovsky, A. (1995). Neutralizing anti-p21 ras Fabs suppress rat sympathetic neuron survival induced by NGF, LIF, CNTF and cAMP. *Eur. J. Neurosci.* **7**, 344–350.

Nobes, C. D., Markus, A., and Tolkovsky, A. M. (1996). Active p21 ras is sufficient for rescue of NGF-dependent rat sympathetic neurons. *Neuroscience* **70**, 1067–1079.

Obeid, L. M., Linardic, C. M., Karolak, L. A., and Hannun, Y. A. (1993). Programmed cell death induced by ceramide. *Science* **259**, 1769–1771.

Ohga, Y., Zirrgiebel, U., Hamner, S., Michaelidis, T. M., Cooper, J., Thoenen, H., and Lindholm, D. (1996). Cell density increases bcl-2 and bcl-x expression in addition to survival of cultured cerebellar granule neurons. *Neuroscience* **73**, 913–917.

Oltvai, Z. N., Milliman, C. L., and Korsmeyer, S. J. (1993). Bcl-2 heterodimerizes in vivo with a conserved homolog, Bax, that accelerates programmed cell death. *Cell (Cambridge, Mass.)* **74**, 609–619.

Oppenheim, R. W. (1981). Neuronal cell death and some related regressive phenomena during neurogenesis. A selective historical review and pregress report. *In* "Studies in Developmental Neurobiology, Essays in Honor of Viktor Hamburger" (W. M. Cowan, ed.). Oxford University Press, New York.

Oppenheim, R. W. (1991). Cell death during the development of the nervous system. *Annu. Rev. Neurosci.* **14**, 453–501.

Park, D. S., Farinelli, S. E., and Greene, L. A. (1996a). Inhibitors of cyclin-dependent kinases promote survival of post-mitotic neuronally differentiated PC12 cells and sympathetic neurons. *J. Biol. Chem.* **271**, 8161–8169.

Park, D. S., Stefanis, L., Yan, C. Y. I., Farinelli, S. E., and Greene, L. A. (1996b). Ordering the death pathway. *J. Biol. Chem.* **271**, 21898–21905.

Payne, D. M., Rossamondo, A., Martino, P., Erickson, A. K., Her, J. H., Shabanowitz, H., Hunt, D. F., Weber, M. J., and Sturgill, T. W. (1991). Identification of the regulatory phosphorylation sites in pp42/Mitogen-activated protein kinase (MAP kinase). *EMBO J.* **10**, 885–892.

Peppelenbosch, M. P., Tertoolen, L. G. J., de Vries-Smits, A. M. M., Qiu, R.-G., M'Rabet, L., Symons, M. H., de Laat, S. W., and Bos, J. L. (1996). Rac-dependent and independent pathways mediate growth factor-induced Ca2+ influx. *J. Biol. Chem.* **271**, 7883–7886.

Pincus, D. W., DiCicco-Bloom, E. M., and Black, I. B. (1990). Vasoactive intestinal peptide regulates mitosis, differentiation and survival of cultured sympathetic neuroblasts. *Nature (London)* **343**, 564–566.

Pinon, L. G. P., Minichiello, L., Klein, R., and Davies, A. M. (1996). Timing of neuronal death in trkA, trkB and trkC mutant embryos demonstrates neurotrophin switching in developing trigeminal neurons. *Development (Cambridge, UK)* **264**, 449–479.

Pittman, R. N., Wang, S., DiBenedetto, A. J., and Mills, J. C. (1993). A system for characterizing cellular and molecular events in programmed neuronal cell death. *J. Neurosci.* **13**, 3669–3680.

Plyte, S. E., Hughes, K., Nikolakaki, E., Pulverer, B. J., and Woodgett, J. R. (1992). Glycogen synthase kinase-3: Functions in oncogenesis and development. *Biochim. Biophys. Acta* **1114**, 147–162.

Pons, S., Asano, T., Glasheen, E., Miralplex, M., Zhang, Y., Fisher, T. L., Myers, M. G., Jr., Sun, X. J., and White, M. F. (1995). The structure and function of p55PIK reveal a new regulatory subunit for phosphatidylinositol 3-kinase. *Mol. Cell. Biol.* **15**, 4453–4465.

Poulsen, K. T., Armanini, M. P., Klein, R. D., Hynes, M. A., Phillips, H. S., and Rosenthal, A. (1994). TGF-beta2 and TGF-beta3 are potent survival factors for midbrain dopaminergic neurons. *Neuron* **13**, 1245–1252.

Prehn, J. H. M., Bindokas, V. P., Marcuccilli, C. J., Krajewski, S., Reed, J. C., and Miller, R. J. (1994). Regulation of neuronal bcl2 protein expression and calcium homeostatis by transforming growth factor-β confers wide-ranging protextion on rat hippocampal neurons. *Proc. Natl. Acad. Sci. U.S.A.* **91**, 12599–12603.

Rabizadeh, S., Oh, J., Zhong, L.-T., Yang, J., Bitler, C. M., Butcher, L. L., and Bredesen, D. E. (1993). Induction of apoptosis by the low-affinity NGF receptor. *Science* **261**, 345–348.

Raines, M. A., Kolesnick, R. N., and Golde, D. W. (1993). Sphingomyelinase and ceramide activate mitogen-activated protein kinase in myeloid HL-60 cells. *J. Biol. Chem.* **268**, 14572–14575.

Raingeaud, J., Gupta, S., Rogers, J. S., Dickens, M., Han, J., Ulevitch, R. J., and Davis, R. J. (1995). Pro-inflammatory cytokines and environmental stress cause p38 mitogen-activated protein kinase activation by dual phosphorylation on tyrosine and threonine. *J. Biol. Chem.* **270**, 7420–7426.

Rosen, L., and Greenberg, M. E. (1996). Stimulation of growth factor receptor signal transduction by aactivation of voltage-sensitive calcium channels. *Proc. Natl. Acad. Sci. U.S.A.* **93**, 1113–1118.

Rosen, L. B., Ginty, D. D., Weber, M. J., and Greenberg, M. E. (1994). Membrane depolarization and calcium influx stimulate MEK and MAP kinase via activation of Ras. *Neuron* **12**, 1207–1221.

Rosette, C., and Karin, M. (1996). Ultraviolet light and osmotic stress: Activation of the JNK cascade through multiple growth factor and cytokine receptors. *Science* **274**, 1194–1197.

Roy, N., Mahadevan, M. S., McLean, M., Shutler, G., Zahra, Y., Farahani, R., Baird, S., Besner-Johnson, A., Lefebvre, C., Kang, X., Salih, M., Aubry, H., Tamai, K., Guan, X., Ioannou, P., Crawford, T. O., de Jong, P. J., Surh, L., Ikeda, J.-E., Korneluk, R. G., and MacKenzie, A. (1995). The gene for neuronal apoptosis inhibitory protein is partially deleted in individuals with spinal muscular atrophy. *Cell (Cambridge, Mass.)* **80**, 167–178.

Rozakis-Adcock, A., McGlade, J., Mbamalu, G., Pelicci, G., Daly, R., Li, W., Batzer, A., Thomas, S., Brugge, J., Pelicci, P. G., Schlessinger, J., and Pawson, T. (1992). Association of the SHC and Grb2/sem-5 SH2 containing proteins is implicated in activation of the ras pathway by tyrosine kinases. *Nature (London)* **360**, 689–692.

Rubin, L. L., Philpott, K. L., and Brooks, S. F. (1993). The cell cycle and cell death. *Curr. Biol.* **3**, 391–394.

Rukenstein, A., Rydel, R. E., and Greene, L. A. (1991). Multiple agents rescue PC12 Cells from serum-free cell death by translation and transcription-independent mechanisms. *J. Neurosci.* **11**, 2552–2563.

Rusanescu, G., Qi, H., Thomas, S. M., Brugge, J. S., and Halegoua, S. (1995). Calcium influx induces neurite growth through a src-ras signaling cassette. *Neuron* **15**, 1415–1425.

Rydel, R. E., and Greene, L. A. (1987). Acidic and basic fibroblast growth factors promote stable neurite outgrowth and neuronal differentiation in cultures of PC12 cells. *J. Neurosci.* **7**, 3639–3653.

Scheid, M. P., Lauener, R. W., and Duronio, V. (1995). Role of phosphatidylinositol 3-OH-kinase activity in the inhibition of apoptosis in haemopoetic cels: Phosphatidylinositol 3-OH0kinase inhibitors reveal a difference in signalling betweeb interleukin-3 and granulocyte-macrophage colony stimulating factor. *Biochem. J.* **312**, 159–162.

Scott, B. S. (1971). Effect of potassium on neuron survival in cultures of dissociated human nervous tissue. *Exp. Neurol.* **30**, 297–308.

Scott, B. S., and Fisher, K. C. (1970). Potassium concentrations and number of neurons in cultures of dissociated ganglia. *Exp. Neurobiol.* **27**, 16–22.

Segal, R. A., and Greenberg, M. E. (1996). Intracellular signaling pathways activated by neurotrophic factors. *Annu. Rev. Neurosci.* **19**, 463–489.

Segal, R. A., Takahashi, H., and McKay, R. D. G. (1992). Changes in neurotrophin responsiveness during the development of cerebellar granule neurons. *Neuron* **9**, 1041–1052.

Sendtner, M., Holtmann, B., Kolbeck, R., Thoenen, H., and Barde, Y.-A. (1992). Brain-derived neurotrophic factor prevents the death of notoneurons in newborn rats after nerve section. *Nature (London)* **360**, 757–759.

Sharkey, J., and Butcher, S. P. (1994). Immunophilins mediate the neuroprotective effects of FK506 in focal cerebral ischaemia. *Nature (London)* **371**, 336–339.

Shou, C., Farnsworth, C. L., Neel, B. G., and Feig, L. A. (1992). Molecular cloning of cDNAs encoding a guanine-nucleotide-releasing factor for Ras p21. *Nature (London)* **358**, 351–354.

Siciniski, P., Donaher, J. L., Parker, S. B., Li, T., Fazeli, A., Gardner, H., Haslam, S. Z., Bronson, R. T., Elledge, S. J., and Weinberg, R. A. (1995). Cyclin D1 provides a link between development and oncogenesis in the retina and the breast. *Cell (Cambridge, Mass.)* **82**, 621–630.

Sjölander, A., Yamamoto, K., Huber, B. E., and Lapetina, E. G. (1991). Association of p21ras with phosphatidylinositol 3-kinase. *Proc. Natl. Acad. Sci. U.S.A.* **88**, 7908–7912.

Smeyne, R. J., Vendrell, M., Hayward, M., Baker, S. J., Miao, G. G., Schilling, K., Robertson, L. M., Curran, T., and Morgan, J. I. (1993). Continuous c-fos expression preceeds programmed cell death in vivo. *Nature (London)* **363**, 166–169.

Smeyne, R. J., Klein, R., Schnapp, A., Long, L. K., Bryant, S., Lewin, A., Lira, S. A., and Barbacid, M. (1994). Severe sensory and sympathetic neuropathies in mice carrying a disrupted Trk/NGF receptor gene. *Nature (London)* **368**, 246–249.

Smith, C. A., Farrah, T., and Goodwin, R. G. (1994). The TNF receptor superfamily of celluolar and viral proteins' Activation, costimulation and death. *Cell (Cambridge, Mass.)* **76**, 959–962.

Smits, A., Kato, M., Westermark, B., Nister, M., Heldin, C.-H., and Funa, K. (1991). Neurotrophic activity of platelet-derived growth factor (PDGF): Rat neuronal cells possess functional PDGF β-type receptors and respond to PDGF. *Proc. Natl. Acad. Sci. U.S.A.* **88**, 8159–8163.

Snider, W. D. (1994). Functions of the neurotrophins during nervous system development: What the knockouts are telling us. *Cell (Cambridge, Mass.)* **77**, 627–638.

Snider, W. D., and Johnson, E. M., Jr. (1989). Neurotrophic molecules. *Ann. Neurol.* **26**, 489–506.

Soppet, D., Escandon, E., Maragos, J., Middelmas, D. S., Reid, S. W., Blair, J., Burton, L. E., Standon, B. R., Kaplan, D. R., Hunter, T., Nikolics, K., and Parada, L. F. (1991). The neurotrophic factors brain-derived neurotrophic factor and neurotrophin-3 are ligands for the trkB tyrosine kinase receptor. *Cell (Cambridge, Mass.)* **65**, 895–903.

Squinto, S. P., Stitt, T. N., Aldrich, T. H., Davis, S., Bianco, S. M., Radziejewski, C., Glass, D. J., Masiakowski, P., Furth, M. E., Valenzuela, D. M., DiStephano, P. S., and Yancopoulos, G. D. (1991). trkB encodes a functional receptor for brain-derived neurotrophic factor and neurotrophin-3 but not nerve growth factor. *Cell (Cambridge, Mass.)* **65**, 885–893.

Steller, H. (1995). Mechanisms and genes of cellular suicide. *Science* **267**, 1445–1448.

Tepper, C. G., Jayadev, S., Liu, B., Bielawska, A., Wolff, R., Yonehara, S., Hannun, Y., and Seldin, M. F. (1995). Role for ceramide as an endogenous nediator of Fas-induced cytotoxicity. *Proc. Natl. Acad. Sci. U.S.A.* **92**, 8443–8447.

Thoenen, H. (1991). The changing scene of neurotrophic factors. *Trends Neurosci.* **14**, 165–170.

Thomas, K. R., and Capecchi, M. R. (1990). Targeted disruption of the murine int-1 protooncogene resulting in severe abnormalities in midbrain and cerebellar development. *Nature (London)* **346**, 847–850.

Thompson, C. B. (1995). Apoptosis in the pathogenesis and treatment of disease. *Science* **267**, 1456–1462.

Tokiwa, G., Dikic, I., Lev, S., and Schlessinger, J. (1996). Actiavtion of Pyk2 by stress signals and coupling with JNK signaling pathway. *Science* **273**, 792–794.

Tolkovsky, A. M., Walker, A. E., Murrell, R. D., and Suidan, H. S. (1990). Ca2+ transients are not required as signals for long-term neurite outgrowth from cultured sympathetic neurons. *J. Cell Biol.* **110**, 1295–1306.

Torres-Aleman, I., Pons, S., and Santos-Benito, F. F. (1992). Survival of Purkinje cells in cerebellar cultures is increased by insulin-like growth factor I. *Eur. J. Neurosci.* **4**, 864–869.

Van Antwerp, D. J., Martin, S. J., Kafri, T., Green, D. R., and Verma, I. M. (1996). Suppression of TNF-alpha induced apoptosis by NF-kappaB. *Science* **274**, 787–789.

van der Geer, P., Hunter, T., and Lindberg, R. A. (1994). Receptor protein-tyrosine kinases and their signal transduction pathways. *Annu. Rev. Cell Biol.* **10**, 251–337.

Veis, D. J., Sorenson, C. M., Shutter, J. M., and Korsmeyer, S. J. (1993). Bcl-2 deficient mice demonstrate fulminant lymphoid apoptosis, polycystic kidneys and hypopigmented hair. *Cell (Cambridge, Mass.)* **75**, 229–240.

Vemuri, G. S., and McMorris, F. A. (1996). Oligodendrocytes and their precursors require phosphatidylinositol 3-kinase signaling for survival. *Development (Cambridge, Mass.)* **122**, 2529–2537.

Verheij, M., Bose, R., Lin, X. H., Yao, B., Jarvis, W. D., Grant, S., Birrer, M. J., Szabo, E., Zon, L. I., Kyriakis, J. M., Haimovitz-Friedman, A., Fuks, Z., and Kolesnick, R. N. (1996). Requirement for ceramide-initiated SAPK/JNK signalling in stress-induced apoptosis. *Nature (London)* **380**, 75–79.

Virdee, K., and Tolksovsky, A. M. (1995). Activation of p44 and p42 MAP kinases is not essential for the survival of rat sympathetic neurons. *Eur. J. Neurosci.* **7**.

Virdee, K., and Tolksovsky, A. M. (1996). Inhibition of p42 and p44 mitogen-activated protein kinase activity by PD98059 does not suppress nerve growth factor-induced survival of sympathetic neurons. *J. Neurochem.* **67**, 1801–1805.

Vogel, K. S., Brannan, C. I., Jenkins, N. A., Copeland, N. G., and Parada, L. F. (1995). Loss of neurofibromin results in neurotrophin-independent survival of embryonic sensory and sympathetic neurons. *Cell (Cambridge, Mass.)* **82**, 733–742.

Vojtek, A., Hollenberg, S., and Cooper, J. A. (1993). Mammalian Ras interacts directly with the serine/threonine kinase Raf. *Cell (Cambridge, Mass.)* **74**, 205–214.

von Bartheid, C. S., Kinoshita, Y., Prevette, D., Yin, Q. W., Oppenheim, R. W., and Bothwell, M. (1994). Positive and negative effects of neurotrophins on the isthmo-optic nucleus in chick embryos. *Neuron* **12**, 639–654.

Wakade, A. R., Wakade, T. D., Malhotra, R. K., and Bhave, S. V. (1988). Excess K+ and phorbol ester activate protein kinase C and support the survival of chick sympathetic neurons in culture. *J. Neurochem.* **51**, 975–983.

Wang, C.-Y., Mayo, M. W., and Baldwin, A. S., Jr. (1996). TNF- and cancer therapy induced apoptosis: Potentiation by inhibition of NF-kappaB. *Science* **274**, 784–787.

Weskamp, G., and Reichardt, L. F. (1991). Evidence that biological activity of NGF is mediated through a novel subclass of high affinity receptors. *Neuron* **6**, 649–663.

Westwick, J. K., Bielawska, A. E., Dbaibo, G., Hannun, Y. A., and Brenner, D. A. (1995). Ceramide activates the stress-activated protein kinases. *J. Biol. Chem.* **270**, 22689–22692.

Wiesner, D. A., and Dawson, G. (1996a). Staurosporine induces programmed cell death in embruonic neurons and activation of the ceramide pathway. *J. Neurochem.* **66**, 1418–1425.

Wiesner, D. A., and Dawson, G. (1996b). Programmed cell death in neurotumor cells invovles the generation of ceramide. *Glycoconjugate J.* **13**, 327–333.

Williams, G. T. (1991). Programmed cell death: Apoptosis and oncogenesis. *Cell (Cambridge, Mass.)* **65**, 1097–1098.

Williams, R. W., and Herrup, K. (1988). Control of neuron number. *Annu. Rev. Neurosci.* **11**, 423–453.

Willumsen, B. M., Norris, K., Papageorge, A. G., Hubbert, N. L., and Lowy, D. R. (1984). Harvey murine sarcoma virus p21 ras protein: Biological and biochemical significance of the cysteine nearest the carboxyl terminus. *EMBO J.* **3**, 2581–2585.

Wood, K., Sarnecki, C., Roberts, T. M., and Blenis, J. (1992). Ras mediates nerve growth factor receptor modulation of three signal-transducing protein kinases: MAP kinase, Raf-1, and RSK. *Cell (Cambridge, Mass.)* **68**, 1041–1050.

Wood, K. A., Dipasquale, B., and Youle, R. J. (1993). In situ labeling of granule cells for apoptosis-associated DNA fragmentation reveals different mechanisms of cell loss in developing cerebellum. *Neuron* **11**, 621–632.

Wright, L. (1981). Cell survival in chick embryo ciliary ganglion is reduced by chrinic ganglionic blockade. *Brain Res.* **227**, 283–286.

Wyllie, A. H., Morris, R. G., Smith, A. L., and Dunlop, D. (1984). Chromatin cleavage in apoptosis: Association with condensed chromatin morphology and dependence on macromolecular synthesis. *J. Pathol.* **142**, 67–77.

Xia, Z., Dickens, M., Raingeaud, J., Davis, R. J., and Greenberg, M. E. (1995). Opposing effects of ERK and JNK-p38 MAP kinases on apoptosis. *Science* **270**, 1326–1331.

Xing, J., Ginty, D. D., and Greeberg, M. E. (1996). Coupling of the RAS-MAPK pathway to gene activation by RSK2, a growth factor-regulated CREB kinase. *Science* **273**, 959–963.

Yao, R., and Cooper, G. M. (1995). Requirement for phosphatidylinositol-3 kinase in the prevention of apoptosis by nerve growth factor. *Science* **267**, 2003–2006.

Yao, R., and Cooper, G. M. (1996). Growth factor-dependent survival of rodent fibroblasts requires phosphtidylinsitol 3-kinase but is independent of pp70s6k activity. *Oncogene* **13**, 343–351.

Zafra, F., Hengerer, B., Leibrock, J., Thoenen, H., and Lindholm, D. (1990). Activity dependent regulation of BDNF and NGF mRNAs in the rat hippocampus is mediated by non-NMDA glutamate receptors. *EMBO J.* **9**, 3545–3550.

Zafra, F., Castren, E., Thoenen, H., and Lindholm, D. (1991). Interplay between glutamate and gamma-aminobutyric transmitter systems in the physiological regulation of brain-derived neurotrophic factor and nerve growth factor synthesis in hippocampal neurons. *Proc. Natl. Acad. Sci. U.S.A.* **88**, 10037–10041.

Zafra, F., Lindholm, D., D. Castren, E., Hartikka, J., and Thoenen, H. (1992). Regulation of brain-derived neurotrophic factor and nerve growth factor mRNA in primary cultures of hippocampal neurona and astrocytes. *J. Neurosci.* **12**, 4793–4799.

Zhong, L.-T., Sarafian, T., Kane, D. J., Charles, A. C., Mah, S. P., Edwards, R. H., and Bredesen, D. E. (1993). bcl-2 inhibits death of central neural cells induced by multiple agents. *Proc. Natl. Acad. Sci. U.S.A.* **90**, 4533–4537.

Zirrgiebel, U., Ohga, Y., Carter, B., Berninger, B., Inagaki, N., Thoenen, H., and Lindholm, D. (1995). Characterization of TrkB receptor-mediated signaling pathways in rat cerebellar Granule neurons: Involvement of protein kinase C in neuronal survival. *J. Neurochem.* **65**, 2241–2250.

Patricia J. Willy
David J. Mangelsdorf
Department of Pharmacology and Howard Hughes Medical Institute
University of Texas Southwestern Medical Center
Dallas, Texas 75235

Nuclear Orphan Receptors: The Search for Novel Ligands and Signaling Pathways

This overview focuses on recent accomplishments and future directions in the nuclear orphan receptor field. Since their initial identification in 1988, research on this family of transcription factors has grown exponentially, and in the past 10 years over 30 different vertebrate orphan receptors have been characterized. The discovery of 9-*cis* retinoic acid as the high-affinity ligand for the retinoid X receptor (RXR) set the precedent that previously uncharacterized ligands exist for some of these receptors. The subsequent identification of ligands and activators for other orphan receptors has implicated the involvement of these novel regulators in a myriad of signaling pathways not previously known to involve nuclear receptors. Because many of these orphan receptors are also associated with metabolic and inherited disorders, they may be ideal targets for the development of improved pharmacologic agents for therapeutic use. Thus, the continued characterization of these proteins and their mechanisms of action is likely to have a significant future influence on many scientific disciplines.

I. Introduction

The nuclear receptors are a superfamily of transcription factors that are regulated in many cases by the binding of small, lipid-soluble ligands. This superfamily includes the known receptors for steroid and thyroid hormones, retinoids, and vitamin D, as well as a large number of newly discovered orphan receptors for which ligand activators are initially unknown. In the last decade, the study of nuclear orphan receptors has resulted in several important new discoveries that have had major impacts on the fields of endocrinology, pharmacology, and molecular biology. These discoveries have revealed the existence of previously unknown hormone signaling pathways and created new paradigms to study transcriptional regulation of gene expression. The purpose of this review is to focus on recent progress in the study of vertebrate nuclear orphan receptors. It is the authors' desire that this review also provide a useful reference source and bibliography. Where appropriate, the reader will be directed to one of several recent reviews that covers a particular topic in more detail or contains related ancillary material.

A. Background

Nuclear receptors are classically defined as ligand-activated transcription factors which share several common structural features that allow for DNA binding and transcriptional activation. These receptors are composed of several regions that can be delineated by function (Fig. 1A; see Mangelsdorf *et al.*, 1995). The amino-terminal A/B region is not well conserved and, in most receptors, contains a transcriptional activation function (AF-1) that works independently of ligand binding. The central DNA-binding domain (region C) is highly conserved and contains two zinc fingers that make critical contacts with specific nucleotide sequences called hormone response elements. The carboxy-terminal portion (regions D, E, and F) is required for ligand binding and receptor dimerization. In most receptors, this region also contains a second transcriptional activation domain (AF-2), which is ligand-dependent and highly conserved.

Members of the nuclear receptor superfamily fall into two distinct evolutionary and functional groups, the steroid hormone receptors and the nonsteroid hormone receptors (reviewed in detail by Beato *et al.*, 1995; Mangelsdorf and Evans, 1995). Steroid hormone receptors are found only in vertebrates, while nonsteroid receptors are phylogenetically diverse and found in virtually every animal species including the simplest metazoans (Mangelsdorf *et al.*, 1995). To date, nuclear receptors have not been found in plants, unicellular protozoans, or yeast (for reviews on the evolution of nuclear receptors, see Laudet *et al.*, 1992; Gronemeyer and Laudet, 1995). As their name implies, steroid hormone receptors bind steroidal ligands that are biosynthetically derived from pregnenolone and share similar structural/

A

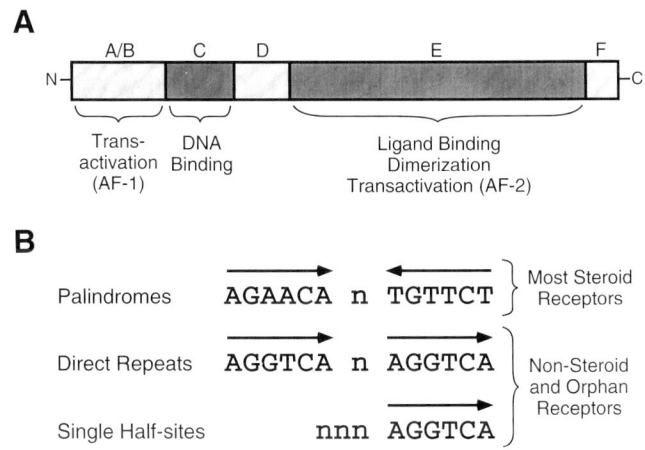

```
     A/B   C    D         E          F
N -[     |███|   |████████████████|    ]- C
    \___/ \_/         _____/
    Trans- DNA         Ligand Binding
  activation Binding   Dimerization
    (AF-1)             Transactivation (AF-2)
```

B

Palindromes	→ ← AGAACA n TGTTCT	Most Steroid Receptors
Direct Repeats	→ → AGGTCA n AGGTCA	Non-Steroid and Orphan Receptors
Single Half-sites	→ nnn AGGTCA	

FIGURE 1 Nuclear receptor structure and consensus DNA response elements. (A) General diagram of a typical nuclear hormone receptor as delineated by its functional domains (regions A through F). Regions C and E contain the conserved DNA- and ligand-binding domains that are the hallmark features of the nuclear receptor superfamily. The ~70-amino-acid DNA-binding domain is connected to the ligand-binding domain (~200–250 amino acids) by a short, flexible hinge (region D). In addition, structure/function studies have revealed the existence of two dimerization regions, one in the DNA-binding domain (not shown) and one in the ligand-binding domain. Nuclear receptors also contain two transactivation domains, designated AF-1 and AF-2. (B) Sequences of consensus DNA-binding sites for steroid, nonsteroid, and orphan receptors. Most steroid receptors bind as homodimers to inverted (palindromic) repeats of the consensus sequence AGAACA, while nonsteroid receptors and the majority of orphan receptors bind as hetero- or homodimers to direct repeats of the consensus sequence AGGTCA. Receptor specificity, and thus hormone responsiveness, is determined in large part by the number of nucleotides (n) between each repeat. For steroid receptors, $n = 3$; for nonsteroid receptors $n = 0$ to 5. In addition to dimeric receptors, several orphan receptors bind as monomers to single hexad consensus sequences. Response element specificity for these receptors is further determined by the nature of the 5-prime flanking nucleotides.

functional properties and mechanisms of action. In their unliganded state, steroid receptors associate with heat shock proteins that prevent their interaction with DNA. Ligand binding induces a conformation change in the receptor that releases heat shock proteins and permits the receptor to bind its cognate response element and interact with a variety of coactivator proteins (reviewed in Horwitz *et al.*, 1996). Steroid receptors bind to DNA exclusively as homodimers on response elements arranged as inverted (palindromic) repeats of two consensus hexanucleotide half-sites separated by three spacer nucleotides (Fig. 1B). This mechanism of steroid receptor DNA binding and activation appears to have evolved more recently than other nuclear receptor activation pathways, since these receptors are found only in vertebrates (Mangelsdorf *et al.*, 1995).

In contrast to the steroid receptors, the nonsteroid receptors include all of the known orphan receptors as well as those receptors that bind to a variety of biosynthetically unrelated, lipophilic ligands (e.g., retinoids, prostanoids, sterols, and thyroxine). Unlike their steroid receptor counterparts, in most cases these receptors do not associate with heat shock proteins and are believed to be bound to their DNA response elements in the absence of ligand. In this way, some nonsteroid receptors are able to function in the absence of ligand as either transcriptional repressors or activators of their target genes (Damm et al., 1989; Apfel et al., 1994; Song et al., 1994; Willy et al., 1995; Wong et al., 1995). For many of the nonsteroid receptors that have known ligands, transactivation is a multistep process. In the absence of ligand, the DNA-bound receptor can be associated with corepressor proteins that effectively block basal transcription of target genes (Chen and Evans, 1995; Hörlein et al., 1995). Upon ligand binding these receptors undergo a conformational change that disrupts their interaction with corepressor and restores basal transcription. Concomitantly, the new conformation of the receptor facilitates an interaction with coactivator proteins that leads to the classic ligand-induced transcriptional response (reviewed in Horwitz et al., 1996).

The nonsteroid receptors can be further classified based on their mode of DNA binding (Mangelsdorf and Evans, 1995). In the first class are those receptors that bind to DNA as heterodimers. This class includes the retinoic acid receptor (RAR), thyroid hormone receptor (TR), vitamin D receptor (VDR), retinoid X receptor (RXR), and several orphan receptors which are discussed in detail below. Although many of these receptors have been shown to bind DNA by themselves at high concentrations *in vitro*, at physiological concentrations all of these receptors appear to require RXR as their exclusive dimeric partner for high-affinity binding to their response elements (Yu et al., 1991; Kliewer et al., 1992b; Leid et al., 1992a; Zhang et al., 1992; Marks et al., 1992; Bugge et al., 1992). The DNA sequences bound by RXR heterodimers are generally arranged as inverted repeats (palindromes) or direct repeats (DRs) of two consensus half-sites separated by a variable number of nucleotides (Fig. 1B). The specificity of receptor binding is determined predominantly by the number of nucleotides separating the half-sites (Umesono et al., 1991). In the uninduced state of many of these heterodimers, RXR cannot bind ligand and its AF-2 domain is not absolutely required for transactivation. In these cases, RXR is said to be a silent partner. However, as will be discussed below, several orphan receptors that heterodimerize with RXR have the unique ability to convert RXR from a silent, nonligand binding partner into a ligand-inducible receptor.

Two other classes of nonsteroid receptors are those that bind DNA as homodimers or monomers. The DNA response elements for homodimeric receptors usually consist of consensus half-sites configured as direct repeats, whereas for monomeric receptors the response elements consist of a single half-site that includes several additional 5' nucleotides (Fig. 1B). Thus far, all nonsteroid receptors that bind DNA as homodimers or monomers are orphan receptors.

B. Definition of an Orphan Receptor

The original members of the nuclear receptor superfamily (i.e., the steroid, vitamin D, thyroid hormone, and retinoic acid receptors) were discovered through a systematic biochemical characterization of the proteins that directly bind fat-soluble hormones and mediate their effects. When it was demonstrated that these receptors shared a common structure consisting of highly conserved DNA- and ligand-binding domains, it was reasonable to presume that other members of this protein family might exist as previously uncharacterized receptors for other lipid signaling molecules. In the late 1980s the term "orphan receptor" was used to describe the first of what has become a large number of new gene products that belong to the nuclear receptor superfamily, but for which ligands are initially unknown. Today it is realized that the term orphan receptor may in some cases be a misnomer, since some of these proteins may not be ligand-dependent (see below). Nevertheless, for more or less historical reasons, the term orphan receptor continues to be used and has come by practice to refer to any protein that is related by sequence identity to other members of the superfamily. Most orphan receptors have been identified by low-stringency hybridization screening of cDNA libraries, purification of proteins that bind specific promoter sequences, or using yeast two-hybrid technology. Thus far, over 50 different orphan receptor genes have been characterized from various species. One of the greatest challenges in the nuclear receptor field has been to decipher the function of these proteins and their mechanism of action. This review focuses on the vertebrate orphan receptors, which have been the most extensively studied. For a comprehensive review of insect orphan receptors, see Thummel (1995).

Although nuclear orphan receptors form a group based on a lack of information regarding their potential ligands, it is important to note that in many cases these proteins differ significantly in their putative modes of action. Elucidation of ligand activating potential is one key to understanding orphan receptor function; however, several other distinct properties of these proteins have yielded a great deal of information about how they may regulate gene expression. A summary of this information is provided in Table I. This table lists the orphan receptors, grouped into their gene families, along with a brief description of their mode of DNA binding, tissue distribution, chromosomal location, and *Drosophila* homologues, and whether or not targeted disruption of their gene has been accomplished. An inspection of Table I reveals that several receptors have been given more than one name as a consequence of their independent identification by several laboratories. For the sake of practicality and to diminish the redundancy of orphan

TABLE I Vertebrate Orphan Receptors

Receptor gene family	DNA binding mode	Subtype	Names[a]	Species[b]	Drosophila homolog[c]	Tissue distribution[d]	Chromosomal localization[e]	K/O[f]
COUP-TF	Homodimer/ heterodimer	α	COUP-TFI, ear-3	h,r,a,f,i	SVP	Widespread, developing organs, CNS	h 5q14; m 13	+
		β	COUP-TFII, ARP-1	h,r,a,c		Widespread, developing organs, CNS	h 15q26; m 7	+
		γ	ear-2	h,r		Widespread, fetal liver	h 19	−
DAX-1	Unknown[g]		DAX-1	h,r		Adrenals, gonads, pituitary, hypothalamus	h Xp21; m X	*
ERR	Monomer/homodimer	α	ERRα, ERR1	h,r		Widespread, esp. CNS	h 11q12-13	+
		β	ERRβ, ERR2	h,r		Kidney, heart	h 14q24.3	+
FTZ-F1	Monomer	α	2 isoforms: SF-1, Ad4BP; ELP	h,r,b,i	FTZ-F1	Adrenals, gonads, brain, hypothalamus, pituitary	h 9q33; m 2	+
		β	FTF, LRH-1, PHR-1, FF1rA	h,r,a,i		Liver, pancreas		−
FXR	Heterodimer		FXR, RIP14	r		Kidney, liver, gut, adrenals	m 10	−
GCNF	Homodimer		GCNF, RTR	h,r		Germ cells	h 20q12-q13.1	+
HNF-4	Homodimer		HNF-4	r,i	dHNF4	Liver, kidney, intestine, pancreas	h 11q23.3	−
LXR	Heterodimer	α	LXRα, RLD-1	h,r		Liver, kidney, spleen, fat, intestine, pituitary, adrenals		
		β	LXRβ, UR, NER, OR-1, RIP15	h,r		Widespread	h 19q13.3	−
MB67	Heterodimer		MB67, CAR	h,r		Liver	h 12q13; m 15	−
NGFI-B	Monomer/heterodimer[h]	α	NGFI-B, Nur77, N10, NAK1, TR3, TIS1	h,r,i	DHR38	Thymus, brain, adrenal gland, muscle, testis		+

312

		β	NURR1, NOT, RNR-1	h,r		Brain, regenerating liver	h 2q22-q23	—
		γ	NOR-1, MINOR, TEC	h,r		Fetal brain, lung; adult heart, skel. muscle	h 9q	—
ONR1	Heterodimer		ONR1	a		Xenopus embryos		—
PPAR	Heterodimer	α	PPARα	h,r,a		Liver, heart, kidney, brown fat	h 22q12-q13.1; m 15	+
		β	PPARβ, PPARδ, NUC-1, FAAR	h,r,a		Widespread	h 6p21.1-p21.2; m 17	—
		γ	PPARγ	r,a		Adipose	h 3p25; m 6	—
REV-ERB	Monomer/homodimer	α	Rev-ErbA-α, ear-1	h,r,i		Widespread, esp. skel. muscle, brown fat	h 17q21	—
		β	Rev-Erbβ, RVR, BD73	h,r		Widespread	m 14	*
ROR	Monomer/homodimer	α	RORα, RZRα	h,i		Widespread, esp. peripheral blood leukocytes	h 15q21-q22; m 9	
		β	RZRβ	r		Brain, retina	m 19	—
		γ	RORγ, TOR	h,r,a,c,f,i		Skeletal muscle, thymus	h 1q22-23; m 3	—
RXR	Homodimer/ heterodimer	α	RXRα	h,r,a,c,f,i	USP	Widespread, esp. skin, liver, kidney, lung, muscle	h 9q34; m 2	+
		β	RXR3β, H2RIIBP	h,r,a,f		Widespread	h 6p21.3; m 17	+
		γ	RXRγ	h,r,a,c,f		Heart, brain, lung, adrenal, kidney, muscle, liver	h 1q22-q23; m 1	+
SHP	None[i]		SHF	h,r		Liver, pancreas, heart		—
TLX	Monomer/homodimer		Tlx	r,c,f,i	Tll	Developing brain, eye		—
TR2	Homodimer	α	TR2	h,r		Testis, prostate, seminal vesical	h 12q22; m 10	—
		β	TR4, TAK1, TR2R1	h,r		Widespread, esp. testis, brain, kidney, skel. muscle	h 3p25	—

(continued)

313

TABLE I (*continues*)

[a] Please see text for nomenclature citations.

[b] Species indicate homologs or related receptors in other species: a, amphibian; b, bovine; c, chicken; f, fish; h, human; i, invertebrate; r, rodent.

[c] For references, see Thummel (1995).

[d] Tissue distribution refers to major sites of mRNA expression. References: COUP-TFα (Miyajima et al., 1988; Lu et al., 1994; Qiu et al., 1994; Jonk et al., 1994); COUP-TFβ (Ladias and Karathanasis, 1991; Qiu et al., 1994; Jonk et al., 1994; Lutz et al., 1994); COUP-TFγ (Miyajima et al., 1988; Barnhart and Mellon, 1994); DAX-1 (Jonk et al., 1994); ERRα (Giguère et al., 1988); ERRβ (Giguère et al., 1988); FTZ-F1α (Honda et al., 1993; Ikeda et al., 1993; Morohashi et al., 1994; Ikeda et al., 1994); FTZ-F1β (Galarneau et al., 1996; Becker-André et al., 1993); FXR (Forman et al., 1995); GCNF (Chen et al., 1994a); HNF-4 (Sladek et al., 1990; Miquerol et al., 1994); LXRα (Apfel et al., 1994; Willy et al., 1995); LXRβ (Song et al., 1994); MB67 (Milbrandt, 1988; Ryseck et al., 1989; Law et al., 1992; Nakai et al., 1990); NGFI-Bβ (Law et al., 1992; Scearce et al., 1993); NGFI-Bγ (Ohkura et al., 1994; Hedvat and Irving, 1995); ONR (Smith et al., 1994); PPARα (Issemann and Green, 1990); PPARβ (Schmidt et al., 1992; Kliewer et al., 1994); PPARγ (Tontonoz et al., 1994); Rev-Erbα (Lazar et al., 1989); Rev-Erbβ (Dumas et al., 1994); RORα (Becker-André et al., 1993); RORβ (Carlberg et al., 1994; Becker-André et al., 1994); RORγ (Hirose et al., 1994b; Ortiz et al., 1995); RXRα, RXRβ, and RXRγ (Mangelsdorf et al., 1992); SHP (Seol et al., 1996); TLX (Yu et al., 1994; Monaghan et al., 1995); TR2α (Chang and Kokontis, 1988); TR2β (Chang et al., 1994; Hirose et al., 1994a).

[e] h, human; m, mouse. References: COUP-TFα and COUP-TFβ (Qiu et al., 1995); COUP-TFγ (Miyajima et al., 1988); DAX-1 (Zanaria et al., 1994; Guo et al., 1996); ERRα and ERRβ (V. Giguère, personal communication); FTZ-F1α (V. Giguère, personal communication); FTZ-F1β (Taketo et al., 1995); FXR (Kozak et al., 1996); HNF-4 (Yamagata et al., 1996); LXRα (G. Evans, personal communication); LXRβ (Song et al., 1994); NGFI-Bα (Ryseck et al., 1989); NGFI-Bβ (Mages et al., 1994); NGFI-Bγ (Ohkura et al., 1996); PPARα (Sher et al., 1993; Jones, P. S., et al., 1995); PPARβ (Yoshikawa et al., 1996a; Jones, P. S., et al., 1995); PPARγ (Greene et al., 1995; Jones, P. S., et al., 1995); Rev-Erbα (Miyajima et al., 1989); Rev-Erbβ (V. Giguère, personal communication); RORα (Giguère et al., 1995); RORβ and RORγ (V. Giguère, personal communication; RXRα, RXRβ, and RXRγ (Almasan et al., 1994; Hoopes et al., 1992); TR2α (Lee et al., 1996; Lee. C. H., et al., 1995); TR2β (Yoshikawa et al., 1996b).

[f] K/O refers to targeted gene knock-outs that have (plus) or have not (minus) been reported. An asterisk indicates a natural gene mutation or disruption. Targeted disruption references: COUP-TFα and COUP-TFβ (M.-J. Tsai, personal communication); DAX-1 (Zanaria et al., 1994); ERRα and ERRβ (V. Giguère, personal communication); FTZ-F1α (Luo et al., 1994); HNF-4 (Chen et al., 1994b); NGFI-Bα (Lee, S. L., et al., 1995); PPARα (Lee, S.-T., et al., 1995); RORα (Hamilton et al., 1996); RXRα (Sucov et al., 1994); RXRβ (Kastner et al., 1996); RXRγ (Krezel et al., 1996).

[g] DAX-1 lacks a conventional DNA binding domain, but has been reported to bind to a DR-5 type sequence (Zanaria et al., 1994).

[h] NGFI-B binds to DNA as a monomer; however, on specific response elements, NGFI-B can heterodimerize with RXR (see text for details).

[i] SHP lacks a conventional DNA binding domain, and while it can heterodimerize with RXR, DNA binding has not been demonstrated thus far.

receptor nomenclature, in this review receptors with more than one name will be referred to by their most common gene family name. Receptor subtypes are the products of individual genes, whereas isoforms are derived through alternative splicing or promoter usage.

The following sections provide a detailed discussion of how the study of nuclear orphan receptors has led to the discovery of novel signaling molecules and pathways, and opened up new areas of research in endocrine physiology. Section II describes the approaches used to search for novel ligand activities and the recent exciting successes that have resulted. Section III reviews those orphan receptors that function primarily as RXR heterodimers. Interestingly, the RXR heterodimer partners are the only orphan receptors to date for which definitive ligand activities have been discovered. The ability of these receptors to be activated by their own unique ligands or participate as partners in RXR retinoid signaling pathways is discussed.

Section IV gives an overview of orphan receptors that can function as transcriptional activators or repressors independently of RXR. Unlike the RXR heterodimers, these receptors display a high degree of diversity in their DNA binding and transactivation properties. While ligands may yet be discovered for these orphan receptors, it appears that many, if not all, of these proteins are capable of regulating gene expression in a ligand-independent manner.

While a great deal of information regarding the *in vitro* DNA-binding and transactivation properties of many orphan receptors has been elucidated, in most cases this information offers little insight into how these proteins function *in vivo*. Targeted gene disruption techniques have been used successfully to demonstrate critical roles for several orphan receptors in early mouse development and results from these studies are also provided in the following sections.

II. Discovery of Orphan Receptor Ligands

A. The Hunt for Ligands

Perhaps the most enticing aspect of orphan receptor research is the prospect of discovering a new ligand. The identification of novel signaling molecules generates exciting possibilities for uncovering previously unknown endocrine pathways and also offers the hope of finding new compounds for pharmacological use in treating disease. Thus, determining which receptors are ligand responsive and characterizing their cognate ligands continues to be a major research goal.

A common approach that has been used to successfully identify orphan receptor ligands employs a cell-based cotransfection assay (for example, see Heyman *et al.*, 1992). In this assay, cultured cells are transiently transfected with a plasmid expressing the orphan receptor of interest, along with a second plasmid encoding a reporter gene capable of expressing a quantifiable product. The reporter gene is driven by a promoter containing a response element specific for the orphan receptor, such that expression of the reporter gene is dependent on activation of the orphan receptor by a potential ligand. Using this assay, a large number of compounds can be screened for receptor activation and subsequent reporter gene expression. Thus far, all known ligands for nuclear hormone receptors have been small lipophilic molecules, most of which are derived from mevalonic acid metabolism (Mangelsdorf *et al.*, 1995). Consequently, the search for new orphan receptor ligands generally employs techniques aimed at identifying small molecules with these characteristics. Sources for potential ligands include known natural and synthetic compounds, chemical libraries, and fractionated lipid extracts derived from tissues, cells, and serum. Once a class of compounds capable of activating a receptor has been found, further experiments are done to identify the most potent activator. It is also critical to determine if the observed activation is through direct, high-affinity ligand binding to the receptor at physiologically relevant concentrations, or if activation is through a molecule that is simply chemically similar to, or derived from, the true ligand. Another possibility is that activation may proceed through a secondary pathway such as phosphorylation or by affecting other proteins that interact with the orphan receptor (e.g., coactivators or corepressors). It is important to note that activation through both direct and indirect mechanisms may reflect physiologically relevant pathways. Within the last 10 years, a small but significant number of hormone-like lipids have been identified as activators for several orphan receptors utilizing this cotransfection approach (Fig. 2). The discovery and biological implications of these novel ligands are discussed below.

B. The Retinoid X Receptors

The first successful screening for a novel orphan receptor ligand resulted in the identification of *9-cis* retinoic acid as a high-affinity ligand for the retinoid X receptor (RXR). The RXR family of receptors contains three subtypes in mammals: α (Mangelsdorf *et al.*, 1990), β (also called H-2RIIBP) (Hamada *et al.*, 1989), and γ (Mangelsdorf *et al.*, 1992). Several additional RXR subtypes have also been described in zebrafish, but like the invertebrate RXR homolog, ultraspiracle (USP), these additional subtypes are not responsive to retinoid ligands (B. B. Jones *et al.*, 1995; Oro *et al.*, 1990). RXRs display distinct but overlapping patterns of expression and have been found in virtually every tissue and cell type that has been examined (Mangelsdorf *et al.*, 1992). In addition to their role in retinoid signaling, RXRs serve as obligate heterodimeric partners to nonsteroid receptors, allowing these proteins to function in several signaling pathways (for detailed reviews of

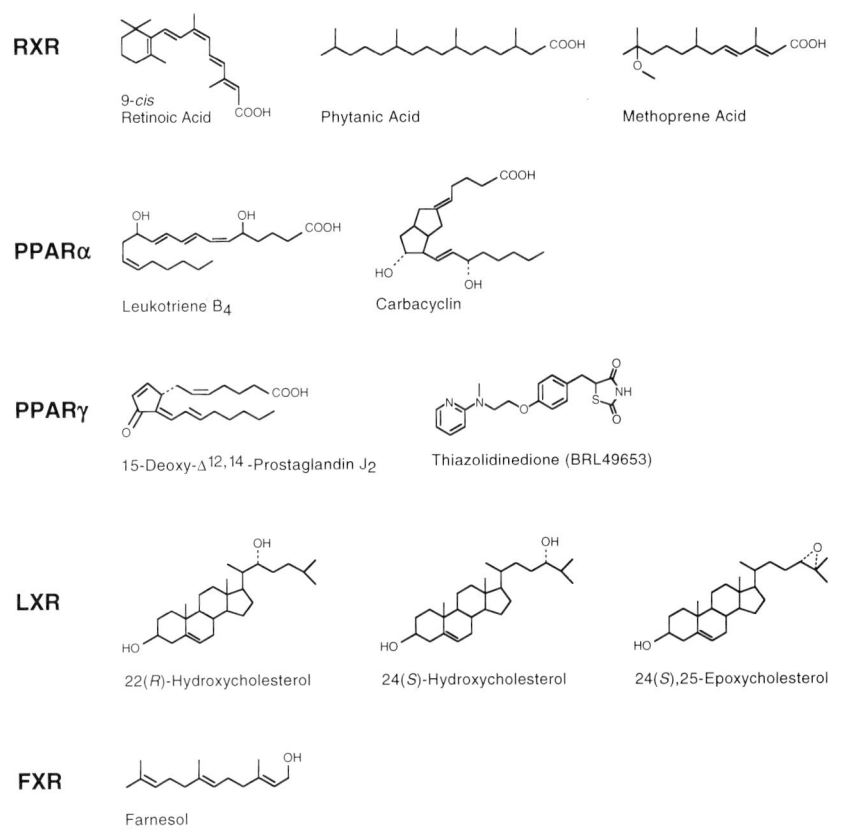

FIGURE 2 Structures for orphan receptor ligands and activators. See text for details on these receptors and their corresponding ligands.

RXR heterodimers, see Leid *et al.*, 1992b; Mangelsdorf and Evans, 1995). RXRα was initially cloned by low-stringency hybridization screening with a cDNA probe to the DNA binding domain of the retinoic acid receptor (RAR). Although its ligand binding domain is only 27% identical to that of RAR, RXRα was found to be transcriptionally responsive to all-*trans* retinoic acid in the cotransfection assay (Mangelsdorf *et al.*, 1990). When binding assays revealed that all-*trans* retinoic acid did not directly bind RXRα, a search for the true ligand, presumed to be a metabolite of all-*trans* retinoic acid (termed retinoid X), was undertaken. This ligand hunt came to fruition when two independent laboratories using different experimental approaches (lipid extract fractionation and ligand trap experiments) identified 9-*cis* retinoic acid (a photo-isomer of all-*trans* retinoic acid) as a high-affinity ligand for all three RXRs (Heyman *et al.*, 1992; Levin *et al.*, 1992; Mangelsdorf *et al.*, 1992). It was subsequently discovered that 9-*cis* retinoic acid is also a high-affinity ligand for RAR (Heyman *et al.*, 1992).

The finding that RXR is both an essential heterodimeric partner and a 9-*cis* retinoic acid receptor has significantly increased our understanding of the complex developmental and regulatory processes controlled by retinoid and other endocrine signaling pathways (reviewed in Chambon, 1996; Mangelsdorf *et al.*, 1994). Targeted disruption of each of the three RXR genes in mice has furthered this understanding and confirmed RXR's role as a master regulator (reviewed in Kastner *et al.*, 1995).

Since the initial discovery of 9-*cis* retinoic acid, two noncyclic terpenoids, methoprene acid and phytanic acid, have also been identified as ligands for RXRs (Harmon *et al.*, 1995; LeMotte *et al.*, 1996; Kitareewan *et al.*, 1996). Unlike 9-*cis* retinoic acid, which can also activate RARs, the noncyclic terpenoids are highly selective for binding and activating RXRs, albeit at much higher concentrations than 9-*cis* retinoic acid. Methoprene acid is a metabolite of methoprene, a synthetic analog of the insect growth regulator, juvenile hormone. Methoprene is an environmental contaminant due to its use as a pesticide. This compound works by conferring a terminal juvenile state on the insect, thus preventing maturation into the adult form (Harmon *et al.*, 1995, and references therein). The ability of a juvenile hormone analog to bind and activate RXR suggests the possibility that a similar receptor exists in insects to mediate the effects of juvenile hormone. To date, however, there is no evidence that USP, the *Drosophila* RXR homolog, or any of the other *Drosophila* nuclear receptors can act as a receptor for juvenile hormone.

The most recently identified RXR ligand, phytanic acid, is a metabolite of phytol, a chlorophyll derivative obtained in the diet and present in human serum (Steinberg *et al.*, 1965). The inability to catabolize phytanic acid has been associated with several disease states (e.g., Refsum's disease, Zellweger syndrome, and neonatal adrenoleukodystrophy), indicating the critical importance of enzymes involved in phytanic acid metabolism (LeMotte *et al.*, 1996; Kitareewan *et al.*, 1996). It has been suggested that RXR's ability to bind this fatty acid may indicate RXR's potential involvement in controlling fatty acid levels through an undefined feedback mechanism (LeMotte *et al.*, 1996). It is also possible that RXR may be involved in some aspect of the disease states associated with faulty phytanic acid breakdown.

As a receptor for 9-*cis* retinoic acid, RXR was originally shown to bind DNA as a homodimer on direct repeat response elements spaced by one nucleotide (DR-1) (Mangelsdorf *et al.*, 1991). Recently, it has been discovered that RXR can also function as a 9-*cis* retinoic acid receptor when heterodimerized with certain orphan receptors (e.g., LXR and NGFI-B) on specific non-DR-1 DNA response elements (Willy *et al.*, 1995; Perlmann and Jansson, 1995). This finding significantly increases the complexity of 9-*cis* retinoic acid signaling in both a tissue- and promoter-specific fashion and is discussed in more detail in Section III.

In addition to the RXR ligands, over the last several years ligands and/or activators have been discovered for three other orphan receptor families (PPAR, FXR, and LXR), all of which form heterodimers with RXR. The structures of these ligands and activators are shown in Fig. 2. The identification of these new ligands and their endocrine and pharmacologic implications are discussed in the following section.

III. Orphan Receptors That Function as RXR Heterodimers

The nonsteroid members of the nuclear receptor superfamily that have ligands are all RXR heterodimers. These members include the vitamin D, thyroid hormone, and retinoic acid receptors, as well as RXR itself. Significantly, the only orphan receptors for which ligands have been discovered are also RXR heterodimers, supporting the contention that these orphans are among the best candidates for novel ligand discovery. Thus far, ligands and/or activators have been identified for three orphan receptor families (PPAR, FXR, and LXR). These receptors are discussed below, in terms of both their own ligand dependent activities and their ability to govern the ligand activity of RXR. In addition to these receptors, there are three others (MB67, ONR1, and SHP) that have been reported to heterodimerize with RXR, but whose ligand activities are unknown. These orphan receptors are also discussed below.

A. RXR Heterodimer Partners with Ligands and/or Activators

The **PPAR** family (peroxisome proliferator activated receptor) consists of three known gene products referred to as PPARα, β (also called PPARδ, NUC1, or FAAR), and γ (Issemann and Green, 1990; Dreyer et al., 1992; Schmidt et al., 1992; Zhu et al., 1993; Chen et al., 1993; Kliewer et al., 1994; Amri et al., 1995). PPARs bind DNA as RXR heterodimers with a specificity for DR-1-type response elements (Kliewer et al., 1992c). The different PPAR subtypes vary in their tissue distribution as well as in their relative ability to be activated by a variety of different compounds (Schmidt et al., 1992; Kliewer et al., 1994; Braissant et al., 1996). The naming of these receptors reflects the initial finding that PPARα could be activated by certain classes of chemicals (e.g., fibrates, plasticizers, and herbicides) that cause peroxisomal proliferation and hepatomegaly (Issemann and Green, 1990). Subsequently, it was discovered that certain long-chain fatty acids, including arachidonic acid and linoleic acid, could also activate PPARα (Göttlicher et al., 1992), and it has since been shown that PPARα can regulate the transcription of several key enzymes involved in fatty acid metabolism (Kliewer et al., 1992c; Dreyer et al., 1992; S. S.-T. Lee et al., 1995; Lemberger et al., 1996). The hypothesis that one function of PPARα is to stimulate lipid metabolism is further supported by its high expression in tissues such as liver and kidney (Kliewer et al., 1994). In addition to long-chain fatty acids, PPARα has also been shown to be activated by eicosanoids, specifically 8(S)-hydroxyeicosatetraenoic acid (8(S)HETE) (Yu et al., 1995), carbacyclin (Hertz et al., 1996), and more recently, leukotriene B_4 (LTB_4), a molecule involved in the inflammatory response pathway (Devchand et al., 1996). It has been proposed that PPARα may act in controlling the duration of an inflammatory response by affecting the expression of genes involved in LTB_4 catabolism (Devchand et al., 1996). Although the physiologic relevance of LTB_4 as a ligand for PPARα is not yet completely understood, this finding offers the potential for development of drugs involved in controlling inflammatory responses.

In contrast to the role of PPARα in the catabolism of fat, the role of PPARγ appears to be in the synthesis of fat. PPARγ is predominantly expressed in adipocytes and is thought to play a key role in adipogenesis through regulation of genes involved in adipocyte differentiation (Tontonoz et al., 1994a,b). Several ligands for PPARγ have recently been identified, including antidiabetic thiazolidinediones (Lehmann et al., 1995) and the prostaglandin metabolite, 15-deoxy-$\Delta^{12,14}$-prostaglandin J_2 (Kliewer et al., 1995; Forman et al., 1995c). More recently it has been discovered that the adipogenic properties of PPARγ can be inhibited through MAP kinase-mediated phosphorylation (Hu et al., 1996). Such results indicate that one mechanism by which mitogenic factors influence the balance of adipocyte differentiation is through their ability to covalently modify PPARγ. Together these findings reveal an unexpected link between prostanoids, adipogenesis, and glucose homeostasis. The discovery that PPARγ is a key regulator of these processes suggests the promise of new pharmacological approaches to combat diseases such as diabetes.

Unlike the α and γ subtypes, the role of PPARβ (PPARδ) is not well understood. PPARβ is more widely expressed, and although its ligand specificity weakly overlaps that of PPARα and PPARγ, its function as well as its endogenous high-affinity ligand are still unknown (Kliewer et al., 1994). Thus far, the most efficacious PPARβ activator is carbacyclin, a stable analog of prostacyclin which is also known to activate PPARα and PPARγ (Brun et al., 1996).

The in vivo function of PPARs is currently being investigated using targeted gene disruption in mice. Thus far, only the knock-out of PPARα has been reported (S. S.-T. Lee et al., 1995). Mice lacking PPARα appear normal, but do not undergo the physiological responses seen in wild-type animals challenged with peroxisome proliferators (S.S.-T. Lee et al., 1995). The relatively subtle phenotype of PPARα-null mice implies that some functional redundancy between PPAR subtypes may exist. Due to the differences in tissue distribution and ligand activation of the different PPARs, the phenotypes of the other subtype knock-outs, specifically PPARγ, may be quite different.

Another orphan receptor for which activators have recently been identified is FXR (farnesoid X-activated receptor) (Forman *et al.*, 1995b), also known as RIP14 (Seol *et al.*, 1995). FXR was cloned from a rat liver cDNA library and is most similar to the *Drosophila* ecdysone receptor (EcR). *In vitro*, FXR binds as an RXR heterodimer to ecdysone response elements (i.e., inverted repeats of half-sites spaced by one nucleotide) and classical DR-4 type sequences (Forman *et al.*, 1995b; C. Weinberger, personal communication). The potential for ligand-dependent transcriptional activation by FXR was tested in transient transfections with a reporter gene containing ecdysone response elements. Using this assay, the insect growth regulator juvenile hormone and its farnesoid precursors were identified as activators of RXR/FXR heterodimers. The significance of FXR activation by an insect hormone remains unclear, although there is evidence that RXR alone can be specifically activated by similar compounds (see Section II) (Harmon *et al.*, 1995). Therefore, it remains a possibility that RXR is the partner in the RXR/FXR heterodimer that is activated by juvenile hormone. In insects, juvenile hormone is metabolically derived from farnesyl pyrophosphate. Since farnesyl pyrophosphate is a product in the mevalonate biosynthetic pathway, metabolites within the mammalian mevalonate pathway were also tested for their ability to activate FXR. Interestingly, several of these metabolites were able to activate FXR, with farnesol being the most potent (Forman *et al.*, 1995b). Although concentrations required for activation were high, they are thought to be within the predicted physiological range for this class of compounds. However, no binding to farnesoids has been demonstrated, suggesting that these compounds may be precursors to the actual FXR ligand or work through a secondary mechanism. Importantly, it was demonstrated that FXR mRNA is expressed in isoprenoidogenic tissues (liver, kidney, gut, and adrenal gland) as would be predicted if FXR is involved in the mevalonate biosynthetic pathway (Forman *et al.*, 1995b). The finding that farnesoids can activate transcription through nuclear receptors was the first demonstration that intermediary metabolites in a crucial biosynthetic pathway may be regulators of gene expression.

An orphan receptor that has recently been implicated in the regulation of cholesterol metabolism is LXRα (originally cloned from rat as RLD-1) (Apfel *et al.*, 1994; Willy *et al.*, 1995), and its related receptor LXRβ (also known as UR, NER, OR-1, and RIP15) (Song *et al.*, 1994; Shinar *et al.*, 1994; Teboul *et al.*, 1995; Seol *et al.*, 1995). The LXRs form heterodimers with RXR on DR-4-type response elements (Apfel *et al.*, 1994; Song *et al.*, 1994; Willy *et al.*, 1995; Teboul *et al.*, 1995; Seol *et al.*, 1995). A unique feature of the RXR/LXR heterodimer is its ability to be activated by both RXR and LXR ligands (discussed in Section III.C) (Willy and Mangelsdorf,

1997). LXRα and β are activated by a select group of stereospecific oxysterols that are key intermediates in cholesterol metabolism (Janowski et al., 1996). The most potent LXR activators identified are 22(R)-hydroxycholesterol, 24(S)-hydroxycholesterol, 24(S),25-epoxycholesterol, 7α-hydroxycholesterol, and FF-MAS (follicular fluid meiosis-activating sterol) (Janowski et al., 1996; Lehmann et al., 1997). Many of these molecules serve as intermediates in the rate-limiting steps of three crucial biosynthetic pathways: the conversion of lanosterol to cholesterol, steroid hormone synthesis, and bile acid synthesis. Concentrations of oxysterols needed to activate LXRs in transient transfections are consistent with their known physiological levels (Kandutsch et al., 1978; Dhar et al., 1973; Javitt et al., 1981; Dixon et al., 1970). Furthermore, the immediate upstream and downstream metabolites of these oxysterols are significantly less potent as LXR activators. These results, in addition to oxysterol-induced protease protection experiments, strongly suggest that these compounds are functioning as LXR ligands (Janowski et al., 1996).

The discovery that oxysterols can positively regulate transcription implies that one function of LXR may be as a sensor of cholesterol. If this is the case, LXR may regulate transcription of genes encoding critical enzymes involved in cholesterol metabolism. This hypothesis is supported by the finding of a potential LXR-dependent oxysterol response element in the promoter of the CYP7A gene, which encodes the enzyme required for the rate-limiting step in bile acid synthesis (Lehmann et al., 1997; B. Janowski and D. Mangelsdorf, unpublished observation). LXRα is expressed in several tissues where cholesterol metabolism occurs (liver, kidney, intestine, adrenals, spleen, and adipose tissue) (Apfel et al., 1994; Willy et al., 1995) and LXRβ is expressed in most tissues tested (Song et al., 1994; Shinar et al., 1994; Seol et al., 1995; Teboul et al., 1995). The discovery of LXR activation by oxysterols demonstrates a novel role for these compounds in transcriptional activation and supports the long-held notion that nuclear receptors are involved in cholesterol signaling pathways.

B. RXR Heterodimer Partners with No Known Ligands or Activators

MB67 (also called mCAR) is an orphan receptor that forms an RXR heterodimer on a subset of DR-5 type response elements, but about which relatively little is known (Baes et al., 1994). This receptor is highly expressed in the liver and was initially identified based on its interaction with RXR in a yeast two-hybrid screen (Baes et al., 1994). Although no activators have been identified, transient transfections with MB67 and a reporter gene containing the DR-5 retinoic acid response element from the RARβ gene (Sucov et al., 1990) demonstrated that this orphan receptor may function as a transcriptional activator in the absence of exogenous ligand. Based on these results, it has been speculated that MB67 may compete with RXR/RAR heterodimer binding to specific DR-5 type response elements and cause activation in the absence of retinoids (Baes et al., 1994).

ONR1 (orphan nuclear receptor 1) is a *Xenopus* orphan receptor expressed during early embryogenesis from both maternal and zygotic sources, that most closely resembles the mammalian vitamin D receptor (VDR). Like VDR, ONR1 binds as an RXR heterodimer to a DR-3 response element; however, ONR1 does not appear to bind the VDR ligand, 1,25-dihydroxyvitamin D_3 (Smith et al., 1994).

SHP (small heterodimer partner) was isolated in a two-hybrid screen as a protein capable of interacting with the orphan receptor MB67 (Seol et al., 1996). Subsequently, SHP was shown to be a potent trans-repressor of several nuclear receptor pathways. SHP has a limited pattern of expression in the liver, pancreas, and heart. Like its close relative, DAX-1, SHP lacks a classical nuclear receptor DNA binding domain, but has a putative ligand binding/dimerization domain similar to other receptors in the superfamily. Through this domain, SHP, like RXR, can dimerize with several receptors including RXR, RAR, TR, and MB67. However, in direct contrast to RXR heterodimerization, the consequence of SHP heterodimerization is inhibition of DNA binding and transactivation. Interestingly, heterodimerization of SHP with some of these receptors requires the partner's ligand, suggesting SHP may function as a ligand-dependent repressor (Seol et al., 1996).

C. Receptors That Mediate Retinoid Signaling through Heterodimerization with RXR

Retinoids have profound effects on the growth and differentiation of hematopoietic and epithelial tissue, development of bone, pattern formation during embryogenesis, and adult metabolism and homeostasis. The two most potent vertebrate retinoids are the acid forms of vitamin A, all-*trans* retinoic acid and 9-*cis* retinoic acid. These retinoids mediate their effects by binding to two classes of receptors: the RARs, which bind to both all *trans* retinoic acid and 9-*cis* retinoic acid, and as discussed above, the RXRs, which bind to only 9-*cis* retinoic acid (Mangelsdorf et al., 1994; Chambon, 1996). Initially, these receptors were shown to modulate ligand-dependent gene expression by interacting as RXR/RAR heterodimers (Yu et al., 1991; Leid et al., 1992a; Kliewer et al., 1992b; Zhang et al., 1992; Marks et al., 1992) or in rare cases as RXR homodimers (Mangelsdorf et al., 1991) on their specific target gene hormone response elements. The discovery that RXRs serve as heterodimeric partners to several different receptors raised the possibility that both RXR and its partner receptor could be activated by their respective ligands within the heterodimer. It is now known that not all RXR heterodimers can respond to 9-*cis* retinoic acid, and that this ability is entirely dependent on which receptor partners with RXR. Three categories of RXR heterodimers have been identified, each defined by RXR's ability to be ligand-activated within the heterodimer complex (Fig. 3). Included in the first two categories are those RXR heterodimers that, in their uninduced state, are unable to respond to RXR's ligand. In the first category (Fig. 3A) RXR is said to be a completely silent partner, since these heterodimers (e.g., RXR/VDR and RXR/TR) are refractory to 9-*cis* retinoic acid activation even in the presence of the partner receptor's ligand (MacDonald *et al.*, 1993; Forman *et al.*, 1995a). In the second category (e.g., RXR/RAR heterodimers), RXR is said to be conditionally silent, since this heterodimer is responsive to the RXR ligand only in the presence of the partner's ligand (Fig. 3B) (Forman *et al.*, 1995a; Minucci *et al.*, 1997). The addition of both ligands can result in an enhanced activation of the heterodimer (Roy *et al.*, 1995; Chen *et al.*, 1996). It should be noted, however, that the ability of RXR to gain ligand responsiveness once RAR is ligand bound remains controversial. Others have reported that RXR is unable to bind ligand when heterodimerized with RAR, regardless of RAR's ligand-binding status (Kurokawa *et al.*, 1994, 1995). In this case RXR would remain a fully silent partner. The difference between these two apparently diametric results may be attributed to experimental conditions, including different response elements and cell types tested (La Vista-Picard *et al.*, 1996).

In contrast to the first two categories, the third category includes several orphan receptors (e.g., NGFI-B and LXR) that when heterodimerized with

---

FIGURE 3 Models for the role of RXR in ligand binding and transcriptional activation by RXR heterodimers. Three classes of RXR heterodimers are thought to exist, exemplified by RXR/TR (A), RXR/RAR (B), and RXR/LXR (C) heterodimers. (A) In the absence of ligand, the RXR/TR heterodimer on DNA is a potent repressor of basal transcription due to the association of a corepressor with TR. Upon ligand binding to TR, the corepressor is released and coactivator proteins associate with TR resulting in transcriptional activation. This heterodimer is not responsive to the RXR ligand and thus RXR is referred to as a silent partner. (B) Similar to the RXR/TR heterodimer, the RXR/RAR heterodimer also associates with a corepressor and in the basal state is not responsive to RXR ligands. Upon ligand binding to RAR, the corepressor is released and coactivator proteins associate. In some cases, once RAR has bound its ligand, RXR is believed to acquire ligand responsiveness. Dual activation of the complex leads to an additive or synergistic response. In this heterodimer, RXR is referred to as conditionally silent because the RXR ligand response is dependent on RAR's ligand binding. (C) Some orphan receptors such as LXR do not appear to associate with a corepressor and therefore the RXR/LXR heterodimer does not exhibit basal repression. The lack of a dominant corepressor binding to RXR's partner receptor may be the reason that this heterodimer is capable of responding to either receptors' ligand alone, or both ligands together to achieve synergistic activation. Within this heterodimer, RXR can be an active, ligand-binding receptor, regardless of the presence of an LXR ligand. An unusual feature of the RXR/LXR heterodimer is that when RXR binds ligand, the AF-2 domain of LXR is required for transcriptional activation. This model suggests that the binding of ligand to RXR induces a conformational change within its unliganded partner, LXR, which in turn leads to activation (see text for details). T3, thyroid hormone; RA, RAR-specific ligand; RX, RXR-specific ligand; LX, LXR ligand.

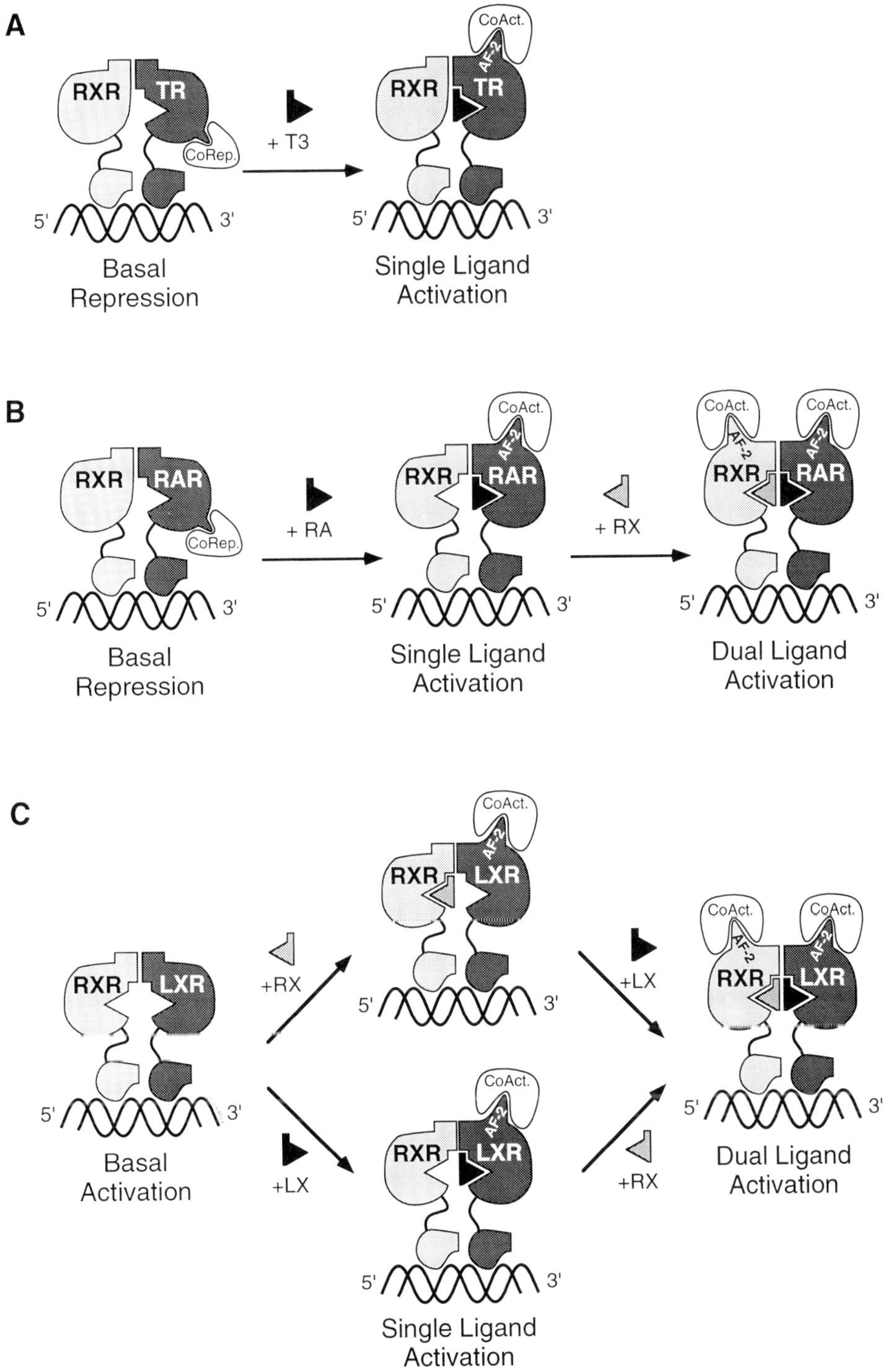
A
RXR
TR
CoRep.
+ T3
CoAct.
AF-2
5'
3'
Basal Repression
Single Ligand Activation
B
RXR
RAR
CoRep.
+ RA
+ RX
CoAct.
AF-2
Basal Repression
Single Ligand Activation
Dual Ligand Activation
C
RXR
LXR
+RX
+LX
CoAct.
AF-2
Basal Activation
Single Ligand Activation
Dual Ligand Activation

RXR, shift RXR's role from a silent partner to a fully active 9-*cis* retinoic acid receptor (Fig. 3C). For those orphan receptors that have recently been shown to have ligands, the heterodimer can be activated by the RXR ligand, the partner's ligand, or synergistically by both receptors' ligands together (Kliewer *et al.*, 1992c; Janowski *et al.*, 1996). The two best studied orphan receptors in this category are NGFI-B and LXR. In these heterodimers, 9-*cis* retinoic acid induction is dependent on both the protein–protein interaction between the two receptors and the precise sequence of their DNA binding site (Perlmann and Jansson, 1995; Willy *et al.*, 1995). NGFI-B and its relative NURR1 were previously thought to bind DNA only as monomers on an extended single half-site response element (i.e., NBRE) on which they were shown to be transcriptionally active in the absence of exogenously added ligand (Wilson *et al.*, 1991). Perlmann and co-workers first showed that NGFI-B can heterodimerize with RXR on specific DR-5 sequences that contain an NBRE as the downstream half-site (Perlmann and Jansson, 1995). In this context NGFI-B permits retinoid signaling through RXR. These findings were supported by studies utilizing chimeric receptors that demonstrated RXR could be activated by retinoids when complexed with NURR1 (Forman *et al.*, 1995a). Interestingly, RXR heterodimerization on DNA and subsequent retinoid signaling are not observed with NOR-1, the most recently identified member in the NGFI-B family (Zetterström *et al.*, 1996) (see Section IV for more information on the NGFI-B family).

The study of the RXR/LXR heterodimer has revealed several unique characteristics that contribute to its ability to respond to retinoids (Willy *et al.*, 1995; Willy and Mangelsdorf, 1997). RXR heterodimerizes with LXR on a novel DR-4 like sequence called an LXRE in which the 5' half-site is degenerate from the canonical sequence. Similar to other RXR heterodimers (Kurokawa *et al.*, 1993; Perlmann *et al.*, 1993; Schräder *et al.*, 1995), the RXR/LXR heterodimer binds to DNA with a polarity that positions RXR over the 5' half-site. However, in contrast to these other heterodimers where only the receptor occupying the 3' half-site is ligand responsive, in the RXR/LXR heterodimer both receptors are independently ligand responsive, indicating that the position of a receptor within the DNA-bound heterodimer is not the sole factor determining its ligand responsiveness.

The different models explaining the mechanism of ligand activation by RXR heterodimers are shown in Fig. 3. In heterodimers where RXR is silent (e.g., RXR/TR and RXR/RAR), ligand binding to the partner receptor results in a conformation change that releases corepressor protein from the partner receptor and recruits coactivator proteins (Kurokawa *et al.*, 1995) (Figs. 3A, 3B). The resulting increase in transcription is dependent on the presence of a functional transactivation (AF-2) domain within the ligand-bound receptor (Durand *et al.*, 1994). In these heterodimers, as the non-ligand-binding partner, RXR's AF-2 domain is not critically required for ligand induction by the partner receptor. In the RXR/LXR heterodimer, however, RXR transactivation proceeds through a novel mechanism (Fig. 3C). In this heterodimer, the AF-2 domain of the non-ligand-binding partner (LXR) is essential for 9-*cis* retinoic acid induction (Willy and Mangelsdorf, 1997). Surprisingly, the presence of the AF-2 domain of RXR (in this case the ligand-binding partner) is not crucial for this heterodimer to be active. Thus, ligand binding to one receptor (RXR) presumably induces a conformation change in its partner (LXR), which in turn recruits coactivator proteins and leads to an increased transcriptional response. This finding suggests the intriguing possibility that this unique mechanism of receptor transactivation through the AF-2 domain of the non-ligand-binding partner may be used by other heterodimers in which RXR is ligand responsive.

Two other orphan receptors, PPAR and FXR, also confer retinoid responsiveness through heterodimerization with RXR on specific DNA sequences (Kliewer *et al.*, 1992c; Forman *et al.*, 1995b). In these cases, the addition of an RXR ligand generally results in a response only in the presence of the partner's ligand. This dual activation is reminiscent of the RXR/RAR heterodimer, which, as described above, can be activated by the RXR ligand only if the RAR ligand is also present.

The studies outlined above demonstrate that when complexed with specific receptors, RXR is able to bind and be activated by its ligand, thereby altering its role from a silent heterodimeric partner to an active, ligand-binding receptor in the heterodimer complex. These discoveries highlight the crucial role played by RXR's partner as well as the DNA response element in determining the ability of RXR to respond to its ligand, and further increase the diversity of retinoid signaling pathways.

IV. Orphan Receptors That Function Independently of Dimerization with RXR

A. Receptors That Generally Function as Transcriptional Activators

In addition to receptor mediated transactivation via ligand binding, several orphan receptors appear to be constitutively active (i.e., capable of activating reporter gene expression in the absence of exogenously added ligand). The mechanisms underlying this regulation are unknown, but could include activation by intracellular ligands or ligands contained in serum used to maintain the cells, or through posttranslational modifications (e.g., phosphorylation). The receptors in this category include HNF-4, NGFI-B, ROR (RZR), and SF-1 (FTZ-F1). The orphan receptor MB67 is also considered constitutively active, but unlike the receptors presented in this section, MB67 heterodimerizes with RXR and is therefore discussed in Section III.B.

HNF-4 (hepatocyte nuclear factor-4) is an orphan receptor that was purified based on its ability to bind liver-specific enhancer sequences (Sladek *et al.*, 1990). In adult animals, HNF-4 expression is limited to the liver, intestine, kidney, and pancreas; however, it is also expressed in the early mouse embryo where it is essential for early embryonic development (see below) (Sladek *et al.*, 1990; Miquerol *et al.*, 1994; Duncan *et al.*, 1994; W. S. Chen, *et al.*, 1994). HNF-4 binds DNA as a homodimer, with a specificity for DR-1-type response elements. Thus far, no evidence of monomeric or heterodimeric interactions involving HNF-4 on DNA have been observed (Sladek *et al.*, 1990; Jiang *et al.*, 1995). HNF-4 has been implicated in the positive regulation of a variety of different genes whose products are involved in lipid transport and metabolism (Costa *et al.*, 1990; Sladek *et al.*, 1990; Mietus-Snyder *et al.*, 1992; Ladias *et al.*, 1992; Metzger *et al.*, 1993; Ochoa *et al.*, 1993; Nakshatri and Chambon, 1994), blood coagulation (Erdmann and Heim, 1995; Crossley *et al.*, 1992; Reijnen *et al.*, 1992; Hung and High, 1996), proliferation and differentiation of red blood cells (Galson *et al.*, 1995), glycolysis (Miquerol *et al.*, 1994), fatty acid β-oxidation (Carter *et al.*, 1993), and xenobiotic detoxification (D. Chen, *et al.*, 1994a,b). HNF-4 has also been implicated in the regulation of a number of liver-specific genes (Sladek *et al.*, 1990; Tian and Schibler, 1991; Miura and Tanaka, 1993; Kimura *et al.*, 1993; Pescini *et al.*, 1994; Hall *et al.*, 1995), as well as Hepatitis B virus through a specific sequence in the viral promoter (Garcia *et al.*, 1993). For most of these genes, HNF-4 is capable of binding enhancer sequences in the promoter regions and transactivating reporter genes containing these sequences in transiently transfected cells. In these cases, HNF-4 acts as a positive regulator of transcription in the absence of added ligands, suggesting that HNF-4 is activated by an endogenous ligand or through a ligand-independent pathway. Interestingly, many of the liver-specific genes that are positively regulated by HNF-4 are repressed by another family of orphan receptors, the COUP-TFs (see Section IV.B). Thus, these two transcription factors may have antagonistic roles in the regulation of many of these gene networks. In this scenario, the relative amount of each receptor as well as its affinity for specific response elements would control whether a target gene is activated or repressed by HNF-4 or COUP-TF, respectively.

HNF-4 also plays a critical role in early mouse development. *In situ* localization of HNF-4 mRNA in developing mouse embryos detected HNF-4 transcripts in the visceral endoderm at Embryonic Day 4.5 and later in specific tissues, suggesting that HNF-4 may be crucial for both postimplantation development and organogenesis (Duncan *et al.*, 1994). Targeted disruption of the HNF-4 gene in mice results in an embryonic lethal phenotype, characterized by cell death in the ectoderm at Embryonic Day 6.5 and a delay in mesoderm formation (W. S. Chen, *et al.*, 1994; Duncan *et al.*, 1994). These results point to an early requirement for HNF-4 expression in the visceral endoderm. In the adult, HNF-4 expression is specific to liver, kidney, intestine, and pancreas, indicating that this orphan receptor is also required later in life for normal metabolic functions (Sladek *et al.*, 1990; Miquerol *et al.*, 1994). Indeed, recent work implicating HNF-4 in certain forms of diabetes supports this contention (see below). Further evidence for the essential role of HNF-4 in development comes from studies with the *Drosophila* homolog dHNF-4. During *Drosophila* embryogenesis the expression of dHNF-4 in midgut, fat bodies, and malpighian tubules is very similar to that of its murine homolog in the corresponding mammalian tissues (Zhong *et al.*, 1993). Interestingly, tissues that normally express dHNF-4 fail to develop normally in *Drosophila* mutants lacking the dHNF-4 gene, again suggesting an essential role for this orphan receptor in gut formation and organogenesis (Zhong *et al.*, 1993).

Information on a possible physiologic role for HNF-4 comes from the finding that a mutation in the human *HNF-4α* gene is associated with an inherited form of diabetes known as maturity-onset diabetes of the young (MODY) (Yamagata *et al.*, 1996b). MODY is an autosomal-dominant inherited form of non-insulin-dependent diabetes mellitus with early onset, usually striking juveniles before age 25 (Fajans, 1989). At least three forms of MODY exist (referred to as MODY1, 2, and 3), each associated with different genetic loci (Bell *et al.*, 1991; Froguel *et al.*, 1993; Vaxillaire *et al.*, 1995). Along with its direct link to MODY1, HNF-4 is believed to regulate expression of the *HNF-1α* gene, which encodes a liver-specific transcription factor that has been associated with MODY3 (Tian and Schibler, 1991; Kuo *et al.*, 1992; Yamagata *et al.*, 1996a). These important findings linking mutations in *HNF-4α* and a potential HNF-4 target gene to diabetes suggest that identification of a ligand for HNF-4 and/or elucidation of its mechanism of action could lead to a greater understanding of diabetes.

The NGFI-B (nerve growth factor-induced) family of receptors includes three distinct genes that have been given many different names reflecting the species, cell type, and method of induction by which these proteins were identified. NGFI-B (Milbrandt, 1988), one of the first orphan receptors isolated, has also been referred to as Nur77, N10, NAK1, TR3, and TIS-1 (Hazel *et al.*, 1988; Ryseck *et al.*, 1989; Nakai *et al.*, 1990; Chang *et al.*, 1989; Lim *et al.*, 1995). The second member of the family, known as NURR1 (Law *et al.*, 1992), is also referred to as NOT and RNR-1 (Mages *et al.*, 1994; Scearce *et al.*, 1993). The most recently identified NGFI-B family member, NOR-1 (Ohkura *et al.*, 1994), is also known as MINOR and TEC (Hedvat and Irving, 1995; Labelle *et al.*, 1995). A unique feature of all NGFI-B family members is that they are immediate-early response genes and their expression can be induced by a variety of stimuli, including mitogens, growth factors, and membrane depolarization (Milbrandt, 1988; Ryseck *et al.*, 1989; Hazel *et al.*, 1991). In addition, all of these receptors are highly expressed in the central nervous system (CNS) (Zetterström *et al.*, 1996). NGFI-B and NOR-1 are also expressed in tissues outside of the CNS, while NURR1 is expressed predominantly in brain (see Table I for specific information on the tissue distribution for these three family members).

NGFI-B family members bind to DNA as monomers and constitutively induce reporter gene expression (Wilson et al., 1991, 1993b). The NGFI-B response element (NBRE) consists of a classical half-site with two additional 5' adenine residues (e.g., aaAGGTCA) (Wilson et al., 1991). NGFI-B activity on an NBRE appears to be regulated by phosphorylation following induction by different stimuli (e.g., growth factor stimulation vs membrane depolarization) (Fahrner et al., 1990; Hazel et al., 1991). Evidence linking phosphorylation to receptor activity comes from experiments demonstrating that phosphorylation of a serine residue in the DNA binding domain of NGFI-B inhibits its DNA binding activity (Hirata et al., 1993; Davis et al., 1993; Davis and Lau, 1994). Recent identification and partial purification of an NGF-activated kinase (NGFI-B kinase) responsible for this phosphorylation in PC12 cells indicates that it may be the cyclic AMP response element-binding protein kinase (Hirata et al., 1995). In addition to preventing binding to an NBRE, it remains possible that phosphorylation also enhances NGFI-B binding to other, as yet unidentified DNA sequences. This may explain why both the receptor and its presumptive inactivating kinase are induced under the same conditions.

NGFI-B has been proposed to influence adrenocortical steroidogenesis based on its expression pattern and its ability to activate transcription through an NBRE sequence in the steroid 21-hydroxylase (CYP21) gene promoter in mouse adrenocortical tumor cells (Wilson et al., 1993a). Targeted disruption of the NGFI-B gene in mice, however, shows no effect on adrenocortical steroidogenesis or the specific expression of CYP21 mRNA, suggesting either that NGFI-B is not critical for regulation of adrenocortical steroidogenic enzymes, or more likely, that redundancy in the family compensates for the loss of one of the receptor subtypes (Crawford et al., 1995). Evidence for redundancy has been seen in response to lipopolysaccharides in NGFI-B knock-out mice, as indicated by the enhanced adrenal expression of the NURR1 gene (Crawford et al., 1995).

A potential role for NGFI-B in T-cell receptor (TCR)-mediated apoptosis has been proposed based on findings that NGFI-B (Nur77) is induced in apoptotic T-cell hybridomas and apoptotic thymocytes, and that overexpression of a dominant-negative form of NGFI-B or inhibition with anti-sense transcripts prevents activation-induced apoptosis in T-cell hybridomas (Woronicz et al., 1994; Liu et al., 1994). It has also been shown that cyclosporin A, a drug known to block TCR-mediated apoptosis, inhibits DNA binding by NGFI-B, thereby preventing apoptosis in these cells (Yazdanbakhsh et al., 1995). The mechanism for NGFI-B-induced apoptosis is unknown, but results from experiments using mice overexpressing NGFI-B in the thymus suggest that the pathway involves the up-regulation of the Fas ligand (Weih et al., 1996).

Unregulated expression of one of the NGFI-B family members, NOR-1, has been associated with a form of soft tissue tumor referred to as extraskeletal myxoid chondrosarcoma (EMC). A chromosomal translocation resulting in a fusion gene containing NOR-1 is associated with many of these tumors, indicating the critical importance of this orphan receptor on the regulation of cell growth. This finding opens up the possibility of potential treatments for this form of cancer, provided the natural mode of NGFI-B activation as well as the mechanism of oncogenic conversion can be elucidated (Labelle *et al.*, 1995).

Accurate assessment of the physiological role of NGFI-B family members in the above-mentioned pathways will likely require targeted disruption of multiple genes, given that the NGFI-B knock-out mice appear to have no identifiable phenotype, including unimpaired T-cell function (S. L. Lee *et al.*, 1995). The absence of a phenotype may be due to functional redundancy with other family members, an idea supported by the overlapping patterns of expression of all three family members both during development and in the adult (Zetterström *et al.*, 1996).

Finally, another interesting finding regarding NGFI-B and NURR1 is their involvement in retinoid signaling through heterodimerization with RXR (Section III.C) (Perlmann and Jansson, 1995; Forman *et al.*, 1995a). Surprisingly, the most recently identified member in the NGFI-B family, NOR-1, does not heterodimerize with RXR and thus does not contribute to retinoid signaling (Zetterström *et al.*, 1996).

The ROR receptor family, also known as RZR, includes α, β, and γ subtypes (Becker-André *et al.*, 1993; Giguère *et al.*, 1994; Carlberg *et al.*, 1994; Hirose *et al.*, 1994b; Ortiz *et al.*, 1995). The name of these receptors (retinoic acid-related orphan receptor, ROR; retinoid Z receptor, RZR) refers to their moderate DNA-binding domain sequence similarity (~70%) to that of the retinoic acid receptor; however, there is no evidence that these receptors respond to retinoids. The different ROR family members vary in their tissue distribution; RORα is expressed in a variety of tissues, RORβ expression is restricted to the brain and retina, and RORγ is highly expressed in skeletal muscle and thymus (Becker André; *et al.*, 1993, 1994; Carlberg *et al.*, 1994; Hirose *et al.*, 1994b; Ortiz *et al.*, 1995). RORα consists of several isoforms that bind as monomers to a response element (called an RORE) consisting of a core half-site of the consensus PuGGTCA preceded by a 6-bp AT-rich sequence (Giguère *et al.*, 1994). The different RORα isoforms vary in their amino-terminal domains which contribute to the distinct DNA-binding specificities exhibited by the various isoforms (Giguère *et al.*, 1994). RORα is considered transcriptionally active when bound to an RORE in the absence of exogenously added ligand, and in this context, it induces a bend in the DNA that may contribute to transcriptional activity (McBroom *et al.*, 1995). It also has been reported that some ROR family members bind as monomers or homodimers to a variety of different response elements, including direct repeat and palindromic sequences (Carlberg et al., 1994; Ortiz et al., 1995; Greiner et al., 1996). The ability of ROR to activate transcription from these different response elements remains controversial and may rely on cell-type specific factors (Greiner et al., 1996). A potential target gene for RORα is murine γF-crystallin. RORα can bind a specific sequence in the promoter of this gene and activate transcription of reporter genes containing this response element. This activation, however, can be repressed by the competitive binding of RXR/RAR heterodimers, suggesting that a balance between RORα and the retinoic acid receptors may play a role in regulating some retinoid signaling pathways (Tini et al., 1995). A similar suggestion has been made regarding RORγ (also called thymus orphan receptor or TOR), which has been proposed to act as a negative regulator of TR and RAR signaling (Ortiz et al., 1995).

Recently, the constitutive activity of ROR has been shown to be downregulated by another orphan receptor, Rev-Erb (Retnakaran et al., 1994; Forman et al., 1994), indicating a possible role for these two receptors in common signaling pathways. The recent identification of a conserved binding site for both RORα and Rev-Erbβ (also called RVR) in a regulatory region of the human and mouse N-myc genes further supports this idea (Dussault and Giguère, 1997). These two receptors could act in a manner similar to that proposed for HNF-4 and COUP-TF, whereby they work antagonistically to regulate the expression of specific genes. As has been suggested for HNF-4 and COUP-TF (see above), the relative amount of each receptor and its affinity for specific response elements could determine if target genes are activated or repressed.

A clue to the potential *in vivo* role of RORα has come from the discovery that a disruption of the *RORα* gene locus in mice is causative for the phenotype *staggerer,* characterized by a defect in Purkinje cell development leading to cerebellar ataxia (Hamilton et al., 1996). In that study, it was postulated that RORα may interact with TRβ to regulate cerebellar Purkinje cell maturation (Hamilton et al., 1996), although it is important to note that no direct interaction between these two receptors can be demonstrated.

Finally, RORβ (also called RZRβ) is highly expressed in brain and has been reported to function as a cell-type-specific transcriptional activator in the absence of exogenous ligands (Greiner et al., 1996). RORβ has also been reported to have a ligand (Becker-André et al., 1994; Missbach et al., 1996). According to these studies, the pineal gland hormone melatonin, as well as a class of thiazolidinediones with anti-arthritic activity, can bind and activate ROR . However, these findings remain controversial and have not been reproduced in several other laboratories (Greiner et al., 1996; Hazlerigg et al., 1996; V. Giguère, personal communication).

SF-1 (steroidogenic factor-1) (Lala et al., 1992), also called Ad4BP (Honda et al., 1993), is the mammalian homolog of the *Drosophila* transcription factor FTZ-F1 that regulates transcription of the fushi tarazu homeobox gene in fly embryos (Lavorgna *et al.*, 1991). A second mammalian FTZ-F1 family member, LRH-1 (Tugwood *et al.*, 1991), is also referred to as PHR-1 and FTF (Becker-André *et al.*, 1993; Galarneau *et al.*, 1996) and a *Xenopus* homolog with similar properties is called FF1rA (Ellinger-Ziegelbauer *et al.*, 1994). A number of biochemical, genetic, and expression studies have provided an overwhelming amount of evidence supporting a role for SF-1 in development and maintenance of the hypothalamic–adrenal–sex axis (see below). SF-1 has at least one alternative splice variant called ELP (embryonal long-terminal repeat binding protein) (Tsukiyama *et al.*, 1992; Ikeda *et al.*, 1993). Both isoforms are expressed in steroidogenic tissues (although SF-1 is expressed at much higher levels than ELP); however, they appear to have different DNA binding and transactivation properties (Morohashi *et al.*, 1994). Interestingly, ELP has been characterized as a transcriptional repressor, while dFTZ-F1 and SF-1 are transcriptional activators (Tsukiyama *et al.*, 1992; Lavorgna *et al.*, 1991; Lala *et al.*, 1992; Honda *et al.*, 1993). SF-1 binds DNA as a monomer to an extended consensus half-site sequence (tcaAGGTCA) (Wilson *et al.*, 1993b). In transient transfections with reporter constructs containing SF-1 binding sites, SF-1 can moderately activate transcription in the absence of exogenously added ligand, although this activation is both promoter- and cell-type-specific (Honda *et al.*, 1993; Morohashi *et al.*, 1993; Lynch *et al.*, 1993). Furthermore, in the context of different cell types and promoter sequences, it has been reported that SF-1 activation is enhanced by cyclic AMP and protein kinase A, indicating that phosphorylation may play a role in activation of SF-1 (Morohashi *et al.*, 1993).

SF-1 is believed to play a major role in the regulation of steroid hydroxylase gene expression, and potential SF-1 binding sites have been identified in the promoters of many of these genes (reviewed in Parker and Schimmer, 1994). Putative SF-1 binding sites have also been identified in promoter regions of the genes encoding Müllerian inhibiting substance (Shen *et al.*, 1994; Hatano *et al.*, 1994), oxytocin (Wehrenberg *et al.*, 1994), luteinizing hormone β (Halvorson *et al.*, 1996), and glycoprotein hormone α-subunit (Barnhart and Mellon, 1994b), making them all potential targets for regulation by SF-1. SF-1 may also regulate the steroidogenic acute regulatory protein (StAR), a protein involved in the rate-limiting step of adrenal and gonadal steroid synthesis (Sugawara *et al.*, 1996). *In situ* analysis of SF-1 expression during mouse embryogenesis demonstrated SF-1 expression in the developing gonads and the adrenal primordium, indicating the potential involvement of SF-1 in early stages of steroidogenic organ development (Ikeda *et al.*, 1994). This study also localized SF-1 transcripts to the embryonic forebrain. In the adult, SF-1 is expressed in adrenals, gonads, pituitary, and hypothalamus (Ikeda *et al.*, 1993, 1995; Shinoda *et al.*, 1997). Based on similar mutant phenotypes and overlapping expression patterns during development, a possible link with the orphan receptor DAX-1 in regulating a common developmental pathway has also been suggested (Ikeda *et al.*, 1996) (see Section IV.B).

Evidence supporting a critical role for SF-1 during development has come from targeted disruption of its gene in mice. These studies have revealed an essential role for this orphan receptor in the development of the hypothalamic–pituitary–adrenal/gonadal axis (Luo *et al.*, 1994; Ingraham *et al.*, 1994). Mice lacking SF-1 are born without adrenals and gonads, and die by Postnatal Day 8, presumably due to adrenocortical insufficiency (Luo *et al.*, 1994). All homozygous mutant mice also have female internal genitalia, indicating that SF-1 functions prior to Müllerian duct regression and supporting the proposed role for SF-1 in regulation of the gene encoding Müllerian inhibiting substance, which is required for sexual differentiation during embryogenesis (Luo *et al.*, 1994; Shen *et al.*, 1994). SF-1 knock-out mice also lack three gonadotroph-specific markers in the pituitary, indicating that SF-1 also has a role in reproductive function (Ingraham *et al.*, 1994). This role is further supported by the finding that SF-1 is critical for normal development of the ventromedial hypothalamic nucleus, a region of the hypothalamus linked to reproductive behavior (Ikeda *et al.*, 1995; Shinoda *et al.*, 1997). Given the critical role of SF-1 in development of steroidogenic tissues and its putative role in the regulation of steroid hydroxylase gene expression, identification of a ligand for this orphan receptor not only would shed light on SF-1's mechanism of action, but also could provide the basis for production of therapeutic agents to control steroid hormone biosynthesis.

B. Receptors That Generally Function as Transcriptional Repressors

A number of orphan receptors have been identified as repressors of other nuclear receptor signaling pathways. The most frequently affected pathway is retinoid signaling through RXR/RAR heterodimers on a natural retinoic acid response element (Sucov *et al.*, 1990). The receptors in this category include COUP-TF, DAX-1, ERR, Rev-Erb, and TR2. Although it remains possible that natural ligands exist for some of these receptors, it is worthwhile noting that many of these proteins (i.e., COUP-TF, Rev-Erb, and TR2) lack the ligand-dependent activation domain (AF-2), which is required for ligand-inducible transcription by all other nuclear receptors with known ligands (Danielian *et al.*, 1992). The orphan receptor SHP is thought to repress transcription through heterodimerization with RXR and other receptors, and is therefore discussed in Section III.B.

The COUP-TF (chicken ovalbumin upstream promoter-transcription factor) family of orphan receptors is believed to function primarily as repressors of multiple signaling pathways. The COUP-TF subfamily includes three members, two of which, COUP-TFI (also known as ear-3) (Wang *et al.*, 1989; Miyajima *et al.*, 1988) and COUP-TFII (also known as ARP-1) (Wang *et al.*, 1991; Ladias and Karathanasis, 1991), are closely related. The third member, called ear-2 (Miyajima *et al.*, 1988), is more distantly related to the other two (Qiu *et al.*, 1996). The members of this family are highly conserved between species, suggesting that they have similar, essential functions in both vertebrates and invertebrates. COUP-TFI was originally characterized as a protein required for expression of the chicken ovalbumin gene (Tsai *et al.*, 1987; Wang *et al.*, 1989). COUP-TFII (ARP-1) was identified as a factor that binds to a regulatory region in the promoter of the apolipoprotein AI gene (Ladias and Karathanasis, 1991). Interestingly, cDNAs for both of these proteins were first isolated by low-stringency hybridization, although their function was unknown (Miyajima *et al.*, 1988). COUP-TFs can bind as homodimers to a wide variety of DNA response elements, although they appear to have greatest affinity for DR-1-type sequences on which they can also heterodimerize with RXR (Kliewer *et al.*, 1992a; Kadowaki *et al.*, 1992; Cooney *et al.*, 1992, 1993).

COUP-TFs have been implicated in the negative regulation of a wide variety of genes, including several that are involved in lipid transport, metabolism, and muscle differentiation (Ladias and Karathanasis, 1991; Ge *et al.*, 1994; Ladias *et al.*, 1992; Mietus-Snyder *et al.*, 1992; Ochoa *et al.*, 1993; Gaudet and Ginsburg, 1995; Muscat *et al.*, 1995; for reviews on the COUP-TF receptor family, see Qiu *et al.*, 1994b, 1996). As negative regulators, COUP-TFs have been shown to repress several other nuclear receptor signaling pathways, including those mediated by RXR (Kliewer *et al.*, 1992a; Widom *et al.*, 1992), RAR (Cooney *et al.*, 1992; Tran *et al.*, 1992), TR (Cooney *et al.*, 1992), VDR (Cooney *et al.*, 1992), PPAR (Miyata *et al.*, 1993), ER (Burbach *et al.*, 1994; Liu *et al.*, 1993), SF-1 (Wehrenberg *et al.*, 1994), and HNF-4 (Mietus-Snyder *et al.*, 1992; Ladias *et al.*, 1992; Galson *et al.*, 1995; Ochoa *et al.*, 1993; Kimura *et al.*, 1993; Carter *et al.*, 1994). Several mechanisms for this negative regulation have been proposed, including competition for common or overlapping binding sites in target gene promoters, titration of RXR from other heterodimeric interactions, and active silencing of transcription through protein–protein interactions involving transcriptional coregulator molecules (Cooney *et al.*, 1993; Leng *et al.*, 1996). The balance between positive and negative gene regulation involving any of these mechanisms is likely to depend on the relative amount of each receptor in a cell, as well as its affinity for DNA response elements and transcriptional regulatory proteins. While the COUP-TFs are generally considered to act as transcriptional repressors, there is some evidence that they can also function as transcriptional activators, depending on promoter context (Kadowaki *et al.*, 1995; Kimura *et al.*, 1993; Gaudet and Ginsburg, 1995) or availability of specific coactivator molecules (Marcus *et al.*, 1996).

COUP-TFI and II are highly expressed in the developing mouse CNS and during organogenesis (Qiu *et al.*, 1994a; Jonk *et al.*, 1994; Lu *et al.*, 1994). In the adult, COUP-TFI and II are expressed in a more widespread pattern, but also continue to be expressed in the CNS. Based on targeted disruption of the COUP-TF genes in mice, both appear to be essential with nonredundant functions. COUP-TFII appears to be required earlier in development than COUP-TFI, as the COUP-TFII knock-out mice die *in utero* at Day 10.5, and the mice lacking the COUP-TFI gene die perinatally within 36 hr of birth (M.-J. Tsai, personal communication). Although the mechanisms underlying these lethal phenotypes are not yet known, it is clear from these studies that both genes have distinct and essential functions in the developing mouse. Like HNF-4, COUP-TF has a well conserved *Drosophila* homolog called seven-up (SVP) that is expressed during fly embryogenesis and is required for development of the central nervous system as well as specific photoreceptor cells in the eye (Mlodzik *et al.*, 1990; Broadus and Doe, 1995). The degree of sequence conservation between *Drosophila* and mammalian COUP-TF is striking (>90%), suggesting a highly conserved role for this orphan receptor family throughout evolution.

DAX-1 (dosage-sensitive sex reversal, adrenal hypoplasia congenita critical region on the X-chromosome, gene 1) was identified as the product of the gene which, when mutated, is responsible for X-linked adrenal hypoplasia congenita (AHC), a disease affecting adrenal development, and hypogonadotropic hypogonadism (HH), a disease often associated with AHC (Zanaria *et al.*, 1994; Muscatelli *et al.*, 1994, and references therein). DAX-1 appears to be expressed most highly in adult adrenal cortex, gonads, hypothalamus, and pituitary (Zanaria *et al.*, 1994; Guo *et al.*, 1995; Swain *et al.*, 1996). These findings suggest a critical role for DAX-1 in development and function of the hypothalamic–pituitary–adrenal/gonadal axis, as well as a possible role in sex determination (Zanaria *et al.*, 1994; Guo *et al.*, 1995; Swain *et al.*, 1996). DAX-1 is an unusual member of the nuclear receptor superfamily that shares homology with other nuclear receptors only in its putative ligand-binding domain (region E in Fig. 1). Unlike several other orphan receptors that appear to act as transcriptional repressors, DAX-1 contains the conserved AF-2 domain that is required for ligand-dependent transcriptional activation in other nuclear receptors. DAX-1 has a unique amino terminus that replaces the typical highly conserved nuclear receptor DNA-binding domain. This N-terminal region consists of four incomplete alanine and glycine-rich repeats, each containing a conserved set of cysteines (Zanaria *et al.*, 1994). Although DAX-1 lacks a conventional nuclear receptor DNA binding domain, it has been reported to bind to a DR-5 type sequence *in vitro*. Interestingly, this DNA binding does not require a partner receptor such as RAR or RXR (Zanaria *et al.*, 1994). In this same study, transient transfection analysis was used to demonstrate that DAX-1 may negatively regulate RAR activation, probably through competition for DNA binding.

A putative binding site for the orphan receptor SF-1 was recently identified in the promoter of the DAX-1 gene, prompting the suggestion that SF-1 may be involved in regulation of DAX-1 (Burris *et al.*, 1995). However, this suggestion is compromised by the finding that DAX-1 expression in the embryonic gonad and hypothalamus is unaffected in SF-1 knock-out mice, indicating that in this context, SF-1 alone does not regulate DAX-1 expression (Ikeda *et al.*, 1996). More data refuting the idea that SF-1 is required for DAX-1 expression come from immunocytochemical studies showing that DAX-1 expression in the fetal testis is not dependent on the presence of SF-1 (Majdic and Saunders, 1996). While it seems clear that SF-1 does not regulate DAX-1 expression in these cases, based on their overlapping expression patterns and the striking similarity of their mutant phenotypes, it has been suggested that SF-1 and DAX-1 may interact to regulate a common endocrine developmental pathway (Ikeda *et al.*, 1996).

ERRα and ERRβ (originally called estrogen-related receptor 1 and 2) were identified based on their similarity to the DNA binding domain of the estrogen receptor with which they share the highest sequence identity (Giguère *et al.*, 1988). There are multiple isoforms of these receptors which can bind a variety of response elements containing inverted and direct repeats of the consensus hexad motif (V. Giguère, personal communication). ERRα has a wide tissue distribution, most notably in the central nervous system (Giguère *et al.*, 1988). ERRα has been implicated in estrogen-controlled regulation of lactoferrin gene expression based on its ability to bind to an extended half-site consensus sequence in the promoter of the human lactoferrin gene (Yang *et al.*, 1996). Evidence suggesting that ERRα acts as a transcriptional repressor comes from its isolation as one of several proteins that bind and cause transcriptional repression of the SV-40 major late promoter (Wiley *et al.*, 1993).

The closely related receptor ERRβ has been identified as a cell-type-specific repressor of several hormone pathways, including those regulated by glucocorticoids, retinoids, and estrogen. Repression of glucocorticoid activity does not appear to involve alterations in DNA binding by either ERRβ or glucocorticoid receptor, but may instead involve titration of a glucocorticoid receptor-specific factor necessary for transcriptional activation (Trapp and Holsboer, 1996). In contrast, ERRβ repression of retinoid signaling requires DNA binding and has been proposed to work either by direct competition or through cell-specific interactions with additional factors (V. Giguère, personal communication). ERRβ has also been reported to associate with heat shock proteins and bind DNA as a homodimer to a palindromic estrogen response element (ERE) *in vitro* (Pettersson *et al.*, 1996). Thus, it appears that both the estrogen receptor and ERRβ may compete for binding to the same target genes, indicating a potential role for ERRβ in the regulation of some aspects of estrogen receptor signaling. Together with their sequence similarity to estrogen receptors, the above results suggest that unlike other orphan receptors, ERRs more closely resemble steroid rather than nonsteroid receptors (Pettersson *et al.*, 1996).

The function of ERRβ during mouse development is under investigation. ERRβ expression begins between 6.5 and 7.5 days postcoitum in the extraembryonic ectoderm and developing chorion (Pettersson et al., 1996; V. Giguère, personal communication). As was predicted from the work of Pettersson and colleagues (1996), targeted disruption of the ERRβ gene results in an early embryonic lethal phenotype due to a placental defect in the formation of the chorion (V. Giguère, personal communication). Along with its role in early development, ERRβ probably has additional functions in the developing and adult mouse, as the mRNA is reexpressed beginning at approximately Day 16 in the heart, kidney, and adrenals (V. Giguère, personal communication).

Rev-ErbAα, also known as ear-1 (Lazar et al., 1989; Miyajima et al., 1989), and Rev-Erbβ, also known as RVR or BD73 (Forman et al., 1994; Enmark et al., 1994; Retnakaran et al., 1994; Dumas et al., 1994), are closely related receptors that appear to act primarily as transcriptional repressors. Rev-ErbAα was originally identified as an orphan receptor encoded on the opposite strand of the genomic DNA coding for the TRα gene (Lazar et al., 1989; Miyajima et al., 1989). Rev-ErbAα is expressed in many tissues, but most highly in skeletal muscle and brown fat where it may be involved in cell-specific differentiation (Lazar et al., 1989). Rev-ErbAα mRNA has been shown to be up-regulated in 3T3-L1 cells during their differentiation into adipocytes, whereas pretreatment of the cells with retinoic acid prevents both Rev-ErbAα mRNA induction and differentiation, suggesting a role for Rev-ErbAα in adipocyte-specific gene expression (Chawla and Lazar, 1993). Rev-ErbAα may also have a negative regulatory role in myogenesis, as its overexpression in myogenic cells can prevent cell differentiation and induction of myogenin mRNA as well as suppress MyoD mRNA levels (Downes et al., 1995).

Rev-Erb family members were first shown to bind DNA as monomers on extended half-site response elements containing an AT-rich 5′ flanking sequence (Harding and Lazar, 1993; Dumas et al., 1994; Retnakaran et al., 1994; Forman et al., 1994). The Rev-Erb receptors are not transcriptionally active on these elements and instead appear to prevent activation by other receptors capable of binding the same element. For example, as mentioned previously, Rev-Erb receptors can interfere with positive signaling by RORα (Retnakaran et al., 1994; Forman et al., 1994), presumably by competition for DNA binding to specific sequences. It has also been demonstrated that Rev-ErbAα acts as a transcriptional repressor when bound as a homodimer to a specific DR-2 type DNA sequence (Harding and Lazar, 1995). This specific DR-2 sequence, referred to as a Rev-DR2, is a natural response element within the cellular retinol binding protein I gene promoter that responds to retinoids through RXR/RAR heterodimers (Smith et al., 1991). In transfections with a reporter containing this response element, Rev-ErbAα prevents retinoid signaling, presumably by competing with RXR/RAR heterodimers for DNA binding (Harding and Lazar, 1995). Rev-ErbAα repression is mediated through interactions with N-CoR, a corepressor originally identified by its association with RAR and TR (Hörlein et al., 1995); however, N-CoR appears to interact with a different region of Rev-ErbAα than it does with TR and RAR (Zamir et al., 1996). As mentioned earlier, Rev-Erb receptors lack the highly conserved AF-2 domain required for ligand-dependent transcriptional activation. Thus, ligands for these proteins may not exist and they may function exclusively as transcriptional repressors.

TR2 (testicular receptor 2) was originally isolated from a human testis cDNA library and its mRNA is highly expressed in testis, prostate, and seminal vesicle (Chang and Kokontis, 1988; C. H. Lee et al., 1995). A second member of this family, TR4 (Chang et al., 1994) (also known as TAK1 and TR2R1 (Hirose et al., 1994a; Law et al., 1994), has a wide tissue distribution with significant expression in both testis and brain (in testis, abundant expression of TR4 is seen mainly in spermatocytes) (Hirose et al., 1994a; Law et al., 1994; Chang et al., 1994). TR2 has been reported to bind to a variety of direct repeat response elements separated by one or more nucleotides, presumably as a homodimer, although this remains controversial (Hirose et al., 1995a; Lin et al., 1995). It has also been demonstrated that TR2 can bind DR-2-type sequences in the SV40 major late promoter and the human erythropoietin gene, repressing transactivation from these elements (Lee and Chang, 1995; Lee et al., 1996). Results from transient transfection experiments involving TR2 family members suggest that these receptors may negatively interfere with retinoid signaling by both RXR homodimers and RXR/RAR heterodimers (Hirose et al., 1995a; Lin et al., 1995). Both of these studies suggest that the mechanism of TR2 repression is due to competitive binding to RXR and RAR response elements and not by titration of RXR or RAR through direct protein–protein interactions.

C. Receptors with Unknown Activation Functions

Even when little is known about the function or potential ligand for an orphan receptor, valuable insight can be gained from determining its temporal and spatial pattern of expression in both embryos and adults. This section reviews what is known about two relatively new members of the nuclear receptor superfamily, GCNF and Tlx, based primarily on their patterns of expression.

An example of a receptor with a limited expression pattern suggesting a potentially interesting role is the gonad-specific receptor GCNF (germ cell nuclear factor) (F. Chen et al., 1994) also known as RTR (Hirose et al., 1995b). GCNF can bind to a DR-0 sequence and an SF-1 site with a critical requirement of the 5' flanking sequence for binding (preliminary evidence suggests GCNF binds as a homodimer) (F. Chen et al., 1994). In situ localization experiments have demonstrated that GCNF is expressed specifically in spermatogenic cells and developing oocytes, giving rise to the suggestion that this orphan receptor may be important for gametogenesis, possibly playing a role in meiosis (F. Chen *et al.*, 1994). However, other studies employing Northern blot analysis from different testis cell populations found GCNF mRNA expression preferentially in haploid germ cells, specifically round spermatids (Hirose *et al.*, 1995b). Expression in this population of haploid cells may suggest a role for this orphan receptor in a postmeiotic phase of spermatogenesis (Hirose *et al.*, 1995b). Although further research will be needed to clarify the exact role of GCNF, these studies strongly imply that this orphan receptor may be important for certain aspects of gamete development.

Tlx is the vertebrate homolog of the *Drosophila* tailless gene (tll) and has been found in mouse, chicken, and zebrafish (Pignoni *et al.*, 1990; Yu *et al.*, 1994; Monaghan *et al.*, 1995). In *Drosophila*, tll is expressed in embryonic termini, developing brain, and peripheral nervous system, and is required for pattern formation in the embryonic poles as well as development of the brain (Pignoni *et al.*, 1990). Results from ectopic expression of tll in fly embryos suggest that tll may have a dual role during development, causing activation of some genes and repression of others (Steingrímsson *et al.*, 1991). Both tll and Tlx differ from other nuclear receptors in their ability to bind DNA as monomers and/or homodimers to the unique half-site sequence AAGTCA (Yu *et al.*, 1994; K. Umesono, personal communication). Based on the similarity in expression patterns and DNA binding specificity, vertebrate Tlx may play a role similar to its *Drosophila* counterpart in early embryonic development. This idea is supported by experiments demonstrating that Tlx is expressed in the forebrain and eye of developing chick and mouse embryos (Yu *et al.*, 1994; Monaghan *et al.*, 1995). In addition, ectopic expression of the vertebrate Tlx in flies is able to mimic the effects observed by overexpression of *Drosophila* tll in flies, suggesting conservation of function between the vertebrate and invertebrate receptors (Yu *et al.*, 1994; Steingrímsson *et al.*, 1991). Future experiments to disrupt the gene in mice should help define the function of this receptor during mammalian brain development.

V. Summary and Perspective

An inspection of Table 1 shows that at least 30 different vertebrate orphan receptor genes have been identified to date, and this list is likely to increase. The nuclear receptors, including the orphan receptors and other known members of the superfamily, now represent the largest family of transcription factors. The importance of these proteins in signal transduction is demonstrated by the impact they have had on the fields of transcription, molecular genetics, and endocrinology. The study of nuclear receptors as transcription factors has led to several discoveries that have become paradigms for transcriptional regulation by enhancer proteins; examples include the discovery of general coactivators such as CBP (Janknecht and Hunter, 1996), the concept of the dimerization on DNA (Gronemeyer and Moras, 1995; Luisi and Freedman, 1995), and the importance of response element specificity (Glass, 1994). In the field of molecular genetics, mutations in the genes for the orphan receptors HNF-4, DAX-1, NOR-1, and RORα have been implicated in four different diseases: diabetes, adrenal hypoplasia congenita, and extraskeletal myxoid chondrosarcoma in humans, and the *staggerer* phenotype in mice, respectively. Finally, in endocrinology, the discovery that 9-*cis* retinoic acid, eicosanoids, and oxysterols mediate their effects through nuclear receptors has opened up three new signaling pathways for investigation, each of which is likely to have an impact on human diseases associated with their biology.

Despite these remarkable advances, the function of many of these proteins is still poorly understood. Thus, while the field has come a long way, much remains to be discovered. The quest for ligands and the development of improved techniques to facilitate this search continues to be at the forefront of future research. In addition to the successful cell-based assays that have been used, recent progress in the disciplines of X-ray crystallography and NMR have begun to provide insights on the mechanism of DNA and ligand binding, as well as transcriptional activation by the nuclear receptors. Thus far, structures have been reported for individual domains of several nonsteroid receptors, including the DNA-binding domains of the RXR homodimer and the RXR/TR heterodimer, and the ligand-binding domains of RXR, RAR, and TR (Lee *et al.*, 1993; Rastinejad *et al.*, 1995; Bourguet *et al.*, 1995; Renaud *et al.*, 1995; Wagner *et al.*, 1995; Wurtz *et al.*, 1996). The conserved structural determinants of the ligand-binding domain suggest that it should be possible in the future to predict or even design ligands for orphan receptors. However, since it is not clear that all orphan receptors will respond to ligands, other specialized techniques are likely to be required to fully understand the function of these proteins. The use of genetic techniques is one of the more powerful tools that may help elucidate the biological role of some of the orphan receptors. Since many receptors may be important during development and in the adult, the ability to generate conditional gene mutations (i.e., knock-outs, knock-ins, or dominant-negatives in a temporally and/or spatially restricted manner) will become increasingly important. Ablating the downstream targets of receptors or the genes involved in metabolism of their ligands is another approach that may help define their specific roles.

In summary, the study of this family of transcription factors continues to offer a number of challenges to investigators posing questions regarding the functions of orphan receptors. The answers to these questions will surely provide us with important information pertaining to many aspects of growth, development, and homeostasis governed by nuclear orphan receptors.

Acknowledgments

We thank Drs. Vincent Giguère, Cary Weinberger, Kazuhiko Umesono, and Ming-Jer Tsai for sharing information prior to publication. We also thank members of the Mango Lab, especially Dr. Daniel Peet, for critical reading of the manuscript. D.J.M. is an investigator of the Howard Hughes Medical Institute. This work was supported by HHMI, a Robert A. Welch Foundation grant to D.J.M., and an NIH Pharmacological Sciences Training Grant to P.J.W.

References

Almasan, A., Mangelsdorf, D. J., Ong, E. S., Wahl, G. M., and Evans, R. M. (1994). Chromosomal localization of the human retinoid X receptors. *Genomics* **20**, 397–403.

Amri, E.-Z., Bonino, F., Ailhaud, G., Abumrad, N. A., and Grimaldi, P. A. (1995). Cloning of a protein that mediates transcriptional effects of fatty acids in preadipocytes. Homology to peroxisome proliferator-activated receptors. *J. Biol. Chem.* **270**, 2367–2371.

Apfel, R., Benbrook, D., Lernhardt, E., Ortiz, M. A., Salbert, G., and Pfahl, M. (1994). A novel orphan receptor specific for a subset of thyroid hormone-responsive elements and its interaction with the retinoid/thyroid hormone receptor subfamily. *Mol. Cell. Biol.* **14**, 7025–7035.

Baes, M., Gulick, T., Choi, H.-S., Martinoli, M. G., Simha, D., and Moore, D. D. (1994). A new orphan member of the nuclear hormone receptor superfamily that interacts with a subset of retinoic acid response elements. *Mol. Cell. Biol.* **14**, 1544–1552.

Barnhart, K. M. and Mellon, P. L. (1994a). The sequence of a murine cDNA encoding Ear-2, a nuclear orphan receptor. *Gene* **142**, 313–314.

Barnhart, K. M. and Mellon, P. L. (1994b). The orphan nuclear receptor, steroidogenic factor-1, regulates the glycoprotein hormone a-subunit gene in pituitary gonadotropes. *Mol. Endocrinol.* **8**, 878–885.

Beato, M., Herrlich, P., and Schütz, G. (1995). Steroid hormone receptors: Many actors in search of a plot. *Cell (Cambridge, Mass.)* **83**, 851–857.

Becker-André, M., André, E., and DeLamarter, J. F. (1993). Identification of nuclear receptor mRNAs by RT-PCR amplification of conserved zinc-finger motif sequences. *Biochem. Biophys. Res. Commun.* **194**, 1371–1379.

Becker-André, M., Wiesenberg, I., Schaeren-Wiemers, N., André, E., Missbach, M., Saurat, J.-H., and Carlberg, C. (1994). Pineal gland hormone melatonin binds and activates an orphan of the nuclear receptor superfamily. *J. Biol. Chem.* **269**, 28531–28534.

Bell, G. I., Xiang, K.-S., Newman, M. V., Wu, S.-H., Wright, L. G., Fajans, S. S., Spielman, R. S., and Cox, N. J. (1991). Gene for non-insulin-dependent diabetes mellitus (maturity-onset diabetes of the young subtype) is linked to DNA polymorphism on human chromosome 20q. *Proc. Natl. Acad. Sci. U.S.A.* **88**, 1484–1488.

Bourguet, W., Ruff, M., Chambon, P., Gronemeyer, H., and Moras, D. (1995). Crystal structure of the ligand-binding domain of the human nuclear receptor RXR-a. *Nature (London)* **375**, 377–382.

Braissant, O., Foufelle, F., Scotto, C., Dauça, M., and Wahli, W. (1996). Differential expression of peroxisome proliferator-activated receptors (PPARs): Tissue distribution of PPAR-α, -β, and -γ in the adult rat. *Endocrinology* **137**, 354–366.

Broadus, J. and Doe, C. Q. (1995). Evolution of neuroblast identity: *Seven-up* and *prospero* expression reveal homologous and divergent neuroblast fates in *Drosophila* and *Schistocerca*. *Development (Cambridge, UK)* **121**, 3989–3996.

Brun, R. P., Tontonoz, P., Forman, B. M., Ellis, R., Chen, J., Evans, R. M., and Spiegelman, B. M. (1996). Differential activation of adipogenesis by multiple PPAR isoforms. *Genes Dev.* **10**, 974–984.

Bugge, T. H., Pohl, J., Lonnoy, O., and Stunnenberg, H. G. (1992). RXRα, a promiscuous partner of retinoic acid and thyroid hormone receptors. *EMBO J.* **11**, 1409–1418.

Burbach, J. P. H., Lopes da Silva, S., Cox, J. J., Adan, R. A. H., Cooney, A. J., Tsai, M.-J., and Tsai, S. Y. (1994). Repression of estrogen-dependent stimulation of the oxytocin gene by chicken ovalbumin upstream promoter transcription factor I. *J. Biol. Chem.* **269**, 15046–15053.

Burris, T. P., Guo, W., Le, T., and McCabe, E. R. B. (1995). Identification of a putative steroidogenic factor-1 response element in the DAX-1 promoter. *Biochem. Biophys. Res. Commun.* **214**, 576–581.

Carlberg, C., Van Huijsduijnen, R. H., Staple, J. K., DeLamarter, J. F., and Becker-André, M. (1994). RZRs, a new family of retinoid-related orphan receptors that function as both monomers and homodimers. *Mol. Endocrinol.* **8**, 757–770.

Carter, M. E., Gulick, T., Raisher, B. D., Caira, T., Ladias, J. A. A., Moore, D. D., and Kelly, D. P. (1993). Hepatocyte nuclear factor-4 activates medium chain acyl-CoA dehydrogenase gene transcription by interacting with a complex regulatory element. *J. Biol. Chem.* **268**, 13805–13810.

Carter, M. E., Gulick, T., Moore, D. D., and Kelly, D. P. (1994). A pleiotropic element in the medium-chain acyl coenzyme A dehydrogenase gene promoter mediates transcriptional regulation by multiple nuclear receptor transcription factors and defines novel receptor-DNA binding motifs. *Mol. Cell. Biol.* **14**, 4360–4372.

Chambon, P. (1996). A decade of molecular biology of retinoic acid receptors. *FASEB J.* **10**, 940–954.

Chang, C. and Kokontis, J. (1988). Identification of a new member of the steroid receptor superfamily by cloning and sequence analysis. *Biochem. Biophys. Res. Commun.* **155**, 971–977.

Chang, C., Kokontis, J., Liao, S., and Chang, Y. (1989). Isolation and characterization of human TR3 receptor: A member of steroid receptor superfamily. *J. Steroid Biochem.* **34**, 391–395.

Chang, C., Da Silva, S. L., Ideta, R., Lee, Y., Yeh, S., and Burbach, J. P. H. (1994). Human and rat TR4 orphan receptors specify a subclass of the steroid receptor superfamily. *Proc. Natl. Acad. Sci. U.S.A.* **91**, 6040–6044.

Chawla, A., and Lazar, M. A. (1993). Induction of Rev-ErbAα, an orphan receptor encoded on the opposite strand of the a-thyroid hormone receptor gene, during adipocyte differentiation. *J. Biol. Chem.* **268**, 16265–16269.

Chen, D., Lepar, G., and Kemper, B. (1994a). A transcriptional regulatory element common to a large family of hepatic cytochrome P450 genes is a functional binding site of the orphan receptor HNF-4. *J. Biol. Chem.* **269**, 5420–5427.

Chen, D., Park, Y., and Kemper, B. (1994b). Differential protein binding and transcriptional activities of HNF-4 elements in three closely related *CYP2C* genes. *DNA Cell Biol.* **13**, 771–779.

Chen, F., Law, S. W., and O'Malley, B. W. (1993). Identification of two mPPAR related receptors and evidence for the existence of five subfamily members. *Biochem. Biophys. Res. Commun.* **196**, 671–677.

Chen, F., Cooney, A. J., Wang, Y., Law, S. W., and O'Malley, B. W. (1994). Cloning of a novel orphan receptor (GCNF) expressed during germ cell development. *Mol. Endocrinol.* **8**, 1434–1444.

Chen, J. D. and Evans, R. M. (1995). A transcriptional co-repressor that interacts with nuclear hormone receptors. *Nature (London)* **377**, 454–457.

Chen, J. Y., Clifford, J., Zusi, C., Starrett, J., Tortolani, D., Ostrowski, J., Reczek, P. R., Chambon, P., and Gronemeyer, H. (1996). Two distinct actions of retinoid-receptor ligands. *Nature (London)* **382**, 819–822.

Chen, W. S., Manova, K., Weinstein, D. C., Duncan, S. A., Plump, A. S., Prezioso, V. R., Bachvarova, R. F., and Darnell, J. E., Jr. (1994). Disruption of the HNF-4 gene, expressed in visceral endoderm, leads to cell death in embryonic ectoderm and impaired gastrulation of mouse embryos. *Genes Dev.* **8**, 2466–2477.

Cooney, A. J., Tsai, S. Y., O'Malley, B. W., and Tsai, M.-J. (1992). Chicken ovalbumin upstream promoter transcription factor (COUP-TF) dimers bind to different GGTCA response elements, allowing COUP-TF to repress hormonal induction of the vitamin D_3, thyroid hormone, and retinoic acid receptors. *Mol. Cell. Biol.* **12**, 4153–4163.

Cooney, A. J., Leng, X., Tsai, S. Y., O'Malley, B. W., and Tsai, M.-J. (1993). Multiple mechanisms of chicken ovalbumin upstream promoter transcription factor-dependent repression of transactivation by the vitamin D, thyroid hormone, and retinoic acid receptors. *J. Biol. Chem.* **268**, 4152–4160.

Costa, R. H., Van Dyke, T. A., Yan, C., Kuo, F., and Darnell, J. E., Jr. (1990). Similarities in transthyretin gene expression and differences in transcription factors: Liver and yolk sac compared to choroid plexus. *Proc. Natl. Acad. Sci. U.S.A.* **87**, 6589–6593.

Crawford, P. A., Sadovsky, Y., Woodson, K., Lee, S. L., and Milbrandt, J. (1995). Adrenocortical function and regulation of the steroid 21-hydroxylase gene in NGFI-B-deficient mice. *Mol. Cell. Biol.* **15**, 4331–4336.

Crossley, M., Ludwig, L., Stowell, K. M., De Vos, P., Olek, K., and Brownlee, G. G. (1992). Recovery from Hemophilia B Leyden: An androgen-responsive element in the factor IX promoter. *Science* **257**, 377–379.

Damm, K., Thompson, C. C., and Evans, R. M. (1989). Protein encoded by v-*erbA* functions as a thyroid-hormone receptor antagonist. *Nature (London)* **339**, 593–597.

Danielian, P. S., White, R., Lees, J. A., and Parker, M. G. (1992). Identification of a conserved region required for hormone dependent transcriptional activation by steroid hormone receptors. *EMBO J.* **11**, 1025–1033.

Davis, I. J. and Lau, L. F. (1994). Endocrine and neurogenic regulation of the orphan nuclear receptors Nur77 and Nurr-1 in the adrenal glands. *Mol. Cell. Biol.* **14**, 3469–3483.

Davis, I. J., Hazel, T. G., Chen, R.-H., Blenis, J., and Lau, L. F. (1993). Functional domains and phosphorylation of the orphan receptor Nur77. *Mol. Endocrinol.* **7**, 953–964.

Devchand, P. R., Keller, H., Peters, J. M., Vazquez, M., Gonzalez, F. J., and Wahli, W. (1996). The PPARα-leukotriene B_4 pathway to inflammation control. *Nature (London)* **384**, 39–43.

Dhar, A. K., Teng, J. I., and Smith, L. L. (1973). Biosynthesis of cholest-5-ene-3β, 24-diol (cerebrosterol) by bovine cerebral cortical microsomes. *J. Neurochem.* **21**, 51–60.

Dixon, R., Furutachi, T., and Lieberman, S. (1970). The isolation of crystalline 22R-hydroxycholesterol and 20α,22R-dihydroxycholesterol from bovine adrenals. *Biochem. Biophys. Res. Commun.* **40**, 161–165.

Downes, M., Carozzi, A. J., and Muscat, G. E. O. (1995). Constitutive expression of the orphan receptor, Rev-erbAα, inhibits muscle differentiation and abrogates the expression of the *myoD* gene family. *Mol. Endocrinol.* **9**, 1666–1678.

Dreyer, C., Krey, G., Keller, H., Givel, F., Helftenbein, G., and Wahli, W. (1992). Control of the peroxisomal β-oxidation pathway by a novel family of nuclear hormone receptors. *Cell (Cambridge, Mass.)* **68**, 879–887.

Dumas, B., Harding, H. P., Choi, H.-S., Lehmann, K. A., Chung, M., Lazar, M. A., and Moore, D. D. (1994). A new orphan member of the nuclear hormone receptor superfamily closely related to Rev-Erb. *Mol. Endocrinol.* **8**, 996–1005.

Duncan, S. A., Manova, K., Chen, W. S., Hoodless, P., Weinstein, D. C., Bachvarova, R. F., and Darnell, J. E., Jr. (1994). Expression of transcription factor HNF-4 in the extraembryonic endoderm, gut, and nephrogenic tissue of the developing mouse embryo: HNF-4 is a marker for primary endoderm in the implanting blastocyst. *Proc. Natl. Acad. Sci. U.S.A.* **91**, 7598–7602.

Durand, B., Saunders, M., Gaudon, C., Roy, B., Losson, R., and Chambon, P. (1994). Activation function 2 (AF-2) of retinoic acid receptor and 9-*cis* retinoic acid receptor: Presence of a conserved autonomous constitutive activating domain and influence of the nature of the response element on AF-2 activity. *EMBO J.* **13**, 5370–5382.

Dussault, I. and Giguère, V. (1997). Differential regulation of the N-*myc* proto-oncogene by RORa and RVR, two orphan members of the superfamily of nuclear hormone receptors. *Mol. Cell. Biol.* **17**, 1860–1867.

Ellinger-Ziegelbauer, H., Hihi, A. K., Laudet, V., Keller, H., Wahli, W., and Dreyer, C. (1994). FTZ-F1-related orphan receptors in *Xenopus laevis*: Transcriptional regulators differentially expressed during early embryogenesis. *Mol. Cell. Biol.* **14**, 2786–2797.

Enmark, E., Kainu, T., Pelto-Huikko, M., and Gustafsson, J.-Å. (1994). Identification of a novel member of the nuclear receptor superfamily which is closely related to Rev-ErbA. *Biochem. Biophys. Res. Commun.* **204**, 49–56.

Erdmann, D., and Heim, J. (1995). Orphan nuclear receptor HNF-4 binds to the human coagulation factor VII promotor. *J. Biol. Chem.* **270**, 22988–22996.

Fahrner, T. J., Carroll, S. L., and Milbrandt, J. (1990). The NGFI-B protein, an inducible member of the thyroid/steroid receptor family, is rapidly modified posttranslationally. *Mol. Cell. Biol.* **10**, 6454–6459.

Fajans, S. S. (1989). Maturity-onset diabetes of the young (MODY). *Diabetes/Metab. Rev.* **5**, 579–606.

Forman, B. M., Chen, J., Blumberg, B., Kliewer, S. A., Henshaw, R., Ong, E. S., and Evans, R. M. (1994). Cross-talk among RORα1 and the Rev-erb family of orphan nuclear receptors. *Mol. Endocrinol.* **8**, 1253–1261.

Forman, B. M., Umesono, K., Chen, J., and Evans, R. M. (1995a). Unique response pathways are established by allosteric interactions among nuclear hormone receptors. *Cell (Cambridge, Mass.)* **81**, 541–550.

Forman, B. M., Goode, E., Chen, J., Oro, A. E., Bradley, D. J., Perlmann, T., Noonan, D. J., Burka, L. T., McMorris, T., Lamph, W. W., Evans, R. M., and Weinberger, C. (1995b). Identification of a nuclear receptor that is activated by farnesol metabolites. *Cell (Cambridge, Mass.)* **81**, 687–693.

Forman, B. M., Tontonoz, P., Chen, J., Brun, R. P., Spiegelman, B. M., and Evans, R. M. (1995c). 15-Deoxy-$\Delta^{12,14}$-prostaglandin J2 is a ligand for the adipocyte determination factor PPARγ. *Cell (Cambridge, Mass.)* **83**, 803–812.

Froguel, P., Zouali, H., Vionnet, N., Velho, G., Vaxillaire, M., Sun, F., Lesage, S., Stoffel, M., Takeda, J., Passa, P., Permutt, M. A., Beckmann, J. S., Bell, G. I., and Cohen, D. (1993). Familial hyperglycemia due to mutations in glucokinase: Definition of a subtype of diabetes mellitus. *N. Engl. J. Med.* **328**, 697–702.

Galarneau, L., Paré, J. F., Allard, D., Hamel, D., Lévesque, L., Tugwood, J. D., Green, S., and Bélanger, L. (1996). The α_1-fetoprotein locus is activated by a nuclear receptor of the *Drosophila* FTZ-F1 family. *Mol. Cell. Biol.* **16**, 3853–3865.

Galson, D. L., Tsuchiya, T., Tendler, D. S., Huang, L. E., Ren, Y., Ogura, T., and Bunn, H. F. (1995). The orphan receptor hepatic nuclear factor 4 functions as a transcriptional activator for tissue-specific and hypoxia-specific erythropoietin gene expression and is antagonized by EAR3/COUP-TF1. *Mol. Cell. Biol.* **15**, 2135–2144.

Garcia, A. D., Ostapchuk, P., and Hearing, P. (1993). Functional interaction of nuclear factors EF-C, HNF-4, and RXRα with hepatitis B virus enhancer I. *J. Virol.* **67**, 3940–3950.

Gaudet, F., and Ginsburg, G. S. (1995). Transcriptional regulation of the cholesteryl ester transfer protein gene by the orphan nuclear hormone receptor apolipoprotein AI regulatory protein-1. *J. Biol. Chem.* **270**, 29916–29922.

Ge, R., Rhee, M., Malik, S., and Karathanasis, S. K. (1994). Transcriptional repression of apolipoprotein AI gene expression by orphan receptor ARP-1. *J. Biol. Chem.* **269**, 13185–13192.

Giguère, V., Yang, N., Segui, P., and Evans, R. M. (1988). Identification of a new class of steroid hormone receptors. *Nature (London)* **331**, 91–94.

Giguère, V., Tini, M., Flock, G., Ong, E., Evans, R. M., and Otulakowski, G. (1994). Isoform-specific amino-terminal domains dictate DNA-binding properties of RORα, a novel family of orphan hormone nuclear receptors. *Genes Dev.* **8**, 538–553.

Giguère, V., Beatty, B., Squire, J., Copeland, N. G., and Jenkins, N. A. (1995). The orphan nuclear receptor RORα (RORA) maps to a conserved region of homology on human chromosome 15q21-q22 and mouse chromosome 9. *Genomics* **28**, 596–598.

Glass, C. K. (1994). Differential recognition of target genes by nuclear receptor monomers, dimers, and heterodimers. *Endocr. Rev.* **15**, 391–407.

Göttlicher, M., Widmark, E., Li, Q., and Gustafsson, J.-Å. (1992). Fatty acids activate a chimera of the clofibric acid-activated receptor and the glucocorticoid receptor. *Proc. Natl. Acad. Sci. U.S.A.* **89**, 4653–4657.

Greene, M. E., Blumberg, B., McBride, O. W., Yi, H. F., Kronquist, K., Kwan, K., Hsieh, L., Greene, G., and Nimer, S. D. (1995). Isolation of the human peroxisome proliferator activated receptorγ cDNA: Expression in hematopoietic cells and chromosomal mapping. *Gene Expression* **4**, 281–299.

Greiner, E. F., Kirfel, J., Greschik, H., Dörflinger, U., Becker, P., Mercep, A., and Schüle, R. (1996). Functional analysis of retinoid Z receptor β, a brain-specific nuclear orphan receptor. *Proc. Natl. Acad. Sci. U.S.A.* **93**, 10105–10110.

Gronemeyer, H., and Laudet, V. (1995). Transcription factors 3: Nuclear receptors. *Protein Profile* **2**, 1173–1308.

Gronemeyer, H., and Moras, D. (1995). Nuclear receptors: How to finger DNA. *Nature (London)* **375**, 190–191.

Guo, W., Burris, T. P., and McCabe, E. R. B. (1995). Expression of DAX-1, the gene responsible for X-linked adrenal hypoplasia congenita and hypogonadotropic hypogonadism, in the hypothalamic-pituitary-adrenal/gonadal axis. *Biochem. Mol. Med.* **56**, 8–13.

Guo, W. W., Lovell, R. S., Zhang, Y. H., Huang, B. L., Burris, T. P., Craigen, W. J., and McCabe, E. R. B. (1996). *Ahch*, the mouse homologue of *DAX1*: Cloning, characterization and synteny with *GyK*, the glycerol kinase locus. *Gene* **178**, 31–34.

Hall, R. K., Sladek, F. M., and Granner, D. K. (1995). The orphan receptors COUP-TF and HNF-4 serve as accessory factors required for induction of phospho*enol*pyruvate carboxykinase gene transcription by glucocorticoids. *Proc. Natl. Acad. Sci. U.S.A.* **92**, 412–416.

Halvorson, L. M., Kaiser, U. B., and Chin, W. W. (1996). Stimulation of luteinizing hormone β gene promoter activity by the orphan nuclear receptor, steroidogenic factor-1. *J. Biol. Chem.* **271**, 6645–6650.

Hamada, K., Gleason, S. L., Levi, B.-Z., Hirschfeld, S., Appella, E., and Ozato, K. (1989). H-2RIIBP, a member of the nuclear hormone receptor superfamily that binds to both the regulatory element of major histocompatibility class I genes and the estrogen response element. *Proc. Natl. Acad. Sci. U.S.A.* **86**, 8289–8293.

Hamilton, B. A., Frankel, W. N., Kerrebrock, A. W., Hawkins, T. L., FitzHugh, W., Kusumi, K., Russell, L. B., Mueller, K. L., Van Berkel, V., Birren, B. W., Kruglyak, L., and Lander, E. S. (1996). Disruption of the nuclear hormone receptor RORα in *staggerer* mice. *Nature (London)* **379**, 736–739.

Harding, H. P., and Lazar, M. A. (1993). The orphan receptor Rev-ErbAα activates transcription via a novel response element. *Mol. Cell. Biol.* **13**, 3113–3121.

Harding, H. P., and Lazar, M. A. (1995). The monomer-binding orphan receptor Rev-Erb represses transcription as a dimer on a novel direct repeat. *Mol. Cell. Biol.* **15**, 4791–4802.

Harmon, M. A., Boehm, M. F., Heyman, R. A., and Mangelsdorf, D. J. (1995). Activation of mammalian retinoid X receptors by the insect growth regulator methoprene. *Proc. Natl. Acad. Sci. U.S.A.* **92,** 6157–6160.

Hatano, O., Takayama, K., Imai, T., Waterman, M. R., Takakusu, A., Omura, T., and Morohashi, K. (1994). Sex-dependent expression of a transcription factor, Ad4BP, regulating steroidogenic P-450 genes in the gonads during prenatal and postnatal rat development. *Development (Cambridge, UK)* **120,** 2787–2797.

Hazel, T. G., Nathans, D., and Lau, L. F. (1988). A gene inducible by serum growth factors encodes a member of the steroid and thyroid hormone receptor superfamily. *Proc. Natl. Acad. Sci. U.S.A.* **85,** 8444–8448.

Hazel, T. G., Misra, R., Davis, I. J., Greenberg, M. E., and Lau, L. F. (1991). Nur77 is differentially modified in PC12 cells upon membrane depolarization and growth factor treatment. *Mol. Cell. Biol.* **11,** 3239–3246.

Hazlerigg, D. G., Barrett, P., Hastings, M. H., and Morgan, P. J. (1996). Are nuclear receptors involved in pituitary responsiveness to melatonin? *Mol. Cell. Endocrinol.* **123,** 53–59.

Hedvat, C. V. and Irving, S. G. (1995). The isolation and characterization of MINOR, a novel mitogen-inducible nuclear orphan receptor. *Mol. Endocrinol.* **9,** 1692–1700.

Hertz, R., Berman, I., Keppler, D., and Bar-Tana, J. (1996). Activation of gene transcription by prostacyclin analogues is mediated by the peroxisome-proliferators-activated receptor (PPAR). *Eur. J. Biochem.* **235,** 242–247.

Heyman, R. A., Mangelsdorf, D. J., Dyck, J. A., Stein, R. B., Eichele, G., Evans, R. M., and Thaller, C. (1992). 9-*Cis* retinoic acid is a high affinity ligand for the retinoid X receptor. *Cell (Cambridge, Mass.)* **68,** 397–406.

Hirata, Y., Kiuchi, K., Chen, H.-C., Milbrandt, J., and Guroff, G. (1993). The phosphorylation and DNA binding of the DNA-binding domain of the orphan nuclear receptor NGFI-B. *J. Biol. Chem.* **268,** 24808–24812.

Hirata, Y., Whalin, M., Ginty, D. D., Xing, J., Greenberg, M. E., Milbrandt, J., and Guroff, G. (1995). Induction of a nerve growth factor-sensitive kinase that phosphorylates the DNA-binding domain of the orphan nuclear receptor NGFI-B. *J. Neurochem.* **65,** 1780–1788.

Hirose, T., Fujimoto, W., Yamaai, T., Kim, K. H., Matsuura, H., and Jetten, A. M. (1994a). TAK1: Molecular cloning and characterization of a new member of the nuclear receptor superfamily. *Mol. Endocrinol.* **8,** 1667–1680.

Hirose, T., Smith, R. J., and Jetten, A. M. (1994b). RORγ: The third member of ROR/RZR orphan receptor subfamily that is highly expressed in skeletal muscle. *Biochem. Biophys. Res. Commun.* **205,** 1976–1983.

Hirose, T., Apfel, R., Pfahl, M., and Jetten, A. M. (1995a). The orphan receptor TAK1 acts as a repressor of RAR-, RXR- and T3R-mediated signaling pathways. *Biochem. Biophys. Res. Commun.* **211,** 83–91.

Hirose, T., O'Brien, D. A., and Jetten, A. M. (1995b). RTR: A new member of the nuclear receptor superfamily that is highly expressed in murine testis. *Gene* **152,** 247–251.

Honda, S., Morohashi, K., Nomura, M., Takeya, H., Kitajima, M., and Omura, T. (1993). Ad4BP regulating steroidogenic P-450 gene is a member of steroid hormone receptor superfamily. *J. Biol. Chem.* **268,** 7494–7502.

Hoopes, C. W., Taketo, M., Ozato, K., Liu, Q., Howard, T. A., Linney, E., and Seldin, M. F. (1992). Mapping of the *Rxr* loci encoding nuclear retinoid X receptors RXRα, RXRβ, and RXRγ. *Genomics* **14,** 611–617.

Hörlein, A. J., Näär, A. M., Heinzel, T., Torchia, J., Gloss, B., Kurokawa, R., Ryan, A., Kamei, Y., Söderström, M., Glass, C. K., and Rosenfeld, M. G. (1995). Ligand-independent repression by the thyroid hormone receptor mediated by a nuclear receptor co-repressor. *Nature (London)* **377,** 397–404.

Horwitz, K. B., Jackson, T. A., Rain, D. L., Richer, J. K., Takimoto, G. S., and Tung, L. (1996). Nuclear receptor coactivators and corepressors. *Mol. Endocrinol.* **10,** 1167–1177.

Hu, E. D., Kim, J. B., Sarraf, P., and Spiegelman, B. M. (1996). Inhibition of adipogenesis through MAP kinase-mediated phosphorylation of PPARγ. *Science* **274**, 2100–2103.

Hung, H. L., and High, K. A. (1996). Liver-enriched transcription factor HNF-4 and ubiquitous factor NF-Y are critical for expression of blood coagulation factor X. *J. Biol. Chem.* **271**, 2323–2331.

Ikeda, Y., Lala, D. S., Luo, X., Kim, E., Moisan, M.-P., and Parker, K. L. (1993). Characterization of the mouse *FTZ-F1* gene, which encodes a key regulator of steroid hydroxylase gene expression. *Mol. Endocrinol.* **7**, 852–860.

Ikeda, Y., Shen, W.-H., Ingraham, H. A., and Parker, K. L. (1994). Developmental expression of mouse steroidogenic factor-1, an essential regulator of the steroid hydroxylases. *Mol. Endocrinol.* **8**, 654–662.

Ikeda, Y., Luo, X., Abbud, R., Nilson, J. H., and Parker, K. L. (1995). The nuclear receptor steroidogenic factor 1 is essential for the formation of the ventromedial hypothalamic nucleus. *Mol. Endocrinol.* **9**, 478–486.

Ikeda, Y., Swain, A., Weber, T. J., Hentges, K. E., Zanaria, E., Lalli, E., Tamai, K. T., Sassone-Corsi, P., Lovell-Badge, R., Camerino, G., and Parker, K. L. (1996). Steroidogenic factor 1 and Dax-1 colocalize in multiple cell lineages: Potential links in endocrine development. *Mol. Endocrinol.* **10**, 1261–1272.

Ingraham, H. A., Lala, D. S., Ikeda, Y., Luo, X., Shen, W.-H., Nachtigal, M. W., Abbud, R., Nilson, J. H., and Parker, K. L. (1994). The nuclear receptor steroidogenic factor 1 acts at multiple levels of the reproductive axis. *Genes Dev.* **8**, 2302–2312.

Issemann, I., and Green, S. (1990). Activation of a member of the steroid hormone receptor superfamily by peroxisome proliferators. *Nature (London)* **347**, 645–650.

Janknecht, R., and Hunter, T. (1996). Transcription—A growing coactivator network. *Nature (London)* **383**, 22–23.

Janowski, B. A., Willy, P. J., Devi, T. R., Falck, J. R., and Mangelsdorf, D. J. (1996). An oxysterol signalling pathway mediated by the nuclear receptor LXRα. *Nature* **383**, 728–731.

Javitt, N. B., Kok, E., Burstein, S., Cohen, B., and Kutscher, J. (1981). 26-Hydroxycholesterol. Identification and quantitation in human serum. *J. Biol. Chem.* **256**, 12644–12646.

Jiang, G., Nepomuceno, L., Hopkins, K., and Sladek, F. M. (1995). Exclusive homodimerization of the orphan receptor hepatocyte nuclear factor 4 defines a new subclass of nuclear receptors. *Mol. Cell. Biol.* **15**, 5131–5143.

Jones, B. B., Ohno, C. K., Allenby, G., Boffa, M. B., Levin, A. A., Grippo, J. F., and Petkovich, M. (1995). New retinoid X receptor subtypes in zebra fish (*Danio rerio*) differentially modulate transcription and do not bind 9-*cis* retinoic acid. *Mol. Cell. Biol.* **15**, 5226–5234.

Jones, P. S., Savory, R., Barratt, P., Bell, A. R., Gray, T. J. B., Jenkins, N. A., Gilbert, D. J., Copeland, N. G., and Bell, D. R. (1995). Chromosomal localisation, inducibility, tissue-specific expression and strain differences in three murine peroxisome-proliferator-activated-receptor genes. *Eur. J. Biochem.* **233**, 219–226.

Jonk, L. J. C., de Jonge, M. E. J., Pals, C. E. J. M., Wissink, S., Vervaart, J. M. A., Schoorlemmer, J., and Kruijer, W. (1994). Cloning and expression during development of three murine members of the COUP family of nuclear orphan receptors. *Mech. Dev.* **47**, 81–97.

Kadowaki, Y., Toyoshima, K., and Yamamoto, T. (1992). Ear3/COUP-TF binds most tightly to a response element with tandem repeat separated by one nucleotide. *Biochem. Biophys. Res. Commun.* **183**, 492–498.

Kadowaki, Y., Toyoshima, K., and Yamamoto, T. (1995). Dual transcriptional control by Ear3/COUP: Negative regulation through the DR1 direct repeat and positive regulation through a sequence downstream of the transcriptional start site of the mouse mammary tumor virus promoter. *Proc. Natl. Acad. Sci. U.S.A.* **92**, 4432–4436.

Kandutsch, A. A., Chen, H. W., and Heiniger, H-J. (1978). Biological activity of some oxygenated sterols. *Science* **201**, 498–501.

Kastner, P., Mark, M., and Chambon, P. (1995). Nonsteroid nuclear receptors: What are genetic studies telling us about their role in real life? *Cell (Cambridge, Mass.)* **83**, 859–869.
Kastner, P., Mark, M., Leid, M., Gansmuller, A., Chin, W., Grondona, J. M., Décimo, D., Krezel, W., Dierich, A., and Chambon, P. (1996). Abnormal spermatogenesis in RXRβ mutant mice. *Genes Dev.* **10**, 80–92.
Kimura, A., Nishiyori, A., Murakami, T., Tsukamoto, T., Hata, S., Osumi, T., Okamura, R., Mori, M., and Takiguchi, M. (1993). Chicken ovalbumin upstream promoter-transcription factor (COUP-TF) represses transcription from the promoter of the gene for ornithine transcarbamylase in a manner antagonistic to hepatocyte nuclear factor-4 (HNF-4). *J. Biol. Chem.* **268**, 11125–11133.
Kitareewan, S., Burka, L. T., Tomer, K. B., Parker, C. E., Deterding, L. J., Stevens, R. D., Forman, B. M., Mais, D. E., Heyman, R. A., McMorris, T., and Weinberger, C. (1996). Phytol metabolites are circulating dietary factors that activate the nuclear receptor RXR. *Mol. Biol. Cell* **7**, 1153–1166.
Kliewer, S. A., Umesono, K., Heyman, R. A., Mangelsdorf, D. J., Dyck, J. A., and Evans, R. M. (1992a). Retinoid X receptor-COUP-TF interactions modulate retinoic acid signaling. *Proc. Natl. Acad. Sci. U.S.A.* **89**, 1448–1452.
Kliewer, S. A., Umesono, K., Mangelsdorf, D. J., and Evans, R. M. (1992b). Retinoid X receptor interacts with nuclear receptors in retinoic acid, thyroid hormone and vitamin D_3 signalling. *Nature (London)* **355**, 446–449.
Kliewer, S. A., Umesono, K., Noonan, D. J., Heyman, R. A., and Evans, R. M. (1992c). Convergence of 9-*cis* retinoic acid and peroxisome proliferator signalling pathways through heterodimer formation of their receptors. *Nature (London)* **358**, 771–774.
Kliewer, S. A., Forman, B. M., Blumberg, B., Ong, E. S., Borgmeyer, U., Mangelsdorf, D. J., Umesono, K., and Evans, R. M. (1994). Differential expression and activation of a family of murine peroxisome proliferator-activated receptors. *Proc. Natl. Acad. Sci. U.S.A.* **91**, 7355–7359.
Kliewer, S. A., Lenhard, J. M., Willson, T. M., Patel, I., Morris, D. C., and Lehmann, J. M. (1995). A prostaglandin J2 metabolite binds peroxisome proliferator-activated receptor γ and promotes adipocyte differentiation. *Cell (Cambridge, Mass.)* **83**, 813–819.
Kozak, C. A., Adamson, M. C., and Weinberger, C. (1996). Genetic mapping of gene encoding the farnesoid receptor, *Fxr,* to mouse Chromosome 10. *Mamm. Genome* **7**, 164–165.
Krezel, W., Dupé, V., Mark, M., Dierich, A., Kastner, P., and Chambon, P. (1996). RXRγ null mice are apparently normal and compound RXRα$^{+/-}$/RXRβ$^{-/-}$/RXRγ$^{-/-}$ mutant mice are viable. *Proc. Natl. Acad. Sci. U.S.A.* **93**, 9010–9014.
Kuo, C. J., Conley, P. B., Chen, L., Sladek, F. M., Darnell, J. E., Jr., and Crabtree, G. R. (1992). A transcriptional hierarchy involved in mammalian cell-type specification. *Nature (London)* **355**, 458–460.
Kurokawa, R., Yu, V. C., Näär, A., Kyakumoto, S., Han, Z., Silverman, S., Rosenfeld, M. G., and Glass, C. K. (1993). Differential orientations of the DNA-binding domain and carboxy-terminal dimerization interface regulate binding site selection by nuclear receptor heterodimers. *Genes Dev.* **7**, 1423–1435.
Kurokawa, R., DiRenzo, J., Boehm, M., Sugarman, J., Gloss, B., Rosenfeld, M. G., Heyman, R. A., and Glass, C. K. (1994). Regulation of retinoid signalling by receptor polarity and allosteric control of ligand binding. *Nature (London)* **371**, 528–531.
Kurokawa, R., Söderström, M., Hörlein, A. J., Halachmi, S., Brown, M., Rosenfeld, M. G., and Glass, C. K. (1995). Polarity-specific activities of retinoic acid receptors determined by a co-repressor. *Nature (London)* **377**, 451–454.
Labelle, Y., Zucman, J., Stenman, G., Kindblom, L. G., Knight, J., Turc-Carel, C., Dockhorn-Dworniczak, B., Mandahl, N., Desmaze, C., Peter, M., Aurias, A., Delattre, O., and Thomas, G. (1995). Oncogenic conversion of a novel orphan nuclear receptor by chromosome translocation. *Hum. Mol. Genet.* **4**, 2219–2226.

Ladias, J. A. A. and Karathanasis, S. K. (1991). Regulation of the apolipoprotein AI gene by ARP-1, a novel member of the steroid receptor superfamily. *Science* **251**, 561–565.

Ladias, J. A. A., Hadzopoulou-Cladaras, M., Kardassis, D., Cardot, P., Cheng, J., Zannis, V., and Cladaras, C. (1992). Transcriptional regulation of human apolipoprotein genes ApoB, ApoCIII, and ApoAII by members of the steroid hormone receptor superfamily HNF-4, ARP-1, EAR-2, and EAR-3. *J. Biol. Chem.* **267**, 15849–15860.

Lala, D. S., Rice, D. A., and Parker, K. L. (1992). Steroidogenic factor I, a key regulator of steroidogenic enzyme expression, is the mouse homolog of fushi tarazu-factor I. *Mol. Endocrinol.* **6**, 1249–1258.

Laudet, V., Hänni, C., Coll, J., Catzeflis, F., and Stéhelin, D. (1992). Evolution of the nuclear receptor gene superfamily. *EMBO J.* **11**, 1003–1013.

La Vista-Picard, N., Hobbs, P. D., Pfahl, M., and Dawson, M. I. (1996). The receptor-DNA complex determines the retinoid response: A mechanism for the diversification of the ligand signal. *Mol. Cell. Biol.* **16**, 4137–4146.

Lavorgna, G., Ueda, H., Clos, J., and Wu, C. (1991). FTZ-F1, a steroid hormone receptor-like protein implicated in the activation of *fushi tarazu*. *Science* **252**, 848–851.

Law, S. W., Conneely, O. M., DeMayo, F. J., and O'Malley, B. W. (1992). Identification of a new brain-specific transcription factor, NURR1. *Mol. Endocrinol.* **6**, 2129–2135.

Law, S. W., Conneely, O. M., and O'Malley, B. W. (1994). Molecular cloning of a novel member of the nuclear receptor superfamily related to the orphan receptor, TR2. *Gene Expression* **4**, 77–84.

Lazar, M. A., Hodin, R. A., Darling, D. S., and Chin, W. W. (1989). A novel member of the thyroid/steroid hormone receptor family is encoded by the opposite strand of the rat c-*erbA*α transcriptional unit. *Mol. Cell. Biol.* **9**, 1128–1136.

Lee, C. H., Copeland, N. G., Gilbert, D. J., Jenkins, N. A., and Wei, L. N. (1995). Genomic structure, promoter identification, and chromosomal mapping of a mouse nuclear orphan receptor expressed in embryos and adult testes. *Genomics* **30**, 46–52.

Lee, H.-J. and Chang, C. (1995). Identification of human TR2 orphan receptor response element in the transcriptional initiation site of the simian virus 40 major late promoter. *J. Biol. Chem.* **270**, 5434–5440.

Lee, H. J., Young, W. J., Shih, C. C. Y., and Chang, C. S. (1996). Suppression of the human erythropoietin gene expression by the TR2 orphan receptor, a member of the steroid receptor superfamily. *J. Biol. Chem.* **271**, 10405–10412.

Lee, M. S., Kliewer, S. A., Provencal, J., Wright, P. E., and Evans, R. M. (1993). Structure of the retinoid X receptor α DNA binding domain: A helix required for homodimeric DNA binding. *Science* **260**, 1117–1121.

Lee, S. L., Wesselschmidt, R. L., Linette, G. P., Kanagawa, O., Russell, J. H., and Milbrandt, J. (1995). Unimpaired thymic and peripheral T cell death in mice lacking the nuclear receptor NGFI-B (Nur77). *Science* **269**, 532–535.

Lee, S. S.-T., Pineau, T., Drago, J., Lee, E. J., Owens, J. W., Kroetz, D. L., Fernandez-Salguero, P. M., Westphal, H., and Gonzalez, F. J. (1995). Targeted disruption of the α isoform of the peroxisome proliferator-activated receptor gene in mice results in abolishment of the pleiotropic effects of peroxisome proliferators. *Mol. Cell. Biol.* **15**, 3012–3022.

Lehmann, J. M., Moore, L. B., Smith-Oliver, T. A., Wilkison, W. O., Willson, T. M., and Kliewer, S. A. (1995). An antidiabetic thiazolidinedione is a high affinity ligand for peroxisome proliferator-activated receptor γ (PPARγ). *J. Biol. Chem.* **270**, 12953–12956.

Lehmann, J. M., Kliewer, S. A., Moore, L. B., Smith-Oliver, T. A., Oliver, B. B., Su, J.-L., Sundseth, S. S., Winegar, D. A., Blanchard, D. E., Spencer, T. A., and Willson, T. M. (1997). Activation of the nuclear receptor LXR by oxysterols defines a new hormone response pathway. *J. Biol. Chem.* **272**, 3137–3140.

Leid, M., Kastner, P., Lyons, R., Nakshatri, H., Saunders, M., Zacharewski, T., Chen, J.-Y., Staub, A., Garnier, J.-M., Mader, S., and Chambon, P. (1992a). Purification, cloning, and RXR identity of the HeLa cell factor with which RAR or TR heterodimerizes to bind target sequences efficiently. *Cell (Cambridge, Mass.)* **68**, 377–395.

Leid, M., Kastner, P., and Chambon, P. (1992b). Multiplicity generates diversity in the retinoic acid signalling pathways. *Trends Biochem. Sci.* **17**, 427–433.

Lemberger, T., Desvergne, B., and Wahli, W. (1996). Peroxisome proliferator-activated receptors: A nuclear receptor signaling pathway in lipid physiology. *Annu. Rev. Cell Biol.* **12**, 335–363.

LeMotte, P. K., Keidel, S., and Apfel, C. M. (1996). Phytanic acid is a retinoid X receptor ligand. *Eur. J. Biochem.* **236**, 328–333.

Leng, X., Cooney, A. J., Tsai, S. Y., and Tsai, M. J. (1996). Molecular mechanisms of COUP-TF-mediated transcriptional repression: Evidence for transrepression and active repression. *Mol. Cell. Biol.* **16**, 2332–2340.

Levin, A. A., Sturzenbecker, L. J., Kazmer, S., Bosakowski, T., Huselton, C., Allenby, G., Speck, J., Kratzeisen, C., Rosenberger, M., Lovey, A., and Grippo, J. F. (1992). 9-*Cis* retinoic acid stereoisomer binds and activates the nuclear receptor RXRα. *Nature (London)* **355**, 359–361.

Lim, R. W., Yang, W.-L., and Yu, H. (1995). Signal-transduction-pathway-specific desensitization of expression of orphan nuclear receptor TIS1. *Biochem. J.* **308**, 785–789.

Lin, T. M., Young, W. J., and Chang, C. S. (1995). Multiple functions of the TR2-11 orphan receptor in modulating activation of two key cis-acting elements involved in the retinoic acid signal transduction system. *J. Biol. Chem.* **270**, 30121–30128.

Liu, Y., Yang, N., and Teng, C. T. (1993). COUP-TF acts as a competitive repressor for estrogen receptor-mediated activation of the mouse lactoferrin gene. *Mol. Cell. Biol.* **13**, 1836–1846.

Liu, Z.-G., Smith, S. W., McLaughlin, K. A., Schwartz, L. M., and Osborne, B. A. (1994). Apoptotic signals delivered through the T-cell receptor of a T-cell hybrid require the immediate-early gene *nur77*. *Nature (London)* **367**, 281–284.

Lu, X. P., Salbert, G., and Pfahl, M. (1994). An evolutionary conserved COUP-TF binding element in a neural-specific gene and COUP-TF expression patterns support a major role for COUP-TF in neural development. *Mol. Endocrinol.* **8**, 1774–1788.

Luisi, B. and Freedman, L. (1995). Nuclear receptors: Dymer, dymer binding tight. *Nature (London)* **375**, 359–360.

Luo, X., Ikeda, Y., and Parker, K. L. (1994). A cell-specific nuclear receptor is essential for adrenal and gonadal development and sexual differentiation. *Cell (Cambridge, Mass.)* **77**, 481–490.

Lutz, B., Kuratani, S., Cooney, A. J., Wawersik, S., Tsai, S. Y., Eichele, G., and Tsai, M.-J. (1994). Developmental regulation of the orphan receptor *COUP-TF II* gene in spinal motor neurons. *Development (Cambridge, UK)* **120**, 25–36.

Lynch, J. P., Lala, D. S., Peluso, J. J., Parker, K. L., and White, B. A. (1993). Steroidogenic factor 1, an orphan receptor, regulates the expression of the rat aromatase gene in gonadal tissues. *Mol. Endocrinol.* **7**, 776–786.

MacDonald, P. N., Dowd, D. R., Nakajima, S., Galligan, M. A., Reeder, M. C., Haussler, C. A., Ozato, K., and Haussler, M. R. (1993). Retinoid X receptors stimulate and 9-*cis* retinoic acid inhibits 1,25-dihydroxyvitamin D_3-activated expression of the rat osteocalcin gene. *Mol. Cell. Biol.* **13**, 5907–5917.

Mages, H. W., Rilke, O., Bravo, R., Senger, G., and Kroczek, R. A. (1994). NOT, a human immediate-early response gene closely related to the steroid/thyroid hormone receptor NAK1/TR3. *Mol. Endocrinol.* **8**, 1583–1591.

Majdic, G. and Saunders, P. T. K. (1996). Differential patterns of expression of DAX-1 and steroidogenic factor-1 (SF-1) in the fetal rat testis. *Endocrinology (Baltimore)* **137**, 3586–3589.

Mangelsdorf, D. J., and Evans, R. M. (1995). The RXR heterodimers and orphan receptors. *Cell (Cambridge, Mass.)* **83**, 841–850.

Mangelsdorf, D. J., Ong, E. S., Dyck, J. A., and Evans, R. M. (1990). Nuclear receptor that identifies a novel retinoic acid response pathway. *Nature (London)* **345**, 224–229.

Mangelsdorf, D. J., Umesono, K., Kliewer, S. A., Borgmeyer, U., Ong, E. S., and Evans, R. M. (1991). A direct repeat in the cellular retinol-binding protein type II gene confers differential regulation by RXR and RAR. *Cell (Cambridge, Mass.)* **66**, 555–561.

Mangelsdorf, D. J., Borgmeyer, U., Heyman, R. A., Zhou, J. Y., Ong, E. S., Oro, A. E., Kakizuka, A., and Evans, R. M. (1992). Characterization of three RXR genes that mediate the action of 9-*cis* retinoic acid. *Genes Dev.* **6**, 329–344.

Mangelsdorf, D. J., Umesono, K., and Evans, R. M. (1994). The retinoid receptors. *In* "The Retinoids: Biology, Chemistry, and Medicine" (M. B. Sporn, A. B. Roberts, and D. S. Goodman, eds.), 2nd ed., pp. 319–349. Raven Press, New York.

Mangelsdorf, D. J., Thummel, C., Beato, M., Herrlich, P., Schütz, G., Umesono, K., Blumberg, B., Kastner, P., Mark, M., Chambon, P., and Evans, R. M. (1995). The nuclear receptor superfamily: The second decade. *Cell (Cambridge, Mass.)* **83**, 835–839.

Marcus, S. L., Winrow, C. J., Capone, J. P., and Rachubinski, R. A. (1996). A p56^{lck} ligand serves as a coactivator of an orphan nuclear hormone receptor. *J. Biol. Chem.* **271**, 27197–27200.

Marks, M. S., Hallenbeck, P. L., Nagata, T., Segars, J. H., Appella, E., Nikodem, V. M., and Ozato, K. (1992). H-2RIIBP (RXRβ) heterodimerization provides a mechanism for combinatorial diversity in the regulation of retinoic acid and thyroid hormone responsive genes. *EMBO J.* **11**, 1419–1435.

McBroom, L. D. B., Flock, G., and Giguère, V. (1995). The nonconserved hinge region and distinct amino-terminal domains of the RORα orphan nuclear receptor isoforms are required for proper DNA bending and RORα-DNA interactions. *Mol. Cell. Biol.* **15**, 796–808.

Metzger, S., Halaas, J. L., Breslow, J. L., and Sladek, F. M. (1993). Orphan receptor HNF-4 and bZip protein C/EBPα bind to overlapping regions of the apolipoprotein B gene promoter and synergistically activate transcription. *J. Biol. Chem.* **268**, 16831–16838.

Mietus-Snyder, M., Sladek, F. M., Ginsburg, G. S., Kuo, C. F., Ladias, J. A. A., Darnell, J. E., Jr., and Karathanasis, S. K. (1992). Antagonism between apolipoprotein AI regulatory protein 1, Ear3/COUP-TF, and hepatocyte nuclear factor 4 modulates apolipoprotein CIII gene expression in liver and intestinal cells. *Mol. Cell. Biol.* **12**, 1708–1718.

Milbrandt, J. (1988). Nerve growth factor induces a gene homologous to the gludocorticoid receptor gene. *Neuron* **1**, 183–188.

Minucci, S., Leid, M., Toyama, R., Saint-Jeannet, J. P., Peterson, V. J., Horn, V., Ishmael, J. E., Bhattacharyya, N., Dey, A., Dawid, I. B., and Ozato, K. (1997). Retinoid X receptor (RXR) within the RXR-retinoic acid receptor heterodimer binds its ligand and enhances retinoid-dependent gene expression. *Mol. Cell. Biol.* **17**, 644–655.

Miquerol, L., Lopez, S., Cartier, N., Tulliez, M., Raymondjean, M., and Kahn, A. (1994). Expression of the L-type pyruvate kinase gene and the hepatocyte nuclear factor 4 transcription factor in exocrine and endocrine pancreas. *J. Biol. Chem.* **269**, 8944–8951.

Missbach, M., Jagher, B., Sigg, I., Nayeri, S., Carlberg, C., and Wiesenberg, I. (1996). Thiazolidine diones, specific ligands of the nuclear receptor retinoid Z receptor retinoid acid receptor-related orphan receptor a with potent antiarthritic activity. *J. Biol. Chem.* **271**, 13515–13522.

Miura, N., and Tanaka, K. (1993). Analysis of the rat hepatocyte nuclear factor (HNF) 1 gene promoter: Synergistic activation by HNF4 and HNF1 proteins. *Nucleic Acids Res.* **21**, 3731–3736.

Miyajima, N., Kadowaki, Y., Fukushige, S.-I., Shimizu, S.-I., Semba, K., Yamanashi, Y., Matsubara, K.-I., Toyoshima, K., and Yamamoto, T. (1988). Identification of two novel members of *erbA* superfamily by molecular cloning: The gene products of the two are highly related to each other. *Nucleic Acids Res.* **16**, 11057–11074.

Miyajima, N., Horiuchi, R., Shibuya, Y., Fukushige, S.-I., Matsubara, K.-I., Toyoshima, K., and Yamamoto, T. (1989). Two *erbA* homologs encoding proteins with different T$_3$

binding capacities are transcribed from opposite DNA strands of the same genetic locus. *Cell (Cambridge, Mass.)* **57**, 31–39.

Miyata, K. S., Zhang, B., Marcus, S. L., Capone, J. P., and Rachubinski, R. A. (1993). Chicken ovalbumin upstream promoter transcription factor (COUP-TF) binds to a peroxisome proliferator-responsive element and antagonizes peroxisome proliferator-mediated signaling. *J. Biol. Chem.* **268**, 19169–19172.

Mlodzik, M., Hiromi, Y., Weber, U., Goodman, C. S., and Rubin, G. M. (1990). The Drosophila *seven-up* gene, a member of the steroid receptor gene superfamily, controls photoreceptor cell fates. *Cell (Cambridge, Mass.)* **60**, 211–224.

Monaghan, A. P., Grau, E., Bock, D., and Schütz, G. (1995). The mouse homolog of the orphan nuclear receptor *tailless* is expressed in the developing forebrain. *Development (Cambridge, UK)* **121**, 839–853.

Morohashi, K., Zanger, U. M., Honda, S., Hara, M., Waterman, M. R., and Omura, T. (1993). Activation of CYP11A and CYP11B gene promoters by the steroidogenic cell-specific transcription factor, Ad4BP. *Mol. Endocrinol.* **7**, 1196–1204.

Morohashi, K., Iida, H., Nomura, M., Hatano, O., Honda, S., Tsukiyama, T., Niwa, O., Hara, T., Takakusu, A., Shibata, Y., and Omura, T. (1994). Functional difference between Ad4BP and ELP, and their distributions in steroidogenic tissues. *Mol. Endocrinol.* **8**, 643–653.

Muscat, G. E. O., Rea, S., and Downes, M. (1995). Identification of a regulatory function for an orphan receptor in muscle: COUP-TF II affects the expression of the *myoD* gene family during myogenesis. *Nucleic Acids Res.* **23**, 1311–1318.

Muscatelli, F., Strom, T. M., Walker, A. P., Zanaria, E., Récan, D., Meindl, A., Bardoni, B., Guioli, S., Zehetner, G., Rabl, W., Shwarz, H. P., Kaplan, J.-C., Camerino, G., Meitinger, T., and Monaco, A. P. (1994). Mutations in the *DAX-1* gene give rise to both X-linked adrenal hypoplasia congenita and hypogonadotropic hypogonadism. *Nature (London)* **372**, 672–676.

Nakai, A., Kartha, S., Sakurai, A., Toback, F. G., and DeGroot, L. J. (1990). A human early response gene homologous to murine nur77 and rat NGFI-B, and related to the nuclear receptor superfamily. *Mol. Endocrinol.* **4**, 1438–1443.

Nakshatri, H., and Chambon, P. (1994). The directly repeated RG(G/T)TCA motifs of the rat and mouse cellular retinol-binding protein II genes are promiscuous binding sites for RAR, RXR, HNF-4, and ARP-1 homo- and heterodimers. *J. Biol. Chem.* **269**, 890–902.

Ochoa, A., Bovard-Houppermans, S., and Zakin, M. M. (1993). Human apolipoprotein A-IV gene expression is modulated by members of the nuclear hormone receptor superfamily. *Biochim. Biophys. Acta* **1210**, 41–47.

Ohkura, N., Hijikuro, M., Yamamoto, A., and Miki, K. (1994). Molecular cloning of a novel thyroid/steroid receptor superfamily gene from cultured rat neuronal cells. *Biochem. Biophys. Res. Commun.* **205**, 1959–1965.

Ohkura, N., Ito, M., Tsukada, T., Sasaki, K., Yamaguchi, K., and Miki, K. (1996). Structure, mapping and expression of a human NOR-1 gene, the third member of the Nur77/NGFI-B family. *Biochim. Biophys. Acta Gene Struct. Expression* **1308**, 205–214.

Oro, A. E., McKeown, M., and Evans, R. M. (1990). Relationship between the product of the *Drosophila ultraspiracle* locus and the vertebrate retinoid X receptor. *Nature (London)* **347**, 298–301.

Ortiz, M. A., Piedrafita, F. J., Pfahl, M., and Maki, R. (1995). TOR: A new orphan receptor expressed in the thymus that can modulate retinoid and thyroid hormone signals. *Mol. Endocrinol.* **9**, 1679–1691.

Parker, K. L. and Schimmer, B. P. (1994). The role of nuclear receptors in steroid hormone production. *Semin. Cancer Biol.* **5**, 317–325.

Perlmann, T., and Jansson, L. (1995). A novel pathway for vitamin A signaling mediated by RXR heterodimerization with NGFI-B and NURR1. *Genes Dev.* **9**, 769–782.

Perlmann, T., Rangarajan, P. N., Umesono, K., and Evans, R. M. (1993). Determinants for selective RAR and TR recognition of direct repeat HREs. *Genes Dev.* **7**, 1411–1422.

Pescini, R., Kaszubska, W., Whelan, J., DeLamarter, J. F., and Hooft van Huijsduijnen, R. (1994). ATF-a0, a novel variant of the ATF/CREB transcription factor family, forms a dominant transcription inhibitor in ATF-a heterodimers. *J. Biol. Chem.* **269**, 1159–1165.

Pettersson, K., Svensson, K., Mattsson, R., Carlsson, B., Ohlsson, R., and Berkenstam, A. (1996). Expression of a novel member of estrogen response element-binding nuclear receptors is restricted to the early stages of chorion formation during mouse embryogenesis. *Mech. Dev.* **54**, 211–223.

Pignoni, F., Baldarelli, R. M., Steingrimsson, E., Diaz, R. J., Patapoutian, A., Merriam, J. R., and Lengyel, J. A. (1990). The drosophila gene *tailless* is expressed at the embryonic termini and is a member of the steroid receptor superfamily. *Cell (Cambridge, Mass.)* **62**, 151–163.

Qiu, Y., Cooney, A. J., Kuratani, S., DeMayo, F. J., Tsai, S. Y., and Tsai, M.-J. (1994a). Spatiotemporal expression patterns of chicken ovalbumin upstream promoter-transcription factors in the developing mouse central nervous system: Evidence for a role in segmental patterning of the diencephalon. *Proc. Natl. Acad. Sci. U.S.A.* **91**, 4451–4455.

Qiu, Y., Tsai, S. Y., and Tsai, M.-J. (1994b). COUP-TF, an orphan member of the steroid/thyroid hormone receptor superfamily. *Trends Endocrinol. Metab.* **5**, 234–239.

Qiu, Y., Krishnan, V., Zeng, Z., Gilbert, D. J., Copeland, N. G., Gibson, L., Yang-Feng, T., Jenkins, N. A., Tsai, M. J., and Tsai, S. Y. (1995). Isolation, characterization, and chromosomal localization of mouse and human COUP-TF I and II genes. *Genomics* **29**, 240–246.

Qiu, Y., Krishnan, V., Pereira, F. A., Tsai, S. Y., and Tsai, M. J. (1996). Chicken ovalbumin upstream promoter-transcription factors and their regulation. *J. Steroid Biochem. Mol. Biol.* **56**, 81–85.

Rastinejad, F., Perlmann, T., Evans, R. M., and Sigler, P. B. (1995). Structural determinants of nuclear receptor assembly on DNA direct repeats. *Nature (London)* **375**, 203–211.

Reijnen, M. J., Sladek, F. M., Bertina, R. M., and Reitsma, P. H. (1992). Disruption of a binding site for hepatocyte nuclear factor 4 results in hemophilia B Leyden. *Proc. Natl. Acad. Sci. U.S.A.* **89**, 6300–6303.

Renaud, J. P., Rochel, N., Ruff, M., Vivat, V., Chambon, P., Gronemeyer, H., and Moras, D. (1995). Crystal structure of the RAR-gamma ligand-binding domain bound to all-*trans* retinoic acid. *Nature (London)* **378**, 681–689.

Retnakaran, R., Flock, G., and Giguère, V. (1994). Identification of RVR, a novel orphan nuclear receptor that acts as a negative transcriptional regulator. *Mol. Endocrinol.* **8**, 1234–1244.

Roy, B., Taneja, R., and Chambon, P. (1995). Synergistic activation of retinoic acid (RA)-responsive genes and induction of embryonal carcinoma cell differentiation by an RA receptor a (RARα)-, RARβ-, or RARγ-selective ligand in combination with a retinoid X receptor-specific ligand. *Mol. Cell. Biol.* **15**, 6481–6487.

Ryseck, R.-P., Macdonald-Bravo, H., Mattéi, M.-G., Ruppert, S., and Bravo, R. (1989). Structure, mapping and expression of a growth factor inducible gene encoding a putative nuclear hormonal binding receptor. *EMBO J.* **8**, 3327–3335.

Scearce, L. M., Laz, T. M., Hazel, T. G., Lau, L. F., and Taub, R. (1993). RNR-1, a nuclear receptor in the NGFI-B/Nur77 family that is rapidly induced in regenerating liver. *J. Biol. Chem.* **268**, 8855–8861.

Schmidt, A., Endo, N., Rutledge, S. J., Vogel, R., Shinar, D., and Rodan, G. A. (1992). Identification of a new member of the steroid hormone receptor superfamily that is activated by a peroxisome proliferator and fatty acids. *Mol. Endocrinol.* **6**, 1634–1641.

Schräder, M., Nayeri, S., Kahlen, J.-P., Müller, K. M., and Carlberg, C. (1995). Natural vitamin D$_3$ response elements formed by inverted palindromes: Polarity-directed ligand sensitivity

of vitamin D$_3$ receptor-retinoid X receptor heterodimer-mediated transactivation. *Mol. Cell. Biol.* **15**, 1154–1161.
Seol, W., Choi, H.-S., and Moore, D. D. (1995). Isolation of proteins that interact specifically with the retinoid X receptor: Two novel orphan receptors. *Mol. Endocrinol.* **9**, 72–85.
Seol, W., Choi, H. S., and Moore, D. D. (1996). An orphan nuclear hormone receptor that lacks a DNA binding domain and heterodimerizes with other receptors. *Science* **272**, 1336–1339.
Shen, W.-H., Moore, C. C. D., Ikeda, Y., Parker, K. L., and Ingraham, H. A. (1994). Nuclear receptor steroidogenic factor 1 regulates the Müllerian inhibiting substance gene: A link to the sex determination cascade. *Cell (Cambridge, Mass.)* **77**, 651–661.
Sher, T., Yi, H.-F., McBride, O. W., and Gonzalez, F. J. (1993). cDNA cloning, chromosomal mapping, and functional characterization of the human peroxisome proliferator activated receptor. *Biochemistry* **32**, 5598–5604.
Shinar, D. M., Endo, N., Rutledge, S. J., Vogel, R., Rodan, G. A., and Schmidt, A. (1994). NER, a new member of the gene family encoding the human steroid hormone nuclear receptor. *Gene* **147**, 273–276.
Shinoda, K., Lei, H., Yoshii, H., Nomura, M., Nagano, M., Shiba, H., Sasaki, H., Osawa, Y., Ninomiya, Y., Niwa, O., Morohashi, K.-I., and Li, E. (1997). Developmental defects of the ventromedial hypothalamic nucleus and pituitary gonadotroph in the Ftz-F1 disrupted mice. *Dev. Dyn.* **204**, 22–29.
Sladek, F. M., Zhong, W., Lai, E., and Darnell, J. E., Jr. (1990). Liver-enriched transcription factor HNF-4 is a novel member of the steroid hormone receptor superfamily. *Genes Dev.* **4**, 2353–2365.
Smith, D. P., Mason, C. S., Jones, E. A., and Old, R. W. (1994). A novel nuclear receptor superfamily member in *Xenopus* that associates with RXR, and shares extensive sequence similarity to the mammalian vitamin D3 receptor. *Nucleic Acids Res.* **22**, 66–71.
Smith, W. C., Nakshatri, H., Leroy, P., Rees, J., and Chambon, P. (1991). A retinoic acid response element is present in the mouse cellular retinol binding protein I (mCRBPI) promoter. *EMBO J.* **10**, 2223–2230.
Song, C., Kokontis, J. M., Hiipakka, R. A., and Liao, S. (1994). Ubiquitous receptor: A receptor that modulates gene activation by retinoic acid and thyroid hormone receptors. *Proc. Natl. Acad. Sci. U.S.A.* **91**, 10809–10813.
Steinberg, D., Avignan, T., and Mize, C. (1965). Conversion of U-C^{14}-phytol to phytanic acid and its oxidation in heredopathia ataxica polyneuritiformis. *Biochem. Biophys. Res. Commun.* **19**, 783–789.
Steingrímsson, E., Pignoni, F., Liaw, G.-J., and Lengyel, J. A. (1991). Dual role of the *Drosophila* pattern gene *tailless* in embryonic termini. *Science* **254**, 418–421.
Sucov, H. M., Murakami, K. K., and Evans, R. M. (1990). Characterization of an autoregulated response element in the mouse retinoic acid receptor type β gene. *Proc. Natl. Acad. Sci. U.S.A.* **87**, 5392–5396.
Sucov, H. M., Dyson, E., Gumeringer, C. L., Price, J., Chien, K. R., and Evans, R. M. (1994). RXRα mutant mice establish a genetic basis for vitamin A signaling in heart morphogenesis. *Genes Dev.* **8**, 1007–1018.
Sugawara, T., Holt, J. A., Kiriakidou, M., and Strauss, J. F., III (1996). Steroidogenic factor 1-dependent promoter activity of the human steroidogenic acute regulatory protein (StAR) gene. *Biochemistry* **35**, 9052–9059.
Swain, A., Zanaria, E., Hacker, A., Lovell-Badge, R., and Camerino, G. (1996). Mouse *DAX-1* expression is consistent with a role in sex determination as well as adrenal and hypothalamus function. *Nat. Genet.* **12**, 404–409.
Taketo, M., Parker, K. L., Howard, T. A., Tsukiyama, T., Wong, M., Niwa, O., Morton, C. C., Miron, P. M., and Seldin, M. F. (1995). Homologs of Drosophila Fushi-Tarazu factor 1 maps to mouse chromosome 2 and human chromosome 9q33. *Genomics* **25**, 565–567.

Teboul, M., Enmark, E., Li, Q., Wikström, A. C., Pelto-Huikko, M., and Gustafsson, J.-Å. (1995). OR-1, a member of the nuclear receptor superfamily that interacts with the 9-*cis*-retinoic acid receptor. *Proc. Natl. Acad. Sci. U.S.A.* **92**, 2096–2100.

Thummel, C. S. (1995). From embryogenesis to metamorphosis: The regulation and function of Drosophila nuclear receptor superfamily members. *Cell (Cambridge, Mass.)* **83**, 871–877.

Tian, J.-M. and Schibler, U. (1991). Tissue-specific expression of the gene encoding hepatocyte nuclear factor 1 may involve hepatocyte nuclear factor 4. *Genes Dev.* **5**, 2225–2234.

Tini, M., Fraser, R. A., and Giguère, V. (1995). Functional interactions between retinoic acid receptor-related orphan nuclear receptor (RORα) and the retinoic acid receptors in the regulation of the γF-crystallin promoter. *J. Biol. Chem.* **270**, 20156–20161.

Tontonoz, P., Hu, E., Graves, R. A., Budavari, A. I., and Spiegelman, B. M. (1994a). mPPARγ2: Tissue-specific regulator of an adipocyte enhancer. *Genes Dev.* **8**, 1224–1234.

Tontonoz, P., Hu, E., and Spiegelman, B. M. (1994b). Stimulation of adipogenesis in fibroblasts by PPARγ2, a lipid-activated transcription factor. *Cell (Cambridge, Mass.)* **79**, 1147–1156.

Tran, P., Zhang, X.-K., Salbert, G., Hermann, T., Lehmann, J. M., and Pfahl, M. (1992). COUP orphan receptors are negative regulators of retinoic acid response pathways. *Mol. Cell. Biol.* **12**, 4666–4676.

Trapp, T., and Holsboer, F. (1996). Nuclear orphan receptor as a repressor of glucocorticoid receptor transcriptional activity. *J. Biol. Chem.* **271**, 9879–9882.

Tsai, S. Y., Sagami, I., Wang, H., Tsai, M.-J., and O'Malley, B. W. (1987). Interactions between a DNA-binding transcription factor (COUP) and a non-DNA binding factor (S300-11). *Cell (Cambridge, Mass.)* **50**, 701–709.

Tsukiyama, T., Ueda, H., Hirose, S., and Niwa, O. (1992). Embryonal long terminal repeat-binding protein is a murine homolog of FTZ-F1, a member of the steroid receptor superfamily. *Mol. Cell. Biol.* **12**, 1286–1291.

Tugwood, J. D., Issemann, I., and Green, S. (1991). LRH-1: A nuclear hormone receptor active in the absence of exogenous ligands. GenBank accession number M81385.

Umesono, K., Murakami, K. K., Thompson, C. C., and Evans, R. M. (1991). Direct repeats as selective response elements for the thyroid hormone, retinoic acid, and vitamin D_3 receptors. *Cell (Cambridge, Mass.)* **65**, 1255–1266.

Vaxillaire, M., Boccio, V., Philippi, A., Vigouroux, C., Terwilliger, J., Passa, P., Beckmann, J. S., Velho, G., Lathrop, G. M., and Froguel, P. (1995). A gene for maturity onset diabetes of the young (MODY) maps to chromosome 12q. *Nat. Genet.* **9**, 418–423.

Wagner, R. L., Apriletti, J. W., McGrath, M. E., West, B. L., Baxter, J. D., and Fletterick, R. J. (1995). A structural role for hormone in the thyroid hormone receptor. *Nature (London)* **378**, 690–697.

Wang, L.-H., Tsai, S. Y., Cook, R. G., Beattie, W. G., Tsai, M.-J., and O'Malley, B. W. (1989). COUP transcription factor is a member of the steroid receptor superfamily. *Nature (London)* **340**, 163–166.

Wang, L.-H., Ing, N. H., Tsai, S. Y., O'Malley, B. W., and Tsai, M.-J. (1991). The COUP-TFs compose a family of functionally related transcription factors. *Gene Expression* **1**, 207–216.

Wehrenberg, U., Ivell, R., Jansen, M., von Goedecke, S., and Walther, N. (1994). Two orphan receptors binding to a common site are involved in the regulation of the oxytocin gene in the bovine ovary. *Proc. Natl. Acad. Sci. U.S.A.* **91**, 1440–1444.

Weih, F., Ryseck, R. P., Chen, L. H., and Bravo, R. (1996). Apoptosis of *nur77/N10*-transgenic thymocytes involves the Fas/Fas ligand pathway. *Proc. Natl. Acad. Sci. U.S.A.* **93**, 5533–5538.

Widom, R. L., Rhee, M., and Karathanasis, S. K. (1992). Repression by ARP-1 sensitizes apolipoprotein AI gene responsiveness to RXRα and retinoic acid. *Mol. Cell. Biol.* **12**, 3380–3389.

Wiley, S. R., Kraus, R. J., Zuo, F., Murray, E. E., Loritz, K., and Mertz, J. E. (1993). SV40 early-to-late switch involves titration of cellular transcriptional repressors. *Genes Dev.* **7**, 2206–2219.

Willy, P. J., and Mangelsdorf, D. J. (1997). Unique requirements for retinoid-dependent transcriptional activation by the orphan receptor LXR. *Genes Dev.* **11**, 289–298.

Willy, P. J., Umesono, K., Ong, E. S., Evans, R. M., Heyman, R. A., and Mangelsdorf, D. J. (1995). LXR, a nuclear receptor that defines a distinct retinoid response pathway. *Genes Dev.* **9**, 1033–1045.

Wilson, T. E., Fahrner, T. J., Johnston, M., and Milbrandt, J. (1991). Identification of the DNA binding site for NGFI-B by genetic selection in yeast. *Science* **252**, 1296–1300.

Wilson, T. E., Mouw, A. R., Weaver, C. A., Milbrandt, J., and Parker, K. L. (1993a). The orphan nuclear receptor NGFI-B regulates expression of the gene encoding steroid 21-hydroxylase. *Mol. Cell. Biol.* **13**, 861–868.

Wilson, T. E., Fahrner, T. J., and Milbrandt, J. (1993b). The orphan receptors NGFI-B and steroidogenic factor 1 establish monomer binding as a third paradigm of nuclear receptor-DNA interaction. *Mol. Cell. Biol.* **13**, 5794–5804.

Wong, J. M., Shi, Y. B., and Wolffe, A. P. (1995). A role for nucleosome assembly in both silencing and activation of the *Xenopus* TRβA gene by the thyroid hormone receptor. *Genes Dev.* **9**, 2696–2711.

Woronicz, J. D., Calnan, B., Ngo, V., and Winoto, A. (1994). Requirement for the orphan steroid receptor Nur77 in apoptosis of T-cell hybridomas. *Nature (London)* **367**, 277–281.

Wurtz, J. M., Bourguet, W., Renaud, J. P., Vivat, V., Chambon, P., Moras, D., and Gronemeyer, H. (1996). A canonical structure for the ligand-binding domain of nuclear receptors. *Nat. Struct. Biol.* **3**, 87–94.

Yamagata, K., Oda, N., Kaisaki, P. J., Menzel, S., Furuta, H., Vaxillaire, M., Southam, L., Cox, R. D., Lathrop, G. M., Boriraj, V. V., Chen, X. N., Cox, N. J., Oda, Y., Yano, H., Le Beau, M. M., Yamada, S., Nishigori, H., Takeda, J., Fajans, S. S., Hattersley, A. T., Iwasaki, N., Hansen, T., Pedersen, O., Polonsky, K. S., Turner, R. C., Velho, G., Chèvre, J.-C., Froguel, P., and Bell, G. I. (1996a). Mutations in the hepatocyte nuclear factor-1α gene in maturity-onset diabetes of the young (MODY3). *Nature (London)* **384**, 455–458.

Yamagata, K., Furuta, H., Oda, N., Kaisaki, P. J., Menzel, S., Cox, N. J., Fajans, S. S., Signorini, S., Stoffel, M., and Bell, G. I. (1996b). Mutations in the hepatocyte nuclear factor-4α gene in maturity-onset diabetes of the young (MODY1). *Nature (London)* **384**, 458–460.

Yang, N. Y., Shigeta, H., Shi, H. P., and Teng, C. T. (1996). Estrogen-related receptor, hERR1, modulates estrogen receptor-mediated response of human lactoferrin gene promoter. *J. Biol. Chem.* **271**, 5795–5804.

Yazdanbakhsh, K., Choi, J.-W., Li, Y., Lau, L. F., and Choi, Y. (1995). Cyclosporin A blocks apoptosis by inhibiting the DNA binding activity of the transcription factor Nur77. *Proc. Natl. Acad. Sci. U.S.A.* **92**, 437–441.

Yoshikawa, T., DuPont, B. R., Leach, R. J., and Detera-Wadleigh, S. D. (1996a). New variants of the human and rat nuclear hormone receptor, TR4: Expression and chromosomal localization of the human gene. *Genomics* **35**, 361–366.

Yoshikawa, T., Brkanac, Z., DuPont, B. R., Xing, G. Q., Leach, R. J., and Detera-Wadleigh, S. D. (1996b). Assignment of the human nuclear hormone receptor, NUC1 (PPARD), to chromosome 6p21.1-p21.2. *Genomics* **35**, 637–638.

Yu, K., Bayona, W., Kallen, C. B., Harding, H. P., Ravera, C. P., McMahon, G., Brown, M., and Lazar, M. A. (1995). Differential activation of peroxisome proliferator-activated receptors by eicosanoids. *J. Biol. Chem.* **270**, 23975–23983.

Yu, R. T., McKeown, M., Evans, R. M., and Umesono, K. (1994). Relationship between *Drosophila* gap gene *tailless* and a vertebrate nuclear receptor Tlx. *Nature (London)* **370**, 375–379.

Yu, V. C., Delsert, C., Andersen, B., Holloway, J. M., Devary, O. V., Näär, A. M., Kim, S. Y., Boutin, J.-M., Glass, C. K., and Rosenfeld, M. G. (1991). RXRβ: A coregulator that enhances binding of retinoic acid, thyroid hormone, and vitamin D receptors to their cognate response elements. *Cell (Cambridge, Mass.)* **67**, 1251–1266.

Zamir, I., Harding, H. P., Atkins, G. B., Hörlein, A., Glass, C. K., Rosenfeld, M. G., and Lazar, M. A. (1996). A nuclear hormone receptor corepressor mediates transcriptional silencing by receptors with distinct repression domains. *Mol. Cell. Biol.* **16**, 5458–5465.

Zanaria, E., Muscatelli, F., Bardoni, B., Strom, T. M., Guioli, S., Guo, W., Lalli, E., Moser, C., Walker, A. P., McCabe, E. R. B., Meitinger, T., Monaco, A. P., Sassone-Corsi, P., and Camerino, G. (1994). An unusual member of the nuclear hormone receptor superfamily responsible for X-linked adrenal hypoplasia congenita. *Nature (London)* **372**, 635–641.

Zetterström, R. H., Solomin, L., Mitsiadis, T., Olson, L., and Perlmann, T. (1996). Retinoid X receptor heterodimerization and developmental expression distinguish the orphan nuclear receptors NGFI-B, Nurr1, and Nor1. *Mol. Endocrinol.* **10**, 1656–1666.

Zhang, X., Hoffmann, B., Tran, P. B.-V., Graupner, G., and Pfahl, M. (1992). Retinoid X receptor is an auxiliary protein for thyroid hormone and retinoic acid receptors. *Nature (London)* **355**, 441–446.

Zhong, W., Sladek, F. M., and Darnell, J. E., Jr. (1993). The expression pattern of a *Drosophila* homolog to the mouse transcription factor HNF-4 suggests a determinative role in gut formation. *EMBO J.* **12**, 537–544.

Zhu, Y., Alvares, K., Huang, Q., Rao, M. S., and Reddy, J. K. (1993). Cloning of a new member of the peroxisome proliferator-activated receptor gene family from mouse liver. *J. Biol. Chem.* **268**, 26817–26820.

Index

Acrosome reaction, 185, 188–192
Activation mechanism, 130
Adenylyl cyclase, sperm, 174–175
Adipogenesis, 253
Advanced glycation endproducts, cell binding site, 43–45
AGE-β_2-microglobulin, binding to monocytes, 51–52
Akt, role in neuronal survival, 272
Alzheimer's disease, RAGE and, 56–59
Amphiregulin, expression and biological role, 126–127
Amphoterin, RAGE and, 55–56
Amyloid-β peptide, RAGE as receptor, 56–59
Antisense strategies, epidermal growth factor family receptors, 143–144
Apoptosis, 7–12
  calcium flux, 8
  cell shrinkage, 10
  CEM cells, oxysterol path, 27–30
  cysteine proteases, 11–12
  DNA lysis, 8
  Fas, 11
  glucocorticoid pathway, 16–20
  induced lethality genes, 10–11
  neurons, 261–262
  pathologic, 258
  proto-oncogenes, 8–10
Arachidonic acid
  catabolism, pathways, 246–248
  metabolites, 247–249

peroxisome proliferator-activated receptors activation, 245–249
Arrestins, binding, 67–69

Bcl-2 family, 10
  apoptosis, 288–291
Betacellulin, expression and biological role, 126
Bindin, 192–193

Caenorhabitis elegans, MAP kinase, 226
Calcium
  interaction with neurotrophin signaling, 276–278
  intracellular, neuronal survival and apoptosis, 262–263
Calcium-calmodulin dependent kinase family, 275–276
Calcium-dependent kinase pathways, 274–276
Calcium flux, 8
Calcium set-point hypothesis, 263–264
Capacitation, 169–170
Caspases, apoptosis, 288–291
cDNA, estrogen receptor β, cloning, 91–95
Cellular signaling, regulation, 66–69
CEM cell
  apoptosis
    glucocorticoid role, 20–27
    oxysterol path, 27–30

CEM cell (continued)
  clones, 12–13
  response to glucocorticoids, 16
Ceramide, neuronal death, 281–285
Chemokineses, see Motility, stimulation
Cholesterol, synthesis and uptake, inhibition, 28
Congestive heart failure, G-protein-coupled receptor kinases, 81–82
COUP-TF, 334–336
Cripto, expression and biological role, 127–128
Cyclin D1, 288
Cysteine proteases, 11–12

DAX-1, 336–337
Desensitization, receptor, 67–68
Diabetes, RAGE and, 51–54
DNA
  lysis, 8
  during glucocorticoid-evoked apoptosis, 16–17
  triplex, epidermal growth factor family receptors, 143
Drosophila, MAP kinase pathways, 226–227

Egg, activation, 194–195
Egg peptide receptors, 175–180
Egg plasma membrane interactions, 192–195
Elk-1, regulation of activity, 223
Endothelial cells, RAGE expression and function, 47–49
Environmental chemicals, estrogenic, 106
Epidermal growth factor, expression and biological role, 123–124
Epidermal growth factor family receptors, 113–145
  activation, 130–135
    heterodimerization, 134–135
    interaction with ligands, 130
    mechanism, 130–134
    transphorylation, 134–135
  antisense strategies, 143–144
  clinical applications, targeting members and ligands, 139–145

  clinical significance, 136–139
  EGF-like growth factors
    amphiregulin, 126–127
    betacellulin, 126
    common structure, 121–122
    Cripto, 127–128
    function, 122–123
    heparin-binding EGF-like growth factor, 127
    heregulin, 128–130
    Neu differentiation factor, 128–130
    transforming growth factor α, 124–126
  ErbB-2
    activation, 132–133
    clinical significance, 137–138
  ErbB-3, 119–120
    activation, 132–133
    clinical significance, 138
    structure, expression, and transforming potential, 119–120
  ErbB-4, 120
    activation, 133–134
    clinical significance, 139
    structure, expression, and transforming potential, 120
  ErbB-2/Neu
    immunotherapy, 140–141
    overview, 114–116
    structure and sequence homology, 115–116
    structure, expression, and transforming potential, 116–118, 120
    tyrosine kinase inhibitors, 144–145
ERK, activation, signaling pathway leading to, 218–219
Estrogen receptor α, amino acid sequences, 94, 337
Estrogen receptor β, 89–107, 337–338
  amino acid sequence, 90
  binding affinity, 96, 98
  cDNA, cloning, 91–95
  comparison with α proteins, 91, 93
  environmental endocrine disruptors, 106
  ERRα knock-out mouse, 103–105
  functional characteristics, 91, 93
  mRNA, expression, 101–103
  phosphorylation, 94
  prostate, 105–106
  protein
    ligand-binding characteristics, 95–99

sucrose density gradient analysis, 95–96
transactivation function, 99
Exocytosis, acrosomal, 190–191

Farnesoid X-activated receptor, 321
Fas, 11
Fatty acids, peroxisome proliferator-activated receptors activation, 245
Fertilin, 193–194
Fertilization, 167–195
   acrosome reaction, 185, 188–192
   egg activation, 192–195
   egg peptide receptors, 175–180
   egg plasma membrane interactions, 192–195
   gamete adhesion, 183–187
   motility stimulation, 170–175
   oviductal transport, 168–170
   sperm, chemotaxis, 180–183

Gamete, adhesion, 183–187
GCNF, 339–340
Gene deletion, G-protein-coupled receptor kinases animal studies, 80–81
Genes, induced lethality, 10–11
Glucocorticoid pathway, apoptosis, 16–20
Glucocorticoid receptor, 2
   constitutively lethal fragments, 23, 25–27
   ligand binding domain, regulation when intact, 20–24
   mechanism of action, 3–6
      growth factor regulation, 5–6
      posttranscriptional control, 5
      transcription control, 3–5
   role in apoptosis, 20–27
G-protein-coupled receptor kinases, 65–83
   animal studies
      gene deletion, 80–81
      transgenic gene expression, 79–80
   cloning and structure, 69–70
   enzymology, 76
   expression, 70–71
   physiological significance, 81–83
   by protein kinase C, 75
   receptor–kinase specificity, 76–79
   regulation
      by membrane association, 71–73
      by membrane lipids, 73–75
   structure, 71–72

G-protein-coupled receptors, kinase specificity, 76–79
G proteins, sea urchin spermatozoa, 191
Growth factors, regulation, by glucocorticoids, 5–6

Heparin-binding EGF-like growth factor, expression and biological role, 127
Hepatocyte nuclear factor-4, 327–329
Heregulin, expression and biological role, 128–130
Heterodimerization
   EGFR/ErbB-2, 134–135
   EGFR/ErbB-3, 135
   EGFR/ErbB-4, 135
   ErbB-2/ErbB-3, 135
   ErbB-2/ErbB-4, 135

Immunotoxins, epidermal growth factor family receptors, 142–143

JNK family, 209–228
   activation, signal transduction pathways leading to, 216–222
   biological functions, 224–227
   c-Jun amino-terminal site phosphorylation, 213–216
   neuronal death, 281–285
   substrates for, 222–224
c-Jun
   amino-terminal site phosphorylation by JNK, 213–216
   induction, apoptosis and, 284–285
      glucocorticoid-evoked CEM-C7 cell, 19–20
   regulation of expression and activity, 212–213

Ligand antagonists, epidermal growth factor family receptors, 141
Lipid
   homeostasis, 237–242
      cell and lipid type, 238
      cross-talk at transcription factor level, 241–242
      peroxisomes as multifunctional organelles, 238–241
   lowering, peroxisome proliferator-activated receptors, 253–254

LXR, 321–322
Lymphoid cells, oxysterol actions, 6–7
Lysis, DNA, 8
  during glucocorticoid-evoked apoptosis, 16–17

MB67, 322–323
Membrane association
  G-protein-coupled receptor kinases regulation, 71–73
  microsomal, G-protein-coupled receptor kinases regulation, 75
Membrane lipids, G-protein-coupled receptor kinases regulation, 73–75
Methoprene acid, 318
Mice, knock-out
  estrogen receptor α, estrogen receptor β in, 103–105
  peroxisome proliferator-activated receptors, 253–255
Mitogen-activated protein kinases, see JNK family
Mitogen-activated protein kinase cascades, 283
Mitogen-activated protein kinase kinase, 210–211, 219–220
Mitogen-activated protein kinase pathways, 224–227
Monoclonal antibodies, against epidermal growth factor receptor, 140–141
Mononuclear phagocytes, RAGE expression and function, 49–53
Motility, stimulation, 170–175
mRNA, estrogen receptor β, expression, 101–103
Myc, overexpression, 9
c-myc, 8–9
  repression, CEM cell apoptosis, 18–19

Nerve growth factor, neuronal survival, 265
Neu differentiation factor, expression and biological role, 128–130
Neurons
  activity, intracellular calcium, and survival, 262–263
  apoptosis, 261–262
    caspases and Bcl-2 family, 288–291
    mechanisms, 278
    second messengers, 281–285
    upstream events and p75, 278–281
    cell-extrinsic signaling meets cell-intrinsic survival and death mechanisms, 291–292
  signal integration, 285–288
  survival, 257–292
    calcium-dependent kinase pathways, 274–276
    calcium set-point hypothesis, 263–264
    interaction of calcium and neurotrophin signaling, 276–278
    neurotrophic theory, 259–261
    neurotrophin-dependent pathway crosstalk, 274
    neurotrophin receptors, 264–266
    pathologic apoptosis, 258
    phosphatidylinositide-3'-OH kinase/Akt pathway, 270–273
    protein kinase C, 273–274
    Ras–MAPK pathway, 266–270
    second messengers, 264–266
Neurotrophic theory, 259–261
Neurotrophin-dependent pathway crosstalk, 274
Neurotrophin receptors, 264–266
Neurotrophins
  functional roles, 260–261
  neuronal death promotion, 279
  signaling, calcium and, 276–278
NGFI-B, 324, 326, 329–331
Nonsteroid hormone receptor, 308–311, see also Nuclear orphan receptors
  classification, 310
Nuclear eicosanoid receptor, inflammation control, 254–255
Nuclear orphan receptors, 307–342, see also Peroxisome proliferator activated receptor; Retinoid X receptors
  COUP-TF, 334–336
  DAX-1, 336–337
  definition, 311–315
  ERRα, 337
  ERRβ, 337–338
  farnesoid X-activated receptor, 321
  GCNF, 339–340
  hepatocyte nuclear factor-4, 327–329
  ligand discovery, 315–319
  LXR, 321–322
  MB67, 322–323
  NGFI-B, 324, 326, 329–331
  ONR1, 323
  retinoid signaling mediation through heterodimerization with RXR, 323–327

Rev-ErbAα, 338–339
ROR, 331–332
as RXR heterodimers, 319–327
SHP, 323
steroidogenic factor-1, 332–334
structures for ligands and activators, 317
testicular receptor 2, 339
Tlx, 340
as transcriptional activators, 327–334
as transcriptional repressors, 334–339
with unknown activation functions, 339–340
vertebrate, 312–314
Nuclear receptor, 90, 308
comparison with estrogen receptor β protein, 94
structure and consensus DNA response elements, 309

Oligonucleotides, antisense, epidermal growth factor family receptors, 143–144
ONR1, 323
Oviductal transport, 168–170
Oxidant stress, induction as consequence of AGE–RAGE interaction, 52–53
β-Oxidation, inducible, peroxisome proliferator-activated receptors, 251, 253
Oxysterols, actions on lymphoid cells, 6–7

p53, 9–10
p38, 222
neuronal death, 281–285
p75, upstream events in neuronal death, 278–281
Peptides, sperm activating, 171–172
Peroxisomal β-oxidation pathway, 239–241
Peroxisome proliferator-activated receptors, 235–255, 319–321
activation profile, 250
activators
arachidonic acid, 245–249
assays, 243–244
fatty acids, 245
species differences, 250
thiazolidinediones, 249–250
expression during development and adulthood, 242–243
functions, 251, 253–255, 320
adipogenesis, 253

inducible β-oxidation, 251, 253
knock-out mice, 253–255
as ligand-activated transcription factor, 236–237
ligands, 251–252
lipid homeostasis, 237–242
nuclear eicosanoid receptor, inflammation control, 254–255
peroxisome proliferators, 249, 253
Peroxisome proliferators, 249, 253
Peroxisomes, multifunctional organelles, 238–241
Phosphatidylinositide-3′-OH kinase/Akt pathway, 270–273
Phosphoinositide-3-kinase, 221
Phosphorylation, c-Jun amino-terminal sites by JNK, 213–216
Phosphotyrosine–SH2 interaction, specificity, 266
Phytanic acid, 318
Polyunsaturated fatty acids, 236
Prostate, estrogen receptor β, 105–106
Protein kinase
cascades, 210
regulation by products of phosphatidylinositol 3-OH kinase, 271
Protein kinase C
G-protein-coupled receptor kinases regulation, 75
neuronal survival, 273–274

RAGE, 41–60
Alzheimer's disease, 56–59
amphoterin and, 55–56
diabetes and, 51–54
effect of diabetic red blood cells, 48–49
expression and functions
endothelial cells, 47–49
mononuclear phagocytes, 49–53
hydrophilicity plot, 45–46
hypothesis, 60
identification and characterization, 42–47
phagokinetic track assay, 50–51
soluble, 47–49
Ras–MAPK pathway, 266–270, 287–288
Ras superfamily, 220–221
sufficiency to promote neuronal survival, 268–269
Receptor for advanced glycation endproducts, see RAGE
Receptors, death-regulating, 279

REJ, 183–184
9-*cis* Retinoic acid, 316–318
Retinoic acid receptor, RXR heterodimer, 323
Retinoid X receptors, 316–319
  heterodimers, see also Nuclear orphan receptors
    AF-2 domain, 326–327
    partners with ligands and/or activators, 319–322
    partners with no known ligands or activators, 322–323
    retinoid signaling mediation, 323–327
    models for role in ligand binding and transcriptional activation, 324–326
  subtypes, 316
Rev-ErbAα, 338–339
Ribozymes, epidermal growth factor family receptors, 144
ROR, 331–332

Saccharomyces cerevisiae, signaling cascades, 216–217
Sea urchin
  acrosomal exocytosis, 191
  acrosome reaction, 188
  egg peptide signal transduction, 172–173
  gamete adhesion, 183–184
  guanylyl cyclases, 176–178
  intercellular plasma membrane adhesion, 192–193
  sperm chemotaxis, 182–183
  sperm motility and respiration, 170–171
Second messengers, 264–266
  neuronal death, 281–285
SHP, 323
Signal transduction pathways
  functions, 225–226
  MAP kinase activation, 210–211, 216–222
Sperm
  adenylyl cyclase, 174–175
  chemotaxis, 180–183
  membrane hyperpolarization, 173
  motility stimulation, 170–175
Sperm activating peptides, 171–172
Steroid hormone receptor, 308–309
Steroidogenic factor-1, 332–334

Testicular receptor 2, 339
Thiazolidinediones, peroxisome proliferator-activated receptors activation, 249–250
THymic cell, subgroups, killed by glucocorticoids, 2–3
Tlx, 340
Toxin molecules, targeted, epidermal growth factor family receptors, 141–143
Toxins, chimeric, epidermal growth factor family receptors, 142
Transactivation, estrogen receptor β protein, 99–100
Transcriptional activators, nuclear orphan receptors functioning as, 327–334
Transcriptional regulatory properties, estrogen receptor β protein, 99–101
Transcriptional repressors, nuclear orphan receptors as, 334–339
Transforming growth factor α, expression and biological role, 124–126
Transgenic gene expression, G-protein-coupled receptor kinases, animal studies, 79–80
Trk receptors, 264–265
Tyrosine kinase inhibitors, epidermal growth factor family receptors, 144–145

VCAM-1, expression, in diabetes, 53–54

Zona pellucida, 184–187
  acrosome reaction, 188–190